Second Language Learning and Teaching

Series Editor

Mirosław Pawlak

For further volumes:
http://www.springer.com/series/10129

About the Series

The series brings together volumes dealing with different aspects of learning and teaching second and foreign languages. The titles included are both monographs and edited collections focusing on a variety of topics ranging from the processes underlying second language acquisition, through various aspects of language learning in instructed and non-instructed settings, to different facets of the teaching process, including syllabus choice, materials design, classroom practices and evaluation. The publications reflect state-of-the-art developments in those areas, they adopt a wide range of theoretical perspectives and follow diverse research paradigms. The intended audience are all those who are interested in naturalistic and classroom second language acquisition, including researchers, methodologists, curriculum and materials designers, teachers and undergraduate and graduate students undertaking empirical investigations of how second languages are learnt and taught.

Jacek Fabiszak · Ewa Urbaniak-Rybicka
Bartosz Wolski
Editors

Crossroads in Literature and Culture

Springer

Editors
Jacek Fabiszak
Faculty of English
Adam Mickiewicz University
Poznan
Poland

Ewa Urbaniak-Rybicka
Faculty of Pedagogy and Fine Arts
Department of English Studies
Adam Mickiewicz University
Kalisz
Poland

Bartosz Wolski
Faculty of Pedagogy and Fine Arts
Department of English Studies
Adam Mickiewicz University
Kalisz
Poland

ISSN 2193-7648 ISSN 2193-7656 (electronic)
ISBN 978-3-642-21993-1 ISBN 978-3-642-21994-8 (eBook)
DOI 10.1007/978-3-642-21994-8
Springer Heidelberg New York Dordrecht London

Library of Congress Control Number: 2012951140

Printed on acid-free paper

Springer is part of Springer Science+Business Media (www.springer.com)

Preface

One of the foundations of science and scholarship is probing into areas unknown or barely chartered as well as crossing the borders of the known. Scholars are also expected to shed a new light on what is already apparently known and researched. Furthermore, they often venture to combine different approaches in order to achieve novelty in their investigations. Commonplaces as these observations are, they nevertheless remain a binding prescriptions for any researcher. No surprising then that the present volume attempts to, on the one hand, discuss literary and cultural phenomena which have not enjoyed a wide scholarly attention so far and, on the other, look at canonical works of literature in English and acknowledged, groundbreaking cultural events from new perspectives. In a nutshell, this is what constitutes the present crossroads in literature and culture, a space where different approaches and different walks of scholarly activity cross and enter into a dialogue with one another.

This volume consists of eight chapters of various lengths, each devoted to a specific aspect of crossing, traversing and mingling in the area of literary and culture studies. Consequently, Part I explores the genuinely unfathomable and forking paths of theoretical considerations, opening with a brilliant essay by Małgorzata Grzegorzewska on prosopopeia and its deployment in Renaissance literature, illustrating her discussion with references to Shakespeare's *A Midsummer Night's Dream*. In turn, Tymon Adamczewski in a somewhat self-reflexive manner treats literary criticism as yet another … literary genre practiced by clandestine writers, or perhaps critics who have always yearned to achieve the status of writers? Grzegorz Koneczniak and Robert Kielawski employ in their readings of, respectively, Irish literature and Ravenhill's drama quite recent and critically acknowledged methodologies, revising their usefulness in critical practice: the former reaches for postcolonial studies, the latter for Slavoj Žižek's ideas. The chapter concludes with Katarzyna Więckowska's interpretation of two works by Doris Lessing, which adopts and adapts psychoanalytical tools, posing significant questions about their presence in scholarly debate.

Part II focuses on the issue of female writers and characters crossing moral and/ or social borders and expanding the limitations imposed by patriarchal order. The works discussed range from Victorian fiction (essays by Aleksandra Budrewicz-Beratan, Agnieszka Setecka, Małgorzata Łuczyńska-Hołdys, and, to a degree, Magdalena Pypeć), to American popular fiction of the 1920s and 1930s (Barbara Leftih), to a modern rewriting of Victorian prose, be it in the Canadian context (Ewa Urbaniak-Rybicka) or British one (Barbara Braid and Malwina Degórska) as well as a contemporary novel addressing sexual anxieties in the era of Margaret Thatcher (Przemysław Uściński). Thus, Budrewicz-Beratan and Setecka explore the ways in which nineteenth-century female characters were trying to go beyond the prescribed social role, characteristically in the works written by canonical Victorian men-writers (Dickens and Trollope, respectively). Łuczyńska-Hołdys investigates how Victorian female gender identity undergoes subversion and transformation in the works of Dante Gabriel Rossetti, whereas Pypeć relates Victorian gender issues contained in a specific genre—dramatic monologue—to its contemporary equivalent, Carol Ann Duffy's volume. Likewise, vibrant problems of the nineteenth-century presentation and treatment of women find their continuation in how the characters are revised and redesigned in contemporary times, which has long challenged the traditional (?) sex/gender division. Urbaniak-Rybicka poses an important question of who the real 'outlander' is in Gil Adamson's novel, the heroine, capable of committing worst crimes and coming to terms with her life, thanks to the Gothic convention, or the crumbling, divided, patriarchal society in which she has to live? Braid further develops the exploration of Gothicism, this time without delving into gender-related issues, in a contemporary novel by Sarah Waters, while Degórska—also looking at a novel by Sarah Waters—approaches the crossing of *the gender frontier* from a male rather than female perspective. The essays by Leftih and Uściński avoid overt references to the Victorian era, although they appear inevitable in a discussion concerning the liberation of rigid social roles of American women conquering the Wild West and their literary counterparts in popular fiction (the former) and the problems of sexuality in Thacherite Britain (the latter).

Part III is the shortest chapter of the volume as it includes only one article—Marcin Jurkowicz's—which, however, addresses a significant issue of translation practices whereby translators, to a better or worse effect, attempt to cross languages and cultures linked with them. Jurkowicz illustrates such strategies in a discussion of a Polish rendering of Alice Walker's *The colour purple*. In Part IV linguistic quibbles yield way to a more 'spatial' issue: the liminal area and constant traffic between variously conceived ideas of the 'Old' and the 'New' worlds. Quite a number of articles in this section are devoted to 'new' literatures—the literatures of the post-colonial cultures remaining in the shade of the 'old'? world. As a result, Anna Branach-Kallas and Dagmara Drewniak discuss the problem of Canadian identity in the context of, respectively, Canadian ethos of pioneers and the space for women in it, and the survivors of Holocaust. Brygida Gasztold addresses similar issues from a Dominican perspective: how to establish one's sense of belonging in an immigrant community in the U.S. Ryszard Wolny approaches from the post-colonial perspective India's national problems on the basis of Vikas Swaruo's *Slumdog millionaire*. Maciej Sulmicki adopts an analogous stance in looking at the relationship between the colonizers and the colonized in *English passengers*.

The next section of the volume contains essays exploring the borderline and the sphere between the private and the public or the social. The articles range from scrutinies of (semi)biographical texts to analyses of characters finding it difficult to draw a line between their private and social selves. Teresa Bruś's work definitely belongs to the former: she is looking at Louis MacNeice's diary in the context of Philip Lejeune's treatment of such liminal forms of writing. A similarly amorphous genre is scrutinised by Joanna Jodłowska, who delves into Aldous Huxley's ideas about social changes contained in his essays. In turn, Dominika Lewandowska looks at the way Iain Sinclair represents London as an urban text, which quite drastically changes the perspective on what is private and public. Artur Skweres's and Natalia Brzozowska's articles touch upon the complex issue of characters' relations with the outer world, of how the social/real affects their private/mental identities. Skweres, importing Philip K. Dick's own anxieties, illustrates the problem with reference to fictitious figures from the writer's oeuvre, whereas Brzozowska looks at the social ramifications of excessive ambition in the case of two Shakespearean tragic characters: Coriolanus and Lady Macbeth.

If Part IV was devoted to spatial 'travels', Part VI is a kind of time travel. Articles collected in this section attempt to compare writers from different, sometimes quite distant, literary periods showing the significance of intertextuality. Anna Budziak considers the ups and downs of Blake's poetry in the eyes of other poets-turn-critics, Dante Gabriel Rossetti and William Butler Yeats, respectively, both trying to explain the 'genius' behind Blake's verse. Joanna Bukowska explores a specific dialogue between the canonical text of English Medieval letters—Chaucer's *The Canterbury Tales*—and a contemporary novel by Peter Ackroyd: *Clerkenwell Tales*, tracing the treatment of literary conventions 600 years after Chaucer. A similar dialogue between a canonical work and its modern version is investigated by Edyta Lorek-Jezińska in her analysis of Deborah Levy's post-modern, post-twentieth-century *Macbeth–False Memories*. Contrary to these three essays, Wacław Grzybowski does not compare texts temporally distant, although he does bring together—like Budziak, Bukowska and Lorek-Jezińska—two artists: Thomas MacGreevy and Samuel Beckett, shedding a new light on their sometimes stormy artistic relationship.

In Part VII borders of literary conventions and genres are reshuffled. Thus, Wojciech Drąg discusses a most important issue of realism/non-realism/surrealism in Kazuo Ishiguro's *When We Were Orphans*. Anna Kędra-Kardela looks at a rather obsolete and critically neglected genre of the eighteenth-century 'fragment', employing a respectable variety of tools, the most significant of which is cognitive poetics. Anna Krawczyk-Łaskarzewska considers the afterlife of Edgar Allan

Poe's arguably most famous short story: *The Fall of the House of Usher* in popular culture and present-day media, tracing the ways in which adaptation ushers in new formats for modern conventions and genres. Adaptation is also the subject of Jadwiga Uchman's essay, albeit in a rather unorthodox way since the scholar includes in her investigations the issue of authorial rights to Beckett's works and what happens in the process of adaptation with both the rights and the works. Finally, Ewa Kębłowska-Ławniczak ruminates on the self-referential issue of art as a conventional subject matter of (non) experimental contemporary English plays.

The concluding part of the volume can be treated as a résumé of what has been explored so far: art extending its limitations across space and time in broadly understood culture. Krzysztof Fordoński interprets the poetic works in Latin by a Polish Jesuit, Maciej Sarbiewski, as a vehicle for political propaganda in English late Renaissance, appropriated by the Catholically-oriented Metaphysical poets. Olga Grądziel contextualizes Thomas More's *Utopia* in the receptive culture of the sixteenth-century England, while Bożena Kucała analyses a modern version of what might have been memoirs of the members of Robert Scott's expedition to the South Pole as rendered in Beryl Bainbridge's novel. Aniela Korzeniowska attempts to face the problem of one's complex, contemporary identity in her search of an answer to the question of what it takes to be a Black Scot (woman) in present-day Glasgow on the basis of the oeuvre of the poetess Jackie Kay. Joanna Mstowska in her essay assumes a historical perspective on the myth of the Flying Dutchman, scrutinising it in the work of a nineteenth-century writer, Frederick Marryat, in light of both the Faustian myth and René Girard's twentieth-century theory. Stankomir Nicieja's essay addresses the issue of the relationship between science and humanities as rendered in contemporary British fiction, illustrated in two novels by the celebrated writer Ian McEwan. A relationship between the country of one's origin and the manner one is framed in the country of one's choice is the focus of Anna Tomczak's article centering on Rose Tremain's novel, *The road home*. Justyna Stępień discusses in her article the crossroads of British popular art, tracing the relationship between the 'high brow' and 'low brow' forms of art. Katarzyna Kociołek, too, looks at British art, albeit from a different perspective: adopting postmodern discourse she analyses documents regulating art in Britain in the 1980s. Postmodern lens is also deployed in Jakub Ligor's reading of the Coens' films, in which he applies the idea of McHalean 'zones'. Agnieszka Rasmus, too, considers films in her essay, focusing however on the borderline between an original movie (a British one, in this case) and its remake (a Hollywood one). Michał Różycki is yet another scholar who in his article looks at films, but he focuses on the concept and figure of conspiracy theorist, both a fictitious character and a historical person. Finally, Urszula Terentowicz-Fotyga's essay is devoted to a different area of cultural studies: she considers the future of humanities in Poland in the context of the 'cultural turn'.

The articles collected in this volume analyse liminal spaces of widely understood cultural studies, showing various intersections and forking paths of research into English letters and culture. Naturally, we can only hope that the essays will provoke and inspire further scholarly debate in the area of the crossroads of literature and culture.

Kalisz–Poznan, 2012

Jacek Fabiszak
Ewa Urbaniak-Rybicka
Bartosz Wolski

Contents

Part VIII Countries and Cultures at Crossroads

Part I
Crossing Thresholds of Literary Theories and Critical Approaches

I See a Voice...

Małgorzata Grzegorzewska

Abstract The text discusses the uses of poetic prosopopeia as a figure of counterfeit voice. Since the term prosopopeia derives from the Greek noun *prosopos*, meaning both face and theatrical mask, the trope invites us to take up the debate concerning the problem of authorial "presence", a promise of authentic meeting "face to face" while at the same time we are confronted with a possibility of deceit: for the actor who puts on a mask may appear to be someone whom in reality he is not. Viewed in this light, prosopopeia leads us towards an idiosyncratic play of sincerity and hypocrisy: in poetry, drama and even in literary criticism. The argument is illustrated with references to William Shakespeare's treatment of rhetorical prosopopeia in *A Midsummer Night's Dream*.

1 Introduction

The observer standing "at the cross-roads of literature and culture" feels encouraged to take risks, embarking on new discoveries and calculating the stakes involved in such an enterprise. I wish to approach this challenge through the perspective of the phenomenon of the voice; that which establishes and negotiates the relationship between us and the outer world. When it reaches out, it forms an extension of our body in space and time. But at the same time, when we speak, we

A Polish version of this text is a part of the Introduction to my book on the rhetorical potential of prosopopeia (*Trop innego głosu w angielskiej poezji religijnej epok dawnych*), Kraków 2011, UNIVERSITAS.

M. Grzegorzewska (✉)
Warsaw University, Warsaw, Poland
e-mail: malgorzatagrzegorzewska@onet.eu

J. Fabiszak et al. (eds.), *Crossroads in Literature and Culture*, Second Language Learning and Teaching, DOI: 10.1007/978-3-642-21994-8_1,

are always left behind by the voice, which must abandon us in order to enter the world. Yet even as it leaves us, it always *re*-turns to us, *re*-sounds, resonates as something that comes from outside (Connor 2000, pp. 3–45). “To hear oneself speak—or, simply, just to hear oneself—can be seen as a minimal formula of narcissism that is needed to produce the minimal form of a self”, says Malden Dolar in his influential study entitled *A Voice and Nothing More.* He also proposes that voices of “other people, of music, of media, as well as our own voice intermingled with the lot; which rise over the multitude of sounds and noises, and interspersed with moments of silence” are “the very texture of the social, as well as the intimate kernel of the subjectivity” (2006, p. 39).

In his little-quoted, country-house poem *Upon Appleton House*, Andrew Marvell, a poet of the seventeenth century, reflected upon the vanity of human endeavors in the following manner:

The Beasts are by their Denns exprest:
And Birds contrive an equal Nest;
The low roof’d Tortoises do dwell;
In cases fit of Tortoise-shell:
No Creature loves an empty space;
Their Bodies measure out their Place (1984, p. 62).

Marvell contrasted this natural measure with “superfluously spread” human pride. My esteem for Marvell’s poetic imagination notwithstanding, I wish to follow the path set by Dolar and propose a revised version of his articulate concept. If man indeed appears a “superfluously spread” creature, it is not because of his attempts to conquer space, but thanks to speech and its attendant, voice. Voice measures out space and stakes out new territories. Our daily experience of soundwaves “on the air” and of soundtracks of speeches delivered by people who passed away years ago have certainly stretched the impact of voice beyond the imagination of past generations. But I venture to say that we still have not paid enough attention to this fascinating phenomenon. We shall try to remedy this omission by addressing the figure of voice, which in the manuals of rhetoric receives a Greek name: *prosopopeia.*

2 The Figure of the Voice

“The trope of voice recurs in our discursive practices with remarkable persistence”, observes Elizabeth Freund, and she adds that “it is as old as poetry or the arts of language and as new as the latest endeavour to recover and represent the absent voices of the living as well as the dead” (unpublished conference paper). In part, this claim is substantiated by the unforgettable one-sentence paragraph that opens Stephen Greenblatt’s *Shakespearean negotiations*, which proves that in principle literary criticism is a domain governed by the figure of voice: “I began with the desire to speak with the dead” (1988, p. 1). This is followed by an explanation which shows that prosopopeia creates a space of dialogue where our voice traverses the voices of those who cannot speak:

> If I never believed that the dead could hear me, and I knew that the dead could not speak, I was nevertheless certain that I could recreate a conversation with them. Even when I came to understand that in most intense moments of straining to listen all I could hear was my own voice, even then I did not abandon my desire. It was true I could hear only my voice, but my voice was the voice of the dead, for the dead had contrived to leave textual traces of themselves, and those traces make themselves heard in the voices of the living (1988, p. 1).

Thus, the figure of prosopopeia can be read two ways. Stephen Greenblatt, in his usual, sharp-witted manner, points out that whenever we lend our voice to the dead or the absent, our manner of speaking inevitably yields to that imagined voice. In literature and literary criticism alike, we therefore constantly encounter phantoms which not only impinge on our writing and speech, but also affect our inner sight and hearing.

Paul de Man once wrote that prosopopeia is the fundamental, constituting characteristic of every poetic utterance, without exception (1984, p. 76). Perhaps he was right, and every poem is an extended prosopopeia. De Man's view on this subject is clear in the following extract from his analysis of Nietzsche's text *On Truth and Lies:*

> What we call the lyric, the instance of represented voice ... includes the grammatical transformation of the declarative into the vocative modes of question, exclamation, address, hypothesis, etc., the tropological transformation of analogy into apostrophe or the equivalent, more general ... transformation of trope into anthropomorphism (1984, p. 261).

So we may ask: Where does the trope of the voice lead? What does it reveal and what does it conceal? For the word prosopopeia derives from the Greek noun *prosopos*, meaning both face and theatrical mask. This name contains both a promise of authentic meeting, 'face to face', and a warning against deceit: for the actor who puts on a mask may appear to be someone whom in reality he is not. The mask may thus serve the purpose of defensive camouflage or deception, unless we treat seriously the vision of the theatre in which the actor gives himself entirely to the play and embodies himself in the role, so that for the duration of the performance he becomes someone whom off-stage he is not. Viewed in this light, prosopopeia is an idiosyncratic play of sincerity and hypocrisy. At the same time, it questions everyday understanding of the subject-object relation, since it presents the face as a product of the mask. By the same token, prosopopeia denies identity: by imparting a face to someone or something, it reveals a primordial lack, evoking a being without a face. The voice that we hear does not come from a mouth, but emerges from the shapeless, hungry chasm of the belly. In consequence, prosopopeia unmasks the emptiness that it attempts to conceal by putting on a mask. The revealing of a non-person precedes and determines the personifying function of prosopopeia. It is precisely this discovery that led de Man to attribute a privileged role in his rhetorical criticism to prosopopeia.

The figures of the voice and the face thus inscribe themselves in the post-modern debate on the theme of identity, which, contrary to what is traditionally thought, is by no identifiable with our inner being. Instead, it comes to us as a closed repertoire of roles that are imposed on us from the outside. The language in which we describe feelings and formulate judgments, precedes and shapes our utterances; it is not I that speak, but words, whose meaning I never define myself, that speak me. Similarly, the face is treated as a repertoire of theatrical masks (in this sense in post-modern discourse it is therefore difficult to employ the concept of hypocrisy, which assumes the possibility of distinguishing between face and mask, truth and falsehood). Finally, the voice appears as a complex structure, like a stream of white, transparent light, which in reality is a composition of many-coloured bands. As a prism diffracts white light, so the concept of heteroglossia serves to make visible the (ideological) colouring of what appears innocently transparent.

3 The Emergence of the Face

In a sense, prosopopeia confirms the postmodernist paradigm, reversing the order of appearance of 'voice' and 'face'. Until the possibility of transmitting and recording voice, and thus overcoming time and space, was discovered, the appearance of a voice was bound up with the experience of physical presence. This is why Prospero's island, filled with invisible voices, as described in Shakespeare's *The Tempest*, makes such an unearthly impression on the castaways; when they hear the singing of Ariel they recognise that they are under the power of enchantment. "Where should this music be? I'th'air or th'earth?/It sound no more; and sure it waits upon/Some god o'th' island", says the son of the King of Naples, Ferdinand [1. 2, 390–393]. The figure of prosopopeia was thus ascribed a visualizing function: whereas everyday experience teaches us that the appearance of a voice must be preceded by the presence of a face and—still more definitely—by a mouth that pronounces the words that we hear, the figure of prosopopeia makes it possible for mouths, eyes and faces to emerge from the uttered words…

Loquere ut videamte. Speak, so that you can be seen. The English poet and dramatist, Ben Jonson, active at the turn of the sixteenth and seventeenth centuries, included this *sententia* in his *silva rerum* (1974, vol. 8, p. 60). In the context of our discussion of prosopopeia, it is worth recalling the historically conditioned changes in that topos, already known to the ancients, which makes the "audible" become "visible". On the one hand, an excellent example of the emergence of a character through the utterance of words may be found in the meeting, described in a fourteenth-century English elegiac poem, between a father deep in mourning and his little dead daughter, who visits him in a dream. The title of this work, *The Pearl*, indicates several possibilities of interpretation; the most literal reading would link it with the little girl's name (Margaret), but medieval allegorical interpretation suggests three other lines of approach, referring to the ethical (an immaculate pearl means a virtuous soul); to faith (in the gospel parable, the pearl is a symbol of the Kingdom of Heaven); and finally, to the eschatological, in the promise of the mystical Wedding Feast of the Lamb (at a certain moment the heroine of the poem is presented as the Bride of the Lamb). However, each of these possibilities is realised only when it is uttered, that is to say revealed to the narrator by the figure whom he meets in his dream. It may be said that in this way the allegory speaks itself. Although visible, Pearl remains anonymous and thus also in a certain sense invisible until the narrator hears her voice.

Considerably later, a nineteenth-century interpretation of the summons *loquere ut videam te* can be found in the parts of Soeren Kierkegaard's *Journal* devoted to the treatise *The Concept of Anxiety*. The Danish philosopher gave a perverse but extremely exact name to these sketches: "Vocalizations of *The Concept of Anxiety*" (1909–1948, vol. 5, p. 104; cf. also Garrf 2005, pp. 348–349). According to Kierkegaard, man is a concealed being, strange and opaque to his own self; thus he can never fully 'utter himself', if that utterance is taken to mean revealing himself. 'Vocalization' thus means only the manifestation of presence, not the revealing of mystery, which would entail becoming accustomed to one's existential strangeness. Kierkegaard's version makes us aware at the same time that the 'face' (or 'mask'), that we find in prosopopeia is a function of discourse. Michele Rifaterre confirms this:

> It [*prosopopeia*] is indeed no more than a mask put on something that may not even have a face. The mask's simplicity gives free play to the viewer's imagination. All the mask does is put the reader in a specific interpretative mood, very much as does the donning of a comic or a tragic mask in Greek and Roman plays. No mimesis, no restriction justified by referentiality may interfere with the reader's constructs on so simple and so abstract a given. If there remains any mimetic component, it must be secondary, and its normal functioning as representation must be reoriented by the given (1985, p. 108).

Here, however, we return to the question of the selfness connected with voice, a question enriched but by no means invalidated by these reflections on the theme of the multiplicity of masks and the variety of discourses that shape our utterances. For as colloquially understood, 'voice', just like 'face', allows us to recognise and authenticate the person of the speaker. When we hear a knocking at the door, we ask "Who's there?", not only in order to hear a known name (without this, we do not open the door), but also, and perhaps above all, in order to recognise the voice that presents itself along with the name. Prosopopeia requires of us exactly this ability to "hear with our eyes" that is the art of interpretation.

4 Pro(zoo)popeia

"The eye of man hath not heard, the ear of man hath not seen, man's hand is not able to taste, his tongue to conceive, nor his heart to report, what my dream was" [4. 1, 218–221], announces simple Bottom in Shakespeare's *A Midsummer Night's Dream*. Thus he strives to convey with the help of a clumsy description the extraordinary story in which he has accidentally become entangled and in which it falls to him to play the main, though not necessarily very heroic, part. Everything, as we recall, starts with the quarrel between a husband and wife, the King and Queen of the Elves, caused by the appearance at the court of a charming Indian boy. When the irate Oberon decides to punish his wife, the malicious fairy Puck helps him in his small revenge, causing Titania by the power of enchantment to fall helplessly in love with a simple artisan turned into an ass. It is Puck who changes Bottom into an ass, by putting an animal mask on his face [what a wonderful example of prosopopeia working backwards, or rather, perhaps, of pro(zoo)popeia], and then restoring his human figure to him. It is significant that the scene in which the unhappy Bottom is disenchanted has an unequivocally theatrical character: for the audience witnesses Puck's removal of Bottom's ass's head, and this of course is no spell but a mere ordinary change of costume, a game with theatre masks. Incidentally to this, the noble figure of prosopopeia falls victim to word games in a world turned upside down. The jest of the spiteful goblin has nothing in it of subtlety; quite the contrary, it calls up the most crude associations. The figure named Bottom essentially deserves his ignoble name still more after his transformation: for *ass*, which in English means "donkey", is a homonym of a vulgar four-letter word referring to what everyone has on their "bottom".

The audience gathered in the theatre looks on Bottom with amusement as he scratches his ass's face and declares in astonishment: "I must to the barber's, mounsieur, for methinks I am marvelous hairy about my face" [4. 1, 25–27]. No wonder that after returning to the normal world, Bottom cannot find the proper words and entertains us yet again by muddling up the learned words of Scripture. "The eye of man hath not heard, the ear of man hath not seen" is, however, no ordinary *lapsus*, and although these words really, as it were, 'happen' to the unlearned artisan, an important lesson concerning the figure of voice derives from them.[1] The transformation that we observe on the stage turns the mind to that very thing against which de Man's warning was directed. For we are witnesses of the giving and taking away of face, of humanizing and dehumanizing, the transformation of a man into a talking donkey, and then of a talking donkey into a man. It is no accident, then, that in the play performed by the artisans, it is to Bottom that it falls to utter the lines that confirm the strange relationship between voice and face. Hearing Thisby, divided from her beloved by a hostile wall, Pyramus-Bottom cries out with joy: "*I see a voice*: now will I to the chink,/To spy an I can hear my Thisby's face" (emphasis added) [5. 1, 195–196].

[1] Critics have found in Bottom's words a warning by the author against the traps of naive literalism (Fergusson 1985, p. 300) or a mockery of bombastic but unskillful rhetoric (E. Freund, "I see a voice': The Desire for Representation and the Rape of Voice"). The present interpretation, I think, opens up a third possibility, that of reading *A midsummer night's dream* as a serious treatise on the theme of the figure of voice, though written in a humorous form.

5 Conclusion

It is worth remembering this lesson of Shakespeare, taking its context into account, of course. Even the very title of the play performed by the artisans, in which Bottom takes the part of Pyramus, sounds grotesque: A tedious brief scene of young Pyramus/And his love Thisby; very tragical mirth. The sentence, "I see a voice, now will I to the chink,/To spy an I can hear my Thisby's face" contains a double catachresis, that is the kind of metaphor in which the elements are in semantic dissonance; it is thus that catachresis is felt as a misuse of language (malapropism). This, in any case, is how the audience of *Pyramus and Thisbe* respond to it. Nevertheless, although we agree with the criticism that Theseus and his guests do not spare the unskillful actors, a moment's reflection makes us acknowledge that Bottom's words may not only amuse, but also arouse a justified unease. The distortion that catachresis brings about inscribes itself disturbingly in the play inspired by Ovid's *Metamorphoses*. It is after all a tale of changing feelings, and its action is player out on a stage bathed in moonlight, the symbol of inconstancy. The dominant theme is already announced in the first scene, when the ruler of Athens reminds the disobedient daughter of one of his subjects that her father is "One that compos'd your beauties, yea, and one/To whom you are but as a form in wax/By him imprinted, and within his power/To leave the figure or disfigure it" [1. 1, 48–51]. One shudders to think what these words might herald if they were spoken by a character in one of Webster's bloody tragedies. But in Shakespeare's comedies, too, it is often a menacing wind that blows. Theseus' words confirm Paul de Man's intuition, uncovering for us the dark side of prosopopeia: a trope that makes it possible not only to "give shape", but also to "disfigure". In this dream, prosopopeia reveals not only an amusing but also a threatening face.

Translated by Prof. Jean Ward

References

Connor, S. 2000. *Dumbstruck. A cultural history of ventriloquism.* Oxford: Oxford University Press.

de Man, P. 1984. *The rhetoric of romanticism.* New York: Columbia University Press.

Dolar, M. 2006. *A voice and nothing more.* Cambridge Ma.: Massachusetts Institute of Technology Press.

Fergusson, M. W. 1985. *Hamlet*: Letters and Spirits. In *Shakespeare and the Question of Theory*, ed. Patricia Hartman and Geofffrey Hartman, New York and London: Methuen.

Freund, E. "I see a voice': the desire for representation and the rape of voice", unpublished conference paper.

Garrf, J. 2005. *Soeren Aabye Kierkegaard. A biography*, trans. Bruce H. Kirmse. Princeton: Princeton University Press.

Greenblatt, S. 1988. *Shakespearean negotiations.* Berkeley and Los Angeles: University of California Press.

Jonson, B. 1974. Timber or Discoveries. In *The poems. The prose works*, ed. C. H. Herford, Percy and Evelyn Simpson, vol. 8. Oxford: Clarendon Press.
Kierkegaard, S. A. 1909-1948. *Papirer*, ed. Peter Andreas Heiberg et al., vol. 5. Copenhagen Gyldendalske Boghandel, Nordisk Forlag.
Marvell, A. 1984. *The complete poems*, ed. George deF. Lord, London: Everyman's Library.
Rifaterre, M. 1985. Prosopopeia. *Yale French studies* 69: 103-127.

A New Territory? Literary Criticism as a Literary Genre

Tymon Adamczewski

Abstract The purpose of the present paper is to study the developments within the field of literary study and literary theory from the perspective of the notion of literature. Since the changes in theoretical discourse at the turn of the 1960 and 1970s literary criticism has become more essayistic in style and descriptive in nature, covering a wide spectrum of cultural phenomena and sharing various features previously reserved for literary or artistic activity. The initial reactions to the omnipresence of structuralism—together with its methodological limitations—resulted in original positions countering the objective modes of literary study. What is more, such standpoints, initially reflecting the counter-cultural vehemence of the times, fit into the anti-systematic tradition (Kierkegaard, Nietzsche, Derrida) which questions the borders between philosophy/theory and art/literature. In the work of such thinkers as Roland Barthes or Jacques Derrida one might observe transformations of the approach to understanding the object of study, the purpose of literary criticism or the way of reading literature. As a result, it is interesting to analyse the selected examples of post-1960s critical discourses in order to determine their traits and possible similarities or differences shared with literary discourse and verify their usefulness for literary study. Particular attention is paid to the work of Barthes, Derrida and post-modern thinking in the light of crossing borders and opening up new territories for interpretation in literary and culture studies. Therefore, such concepts as literature, style, and theory are analysed in the context of contemporary theoretical discourse.

T. Adamczewski (✉)
Nicolaus Copernicus University, Toruń, Poland
e-mail: tymon.a@wp.pl

J. Fabiszak et al. (eds.), *Crossroads in Literature and Culture,* Second Language Learning and Teaching, DOI: 10.1007/978-3-642-21994-8_2,

1 Introduction

Accompanying the social, cultural and political changes at the turn of the late 1960s and early 1970s, theoretical discourse underwent radical reformulations resulting in new possibilities of approaching literature. The erstwhile dominant structuralist critical positions were gradually supplanted by more philosophically inclined approaches, especially by the ones aiming to revise the considerably more traditional tenets, and frequently grouped under such terms as 'post-structuralism', or 'post-modernism'. The 1970s also brought in significant modifications of the nature and function of academic disciplines together with a refashioned role and shape of universities, which suddenly, in the aftermath of the civil-rights, leftist, or feminist movements' initial momentum, had to cater for newly-emerged political issues simultaneously opening a novel dimension for academic discourses. Among the turbulent transformations of the time considerable methodological permissiveness entered literary studies, not least with the introduction and subsequent proliferation of the notion of 'text'.

Viewed as a part of the communicative system, the concept exerted a profound influence on the relationships and functions of the author, the reader and the 'literary'. Contrasted with the idea of the 'text', literary work lost its meaning as an autonomous object of study. Defined by Barthes (1971) as an entity held in language and as an embodiment of a subversive force with regard to the old classifications, it became one of the key concepts in the post-1970s theoretical debate. The subsequent post-structuralist emphasis on ubiquitous textuality, despite its gradual waning throughout the latter part of the twentieth century, resulted not only in a particular blurring of boundaries between once separate discourses but also instigated a change in the ontological and epistemological status of the theoretical discourse, which, among its other effects, gave rise to numerous speculative positions such as 'culture studies' and the establishment of a set of key terms used by theorist, philosophers, readers, writers and critics alike. Once reserved for the discussion of 'the literary', criticism seems to perfectly reflect these changes and constitutes a record of the reshufflings marking the development of contemporary literary studies.

In "From Work to Text", the essay founding the textual studies, Barthes (1971, p. 57) himself relies on a literary device: the proposition of the text's traits concerning method, genres, signs, plurality, filiation, reading and pleasure is indeed 'metaphorical'. His meditations, typifying the emblematic post-structuralist preoccupation with language and signification, offer a seemingly provisional definition of the whole concept of the text as "that space where no language has a hold over any other, where languages circulate" (Barthes 1971, p. 64). By stating further that "the discourse on the Text should itself be nothing other than text, research, textual activity, since the Text is that social space which leaves no language safe, outside, nor any subject of the enunciation in position as judge, master, analyst, confessor, decoder", Bathes (1971, p. 64) ultimately obliterates the traditional boundaries between the eponymous concepts of a 'work' and 'text'.

Since "the theory of the Text can coincide only with the practice of writing" (Barthes 1971, p. 57), what is advocated is a novel, less-objectifying meta-language producing a more creative approach to reading and writing.

Moreover, Barthes' textual study seems to inscribe this part of his work into the heritage of a socially conscious Critical Theory. What the French intellectual shares with Max Horkheimer—one of the leading Frankfurt School's theorists—is a particular awareness of the social, i.e. political, implications of textuality, and, in fact, the view of theory as a liberating discourse, which is classified as *polemical, corrective,* and *oppositional* and which mobilizes the subjective relationships involving the scientist as a social being (Horkheimer 2002, p. 197). In addition, the textual-critical practice acquires the status of a creative production at least partly similar to the traditional artistic literary creations as a record of a particular encounter with a text, and producing, in effect, yet another text for analysis. Marking the beginnings of a new territory, Barthes' work not only describes a new, 'open-ended' object of study but also affects criticism as a theoretical practice distorting the boundaries between formerly established distinctions, genres or fields. In fact, such was the concept's richness that it stimulated further inquiry in other fields with the works of such authors as Michel Foucault, or Julia Kristeva.

In stylistic terms, the post-structuralist textuality and interest in language as a means of communication and signification overlap with the interest in etymology, the emphasis on a somewhat personal declamatory style, or unusually extravagant methodological and typographical practices. In fact, the coinages of new words and terms, the introduction of the first-person pronoun (see e.g. *S/Z* by Barthes)—contrasted with the more objective and impersonal rhetoric of the frequently passive voice endemic to the hitherto dominant structuralist discourses—and the excessive textual fireworks of Jacques Derrida's latter work correspond to an effort of finding new means of expression, perhaps similar in vein to the efforts of modernist artists.

With its disparate discourses and modes, post structuralism—frequently referred to as French Theory—exhibits the radical forces of transformations at work in the practices of literary critics. It traces the movement from the objective, detached structuralist meta-discourse to the altogether different methodology which centres on the intricacies of language as a medium of communication and draws on its self-referential qualities, such as the constant play of meanings, while also accommodating a degree of political significance. Interestingly, the term French Theory represents the frequently reductive tendency of many present day discourses attempting to represent the originally diverse, rich and otherwise disparate theoretical positions, e.g. of such *maîtres à penser* as Jacques Derrida, Michel Foucault, Jacques Lacan, or Roland Barthes, in a narrow and simplified form. Such is the case, for example, with deconstruction, which, due to its American filtering in the 1970s, often boils down to political practice. Nevertheless, the impact of the French thinker's works on the theoretical discourse in general turned it into a less rigid form of scholarly practice, simultaneously producing as many controversies as supporters of the new modes of theoretical meditations. In spite of the critical voices or the charges of deliberate haziness of argumentation—especially in the case of deconstruction—it can be argued that the theoretical developments of the turn of the 1960 and 1970s can be looked at from the perspective of literature, or studied, as it were, in their own right. Indeed, the theoretical texts of the French thinkers, dealing not solely with literary production, exhibit striking affinities to literary works: not only with respect to the use of particular aesthetic tropes, rhetorical figures or structural devices present in the critical discourse, but even more as an embodiment of an often personal experience, or a document of an encounter with a text, which shares various features with many areas of interests of the traditional literary production—whatever its definition for the time being might be.

Consequently, the present paper focuses on this relationship between literary criticism, theory and literature and the possibilities of looking at literary meta-discourse from the perspective of the notion of literature. It deliberately centres on the works of selected French thinkers due to the most explicit degree in which their critical texts illustrate the issues at hand.

2 Theory's Limits

The possibility of studying literary criticism from the point of view of literature is invariably related to the opposition between theory and practice. Problematic as the distinction might currently be, one may look at the development of the theoretical ponderings—including the ones concerning literature—as a reaction to this pair elaborated on already by Plato and Aristotle. Presently however, as numerous post-war literary texts and critical orientations point out, the split between theory and practice is never complete and, in fact, the contemporary Western philosophy partly constitutes an effort of showing that such a clear-cut distinction implies rigid hierarchical differentiation. Jacques Derrida's work is immensely informative in this respect—also in terms of his fundamental influence on literary studies—since it aims to show, among other issues, that such binary oppositions are implicitly interrelated and that in such a contrasting framework one discourse is given precedence over another. Therefore, even though it might seem that theory underlies practice, the latter never constitutes a simple and passive reflection of the former. Actually, one does not need to look very far in order to notice the reciprocal nature of this relationship: practice never consists in a mere application of theoretical premises and the two notions constantly interact, since in order to create literature one needs to ponder on its mechanisms in the first place. All human practice is, in this light, implicitly theoretical and the reflection about literature might be either silently included in the practice of creating it, or constitutes an overt meta-discourse commenting, accompanying and explaining literature (Ayers 2008, p. 4). Traditionally literature was seen not solely as an expression of an author's creativity, but at the same time, as a comment on the nature of the whole discipline. This feeling is expressed in the bulk of contemporary so-called postmodern fiction, which incorporates the meditations on the status of literary discourse into the actual writing together with the observations about the perception of reality. Interwoven into the literary practice, the metafictive or theoretical dimension emphasises the way in which this type of fiction parallels the problems facing today's theoretical and literary ponderings such as the debatable status of the very notion of theory, the growing anxiety and awareness about language as a means of communication, literary forms, the very act of writing, and their relationship with reality. In this way fiction mimics the changes in the areas of contemporary intellectual inquiry and evidences the link with the theoretical and philosophical traditions questioning the legitimacy of human inquiry and its status and relation to the material world.

Accordingly, the main problem with the exact definition of literary theory is the difficulty in determining its boundaries. On the one hand, theory as a meta--discourse can signify a type of an external, detached contemplation devoted to epistemological inquiry involving learning, systematising and aiming at a formulaic description of the *praxis*. In this respect, theory fits into a long-standing tradition of the systematic strain in Western philosophy exemplified by the work of Aristotle, or Immanuel Kant which ascribes to theory the place of an objective and explanatory discourse. Practice is thus located in a subservient post and further demonstrates the links between the philosophical and literary meditations. Hence, such thinking locates theory in a space beyond practice, which, in turn, is governed from without, and believed to contain distinct features or characteristics. However, the bulk of ongoing critical debate demonstrates the opposite, i.e. the effacement of boundaries, persistent questioning the notion of style, or the insistence on interdisciplinarity, all of which work to dissolve the original clear-cut distinctions. In fact, most of contemporary thought devoted to theory concerns its repeated entanglement with practice. Many critics stress that theory is not a notion neatly described by a set of rules, or characteristics, to be found in some official documents about 'Literary Theory', but can also be located in criticism and commentary, and "may be silently embodied in literary works, in the practice of their creation, distribution and reception, and—if we are to assume any relationship between 'literature' and 'life' at all—theory as discourse and as practice will reveal itself as to belong in large part to the cultural, social and political discourses, institutions and practices of which it is part" (Ayers 2008, p. 4). It should be clear by now that recent discussions stress theory's *immanence*, or the fact of its continuous involvement with institutions other than University (Ayers 2008, p. 5). Thus, theory does no longer occupy the external, abstract dimension of a master discourse but, as it is aptly characterised by Jameson (1991), serves to include a plethora of discourses, also working within the domain of practice additionally shaped by its relation to and concern with the social or cultural fields of which it is part. In addition, the notion of theory is entangled in criticism and literary theory, all of which have undergone significant transformations merging into a "new discursive genre" of a "theoretical discourse" (Jameson 1991, pp. x–xvi), which, according to this American critic, performs a political function of discrediting the autonomy of academic disciplines, and thereby the classification of texts they perpetuate, into political philosophies, historical and social speculation, novels and plays, philosophy, and autobiographical writing, each of which is claimed by a separate tradition (Jameson 1991, pp. xvi, 240). Originally dating back to the 1960s, theoretical discourse is endemic to the postmodern era, which effaces the former distinctions between particular disciplines, and, at least partly, preserves its subversive force against the more traditional academic sites of knowledge and regular forms of intellectual inquiry, attempting simultaneously to translate this into concrete practice and politics.

Inherently linked with the practice of criticism, the field of theory is thus subject to little agreement about its content, meaning and boundaries. Unfortunately, theory's heterogeneity, which includes numerous disparate discourses and speaks with various voices, contributes to the problematic nature of the essence and omnipresence of the concept. Theory is in fact ubiquitous: there is just so much of it about that it can be easily found in academic bookshops and scholarly journals (Macey 2000, p. v); and the students in departments of literary–and cultural studies are increasingly required to read 'theory' rather then what was once termed 'literary criticism' or even 'theory of literature' (Macey 2000, p. 379). The contemporary insistence on interdisciplinarity and plurality of discourses within the modern humanities stems partly from the decline of the power and importance of traditional disciplines, as well as from the effort to question their status and influence. The fact that 'theory' has poured over numerous fields of human activity testifies to important changes in the way in which the role, as well as the status of theory is perceived in contemporary culture. No longer seen as a master-explanatory discourse, theory, its scope of interest, and the spectrum of the objects of its study has certainly widened and diversified. Various forms of the present day theory might be thus identified among the most distant domains of contemporary intellectual activity such as film, literature, culture, women studies, arts, or politics in general. Doubtlessly, the contemporary theoretical discourse incorporates numerous diverse aspects of theoretical ponderings, including the questions of methodology, interpretation as well at the aims to which they are to lead.

Literary criticism, as a seemingly more restricted branch of 'theory', may serve as a good example of the current problems affecting the concept under consideration because reading literature nowadays involves much more than it used to. Instead of limiting the understanding of this activity to a simple discussion of the way texts portray class or gender, Habib (2008), for example, acknowledges the significance of modern criticism and theory in its ability to recognise the full consequences of the practice of reading literary texts. Reading literature today entails something significantly different than it used to do still half a century ago and has very tangible consequences in the light of the present socio-political situation. According to the critic, not only the social institutions, traditionally perceived as harbouring widely understood positive intellectual values and the possibility of free-thinking, are now threatened but because of the transformations in the humanities, church, public sphere, media and the impact of technology and capitalism on life in general, theory touches upon the democratic process itself and the very values endemic to it (Habib 2008, p. 1). Indeed, the insistence on the link between the political and cultural phenomena heavily influences the consequences of practising literary criticism, and theory as such. Despite the initial differences in modes or aims of literary theories, what binds the disparate instances of theory together is the fact that they all revolve around the reading of texts. This straightforward conclusion has fundamental consequences for understanding the nature and purpose of theory and criticism and its relation to literature. If the post-war literary theory emphasises theory's immanence, its accompanying features consist in the attempts at refashioning the distinctions between literature and criticism as well as in subversion against the possibilities of a stable and fixed interpretation supplying an objective and final truth about a given text.

Consequently, contemporary literary criticism is still related to critical theory. For Lois Tyson (2006, p. 6) for example, since criticism and theory try "to explain the assumptions and values upon which various forms of literary criticism rest", literary criticism is, in fact, critical theory put into practice. For her, the critics' activity does not rest in the evaluation of literature but its explanation, especially with respect to the theories which involve a desire to change the world—such as, for example, feminism or Marxism—which benefit even from the 'flawed' works, since through them they can expose the operation of oppressive ideologies. Consequently, criticism does not always aim at finding faults with literary works, but focuses on the explanation of literary works with respect to their production, meaning, design and beauty. On the other hand, critics also tend to be more centred on one another's interpretations, rather than literary works themselves.

3 Reading Criticism as Literature

Reading anything *as* literature suggests that what is to be read can be approached at least in two ways: in a literary and non-literary manner. Moreover, it implies an ability to formulate a clear definition of the notion of literature together with the way in which it is, or should be read, assuming the method and forms of attention directed at the texts under scrutiny. This, on the other hand, presupposes the set of qualities and characteristics typical for the texts generally labelled as literary. Consequently, the training in the ability to read texts in such a way, simultaneously producing a proper discourse for their description, results in the production of a class of authorities, or individuals believed to be able to describe, explain and comment on the texts in question. That is exactly what the post-structuralist critics rebelled against, hence their work could also be looked at as an effort to dissent from the domination of authorities—an idea corresponding to the 1968 student upheavals across the Western culture.

On the other hand, this attitude also suggests the idea of a particular, 'literary' attention which is directed at texts and leaves it to the reader to decide which texts are in fact literary, instead of recognising a set of qualities in the texts themselves.[1] The notion of text characterized by Roland Barthes together with the Linguistic Turn in the late 1960s, deconstruction, postmodernism, and the general blurring of borderlines between practice and theory, or philosophy and literature, contributed to a novel view on criticism. As Jacques Derrida comments:

Literature has often been read in terms of a dominant meaning or of a dominant form; although a given critical tradition may emphasise one of these at the expense of the other, or insist on their interdependence, this does not diminish the determining force of the philosophical categories themselves. Or it has been read as understandable in terms of origin (biographical, historical, socio-economic, psychoanalytic) or goal (aesthetic, moral, spiritual, political), or as fundamentally mimetic and therefore answerable to a classical notion of truth. The result has been the representation of literature as itself governed by these oppositions and assumptions, a representation which one cannot simply call "inaccurate" since it responds to a marked tendency in a large part of the Western literary writing. Of course, the literary tradition is far from homogeneous, and some works of literature, criticism, and literary theory have resisted such philosophical categories more than others—not by abolishing mimesis, reference, form, content, genre, origin, intention, and so on, but by staging, suspending, and testing these concepts, showing them to other than the self-consistent, controlling categories they are usually taken to be (Derrida 1992, pp. 3–4).

In this respect, if literature consists in an eclectic heterogeneous and clever teasing out or testing of the traditional concepts used to describe or characterise it, then criticism and theory, especially in its French post-structuralist variety, can be seen as an example of a discourse operating between the fields of literature (artistic creation) and scholarly objectivity, even though simultaneously challenging this tradition.

The notion of text, due to its entanglement in language and simultaneous shifting of the positions of reader and writer, allows for a new perspective on the critical commentary. Traditionally, criticism is a predominantly a parasitical genre—it cannot exist without the original text to which it constantly has to refer to, or feed on. Through the image of a network the semantic textual plurality—etymologically retained within in the original meaning of the word 'text', i.e. 'fabric'—works as a place of meetings for the three positions: the text, the author, or the reader/critic. In this way criticism can be seen, for example, as a form of autobiography, since as readers we are faced with a subject commenting on a yet another (literary) text. Such a perspective also parallels the way Jacques Derrida sees literature: as an "institution which allows one to *say everything, in every way*" (Derrida 1992, p. 36). Such a remark is especially informative in the view of the post-structuralist critique of the classical (New Critical, or Structuralist)

[1] Good sources which discuss this issue are, among others, Jonathan Culler's (1997) *Literary Theory: A Very Short Introduction*, or *Literary Theory: An Introduction* by Terry Eagleton (1996).

theoretical pretence to its status as a privileged metalanguage. In fact, contemporary literary theory and criticism can be viewed from the perspective of borderlines, frontiers and demarcation lines. Even the father of deconstruction himself acknowledges the lack of necessity to distinguish between 'literature' and 'literary criticism', however, making a reservation that all forms of writing or reading should not be assimilated, but these new distinctions "ought to give up on the purity and linearity of the frontiers" (Derrida 1992, p. 52). This, in turn, leads to the questioning the traditional distinction between structuralism and post-structuralism. The former movement expressed epistemological fundamentalism, shared at some point by its most prominent exponents who expressed an explicit belief in the possibility of reaching under the veneer of transitional cultural phenomena or processes in order to obtain positive knowledge about the continuous, ordered base (or deep structures). In contrast, post-structuralism voices a general epistemological scepticism, most visible perhaps in the discussion about interpretation and criticism, and places theory within the domain of discourse. In this way, literary theory can neither provide an objective external perspective on the interpretative practice—because it itself relies on its results—nor on the literary practice—since it is itself entangled in the processes and conditions which were subject to analysis (Nycz 2000, pp. 21–23).

Accordingly, post-structuralist criticism seems closer to literary production since it is centred on the experience of the text rather than on its assessment from an external objective vantage point. In the same vein, criticism can be read as a literary genre, however, with several reservations. The dominant one being that contemporary criticism, or theory for that matter, has been heavily influenced with theoretical positions which had originated in philosophical inquiry (Jacques Derrida, Michel Foucault) coupled with sociological and psychoanalytical observations (Roland Barthes, Jacques Lacan) and resulted in their inclusion into literary studies departments. In consequence, (literary) texts were seen as cultural phenomena, or sites of ideology which, from a critical standpoint, should be analysed and resisted. As time went by, theory seemed to have turned into a meta-discourse focusing on determining the status of the dominant tenets. Therefore, literary theory has also evolved into an essayistic and self-commenting endeavour, employing numerous literary devices and textual strategies which contribute to the possibility of reading it as a (literary) genre, class or type.

4 Examples: Two Important Figures

The selection of examples is always crucial and particularly telling, also by means of exclusion of the aspects, issues or objects *not* chosen to illustrate a given point. In this case the situation is even more problematic given the scope, diversity and manner of the so called post-structuralist works. Nonetheless, the two authors whose critical works perhaps most clearly lend themselves to a literary attention are Roland Barthes and Jacques Derrida and the choice of these thinkers seems to me especially important due to a particular evolution of their respective critical positions, which may help to demonstrate certain points.

The writings of Barthes, the instigator of the "Death of the Author" (1968), who managed to produce a substantial array of various theoretical texts, are universally recognised as illustrating the transition from structuralist formality to the post-structuralist openness. However, whether it is the Barthes who reads into cultural phenomena trying to uncover the latent meaning of soap powders and detergents in *Mythologies* (1972 originally published in 1957), the Barthes (1974) analysing the most minute yet striking textual connotations in *S/Z*, or the Barthes (1975) pursuing *The Pleasure of the Text*, his writings produce a compelling discourse which, apart from its frequently unusual insights, infallibly grasps the reader's attention. Moreover, his now canonical study of Balzac's short realist story "Sarrasine" in *S/Z*, with the length of the whole commentary far exceeding the original narrative, is a perfect illustration of the way in which a textual commentary turns into a text. The critical discourse, through its particular insistence on the ideas of random selection of the lexias, or studied fragments, as well as personal remarks about their connotations results in producing the impression that both the meaning and a text's functions depend on and are created by the very practice of reading. The scant methodology (compare e.g. the codes of narrative) used in his later critical commentaries serves to demonstrate and explain rather than prove a given point. If any method is used, it is rather employed not as a reductive formal tool, but as a suggestion for ways of approaching the text so as to observe the potential disruptive play of meanings in the original text. In this way, literary criticism becomes an informative written record of an almost personal confrontation with a text, doubtlessly enriching the experience of reading which, similarly to literary works, opens new perspectives on further encounters with writing.

Similarly immediately recognisable for his unusual textual strategies, the works of Derrida (1979) exemplify the philosophical tradition in literature and prove valuable in their own right. Literary criticism for Derrida (1992, p. 53) is in general "very philosophical in its form, even if the professionals in the matter haven't been trained as philosophers, or if they declare suspicion of philosophy; literary criticism is perhaps structurally philosophical". His notoriously difficult texts constantly brake away with all possible conventions, borders and frontiers; also in terms of typography as such works as Glas (1986), The Truth in Painting, or "Living On/Borderlines"—among others demonstrate. Apart from the content and form, which are frequently difficult to follow, since Derrida's (1987) text themselves often illustrate the practice of deconstruction, the texts of the author of *De la grammatology* (1967) are akin to a certain extent to some modernist experiments with the powers of linguistic representation.

Needless to say, the writings of these two authors—as the representatives of the most infectious 'French Theory'—with their concepts, forms and style of criticism have had a profound impact on the tradition of contemporary literary criticism and theory, the developments of which are equally associated with the late-1960s revolutions and the resulting theoretical transformations. The two thinkers as such recognise the role which language plays in the reality and acknowledge the instabilities lying at its core but, equally importantly, their work, to an extent, demonstrates the fate of the revolutionary theoretical discourses. Both textuality and deconstruction can be described in the terms of subversion, radicalism and revolution which—despite their original disruptive force—have become, in a way, institutionalised, illustrating the consequences of such critical practice for the status and powers of literary studies as a discipline and its role in the process of education. In fact, both textuality and deconstruction have become a standard on university curricula and now constitute well-established methodologies for working with both—literary and non-literary—texts, in spite of Derrida's reservations that the aim of deconstruction is, actually political and that it "is not a method and cannot be transformed into one" (Derrida 1985, p. 3). Therefore, it seems thoroughly unnecessary to analyse the works of the post-structuralist critics, such as Derrida, Barthes or any other, from the point of view of traditional formal (literary) analysis in order to trace their use of allegory, metaphor, ellipsis, or any other figures, since the application of formalist or stylistic methodological apparatus to the texts would delimit their original explanatory powers and appeal. What is perhaps more important is the fact that yet another way in which one can look at such criticism is its historical significance—not without influence on the present shape of literary studies as well—since it also constitutes the record of responses to literary texts and the manner in which the critics, or philosophers have related themselves to them.

5 Conclusion

Paving the way for numerous contemporary approaches, the post-structuralist developments in the area of literary criticism and theory in general certainly set up a new territory for the theoretical ponderings of the past. Interestingly, the critical writings of especially the French thinkers, demonstrated a new style of theoretical commentary. On the one hand, the downside of such criticism at times consists in the impression of being incoherent, difficult to grasp and sceptical, since such criticism focuses on the negative aspects such as lacks, incongruities and inconsistencies of a text. The origins of such practice lie in its inclusion in the anti-systematic tradition in philosophy, exemplified by the works of Søren Kierkegaard, Frideric Nietzsche, or Ludwig Wittgenstein among others. Owing to this heritage, the type of criticism in question is much more personal (at times autobiographical), aphoristic or lyrical, but is equally capable of penetrating the problems under analysis.

It seems, however, that contemporary literary criticism does no longer constitute a new territory. It can nonetheless still be framed into a category of a genre, especially a historical one (a record of responses to literary texts and critical tendencies), which due to its multifarious content and scope of interdisciplinary interests—regardless of the problems with definition this produces—can still evidence important tendencies for the humanities. Namely, the mechanisms—also visible within literary studies—which seem to govern contemporary Western culture: a certain absorbent power, which incorporates the once subversive and radical positions into standardized responses; methods taught within the same sites of knowledge; and power which these positions initially were critical of. What remains to be studied is perhaps the usefulness of the approaches presented here, and the possibility of viewing the critical texts from the perspective of literature. The solutions which seem most logical is the possibility of criticism's combination with various types of textual hermeneutics which might result in opening up diverse interesting and important contexts for literary study, and which seems to have been already initiated by some contemporaneous orientations. Nevertheless, the merits of the post-structuralist approaches remain in their contribution to democratization processes which work against the supremacy of the authorities. They result in the liberation of the critic from the position of a 'slave' of the text, language, concepts, or terminology. Additionally, the frequent inclusion of the political function, derived from various Marxist approaches, into the textual practices of criticism contributes to its significant position among the discourses of contemporary critical theory, which, on the other hand, have been preoccupied with a much wider set of 'texts' than the original literary ones. Besides, bearing in mind the influence on academic education and the critical business, the methodologies of many textual practices—despite their original hostility towards any forms of formalisation or systematisation—have become so widely employed and discussed in university classes that one needs to remember that such a representation of criticism is highly selective and simplifying, especially with respect to the original richness and heterogeneity of the critical discourses. Consequently, post-structuralist thinking about language and texts still reminds us of the vicissitudes of meaning and makes us more cautious about such issues as interpretation, or the seemingly simple reading practice.

References

Ayers, D. 2008. *Literary theory: a reintroduction*. UK: Blackwell Publishing.

Barthes, R. 1968. The death of the author. In *The rustle of language*, trans. Richard Howard, pp. 49-56. California: University of California Press, 1989.

Barthes, R. 1971. From work to text. Reprinted in *The rustle of language*, trans. Richard Howard, pp. 56- 64. California: University of California Press, 1989.

Barthes, R. 1972. *Mythologies*. New York: The Noonday Press.

Barthes, R. 1974. *S/Z*. USA: Farrar, Strauss and Giroux.

Barthes, R. 1975. *The pleasure of the text*. USA: Farrar, Strauss and Giroux.

Culler, J. 1997. *Literary theory: a very short introduction*. New York: OUP.

Derrida, J. 1979. Living On/Borderlines. In *Deconstruction and criticism*, pp. 75 -177. London and Henley: Routledge and Kegan Paul,

Derrida, J. 1985. Letter to A Japanese Friend. In *Derrida and différance*, ed. David Wood and Robert Bernasconi, pp. 1-5. Warwick: Parousia.

Derrida, J. 1986. *Glas*, trans. John P. Leavey, Jr. & Richard Rand. Lincoln & London: University of Nebraska Press.

Derrida, J. 1987. *The truth in painting,* trans. Geoffrey Bennington and Ian McLeod, Chicago, IL: University of Chicago Press.
Derrida, J. 1992. *Acts of literature*. Edited by Derek Attridge. New York, London: Routledge.
Eagleton, Terry. 1996. *Literary theory: an introduction*. Oxford: Blackwell.
Habib, M. A. Rafey. 2008. *Modern literary criticism and theory. A history*. Oxford: Blackwell.
Horkheimer, M. 2002. *Critical theory: selected essays.* New York: The Continuum Publishing Company.
Jameson, F. 1991. *Postmodernism, or, the cultural logic of late capitalism.* London – New York: Verso.
Nycz, R. 2000. *Tekstowy świat: poststrukturalizm a wiedza o literaturze. Kraków:* TaiWPN Universitas.
Macey, D. 2000. *The penguin dictionary of critical theory*. London: Penguin Books.
Tyson, L. 2006. *Critical theory today. A user-friendly guide*. New York – London: Routledge.

Cutting into a New World: Reading *The Cut* Through Slavoj Žižek

Robert Kielawski

Abstract The art of Mark Ravenhill revolves around gender constructs and relations between people in the consumer society. With his conscience clearly on the left end of the political (as much as philosophical) spectrum, the playwright would not shun the reputation of a critic of the so-called postmodern ideology. The paper aims to analyse Ravenhill's *The Cut* as an attempt on the part of the playwright to engage with cultural theory and psychoanalysis. With its exploration of the phantasmic background of a society at a time of change, the play is interpreted with reference to Slavoj Žižek's numerous deliberations on the reinvention of utopia revolution and the big Other in decline. The analysis sets out to prove that *The Cut* and its representation of the post-revolutionary reality moves Ravehnill into a new territory of engagement with theory and politics, a territory much exceeding that of in-yer-face sensibility. The paper concludes that the play is a warning against post-politics as much as a calling for reconsider political agency. The characters in the play fail to constitute subjective positions capable of anything beyond a politics based either on the fear of negativity, the fear of the cut, or on faith in the university as an apolitical institution driving social change. That said such an attempt to confront philosophical ideas with drama results in new illuminating interpretations which both scrutinize philosophical ideas and shed new light on contemporary drama.

R. Kielawski (✉)
Philological School of Higher Education, Wrocław, Poland
e-mail: robert.kielawski@gmail.com

J. Fabiszak et al. (eds.), *Crossroads in Literature and Culture,* Second Language Learning and Teaching, DOI: 10.1007/978-3-642-21994-8_3,

1 Exploring New Territories of the Political in Drama

The present paper examines Mark Ravenhill's *The Cut* (first published in 2006), a play that confirms his moving away from, rather than extending, the sensibility of in-yer face theatre (Sierz 2008, p. 32). Indeed, as this paper argues, the play provokes a reflection on crossing the frontiers of a certain politics of the 90s relying on provocation. In particular, it examines the playwright's handling of the subject of social and political change. The ideas of Slavoj Žižek dealing with political agency, utopianism, and the idea of revolution, as scattered about in *The sublime object of ideology* (2008), *The ticklish subject* (2000) and *The paralax view* (2006), will be used to provide theoretical scaffolding for the adopted line of argumentation. The very choice of Žižek is by no means accidental. It discloses an assumption that Ravenhill shares and, at the same time, labours to scrutinize a leftist sensibility. Given this affiliation, the following discussion makes numerous references to Lacanian psychoanalysis and, ultimately, seeks to deal with some general questions about its political potential. Thus, on the one hand, the play itself seems to be the playwright's leap into a different territory. On the other hand, the analysis proposed below tries to stake out new territories by making use of political theory and psychoanalysis to analyse drama. Refraining from providing overtly unambiguous conclusions, the paper arranges a meeting of the chosen dramatic text with a selection of texts from the field of cultural theory and philosophy.

1.1 Taking Theoretical Bearings

David Greig once said that political theatre must contain a suggestion that change is possible (Sierz 2001, p. 240). If one takes this somewhat bald statement at face value, adds the fact that it concerns the socio-symbolic sphere, and benchmarks it against the work of Mark Ravenhill, there appear questions about the political message behind his plays. One way of answering them is perhaps best summarized by Clare Wallace who writes that "Ravenhill never presents any convincing alternative" (Wallace 2006, p. 130). On the other hand, Aleks Sierz posits that Ravenhill represents the "naughty 90s" when the political in drama took on the form of individual or private pain and alienation (Sierz 2001, p. 241). This corresponds to what Hans-Thies Lehmann has to say about the political in the epilogue to *Postdramatic theatre*, namely that the political is universal whereas art is the particular *par excellence* (Lehmann 2006, p. 178). Thus, as Lehmann concludes, the politics of drama in multinational capitalism can be resistant but not transgressive. Ravenhill *à la* Wallace would hardly protest here. Such a political agenda is not uncommon among playwrights sharing the sensibility of in-yer face theatre. As theorised and historicised by Sierz, their political agenda boiled down to a shock therapy, both in terms of theatrical form and content, whereby the audiences were to be awaken from their (post)ideological slumber by being exposed to violent images.

In the past decade or so, the theorising of drama and theatre engaged in a rethinking of the political can be traced back to similar debates in philosophy, cultural theory and political science. It was within those fields that some voices proclaimed the revival of a theory of ideology[1] or a reconsideration of universality.[2] Unsurprisingly, the loudest cry came from post-Marxist philosophers and theorists. Given the demise of the so-called Grand Narratives as pronounced by Jean-François Lyotard in *The postmodern condition* (1984), Frederic Jameson was precarious when writing in his *Postmodernism* about a global cognitive mapping as a means to "regain a capacity to act" in the disintegrated socio-symbolic space (quoted in Malpas 2005, p. 121). Again, what Jameson and, another prominent post-Marxist, Ernesto Laclau seem to have stressed is the fact that the politics of postmodernism struggled to move beyond "resistance without foundations" (Malpas 2005, p. 128). Yet one of the more provocative theoretical endeavours to undermine the politics of postmodernism has been made by Slavoj Žižek. The Slovenian cultural theorist pounds away at ideas that fly in the face of post-politics: a term he defines as a pathological form of politics which aims at converting the unpredictability in democracy into apolitical administration. In a post-political world individuals can hardly do more than pursue their consumerist fantasies engineered by experts (Žižek 2008b, p. 325). Such a political deadlock closes the door for what he calls "traversing the fantasy," or, in other words, for an authentic political Act, or a *Truth-Event*[3] capable of shattering "the existing

[1] Cf. The chapter on politics in Malpas' *The postmodern* (2005, pp. 105–132) and Laclau's article "The death and resurrection of the theory of ideology" (1997).

[2] Cf. The manifesto of the left, *Contingency, hegemony, solidarity* (2000) written jointly by Judith Butler, Ernesto Laclau and Slavoj Žižek.

[3] Žižek borrows the term Truth-Event from Alain Badiou. The term is directly linked to Badiou's views on the role of philosophy, which is made clear in his *Manifesto for philosophy*: "The specific role of philosophy is to propose a united conceptual space in which naming takes place of events that serve as the point of departure for truth procedures. Philosophy seeks to gather together all the additional-names. It deals within thought with the compossable nature of the procedures that condition it. It does not establish any truth but it sets a locus of truths. It configurates the generic procedures, through a welcoming, a sheltering, built up with reference to their disparate simultaneity. Philosophy sets out to think its time by putting the state of procedures conditioning it into a common place. Its operations, whatever they may be, always aim to think 'together' to configurate within a unique exercise of thought the epochal disposition of the matheme, poem, political invention and love [...]. In this sense, philosophy's sole question is indeed that of the truth" (1997, p. 37). It seems no wonder why Žižek to a large extent allies with Badiou with regard to locating the political at the very centre of any philosophical project. Both philosophers try to rehabilitate philosophical grounds for effective political agency. Yet the Truth they theorise is only discernable from the standpoint of the Event. Or as Žižek puts it: "The Event emerges ex nihilo: if it cannot be accounted for in terms of the situation, this does not mean that it is simply an intervention from Outside or Beyond—it attaches itself precisely to the Void of every situation, to its inherent inconsistency and/or its excess. The Event is the Truth of the situation that makes visible/legible what the 'official' situation had to 'repress', but it is also always localized—that is to say, the Truth is always the Truth of a specific situation. The French

discursive universe" (Žižek 2000, p. 377). The archetypal political act is a revolution, a collective act that leads one to consider yet another crucial hub of Žižek politically engaged ideas, namely that of his criticism of the liberal division into the private and the public (Cf. Boucher 2005, p. 39). Political agency omits the liberal trap only when the said division is discarded, only when the individual realizes that the antagonism permeates both the private and the public and thus abolishes the division as such (Cf. Dybel, Wróbel 2008, p. 319). Žižek emphasises that the problem is that we do not believe that individual acts can have global consequences and thus view catastrophes as systemic or anonymous (Žižek 2008a, p. 454). Adding all that to the already mentioned lack of cognitive mapping renders the subject battered and bruised, not unlike many of the characters in Ravenhill's plays. The falling down of symbolic walls makes the subject resort to "imaginary *simulacra*" which "trigger the need for violence in the Real of the body itself" (Žižek 2000, p. 369). Žižek sees this consequence as the obverse of the consumerist utopia of social relations viewed as an infinite multitude of free choices and unconstrained enjoyment. His interpretations of sadomasochism and the Master/Slave relations suggest that the consumer society needs to be counterbalanced by a strong disavowed attachment to subjection (Žižek 2000, p. 345). Therefore, both an often desperate quest for subjection to a Master and the need to mutilate the body are considered by the Slovenian philosopher as the symptoms of the consumer society. Yet Žižek is wary to distinguish between the "traditional" and the "postmodern" cut:

> [T]he traditional cut ran in the direction *from the Real to the Symbolic,* while the postmodern cut runs in the opposite direction, *from the Symbolic to the Real.* The aim of the traditional cut was to inscribe the symbolic form on to raw flesh, to 'gentrify' raw flesh, to mark its inclusion into the big Other, its subjection to it; the aim of postmodern sado-maso practices of bodily mutilation is, rather, the opposite one—to guarantee, to give access to, the 'pain of existence', the minimum of the bodily Real in the universe of symbolic simulacra. In other words, the function of today's 'postmodern' cut in the body is to serve not as the mark of symbolic castration but, rather, as its exact opposite: to designate the body's resistance against submission to the socio-symbolic Law (Žižek 2000, p. 372).

2 Political and Psychoanalytical Traces in *The Cut*

What Ravenhill does in *The Cut* is thus engage with the above ideas, namely with the problem of the need for the (postmodern) cut (Scene One), the division into the public and the private (Scene Two), and the idea of social change (Scene Three). Given that interesting terminological coincidence that suggests that Ravenhill is no

---

(Footnote 3 continued)
Revolution, for example, is the Event which makes visible/legible the excess and inconsistencies, the 'lie', of the ancient régime, and it is the Truth of the ancient régime situation, localized, attached to it" (Žižek 2000, p. 130).

stranger to cultural theory and the academic debate of the philosophical left, it seems plausible to consider the cut, as it functions in the play, to be a form of interpellation. The general connotations surrounding the cut are thus quite explicit. One is told that it is painful and barbaric (Ravenhill 2008, pp. 186, 216). It evokes images of ordeals, torture and the like. One also quickly learns that the cut can be read as a (perhaps modernist-like) way of interpellation, which during that "last lot" was hardly a matter of free choice (Ravenhill 2008, p. 194). Now with the "new lot," John wants to undergo the cut, which in turn makes Paul anxious. This anxiety, arising from John's perverse desire to be cut, proves to be a symptom of Paul's hysteria, which shows his uncertainty about his symbolic position of power. This results in his obsession with his own impotence:

PAUL: [...] After the incident with the gun I find trust impossible.
JOHN: Of course. [...]
PAUL: Which has made lovemaking with my wife impossible. It's only when you can't ... when you can no longer close your eyes during, the, the, the ... act that you realise ... lovemaking with the eyes wide open ... impossible. (Ravenhill 2008, pp. 194–195)

It also results in his reassertions about superiority over John, who by no means questions his authority:

JOHN: Beat me away. Beat me to the ground. Beat me to death. I'm weak. You're strong. You can easily beat me
PAUL: Yes, yes, I suppose I can.
JOHN: But I'm not going to strangle you. (Ravenhill 2008, p. 195)

Given that true authority operates without such self-assertions, Paul sees his symbolic mandate and its exercise as a burden. It seems that all this anxiety and hysteria pushes Paul into confiding in John as if he was the big Other itself, as if he was the "natural receptacle of his secrets" (Žižek 2004, p. 207). It is that moment that opens up an interesting twist in the dialogue between the two characters and as if saturates their subjective positions. What then John imputes to Paul in an almost conscious use of psychoanalytic language is the desire to return to darkness, to the pre-symbolic, where there is no body and no language.[4] This return to darkness is clearly equalled in the play with the postmodern cut. Interestingly, Ravenhill gives it a strong anti-Enlightenment edge of moving away from light into darkness.

Therefore, in Scene One Ravenhill tries to dramatise the deadlock of a pervert (John) organising the desire of a hysteric (Paul). Žižek explicates that "beneath the

---

[4] "JOHN: Ah-hah. Total darkness. And you have no body. Your body has dissolved. [...] The cage has vanished. And you are free. Feel the darkness. [...] Where the monsters live. Where the witches live. Where the paedophiles are. [...]. Take a torch into the woods. Lies. All of it lies. The void. It'll eat you up. The chasm that swallows the sailors [...] Take a map, make a rope bridge. Steer clear of the void. Lies lies, all of it lies. They've told you lies and you've kept you eyes open. When all freedom asked of you was to close your eyes. And now you've closed them. And you've made a start. [...] There's no history. All that struggling to move forward, To expand, to progress. That's gone away. And there's no society. [...] Don't try and feel your body. Don't reach for the reports. Don't try and call your wife. Because it's all nothing. There's only truth. There's only you. Darkness is light. Void is everything. You are truth" (Ravenhill 2008, pp. 195–197).

hysteric's rebellion and challenge to paternal authority there is thus a hidden call for a renewed paternal authority/law, for a father who would really be a 'true father' and adequately embody his symbolic mandate" (Žižek 2000, p. 334). Yet for traversing the fantasy, the hysteric needs an analyst, who would confront him with the truth of his desire, and not a pervert, who is the "inherent transgressor," who brings to light the spirit of a diligent consumer preparing for the cut (Žižek 2000, pp. 247–250). What their relation exhibits is mutual dependence. To recapitulate: the hysteric Paul cannot speak the discourse of the Master and therefore is left at the mercy of John, who wants to enter into a Master/Slave contract. Again—the hysterical and pervert positions are presented here as complementary and lead to a deadlock.

Scene Two moves away from the public space into Paul's private homeliness. It is there that Paul's discourse of the hysteric is complemented by the discourse of university represented by his son Stephen. After Lacan, Žižek defines the discourse of university as a discourse which produces subjects occupying the "neutral" position of knowledge, a position disavowing the performative aspect of its own language (Žižek 1998, p. 78). It is precisely the disavowal of politics that Paul recognizes and already in Scene One tells John: "I'll send you straight off to the university and they'll soon put a stop to those political statements" (Ravenhill 2008, p. 188). Yet it is the university that serves here as the hotbed of revolutionary ideas:

> SUSAN: […] Stephen wrote from the university. He's looking forward to his fruit cake. Stephen wrote and he said there's a big push now in the universities. The students mainly. But also the lecturers. And there's a big push against the Cut. (Ravenhill 2008, p. 206)
> […]
> PAUL: Writing. Discussing. They [students] never change anything.
> SUSAN: Not immediately. (Ravenhill 2008, p. 215)

All one knows about Stephen before Scene Three is that he writes letters from the university to his parents and that he enjoys the fruit cake. Both Susan and Paul indulge in reminding themselves that their son has always liked the fruit cake. It harbours their illusion about their knowing what Stephen desires. It may thus suggest that there is some revolutionary potential in desire itself.[5] As quoted above, Paul denies university the potential for driving social change and when the Scene Three brings about the revolution, he indulges in a patronizing tone claiming that "it's only a matter of time this [new] lot falls back on some of the old ways" (Ravenhill 2008, p. 231). Before that happens, however, Paul goes on

[5] It is worth noting that the way desire functions in the play lends itself to a reading that based on the post-Marxist ideas of Deleuze and Guattari, who in *Anti-Oedipus: capitalism and schizophrenia* (1984) proposed "a politics of participation and direct involvement by those who desired change, not representation and the abstraction of power in the interests of an ossified or 'molar' subject ('the working class')" (Tormey and Townshend 2006, p. 45). Interestingly, it is suggested in the play that any social change is triggered by and for those who have voice. The two female characters deprived of language in the play, Gita and Mina, are thus deprived of the most elementary tool of articulating desire.

ranting about the world being too complicated for university to make any definite political claims. His apology of complexity is best expressed in the following lines: "Paul [...] black and white. Goodies and baddies. Us and them. We Cut. They are Cut. [...] Life isn't simple. Things aren't simple. Don't simplify—let Stephen—fine, he's a student—maybe at the university but don't simplify" (Ravenhill 2008, p. 216). Interestingly though, immediately after this apology of complexity and symbolic disempowerment, Susan enacts something that could be called a psychoanalytic insight. Psychoanalysis, as Žižek writes in *The sublime object of ideology*, aims at shifting the emphasis from the symbolic to the Real, from the visible to what have so far seemed meaningless (Žižek 2008b, p. 148). What Susan does is voice the truth about Paul and her eagerness to become socially active. When lying and "looking at the ceiling," Susan "noticed a mark," a small detail which makes her as if peek into the blackout moment at the end of Scene One when Paul was performing the cut on John (Ravenhill 2008, p. 217).

Given the four discourse theory as formulated by Lacan and appropriated by Žižek, what the play leaves out is the discourse of the Master and that of the analyst. On a metatextual level, this conclusion connects with the role of the reader or spectator of the play, the role of organizing the text/spectacle. Lehmann would chime in and claim that it is precisely the political aim of (post)drama: to get rid of such a perspective, redolent of the Master (key of interpretation) that could impose a Master Signifier (or a political message). Postdramatic theatre relies on "the deconstruction of a discourse oriented towards meaning and the invention of a space that eludes the laws of telos and unity" (Lehmann 2006, p. 146). The demise of the Master's discourse, as Žižek writes in *The paralax view*, triggers either an excess of doubt (hysteria) or post-political excess of administration legitimized by objective knowledge (university) (Žižek 2006, p. 296). Only the analyst is capable of appointing a new Master Signifier. It is essential for the revolutionary act to take place. There is no one who is "supposed to know," someone who would represent the "illusory Other Place at which everything is always-already written, at which the (unconscious) meaning of all symptoms is always-already fixed" (Žižek 2006, p. 77). Thus without the analyst on the level of plot, yet with language prone to a psychoanalytic reading, *The cut* moves to Scene Three, where the meeting of Paul and Stephen takes place after some sort of a revolution (Ravenhill 2008, pp. 223–224). The world has changed as Stephen tells us and now "the light goes on at five thirty" Ravenhill 2008, p. 225); there are new prisons as temporary measures, the Ministry of Forgiveness, and perhaps most importantly, there is no punishment, no practices of the Cut. If we consider the cut to represent the negativity, the post-revolutionary scene brings to fore its ultimate erasure and the establishment of "total transparency" (Žižek 2006, p. 306). Ravenhill seems to suggest that the dream of such an erasure leads to a post-political dystopia.

3 Conclusions: *The Cut* and its Engagement with Politics-Laden Theory

Ravenhill's earlier plays, with their in-yer-face aesthetics, are often read as interested in social matters, yet unencumbered by ideology and stripped of any remnants of political agency. Commentators often repeat the platitude that Ravenhill and the young writers of the 90s lacked a firm scaffold for their ideas and claims and struggled to develop a stage language that made "small stories … resonate as widely as grand narratives" (Sierz 2001, p. 241). *The Cut* seems to prove that the in-yer-face aesthetics really "ran out of steam by the end of 1999" (Sierz 2008, p. 32). *The Cut* differs from the playwright previous attempts at political ideas. It is nothing like *Some explicit Polaroids* (Ravenhill 1999), a play boldly addressing the questions of political engagement, which some critics saw as rehashed state-of-the-nation play.[6] With the dissemination of what is considered political (Cf. Dybel, Wróbel 2008, pp. 51–52), as much as of what is politically practicable in theatre, *The Cut* seems to be groping for a dramatic form that would suit smuggling ideas not too unambiguously. Therefore, on the one hand, the play lends itself to be read as a dystopia, whereby Scene Three could be viewed as a cautionary new Order devised by an apolitical university and as a warning against post-politics. On the other hand, beyond this subtle apology of politics, there is scarcely any indication to an authentic act or a Truth-Event that philosophers dream about. The scattered symptoms and subjective positions fail to constitute anything beyond a politics based on the fear of negativity, the fear of the cut. Again, what Ravenhill wants to confront us with is what he leaves out: the analyst as the required subjective position for the authentic act to take place. What he dramatizes is this very lack, thus revealing engagement with cultural theory and psychoanalysis. Although some of Ravenhill's earlier plays could easily be labelled as engaged with theory, it is the kind of theory and its political bent that drive the politics of *The Cut.* Such an approach to writing for theatre, whether conceived of as engaged or not, bears witness to his interest in dramatizing ideas and, to lesser extent, social realities.

References

Boucher, G. 2005. The law as a thing: Žižek and the graph of desire. In *Traversing the fantasy: critical responses to Slavoj Žižek,* ed. G. Boucher et al., 23-31. Adlershot: Ashgate Publishing Ltd.

Butler, J., Laclau, E., Žižek, S. 2000. *Contingency, hegemony, universality*. London: Verso.

[6] For Dan Rebellato the state-of-the-nation play "mirrored the nation-state in its mapping of the political onto the personal, and the general onto the particular." He rightly points out that the genre is a thing of the past, because "the values of nation and state no longer coincide at the territorial level" (Rebellato 2008, p. 252).

Deleuze, G., Guattari, F. 1984. *Anti-Oedipus: capitalism and schizophrenia*. London: Athlone Press.
Dybel. P., Wróbel. Sz. 2008. *Granice polityczności*. Warszawa: Aletheia.
Laclau. E. 1997. The death and resurrection of the theory of ideology. *MLN* 112.3: 297-321.
Lehmann, H. T. 2006. *Postdramatic theatre*. New York: Routledge.
Lyotard, J. F. 1984. *The postmodern condition*. Manchester: Manchester University Press.
Malpas, S. 2005. *The postmodern*. London: Routledge.
Rebellato, D. 2008. From the state of the nation to globalization: shifting political agendas in contemporary British playwriting. In *A concise companion to contemporary British and Irish drama*, eds. N. Holdsworth & M. Luckhurst, 245-262. Malden: Blackwell Publishing.
Ravenhill, M. 1999. *Some explicit Polaroids*. London: Methuen Publishing.
Ravenhill, M. 2008. *The Cut*. In *Plays: 2,* 180-231. London: Methuen Drama.
Sierz, A. 2001. *In-yer-face theatre. British drama today*. London: Faber and Faber.
Sierz, A. 2008. 'We all need stories': The politics of in-yer-face theatre. In *Cool Britannia? British political drama in the 1990s*, eds. R. D'Monte, G. Saunders, 23-37. New York: Palgrave Macmillan.
Tormey, S., Townshend, J. 2006. *Key thinkers from critical theory to post-Marxism*. London: Sage Publications.
Wallace, C. 2006. *Suspect cultures*. Prague: Litteraria Pragensia.
Žižek, S. 1998. Four discourses, four subjects. In *Cogito and the unconscious*, ed. S. Žižek, 74-113. Durham: Duke University Press.
Žižek, S. 2000. *The ticklish subject*. London: Verso.
Žižek, S. 2004. *Revolution at the gates*. London: Verso.
Žižek, S. 2006. *The paralax view*. Cambridge: The MIT Press.
Žižek, S. 2008a. *In defence of lost causes*. London: Verso.
Žižek, S. 2008b. *The sublime object of ideology*. London: Verso.

Internal (Post)Coloniality in Anglo-Irish Literature: Crossing the Boundaries in Postcolonial Comparative Studies

Grzegorz Koneczniak

Abstract This study is the revision and exploration of the author's essay entitled "Ireland as an Internal Colony: the Irish Literary Revival in Search of de-Anglicisation", published in the proceedings of the 2006 PASE conference. As demonstrated in that paper, the Irish colonial and postcolonial status is characterised by its distinctiveness, due to the idea of internal colonialism, which, according to Michael Hechter, has governed the Anglo-Irish economic, political and cultural relations since the beginning of the nineteenth century. In the field of this internal colonialism, literary scholars and critics have already noticed patterns of colonisation, studied, for instance, by Vincent Cheng or Anna Cisło on the basis of the Victorian ethnocentrism, and decolonisation, frequently studied on the example of the Irish Literary Revival. However, these observations were rather omitted in a more general field of postcolonial studies, dealing with the literatures of the Second- and Third-World societies, and it was not until the beginning of the twenty-first century that Irish postcoloniality crossed the boundaries of postcolonial investigation. Developed from both the aforementioned essay and the author's further research projects concerning Irish postocoloniality, the present study offers a review of the most current approaches towards the question of internal colonialism and postcolonialism, discernible in Irish literature, in the context of postcolonial comparative studies.

G. Koneczniak (✉)
Nicolaus Copernicus University, Toruń, Poland
e-mail: grzegorz.koneczniak@gmail.com

J. Fabiszak et al. (eds.), *Crossroads in Literature and Culture,* Second Language Learning and Teaching, DOI: 10.1007/978-3-642-21994-8_4,

1 Introduction

This study is the revision and exploration of my essay entitled "Ireland as an internal colony. The Irish Literary Revival in search of de-Anglicisation," published in the proceedings of the 2006 PASE conference. As demonstrated in that paper, the Irish colonial and postcolonial status is characterised by its distinctiveness, due to the idea of internal colonialism, which, according to Michael Hechter, has governed the Anglo-Irish economic, political and cultural relations since the beginning of the nineteenth century (Hechter 1975; Cisło 2003; Koneczniak 2007, pp. 207–216; Koneczniak 2009), pp. 94–104. In the field of this internal colonialism, literary scholars and critics have already noticed patterns of colonisation, studied, for instance, by Vincent Cheng (1995) or Anna Cisło (2003) on the basis of the Victorian ethnocentrism, and decolonisation, frequently studied on the example of the Irish Literary Revival (Koneczniak 2007, pp. 207–216; Koneczniak 2008, pp. 94–104; Koneczniak 2010).

However, these observations were initially omitted in a more general field of postcolonial studies, dealing with the literatures of the Second- and Third-World societies, and it was not until the end of the twentieth century that Irish postcoloniality crossed the boundaries of postcolonial investigation (Kiberd 1995; Crump 1996; Lennon 2004; Koneczniak 2007, pp. 207–216; Koneczniak 2009, pp. 94–104; Koneczniak 2010). Developed from both the aforementioned essay and my further research projects concerning Irish postocoloniality, the present study offers a review of the most current approaches towards the question of internal colonialism and postcolonialism. It will seek to determine the features of Irish internal (post)coloniality with the focus on comparing or contrasting them with (post)colonial aspects inherent in the literary and cultural developments of more external colonies of the former British Empire, the original scope of postcolonialism.

2 The Question of Internal (Post)Coloniality on the Example of Ireland

Internal colonialism is the term developed by Michael Hechter in his influential study entitled *Internal colonialism: the Celtic fringe in British national development*. The book was published in 1975, and it was the first publication to address the question of the distinctive relations between the imperial England and its home colonies including Ireland, Scotland and Wales. Hechter focused on the reasons for the economic and cultural inequality among the states of the United Kingdom (Cisło 2003, p. 82; Koneczniak 2007, pp. 210–211; Koneczniak 2009, p. 96). The term itself was explained as "a process of unequal exchange between the territories of a given state that occurs either as a result of the free play of market forces or of economic policies of the central state that have intended or unintended distributional consequences for regions" (Hechter 1999, p. xiv). This definition is rather vague and can be applied in order to study any single society in which its centre is responsible for the uneven distribution of economic wealth among its peripheral regions. In fact, Hechter stresses the applicability of internal colonialism with respect to the poor regions of the former Soviet Empire, the Italian states and even the minorities on the American continents (Hechter 1999, pp. xiii–xiv; Koneczniak 2007, p. 210). In reference to the aforementioned nations and groups, the term had already been investigated. By way of illustration, Robert Blauner in his 1969 essay "Internal colonialism and ghetto revolt" considers Afro-American minorities to be internal colonies of the United States, affected firstly by external and secondly by internal colonial experience (Blauner 1969, p. 39).

The initial implication of internal colonialism was rather rooted in its economic and social distinctiveness. However, Hechter's notion of 'exchange' also refers to culture, as in the case of internal colonialism the peripheral territories "are simultaneously economically disadvantaged and culturally distinctive from the core regions of the host state" (Hechter 1999, p. xiv). Cultural distinction is also the key foundation element as regards the theories of internal colonialism offered by other scholars, even prior to Michael Hechter's publication. In the essay "Internal colonialism and national development," published in 1965, Pablo Gonzalez-Casanova emphasizes that "[i]nternal colonialism corresponds to a structure of social relations based on domination and exploitation among culturally heterogeneous, distinct groups. It is the result of an encounter between two races, cultures, or civilizations, whose genesis and evolution occurred without any mutual contact up to one specific moment" (Gonzalez-Casanova 1965, as cited in Walls 1978, p. 4). Both Hechter and Gonzalez-Casanova take into account the cultural distinctiveness that was coerced into co-existence with the moment of internal colonisation.

This distinctiveness is much in question as regards Irish internal coloniality. With the arrival of the English in Ireland in the twelfth century, throughout the period of colonisation, characterised by a varied intensity of colonial relations and imperial dependency (Smyth 1998; Koneczniak 2009, p. 99; Koneczniak 2010), cultural distinctiveness became one of the reasons for imposing inferiority upon the Irish (Koneczniak 2007, p. 211). The colonial relations between England and Ireland, its home colony, were different from the imperial practices in Africa or Asia predominantly on account of the geographical proximity, which had given rise to distinctive political, economic, social, and cultural links between England and Ireland. Yet, the analogy between the domestic and overseas forms of colonisation can be drawn with respect to the connection between the colonial periphery and core, which, as argued by many scholars and demonstrated in colonial literary works and cultural artifacts, is equated with metropolitan London for both the internal and external colonies of the British Empire.

William Butler Yeats, James Joyce, Sean O'Casey or Oscar Wilde, as representatives of literature from the internal colony, were aware of the binary opposition between Ireland, the site of apparent backwardness and degeneration, and England, the place of apparent social advancement and cultural refinement. This opposition is visible in both their oeuvre and biographies (Welch and Stewart 1996). Correspondingly, writers from external colonies of the British Empire perceived England, and especially metropolitan London, in similar terms, as was articulated by George Lamming in his "The occasion for speaking" (Lamming 2006, pp. 14–18).

The historical, social and predominantly cultural narrative of Anglo-Irish relations also offers representations of Ireland and the Irish as the colonial 'Other.' Such studies concerning British ethnocentrism directed against the Irish distinctiveness of an internal colony as *Anglo-Saxon and Celts: a study of anti-Irish prejudice in Victorian England* written by Lewis Perry Curtis or *Joyce, race and empire* by Vincent Cheng unanimously confirm this perception of the Irish. The anthropological, ethnological, and political distinctiveness of Ireland was, or perhaps has been, transformed into the anti-Irish stereotype-formation featuring English writings, especially during the Victorian epoch. (Cisło 2003, pp. 68–86; Koneczniak 2007, pp. 209–210).

As many scholars confirm, the examples of anti-Irish prejudice range from visions of this nation as the most backward ethnic group of Europe, often compared to indigenous Africans, to depictions of the Irish as monsters and animals. Among the political, social, cultural and literary texts analysed by Curtis and Cheng, such derogatory expressions as 'Irish Negro,' 'Frankenstein,' 'ape' and 'chimpanzee' are often used in order to describe the Irish population. Moreover, the use of these particular expressions is seemingly justified on the grounds of philosophical, moral and pseudo-scientific speculations. As regards the lastly-mentioned, Robert Knox is often quoted as an anthropologist who sought to prove the inferiority of certain races, including the Irish and the Jews by means of medical research (Cisło 2003, pp. 68–96; Koneczniak 2007, pp. 207–210).

The concept of internal colonialism, apart from cultural discrimination of the periphery, comprises a reciprocal movement, namely, the counter-response which frequently takes the form of regional nationalism. It stems from the fact that the imposition of cultural inferiority upon the peripheral group precipitates counter-reactions seeking to belie the vision of this minority collectivity propagated by the core central majority. This mechanism is observed, for instance by John Stone, the author of the essay entitled "Internal colonialism in comparative perspective." The scholar stresses that "by defining inter-regional relationships as 'colonial', nationalist leaders have tried to inspire popular support for movements designed to promote greater autonomy, if not outright secession" (as cited in Hicks2010, p. 1). Thus, the concept of internal colonialism bears ideological implications not only as the residue of colonial discrimination but also as the justification for anti-colonial movements.

This dialectic of imperialism and anti-imperialism, colonisation and decolonisation, discourse and counter-discourse is then much implied by the model of internal colonialism (Koneczniak 2007, pp. 207–216). The more intense imperial policies were, the more extensive literary and cultural oppositions seem to have occurred, frequently with the emphasis on the value of common heritage, elements of national identity, and as opposed to those of the imperial core. Such critics, historians and theorists of Irish literary and cultural legacy as Ian Crump or Declan Kiberd point to the political, counter-colonial and anti-imperial aspect as the defining feature of Irish literature (Koneczniak 2009, pp. 94–104; Koneczniak 2010). Even if differences emerge as to the individual poetics of Irish literary tradition, the common denominator frequently appears to have been the explicit or implicit opposition of the Irish periphery against the English core.

This opposition manifests itself both in the literature created during the period of colonisation and the period after the Irish gained its political independence in 1937. The studies in the former literary development predominantly focus on instances of Irish revivals and the most significant, of course, was the Irish Literary Revival formed at the end of the nineteenth century by such figures as Lady Augusta Gregory, John Millington Synge and William Butler Yeats (Cisło 2003, pp. 92–96; Koneczniak 2007, pp. 213–215). As regards common critical approaches to the opposition staged during the Irish Literary Revival, on the one hand, scholars regard the aforementioned figures as the founders of Irish national literature, especially theatre and drama; on the other hand, they also discern the Revivalists' connection with England (Kiberd 1996; Koneczniak 2010). In reference to the later literary developments, the studies predominantly concentrate on the oeuvre of such contemporary authors as Brian Friel, Sebastian Barry, Martin McDonagh or Seamas Heaney, whose works are thought to feature the long-lasting repercussions of the colonial period (Morash 2002; Llewellyn-Jones 2002; Richards 2004).

3 Internal (Post)Coloniality in the Context of Postcolonial Studies

It is quite comprehensible that some features of internal colonialism bear resemblance with the external varieties of colonialism and postcolonialism. This similarity has so far been signalled by theorists and critics of internal colonialism. For instance, Robert Blauner deals with four features that are present in internal as well as external colonialism. The first feature that Blauner enumerates is related to the manner of the early encounter between the coloniser and the colonised, which is often tantamount to infringement and enforcement. The second feature concerns the cultural processes between the two societies, which are far from such natural developments as acculturation. The third feature revolves around administrative and political matters: the colonial relations are close to manipulation by the colonisers who are anthropologically different from the colonised. The last point of convergence is linked to the idea of racism: the ethnic, biological and cultural differences are used as a means of humiliating and subjugating the colonised (Blauner 1969, as cited in Hicks 2010, pp. 4–5).

Common features between the internal and external forms of colonialism are also noticed by Robert J. Hind in his "The internal colonial concept" (1984). According to this scholar, "[t]heories of internal colonialism derive from analogies" (1984: 552). Hind is aware of the fact that in the case of internal colonialism "the colonising and colonised sections of society live in the same country" (1984: 552), which, by definition, "excludes that feature of traditional views of colonisation which assumes geographical separation, and also that feature which rests upon the premise that an entire population has imposed its authority upon an extraterritorial society or group of communities" (1984: 552). Nevertheless, Hind also states that between the internal and external colonisation there are certain points of convergence, comprising "such characteristics of conventional colonialism as political subjection, economic exploitation, cultural domination, and racial conflict" (Hind 1984, p. 552).

Accepting Blauner and Hind's similarities may lead to the assumption that the term internal colonialism should have long been present in the field of postcolonial studies. However, this interesting phenomenon of internal colonialism was initially omitted. Even in the immensely popular collection of postcolonial-oriented essays, *The post-colonial studies reader*, first published in 1995, the term "internal colonialism" is used only once, in the article "What ish my nation?", dealing with the problem of identity experienced by Welch, Scottish and Irish Nations. (Cairns and Richards 2006, pp. 178–180).

The authors of the article mentioned in the previous paragraph may be considered to have been aware of the potential analogy between the studies in Irish internal (post)colonialism and 'external' postcolonial studies, which is reflected in their book *Writing Ireland: colonialism, nationalism, and culture* (1988), in which the article "What ish my nation?" was originally included. Cairns and Richards accept the proposition that "relations of domination and subservience may exist between regions within a state analogous to those between a colony and the metropolitan state" (1988: 169). Thus, as has already been stated, both in internal and external forms of colonialism similar mechanisms can be observed with respect to the processes of imperialism and hegemony as well as anti-imperialism and counter-hegemony (Cairns and Richards 1988, p. 169), which should give rise to investigations into internal (post)coloniality within postcolonial studies.

The boundary between studies in internal and external varieties of postcolonialism was crossed in the second half of the 1990s, and since then research in Irish literature and culture has played the most significant role. In this field, critics and scholars have focused on comparative aspects, and currently they seem well aware that Irish (post)colonial internal condition could facilitate the future development of postcolonial studies. The end of the previous decade witnessed the emergence of Irish studies that have been developing towards the present (Koneczniak 2009), pp. 94–104.

Initially, these studies, supported by the historical account of internal colonialism, focused on direct application of postcolonial methodology, perspective and terminology in order to analyse and interpret the colonial ethnocentrism discernible in English writing in terms of colonial discursiveness and the opposition against it articulated in Irish literature. Vincent Cheng, for instance, states that for that in imperial England Ireland was frequently seen as the Orient because of similar examples of stereo-typology and ethnocentrism (Cheng 1995, p. 24; Cisło

2003, p. 71; Koneczniak 2007, p. 210). In general, this direct translation was the development of the thought offered by Robert Blauner forty years earlier, insisting on the similarities between external and internal forms of colonialism and postcolonialism.

By this analogy in question, the Irish cultural and literary development can be analysed as a counter-narrative that has opposed the English coloniser. In fact, when investigating the Irish literary programs and manifestos, like the famous "Our Irish theatre," written by Lady Augusta Gregory to provide the tenets for the activities of what would become the Irish national theatre (Gregory 1899, pp. 377–386), the idea of postcolonial counter-discursiveness seems applicable. Furthermore, the most essential dramatic, poetic and prose works of Gregory and other Revivalists, especially those of William Butler Yeats, can be approached as examples of textual opposition against the English within the paradigm of postcolonial studies, wherein postcolonial 'terms' can easily be used to interpret the Revivalists' works (Koneczniak 2007, pp. 213–215; Koneczniak 2010).

However, this straight translation of postcolonial theory into Irish literature and culture also poses the threat of superficiality and imposition of foreign terminology on the unique Irish colonial and postcolonial experience. In order to both retain Irish distinctiveness within postcolonial studies, the theory should be redefined and appropriated, which I have already demonstrated with reference to the oeuvre of the Irish Literary Revival in my recent study *Women on stage and the decolonisation of Ireland: counter-discursiveness in the drama of the Irish Literary Revival* (1892–1926). Through appropriating postcolonial theory, Ireland will not be 'swallowed' by the mainstream postcolonial studies and will retain its distinctiveness of a former internal colony, which is emphasised by current critics of Irish postcoloniality, including Declan Kiberd or Joseph Lennon (Koneczniak 2009, pp. 94–104; Koneczniak 2010).

With reference to the question of Irish literature and culture, David Lloyd's *Anomalous states: Irish writing and the postcolonial moment* can be quoted as an example of such appropriation. Lloyd devotes one chapter to the oeuvre of James Joyce from a redefined postcolonial perspective. The scholar considers the possible implications of hybridity, the concept that has recently gained popularity in postcolonial theory, and which initially was applied to study the voices, perspectives and visions of national identity in Joyce's fiction. However, Lloyd argues that in order to avoid the practices of postcolonisation and nationalisation of Irish unique experience, Joyce's oeuvre should rather be seen in terms of 'adultaration', understood as a stylistic device studied on the basis of *Ulysses* (Lloyd 1993, pp. 88–124).

The term adulteration is developed from the notion of adultery that features both as the origin of Irish colonial history, which goes back to the marital unfaithfulness of queen Dervorgilla, and as the important element in Joyce's novel (Lloyd 1993, pp. 88–124). As Lloyd states, adulteration is related to the forbidden, which in the case of Joyce and Irish internal (post)coloniality has a more metaphorical dimension: it "undermines the stable formation of legitimate and authentic identities" (Lloyd 1993, p. 109). According to Lloyd, Joyce's work illustrates lack of the possibility to impose one genuine postcolonial discourse upon Irish literature, which, similar to adultery and adulteration, is characterised by impurity, instability and inauthenticity: the features of which Joyce was aware, and thus refused any involvement in search of Irish homogenous national identity, common among his contemporaries. Lloyd's theory of adulteration was initiated in 1993; however, it was not until the twenty-first century that the concept entered the theoretical and critical paradigms of postcolonial theory.

Similar to the recent theoretical and critical interest in the works of Joyce, other prominent literary figures of Irish origin have gained some scholarly attention within postcolonial comparative studies (Koneczniak 2009, p. 104). By way of illustration, studies devoted to the drama of Brian Friel, especially the text authored by Francis Charles McGrath, *Brian Friel's (post)colonial drama: language, illusion and politics*, cast new light on the importance of language in shaping distinctive Anglo-Irish colonial and postcolonial relations, which Friel's play *Translations* illustrates. As McGrath maintains, "Friel embraces an orientation in which illusion, myth, identity and even history are products of language, discourse and narrative" (McGrath 1999, pp. 18–19), which in fact have affected Irish internal (post)coloniality. As implied in Friel's play, the idea of translation is yet another theoretical and critical concept developed from the Irish version of internal (post)coloniality that has influenced postcolonial theory, or at least it will (Cronin 1996).

4 Conclusion

It is evident that Ireland as an internal colony has been included in the research domain of postcolonial studies. However, the problem arises as regards the place of internal colonialism as a more theoretical concept that can be applied to approach the literary and cultural developments of other societies, or parts of societies which so far have either been omitted or treated homogenously as postcolonial without any further differentiation. In the case of Canada or Australia, which have long been analysed as postcolonial Second-World societies studied in the context of their colonial history, colonial discursiveness, and postcolonial counter-discursiveness of writing back to the empire, the appropriate model of internal colonialism could in fact be useful in order to address the interior cultural and social stratification of these states, perhaps going beyond the common distinction between First, Second, and Third nations, so popular in the current postcolonial discourse.

Shifting the scope of postcolonial studies towards the distinctiveness of internal colonialism could also raise the problems of the states in reference to which the attribute postcolonial has rather been avoided and which do experience the mechanisms of internal colonialism. This postulate is particularly applicable in reference to the societies which in the sixties and seventies were analysed from the perspective of internal colonialism and which in the nineties were totally absent in the research domain of postcolonial studies. The former colonies of the French, Portuguese, and predominantly English empires are frequent research subjects of postcolonial studies; however, the states of the former Russian empire are still rather rare.

I can point to the phenomenon that some single texts dealing with the problems of internal colonialism and postcolonialism of Russia, Canada, Spain, Italy, the United States and Palestine have recently appeared (Todorova 2004; Caprotti 2007; Allatson 2007). Their common feature is emphasis on distinctiveness and individuality, the regional variety of colonialism and postcolonialism, which has been rather suppressed in the epistemological and methodological paradigms offered in the mainstream postcolonial studies. Ireland, initially absent from postcolonial studies is currently a common research subject within this field. The inclusion of this state has significantly enriched the comparative approach to postcolonial literatures, which should also occur as regards other internal colonies and thus further transgress the boundary between the Second- and Third-World societies and the theories initially developed in order to address them.

References

Allatson, P. 2007. *Key terms in Latino/a cultural and literary studies*. Maiden, Oxford and Carlton: Blackwell Publishing.

Blauner, R. 1969. Internal colonialism and ghetto revolt. *Social problems* 16, 396. http://heinonline.org/HOL/LandingPage?collection=journals&handle=hein.journals/socprob16&div=45&id=&page=. Accessed 14 July 2010.

Cairns D., and S. Richards. 1988. *Writing Ireland: colonialism, nationalism, and culture*. Manchester: Manchester University Press.

Cairns D., and S. Richards. 2006 [1995]. What ish my nation. In *The post-colonial studies reader*, ed. B. Ashcroft et al., 134-136. Cornwall: Taylor & Francis Group (Routledge).

Caprotti, F. 2007. *Mussolini's cities: internal colonialism in Italy, 1930-1939*. New York: Cambria Press.

Cheng, V. J. 1995. *Joyce, race, and empire*. Cambridge: Cambridge University Press.

Cisło, A. 2003. *Tożsamość Irlandczyków w prozie odrodzenia literackiego z przełomu XIX i XX wieku*. Wrocław: Wydawnictwo Uniwersytetu Wrocławskiego.

Cronin, M. 1996. *Translating Ireland: translation, languages, culture*. Cork: Cork University Press.

Crump, I. 1996. "A terrible beauty is born": Irish literature as a paradigm for the formation of postcolonial literatures. In *English postcoloniality: literatures from around the world*, ed. M. Radhika and G. Rajan, 31-42. Westport: Greenwood Press.

Gregory, A. I. 1991 [1899]. Our Irish theatre. In *Modern Irish drama*, ed. J. P. Harrington, 377-386. New York and London: W. W. Norton & Company.

Hechter, M. 1999 [1975]. *Internal colonialism: the Celtic fringe in British national development*. New Brunswick: Transaction Publishers.

Hicks, J. On the application of theories of 'internal colonialism' to Inuit societies. http://www.cpsa-acsp.ca/papers-2004/Hicks.pdf. Accessed 14 July 2010.

Hind, R. J. 1984. The internal colonial concept. In *Comparative studies in society and history*, 26, 3: 543-568. Cambridge: Cambridge University Press.

Kiberd, D. 1996. *Inventing Ireland: the literature of the modern nation*. London: Random House.

Koneczniak, G. 2007. Ireland as an internal colony. The Irish Literary Revival in search of de-Anglicisation. In *PASE Papers 2007. Studies in culture and literature*, ed. W. Kalaga et al., 207-216. Katowice: University of Silesia.
Koneczniak, G. 2009. Status Irlandii w studiach postkolonialnych. In *Studia postkolonialne w literaturoznawstwie i kulturoznawstwie anglojęzycznym*, ed. M. Buchholtz, 94-105. Toruń: Wydawnictwo Adam Marszałek.
Koneczniak, G. 2010. *Women on stage and the decolonisation of Ireland: counter-discursiveness in the drama of the Irish Literary Revival (1892-1926)*. Toruń: Wydawnictwo Naukowe Uniwersytetu Mikołaja Kopernika (publication in progress).
Lamming, G. 2006. The occasion for speaking. In *The post-colonial studies reader*, ed. B. Ashcroft et al., 14-18. Cornwall: Taylor & Francis Group (Routledge).
Lennon, J. 2004. *Irish Orientalism: a literary and intellectual history*. Syracuse: Syracuse University Press.
Llewellyn-Jones, M. 2002. *Contemporary Irish drama and cultural identity*. Bristol: Intellect Books.
Lloyd, D. 1993. *Anomalous states: Irish writing and the post-colonial moment*. Durham: Duke University Press.
McGrath, F. C. 1999. *Brian Friel's (post)colonial drama: language, illusion, and politics*. Syracuse: Syracuse University Press.
Morash, C. 2002. *A history of Irish theatre, 1601-2000*. Cambridge: Cambridge University Press.
Richards, S. 2004. *The Cambridge companion to twentieth-century Irish drama*. Cambridge: Cambridge University Press.
Smyth, G. 1998. *Decolonisation and criticism: the construction of Irish literature*. London: Pluto Press.
Todorova, M. N (ed.). 2004. *Balkan identities: nation and memory*. London: C.Hurst & Co. (Publishers) Ltd.
Walls, D. S. 1978. Internal colony or internal periphery? A critique of current models and an alternative formulation. http://sonoma.edu/users/w/wallsd/pdf/Internal-Colony.pdf. Accessed 14 July 2010.
Welch, R. and B. Stewart. 1996. *The Oxford companion to Irish literature*. Oxford: Oxford University Press.

A Post-Battle Landscape: Doris Lessing's *The Golden Notebook* and *The Cleft*

Katarzyna Więckowska

Abstract The central theme of Doris Lessing's *The golden notebook* (1962) is struggle and the heroine's attempts to stake out a new territory where she would be able to re-define herself and gain on a new identity. Similarly, although in a different manner, this desire to find a space beyond the existing social limits is repeated in *The Cleft* (2007) as a dream of an all-female world before the arrival of sexual difference. Thus repetition can be seen as the regulating principle working both within the novels and between them, where it takes the form of a certain mourning, or a process of releasing the desire for a once possessed, but lost unified self and for an imagined sociality without otherness, a desire underlying also certain forms of literary and social criticism. In this essay I refer to the work of Sigmund Freud and Jacques Lacan, and particularly to the psychoanalytic rendering of the concepts of the lost object, the return of the repressed and the woman, to read *The golden notebook* and *The Cleft* as excursions beyond the phallocentric social and literary orders into a space where the woman might be imagined differently than the lacanian "symptom of man".

1 Introduction

When Doris Lessing was awarded the Nobel Prize in Literature in 2007, she was addressed by the committee as "that epicist of the female experience, who with scepticism, fire and visionary power has subjected a divided civilization to scrutiny" (Nobelprize.org). The careful framing of the description of the writer as "the

K. Więckowska (✉)
Nicolaus Copernicus University, Toruń, Poland
e-mail: klew@umk.pl

J. Fabiszak et al. (eds.), *Crossroads in Literature and Culture,* Second Language Learning and Teaching, DOI: 10.1007/978-3-642-21994-8_5,

epicist of the female experience" emphasizes the centrality of women's issues to her fiction and also evades the problem of identifying her as a feminist writer, a label attached to her work since the publication of the notorious *The golden notebook* in 1962. The feminist identification, which the novelist strongly and persistently rejects, is obviously one of the symptoms of the "divided civilization" that her fiction both describes and produces—a civilization divided not only by sexual difference but, as Lessing's relation with feminism proves, also by the various political and ideological interests of women as a divided group.

The question whether Lessing's work in general, and *The golden notebook* in particular, is feminist or not has been the subject of numerous studies and debates, and continues to be asked, despite the writer's repeated objections, such as the one added by her as early as in the 1971 preface to the novel:

> handing the manuscript to publisher and friends, I learned that I had written a tract about the sex war, and fast discovered that nothing I said then could change that diagnosis. Yet the essence of the book, the organization of it, everything in it, says implicitly and explicitly, that we must not divide things off, must not compartmentalize (Lessing 1989, p. 10).

As Gayle Greene's discussion of the critical reception of *The golden notebook* shows, the debate over the feminism of Lessing's novel is itself an interesting document of the critical approaches dominant at a specific point of literary history (Greene 1991, p. 109). In contrast to the ideas of New Criticism that were used to criticize the novel on its publication (interestingly, mostly by male critics), feminist interpretations, including the one by Greene, resort to reader response theory, where the decision on the meaning of the text is shifted onto the reader, and to the Barthesian proclamation of the "death of the author", which annuls the problem of authorial intentions.[1] The critical debate not only shows the contradictory role of the author, who is treated as both the support of the meaning of her work and as simply "the principle of thrift in the proliferation of meaning [....] by which, in our culture, one limits, excludes, and chooses" (Foucault 1994, p. 209), but also unmasks the critic as author and ideologically motivated master of interpretation. Considering its critical fortunes, *The golden notebook* can be seen as a document in cultural, literary and social history, illustrating not only the changing critical approaches, including feminism, but a more general and slow coming to terms with the collapse of the notion of personal identity as stable and knowable. The dissolution of identity is closely related to one of the major problems of contemporary feminism and feminist literary criticism, i.e. the loss of the stable female subject at the time she has finally regained her self, and might be one of the reasons which make it difficult to interpret Lessing's work along feminist lines. In accordance with Lessing's wish to pay attention to the structure of the book and her authorial intentions, I read the novel as an account of breakdown, which is also

---

[1] It is interesting to note that Roland Barthes' seminal essay was published in 1967 and can therefore be seen as coming out of the same critical upheavals from which Lessing's novel originates.

"a way of self-healing" (Lessing 1989, p. 8), staged along a number of textual erasures and absences which hamper the formulation of a finite interpretation. The erasures are, as I argue, repeated and apparently compensated for in *The Cleft* (2007), a novel which can be read as a continuation and complementation of the thematic concerns and problems of *The golden notebook*. The key word in analyzing the novels is repetition, both within and between the books, which is approached here from the psychoanalytic perspective, thus binding the texts to the process of mourning after an object which, although lost, continues to be secretly desired and possessed (Freud 1995a, p. 586). The question of the supposed feminism of Lessing's work is left open—mostly because, as Ros Coward(1992) writes, "it is just not possible to say that women-centred writings have any necessary relationship to feminism" (1992, p. 378; my emphasis)—but the reading of the novels is motivated by a kind of feminist practice which Rosi Braidotti describes as a "strategy of working through the historical notion of 'Woman'" which aims at "unveiling and consuming the different layers of representation of 'Woman'" (1994, p. 168).

2 An Absent Centre

The golden notebook is a novel about Anna Wulf, a writer suffering from a block, and her friend, Molly Jacobs, an actress, and it documents Anna's crisis of identity. Although the crisis spreads over all areas of Anna's life—her disillusionment with the Marxist party, her problems with writing the second novel, and the end of her love affair with Michael, who returns to his wife—its ultimate cause is the fact that Anna is a woman, or, to be more specific, that she represents a new type of a "free woman": as Molly states at the beginning of the novel, "Free. Do you know ... I was thinking about us, and I've decided that we're a completely new type of women. We must be, surely?" (Lessing 1989, p. 26). A single woman with a child, an artist and a political activist, Anna certainly does not fit any of the models of womanhood offered by the English society of the 1950s, represented in the novel as itself a fragmented and divided society in transit. Interestingly, the fragmentation of the society is reflected in its literature, particularly the novel, which, as Anna writes, "has become a function of the fragmented society, the fragmented consciousness. Human beings are so divided, are becoming more and more divided, and more subdivided in themselves"; they are beings for whom reading has become "a blind grasping out for their own wholeness" (Lessing 1989, p. 75). This binding of the recovery of stable identity to writing and reading is reflected by Anna's almost compulsive need to write, to compartmentalize her life into separate spheres so as to finally discern there a recurring pattern:

> I keep four notebooks, a black notebook, which is to do with Anna Wulf the writer; a red notebook, concerned with politics; a yellow notebook, in which I make stories out of my experience; and a blue notebook which tries to be a diary (Lessing 1989, p. 418).

Anna's writerly gesture inside the book mirrors the structure of the whole novel which Lessing describes in the introduction in the following way:

> There is a skeleton, or frame, called *Free Women*, which is a conventional short novel, about 60,000 words long, and which could stand by itself. But it is divided into five sections and separated by stages of the four Notebooks, Black, Red, Yellow and Blue. The Notebooks are kept by Anna Wulf, a central character of *Free Women*. She keeps four, and not one because, as she recognizes, she has to separate things off from each other, out of fear of chaos, of formlessness – of breakdown. Pressures, inner and outer, end the Notebooks: a heavy black line is drawn across the page of one after another. But now that they are finished, from their fragments can come something new, *The golden notebook* (Lessing 1989, p. 7).

Anna the writer, like Lessing the novelist,[2] are thus engaged in recording, documenting, repeating in order to restore the dissolving identity and to reach the lost and absent centre, to encounter what Jacques Lacan calls the Real (Grosz 1995, p. 34).

As Gayle Greene notices, *The golden notebook* is marked by repetition: the same characters appear under different names in the various notebooks, the same situations are re-staged in different contexts, the reactions of male characters within the novel are repeated by the male critics of Lessing's book (Greene 1991, p. 114). The ultimate repetition is the very act of writing, with the notebooks, initially imagined as a writing cure, turning into a frightening and poisonous testimony to one's disappearance:

> It occurs to me that what is happening is a breakdown of me, Anna, and this is how I am becoming aware of it. For words are form, and if I am at a pitch where shape, form, expression are nothing, then I am nothing, for it has become clear to me, reading the notebooks, that I remain Anna because of a certain kind of intelligence. This intelligence is dissolving and I am very frightened (Lessing 1989, p. 419).

But frightened of what?

The golden notebook opens with a scene of a meeting between Anna and Molly in the summer of 1957 after a separation. Not having seen each other for some time, the women try to pick up their friendship where they left it and to restore the old feelings of intimacy and confidence. But their conversation is strained, the familiarity and frankness gone, and therefore Anna's opening comment that "as far as I can see, everything's cracking up" (Lessing 1989, p. 25) can be understood as referring not only to their colleagues and the Party, but also to their friendship and to Anna herself, who is "cracking up" under the burden of the secret of her changed relationship with Richard, Molly's ex-husband. In the relation between the women, where "balance had been struck early on" (Lessing 1989, p. 26), Molly was the confidante, to whom Richard entrusted the problems with his second marriage, and Anna was the enemy, the "free woman" he always disagreed

---

[2] It is indeed hard not to draw analogies between Doris Lessing and Anna Wulf: they are both writers who have successfully published their first books, they are both members of the Communist Party and they are both "free women".

with. In Molly's absence, Anna takes over her role and becomes not only the man's support, but also the object of his sexual desire. That this is a disturbing change which, in fact, damages the balance between the women is visible in this scene in the way their conversation circles around the event, in the various detours they take to talk about Richard through references to his wife and their marital problems without addressing the role he plays in the relationship between Molly and Anna. The strategy of dismissing Richard's presence mirrors the representation of men in general in the novel in which failed encounters with men play so important a role that the text can be described as having men at its hidden centre: they are the common secret, the regulating principle that nobody speaks about.

It is therefore the man (and men in general) who disturbs the women's relationship, who becomes Anna's guilty secret and the seemingly absent centre of their relation. If we see the man as occupying the central position in the triangle, then Anna's comment, dropped just before she makes her confession about Richard to Molly, gains new significance: when Anna states with anger that, although they call themselves free women, "[t]hey still define us in terms of relationships with men, even the best of them" (Lessing 1989, p. 26), she passes a judgment on the structure of society as a whole and recognizes the impossibility of escaping the social reality of sexual difference. A free woman is, therefore, an ironic label and a state impossible to achieve in a society where identity is intersubjectively constituted and where, as psychoanalysis would have it, it is the encounter with the (historically contingent) hierarchy of sexual difference that enables the entry into the social space.

It is not accidental then that Anna resorts to men as the way out of emotional and psychic collapse; towards the end of the novel, Anna "decided the remedy for her condition was a man. She prescribed this for herself like a medicine" (Lessing 1989, p. 562). In a text dramatically documenting the lack of place for women in social, political and critical life, the medical choice of a man, though it may seem surprising, is also, I think, the only choice possible: the man is both dangerous and necessary, at one and the same time a poison and a cure whose presence, although unmasking the constructed nature of the socio-sexual hierarchy, reinforces its validity. The encounter with Milt that takes place just before the novel's closure marks Anna's return to normal life—which begins with her daughter's return from the cinema and with the prospects of Molly's future marriage. But Anna's coming out of the crisis is also her agreeing to the existing divisions of genders based on the woman's submission to the role of the one who supports the system, serving for the men as the other who guarantees social cohesion. As Lacan emphatically claims, "[a] woman is a symptom" of man (1985, p. 168); she is "the place where [male] lack is projected and through which it is simultaneously disavowed" (Rose 1985, p. 48); without her acquiescence to play her role in the social fiction, that "fiction" would collapse. The plea uttered by Milt, an impotent and therefore powerless and emasculated American, discloses the protagonists' awareness of the social mechanism and the necessity for women to continue its game:

> You've got to take us on, you've got to, don't you know that? Don't you see it's all much worse for us than it is for you? I know you are bitter for yourselves and you're right, but if you can't take us on now, and see us through it… […] You're tougher, you're kinder, you're in a position to take it (Lessing 1989, p. 574).

That Anna agrees to "take him on", to comfort him by reflecting him back to himself as complete, testifies to her readiness to join the social masquerade and to perform the supportive role the society ascribes to women in the process of social identification.

In *The golden notebook* psychoanalysis is presented not as a possibility of escaping the social fiction of gender, but as a chance of having it explained and thus making it easier to accept, although not to fight. An explanation of why Anna continues to attend the ineffective sessions with her Jungian psychoanalyst, whom she ironically and tellingly calls "Mother Sugar", can be found in Elizabeth Grosz's statement that:

> Psychoanalysis exerts an appeal for women which can also be seen as a lure or trap, especially for those who want to challenge the social functions and values attributed to women and femininity in our culture (actively affirmed in psychoanalytic theory) (1995, p. 6).

Like the world in which Anna lives, psychoanalysis proves to be a phallocentric trap (Grosz 1995, p. 3); therefore, the only function it can perform for the woman is that of returning her to her place in the social fiction, of "sweetly mothering" her back into the dominant order and to thus rewarding her with a stable identity.

The veiled acquiescence to the inevitability of phallocentrism is repeated in Doris Lessing's preface to the novel where she identifies herself as a writer by the masculine pronoun "he". This may have been a defensive gesture to counter the insistence on her womanly status in the hostile reviews of male critics where she was usually addressed as "Mrs. Lessing", as evidenced by the more sympathetic review of her work by Michael Thorpe who defends *The golden notebook* by writing that Mrs. Lessing's aim was to produce

> a comprehensive work 'describing the intellectual and moral climate' of her time, not to produce […] a feminist broadside. […] She aspired instead to meet the need for the more varied view of the human condition such as a Tolstoy or a Stendhal could provide (1973, p. 25).

By comparing the female writer to a Tolstoy or a Stendhal rather than to an Eliot or a Woolf, the critic repeats Lessing's and her novel's gesture, seemingly disregarding sexual difference, but only through identifying writing and the human condition with the male subject.

3 A Return of the Repressed

In "The Ego and The Id" Sigmund Freud describes the mechanism of loss as resulting in an alteration of the ego "which can only be described as a setting up of the [lost] object inside the ego, as it occurs in melancholia" (Freud 1995b, p. 638).

The ego is re-defined here as "a precipitate of abandoned object-cathexes" where the lost objects survive, conserved through prior identifications (Freud 1995b, p. 638). The process of introjecting the lost object is posited as a "universal road to sublimation" so that all sublimation "take[s] place through the mediation of the ego, which begins by changing sexual object-libido into a narcissistic libido and then, perhaps, goes on to give it another aim" (Freud 1995b, p. 639). A work of art, as an effect of sublimation and mediation of the ego, contains the traces of all that has been lost or repressed; hence the possibility of another relation of sexual difference, of a female identification that has been lost in *The golden notebook* returns "conserved" as a trace in *The Cleft*.

As Doris Lessing writes in the Preface to the book, *The Cleft* (2007) originated from a remark she encountered in "a scientific article" which speculated that "the basic and primal human stock was probably female, and that males came along later, as a kind of cosmic afterthought" (Lessing 2008). Consequently, the creation myth the novel presents begins with an idyllic state disturbed by the appearance of an other, in this case men, and situates the origins of civilization in a war between the sexes. Clefts, an all female race living close to the sea, impregnated with water or the Moon, lead uneventful and happy lives, lacking nothing and wanting nothing, with no sense of time or place, until one day, for reasons never explained, one of them gives birth to a baby boy and thus begins the race of what to the women appear as "[m]onsters. The deformed ones, the freaks, the cripples" (Lessing 2008, p. 8). The first boys are treated as a curious though threatening natural perversion: like toys, they are mercilessly played with, experimented on, mutilated, left to die on the rocks or killed by being thrown into the Cleft. As more of them are born, some are saved by the eagles who transport them to a nearby valley where the boys, suckled by a doe, set up a rival community. Slowly but inevitably, the two communities begin to establish contacts, particularly when Clefts realize that they have lost their ability to self-impregnate and that they need men to have children. It is at this point when the need for mutual dependency becomes obvious to both groups that the sexual relations as we know them begin to be established, leading to the hierarchy prescribed by the novel's epigraph according to which "Man does, woman is" (Robert Graves in Lessing 2008).

The novel represents the appearance of men in the women's world as the beginning of our civilization: the men radically change the women's lives, destroying their communality and the old way of thinking and giving them in exchange the inventions of language, individuality, time and space. It is also the men's presence that forces the women to record their past, although this has to be done in the new language of the colonizers, with the use of concepts alien to the women; as one of the Memories, the women keeping record of the oral history of Clefts, states:

> Even words I use are new, I don't know where they came from, sometimes it seems that most of the words in our mouths are this new talk. I say I, and again I, I do this and I think that, but then we wouldn't say I, it was we. We thought we (Lessing 2008, p. 7).

With the introduction of the new language, there comes the need for history and for a new way of thinking: "I said think but did we think? Perhaps a new kind of thinking began like everything else when the Monsters started being born" (Lessing 2008, p. 8). Above all, the men bring social organization, the division into separate communities, both between men and women, and among the women themselves, who are divided over the question of how to treat the new race. Importantly, what underlies the Clefts' decision to establish relations with Squirts is shame and the feeling of guilt for having killed the first boys.

In his account of the origins of civilization, Freud identifies the beginning of social organization with patricide and with the sons' guilt for having killed the all-powerful patriarch:

> Society was now based on complicity in the common crime; religion was based on the sense of guilt and the remorse attaching to it; while morality was based partly on the exigencies of this society and partly on the penance demanded by the sense of guilt (Freud 1995c, p. 503).

The original patricide and the ensuing guilt are presented also as the reasons for the unequal distribution of power within patriarchy as

> atonement with the father was all the more complete since the sacrifice was accompanied by a total renunciation of the women on whose account the rebellion against the father was started (Freud 1995c, p. 508).

In Lessing's novel, the original crime from which society springs is ironically presented as committed by women driven not by their will to power, but by curiosity and disgust with the boys' (bodily) otherness. But, although the story of female supremacy is "abrasive and may upset certain people" (Lessing 2008, p. 7), its patriarchal effects are the same, even if this time it is the women's choice, dictated by what one might call, again ironically, "the feminine ethics of care", wisdom or simple desire to survive the actions of men who, as Clefts are fond of repeating, act, but do not think or care.

The Cleft emphasizes the accidental and subjective nature of official history, as well as the fact that it is being issued by a man. The story of Clefts is a "mass of material accumulated over ages, originating as oral history, some of it the same but written down later", "a cumbersome, unwieldy mass" (Lessing 2008, p. 6). Moreover, "more than one hopeful historian had been defeated by it, and not only because of its difficulty, but because of its nature" (Lessing 2008, p. 6) which might explain why the material "had at various times been regarded as so inflammatory it had been put with other 'Strictly Secret' documents" (Lessing 2008, p. 7). The Clefts' tale is finally put together by a Roman historian, an elderly man hopelessly in love with his much younger wife, who retreats from the dangers of Nero's burning Rome to immerse himself in writing and in fatherly duties. A lover of peace and family, the historian is contrasted with his wife, an adventurous and clever woman who safely manages her complicated public affairs, leaving to him the care over their private life. In this marriage, it is the woman who acts and the man who is, which contradicts not only the epigraph to the novel, but also the existing hierarchy of sexual difference, origin myths, or historical accounts. As for the latter, their nature is made sufficiently clear by the historian's frequent remarks on the gradual changes introduced to the history of Squirts which finally presents them as the first men who were not mothered by Clefts, but fathered by the Eagle.

The Cleft begins with the historian observing the arrival of Marcus, a slave, and his encounter with Lolla, the beloved who has been waiting for his return. Impatient to spend the night in the company of his male friends, Marcus leaves the girl, who starts crying; still, for the man watching the pair it is obvious the boy will return to spend the night with Lolla. Yet another contradictory comment on the binding of women with being and men with acting, for the historian "the little scene seems [...] to sum up a truth in the relations between men and women" (Lessing 2008, p. 6). The truth, if there is one, might relate to the correspondence between the official account and the lived experience, between what is prescribed and what is practiced, to the variety of relations acted out daily, and to the fact that we are not "One—one Race, one Species, or one People" (Lessing 2008, p. 23), but always already divided.

4 A Lost Object

The work of mourning, as Freud tells us, proceeds through incorporating the lost object into one's ego and working through the relation, gradually de-cathecting it, until it can be substituted for by something new (Grosz 1995, p. 30). The desire for a stable identity, for a complete self before the arrival of any difference—and perhaps particularly the sexual difference—is what *The golden notebook* depicts, while simultaneously, from the start, disbelieving its possibility. "There is nothing in the unconscious that accords with the body" (Lacan 1985, p. 165) or, to put it differently, there is no body as a biological fact, but instead there are the linguistically coded body parts of Clefts and Squirts for whom sexual difference is an invention, a necessary invention of the other (Derrida 1991). To identify oneself beyond social divisions and not to divide things off, not to compartmentalize, as Lessing would like us to live (Lessing 1989, p. 10) is, as Lacan would say, impossible: there is no one without an other, and no two without the Other of the social (Lacan 1985). If the lost object of *The golden notebook* and *The Cleft* is the woman as One—beyond or before sexual difference—then it is an object which must be substituted by something new, possibly by writing.[3] That this writing as a whole has been described as "subject[ing] a divided civilization to scrutiny" proves that the object of the lost unity must be abandoned—in anything but fiction.

[3] To refer to Lacan again, when he identifies the woman as the symptom of man, he also writes that the symptom is what never ceases to be written and that writing is "a means of situating the repetition of the symptom" (Lacan 1985, p. 166).

The desire for One returns, like the repressed, in the critical wish to have a clearly identified writer of a fragmented novel, to arrive at or to secure the one interpretation, to come up with one definition of what it means to be a feminist, a feminist writer or a woman, particularly the woman as other than man. That this desire is recognized as one that cannot be fulfilled is visible in Lessing's preface to *The golden notebook* when she writes about the impossibility and dangers underlying the authorial wish to control her text:

> it is not only childish of a writer to want readers to see what he sees, to understand the shape and aim of a novel as he sees it – his wanting this means that he has not understood a most fundamental point. Which is that the book is alive and potent and fructifying and able to promote thought and discussion *only* when its plane and shape and intention are not understood, because the moment of seeing the shape and plan and intention is also the moment when there is not anything more to be got out of it (Lessing 1989, p. 21).

If, as *The Cleft* illustrates, our stories are incessantly re-constructed and through them, so are we, perhaps it is time to see that "[t]he myth of *Woman* as other is now a vacant lot where different women can play with their subjective becoming" and that "the question for the feminist subject [the critic or the reader] is how to intervene upon *Woman* in this historical context, so as to create new conditions for the becoming-subject of women here and now" (Braidotti 1994, p. 168). After all, as one of the male characters of *The golden notebook* claims, "the real revolution is, women against men" (Lessing 1989, p. 198), a revolution which, as *The Cleft* ironically shows, started a long, long time ago—and doubtless will continue.

References

Braidotti, R. 1994. *Nomadic subjects. Embodiment and sexual difference in contemporary feminist theory*. New York: Columbia University Press.

Coward, R. 1992. 'This novel changes lives': are women's novels feminist novels? In *Feminisms. A reader*, ed. M. Humm, 377–380. London: Harvester Wheatsheaf.

Derrida, J. 1991. Psyche: inventions of the other. In *A Derrida reader: between the blinds*, ed. P. Kamuf, 200-220. New York: Columbia University Press.

Foucault, M. 1994. What is an Author? In *Modern criticism and theory*, ed. D. Lodge, 197–210. London, New York: Longman.

Freud, S. 1995a. Mourning and melancholia. In *The Freud reader*, ed. P. Gay, 584–589. New York, London: W.W. Norton & Company.

Freud, S. 1995b. The Ego and the Id. In *The Freud reader*, ed. P. Gay, 628–658. New York, London: W.W. Norton & Company.

Freud, S. 1995c. Totem and Taboo. In *The Freud reader*, ed. P. Gay, 481–513. New York, London: W.W. Norton & Company.

Greene, G. 1991. *Changing the story: feminist fiction and the tradition*. Bloomington: Indiana University Press.

Grosz, E. 1995. *Jacques Lacan. A feminist introduction*. London and New York: Routledge.

Lacan, J. 1985. Seminar of 21 January 1975. In *Feminine Sexuality. Jacques Lacan and the école fruedienne*, ed. J. Mitchell and J. Rose, Trans. J. Rose. New York, London: W.W. Norton & Company, 162–171.

Lessing, D. 1989. *The golden notebook*. London: Paladin.

Lessing, D. 2008. *The Cleft*. New York, London: Harper Perennial.
Nobelprize.org. The official web wite of the Nobel Prize. http://nobelprize.org/nobel_prizes/literature/laureates/2007/. Accessed 26 January 2010.
Rose, J. 1985. Introduction—II. In *Feminine sexuality. Jacques Lacan and the école fruedienne*, ed. J. Mitchell and J. Rose. New York, London: W.W. Norton & Company, 27–57.
Thorpe, M. 1973. *Doris Lessing*. Burnt Mill, Harlow, Essex: Longman.

Part II
Remapping Women's/Men's Moral and Social Borders

Near the Riverbank: Women, Danger and Place in Dickens

Aleksandra Budrewicz-Beratan

Abstract The paper analyzes the way selected Dickens's female characters attempt to cross the physical and metaphorical borders laid down by Victorian society in England, focusing on Nancy from *Oliver Twist* and Martha's, the heroine's from *David Copperfield*, relationship with and attitudes towards this physical and figurative geography and how it shapes the crucial moments in their lives. Exploring the relationship of these two characters to the many boundaries they encountered, crossed, and could not cross it will become clear that their encounters at these boundaries are shaped by the society's demands and expectations, and their ability or inability to conform to these demands.

1 Introduction

Both marked and unmarked, borders are of great significance in the novels of Charles Dickens. Physical and personal boundaries shape the lives and experiences of many of his characters. The article looks at these borders in two novels by Charles Dickens: *Oliver Twist* and *David Copperfield*. Two female characters, Nancy and Martha, will be analyzed, in particular how and why they approach and

A. Budrewicz-Beratan (✉)
The Pedagogical University of Kraków, Kraków, Poland
e-mail: a.beratan@gmail.com

J. Fabiszak et al. (eds.), *Crossroads in Literature and Culture*, Second Language Learning and Teaching, DOI: 10.1007/978-3-642-21994-8_6,

cross certain real—physical—and metaphorical—personal and social—borders. The main focus of this article will be Nancy's and Martha's relationships with and attitudes towards this physical and figurative geography, as well as the way it shapes the crucial moments in their lives. Special attention will be given to the symbolic meanings of the river.[1] The readers will be offered a bit of the social and historical context of the river Thames in the lives of 19th century Londoners in general, and in particular the importance of the river in the life and work of Charles Dickens. The paper will also discuss the changing perception of prostitution in the Victorian period. The remainder of the article will concentrate on Nancy's and Martha's relationships to the many boundaries they encountered, crossed, and could not cross. It will become clear that their encounters at these boundaries are shaped by their society's demands and expectations, and their ability or inability to conform to these demands.

2 Sweet Thames[2]

The river Thames has been imagined in many ways by a number of writers. William Wordsworth saw the Thames as "proudly bridged" (Wordsworth 1970, p. 108), whereas Edmund Spenser wrote of "the shore of silver-streaming Thames" which he also famously called "sweet Thames" in his *Prothalamion* (Spenser 1989, p. 24). The last phrase is often applied to T. S. Eliot who in fact "borrowed" the whole verse from Spenser: "sweet Thames! Run softly, till I end my song", and used it three times in *The waste land* (Eliot 1969, p. 67). Whatever the perception of the Thames was, the river with its numerous bridges easily stood out as an important element of the city.

2.1 The River of London

Rivers often indicate a variety of boundaries: the most obvious is a geographical one, but there are also social or historical borders as well. The river can be metaphorically seen as a border between the world of the living and the dead (Masłowska 1990, p. 197), just like the Nile was an encounter and the border of two parts of the world: the West (the dead) and the East (the living) (Niemczyk 2002, p. 67). The underground river Styx described in the Greek myth was a

[1] An interesting critical look at the rivers of America presented in literature can be found in a selection of essays called *Rivers and the American experience,* edited by Jerzy Durczak, Lublin: Uniwersytet Marii Curie-Skłodowskiej 2000. Among the most informative are Jiří Flajšar's Epiphanic Transformation of the Self in American River Poems from Whitman to Hugo (pp. 59–70), Schöpp's (2000), and Zapędowska's (2000).

[2] The title comes from Edmund Spenser's poem *Prothalamion* (1989, p. 24).

border between the worlds of the living and the dead. According to Hindu and Buddhist beliefs, sacral powers are concentrated around the junction of the rivers (Niemczyk 2002, p. 68). Rivers have the power to shape and create the landscape, together with the movements of the Earth's tectonic plates (Czaya 1987, p. 58). Great cities were located either on riverbanks, or at the mouth of a river (Piskozub 2001, pp. 87–109), which led scholars to distinguish between four important roles of rivers: functional, compositional, aesthetic, and that related to worship (Niemczyk 2002, p. 68). Certainly, we could take this concept further and mention a number of symbolic meanings which have been applied to water in general: life, purity, fertility, cleansing, liberation from sins—baptism in Christianity, etc.

London's fortunes have long been tied to its river (Porter 2001, p. 188)—the Thames. It is the longest river in England with its 215 miles and has always been the heart of the capital, but it has also fascinated artists as an object of description. The Thames, with its 134 bridges, runs through the borders of nine English counties, both rural and urban regions (Ackroyd 2009, p. 3). For this reason, as Peter Ackroyd believes, some parts of the river evoke calm and forgetfulness, whereas others may provoke anxiety and despair: "it is the river of dreams and the river of suicide" (Ackroyd 2009, p. 6).[3]

2.2 The 19th Century London

For the 19th century Englishman London was the centre of all things (Porter 1995, p. 185).[4] Charles Dickens was its social and a physical topographer (Collins 1973, p. 551) whose depictions of both London's beautiful places and slums (Dyos 1982, pp. 129–153) were suggestive and truthful. However, Dickens was not the only one whose novels make London their most important setting (Collins 1973, pp. 537–557)[5]. He was not even a pioneer in doing so, since Daniel Defoe, Henry Fielding and William Hogarth had exploited urban living before Dickens and Dickens himself cited them as his models in writing about the city. It was also not just London that fascinated people, of course, as other big cities—Paris, Venice—were described, and explored (Pendergrast 1995, p. viii). From Dickens's London though, one could view the world (Bell 1967, p. 43). The term "Dickensian architecture" (Wolfreys 1998, p. 144) was coined to explore the way Dickens saw

[3] The author uses other interesting and apt metaphorical phrases, for example that of the river as "liquid history because within itself it dissolves and caries all epochs and generations. They ebb and flow like water" (Acroyd 2009, p. 6); "A Museum of Englishness" (Acroyd 2009, p. 10); or "a perpetual remembrance of the past", "timeless river" (Acroyd 2009, p. 12).

[4] By 1800 London was the greatest city in Western Europe and possibly the world, with almost a million inhabitants. By the 1850s it was joined by Paris and New York. By 1900 London had a population of 4.5 million. For more information about this period from a contemporary of Dickens see Henry Mayhew's famous *London labor and the London poor* (Mayhew 2010).

[5] On page 551 Collins quotes different writers' fascination with the London of Dickens, like Nathaniel Hawthorne and Henry James. In J. K. Huysmans' famous novel *A rebours* (1884) its hero Des Esseintes talks about how well Dickens's works evoked the feelings of the English life.

and wrote about London. The London Dickens depicted was called "the massive web of intricate interconnections" (Freeman 2007, pp. 21, 43–46).

The question of the way Dickens portrayed London has been the focus of much scholarly attention (Tambling 2009). Detailed topography of the locations of his novels, as well as its symbolism were emphasized. Let us then concentrate on a single element which has not been widely explored: Dickens's use of the river. This element of nature was frequently present in his novels, both as an important symbol, or a place of location. Dickens himself was fascinated with the riverside world. As a child, he crossed London Bridge every day going to Warren Blacking factory to work. The Thames and its bridges play an important role in much of Dickens's fiction,[6] in particular in *Our Mutual Friend* which has been studied critically (Sanders 2000, pp. 348–349; Sanders 2010, pp. 180–193; Litvack 2003; Depa 2009, pp. 139–148).

Dickens's fascination with the river and its bridges in many ways mirrors their importance to London and its inhabitants. A number of new bridges and canals were built in the 19th century, changing the social and economic landscape of London as a whole (Arnold 2005, p. 104). A huge investment in the docks (Porter 1998, p. 188) contributed to London's economic growth. Riverside areas were full of industries and factories, coal delivery started to come by coaster,[7] which was cheaper. Before a civil engineer Joseph Bazalgette introduced a new sewage system, the Thames had brought serious hygienic problems as well, as the Victorians flushed their waste directly into the river. As a result, pollution and a cholera epidemic had to be dealt with. During the hot summer days of 1858 the odour became so unbearable that a session in Parliament was disrupted; this event became known as "The Great Stink" (Halliday 2001; Barnes 2006). There were other issues that began receiving greater attention in the Victorian era than they had had in the past.

2.3 *Found Drowned*[8]

Prostitution had always been around (Valverde 1989),[9] but in Victorian times it came to be seen as a "Great Social Evil" (Slater 1983, p. 339; Walkovitz 1980,

[6] This fascination was passed over to Dickens's son, Charles Dickens. In 1887 he published *Dictionary of the Thames* in order to "give practical information to oarsmen, anglers, yachtsmen, and other directly interested in the river; to serve as a guide to the numerous strangers who annually visit the principal places on its banks" (Dickens 2007, p. 3). The book is an interesting portrait of the Thames at the end of the 19th century, with a number of detailed descriptions of towns, inns, boats renting etc.

[7] Coasters were transport ships that enabled the larger and faster supply of coal from the north than had been previously possible (Porter 1998, p. 191).

[8] This is the title of a popular painting by George Frederic Watts's (ca. 1848–1850)

[9] Mariana Valverde's interest focused on the way prostitutes were perceived by society. She describes the way clothes signified women in the 19th century, the way their fashion and the then social life were connected; also, among others, she talks about Esther, a prostitute from Elisabeth Gaskell's popular novel *Mary Barton.*

p. 4). In mid-19th century there were more than 6,000 brothels in London, and around 80,000 prostitutes (Laurence 2004, p. 39; Liggins 2003, pp. 39–55). Because of the associations with venereal disease, prostitution was seen as a dangerous pollution, simultaneously physical and moral[10] (Ingham 1996, p. 24).

Prostitutes were often seen as ill-fated women because they would never marry or have children (Slater 1983, p. 339). Charles Dickens, who viewed the prostitute as a "Ruined Temple of God" (Slater 1983, p. 443), is believed to have played an important role in arousing public sympathy for "fallen women" (Westland 2000, pp. 476–477). Dickens's earliest allusion to prostitution was made in *Sketches by Boz*, where prostitutes are depicted as victims who struggled to live respectably.

Dickens was especially intrigued by the miserable life of prostitutes. For 11 years (1847–1858) he was actively involved in establishing and administering Urania Cottage (Hartley 2009; Slater 2009, pp. 295–296, 317, 326, 331, 404; Pope 2000, pp. 590–591). Urania Cottage, or "The *Home* for Homeless Women" (Dickens himself put emphasis on the word home; Slater 1983, p. 342) was an organization administered by the millionaire philanthropist Angela Burdett Coutts and significantly supported by Dickens. It was an 'asylum' for homeless women, as well as for women from prisons, workhouses and the streets of London. They were given education and care, in order to prepare them to immigrate to England's colonies. Burdett Coutts financed the place, whereas Dickens provided ideas for organizing it. The word "Urania" meant "celestial", as opposed to earthly or sexual love[11] . Some scholars call this community Dickens's alternative family (Hartley 2009, p. 73).

According to Michael Slater, Dickens imposed his own conceptions on prostitutes: either they were so far "fallen" that they seemed mischievous in his eyes, or they were quite unhappy (Slater 1983, p. 342). Prostitutes in Dickens's novels frequently feel embarrassed, have remorse, and think about changing into respectable and domestic women. They were mostly seen by the society as a moral poison, but also by some as a useful and necessary service that helped to preserve the home (Ayres 1998, p. 130). However, this "forgiving" approach to prostitutes was not accepted by many Victorians, who would rather treat these women as fallen, immoral, and sinful. Many prostitutes ended their lives by drowning themselves in the river Thames, which was often portrayed by painters, like George Frederic Watts in his *Found Drowned*. This is something that Dickens obviously took note of as it is alluded to in both *Little Dorrit* (Chap. 3) and *Bleak*

[10] Greater condemnation was reserved for the women forced into prostitution than for the men who engaged their services.

[11] Urania was the muse of astronomy, which is why nowadays lots of astronomical observatories are named after her. The name had associations with Aphrodite Urania the goddess of heavenly love, as opposed to Aphrodite, the goddess of physical love. The word "cottage" in the House's name aroused a pleasant and domestic atmosphere, unlike other casual names of the London institutions. The main intention for the Urania Cottage was to provide women with "kind treatment" so that they could become good wives and mothers when they emigrate to other British-owned lands, like Australia (Slater 2009, p. 269).

House (Chap. 5). What is more, Dickens came back to this aspect of the experience of prostitutes quite often, for example in *David Copperfield* in which it served as a plot point for Martha. This example will be examined in depth later in this paper.

3 Borders in Dickens

The depiction of the 19th century London given above, in particular its river, bridges and prostitutes, is essential to understanding the way certain "borders" were approached and crossed by two female characters created by Dickens in his *Oliver Twist* and *David Copperfield.* For both of these women the riverbank plays a crucial role, serving as a central backdrop to significant events in their lives. With this in mind it would be helpful to quote Peter Ackroyd again. Seeing the river as something "suspended in time" and poised between two worlds, Ackroyd talks about the act of reading water: the thoughts of a person standing on the bend of the river derive from the necessity to go forwards and backwards (Ackroyd 2009, p. 15). If one passes from land to water, he or she physically and mentally crosses boundaries. Therefore, for the purpose of this article I distinguish between two types of borders:

1. The real/physical ones: both women cross the "real" borders of the districts of London, walking alone, in defiance of Victorian society's norms and proprieties. Those unmarked but implicitly understood boundaries within the space of London often separate the rich and the poor, the attractive and the dirty, as well as the happy and the miserable. These borders can be seen in the dangers both Martha and Nancy encounter in the very act of crossing London.
2. The metaphorical boundaries are more complex and stem from a number of possible interpretations of the novels. Prostitution can be then seen as a metaphor for the boundaries of respectability in the society. Nancy crosses one border by betraying her loyalty to Fagin and Sikes, and another in going from "performing" a role to taking on that role as part of her real life when Nancy adopts the part she was told to play for herself).

3.1 Real Borders for Nancy and Martha

Nancy from *Oliver Twist* and Martha Endell from *David Copperfield* were prostitutes who felt ashamed of their profession. They wandered around the most gloomy and sordid parts of London, almost always covering their faces with shawls, as if they were trying to hide from the world. Both characters had their prototypes in earlier Dickens sketches. Nancy's was a sketch of a 'fallen woman' living on the streets of London in *The Pawnbroker* (Slater 1983, p. 339, 340), combined with a story of a dying girl, tormented by shame and remorse, murdered by her pimp in *The Hospital Patient.* On the other hand, the portrait of Martha was modeled after an inhabitant of Urania Cottage, Isabella Gordon, a high-spirited girl who added to the overall cheerfulness of the house, which Dickens loved, but at some point she was forced to leave (Hartley 2009, pp. 72–74). Nancy and Martha both transcend the traditional depiction of prostitutes. Scholars have usually interpreted Nancy as a victim because she allowed herself to be killed in Oliver's place (Schor 1999, p. 35; Slater 1983, p. 340). Nancy has been seen as an exceptional image of a prostitute who also plays roles of a wife to Sikes and a sister or a mother to Oliver (Westland 2000, pp. 476–477). Her life has been "squandered in the streets and among the most noisome of the stews and dens of London" (Dickens 1994b, p. 369).

Martha's prostitution is the main reason for her marginalization. We see her through the eyes of David, Mr Peggotty, Ham, and Emily. The society thought Martha was the reason for Emily's downfall, whereas in fact she was not to blame. Nonetheless, when she helps to find Emily, Martha earns people's respect. Even though prostitutes were usually denied marriage, Martha does get married after she immigrates to Australia (Poovey 1998, p. 96). To some critics Martha was a version of what Dickens hoped to accomplish at Urania (Dunn 2004, p. 67).

3.2 Nancy: "If There Was More Like You, There Would Be Fewer Like Me"

Dickens 1994b, p. 371 Nancy crosses the physical border of unfamiliar areas of London. She lives near Field Lane which was close to the river and St Paul's cathedral, "a dirtier and more wretched place Oliver had never seen", with narrow and muddy streets, with "the air impregnated with filthy odors" (Dickens 1994b, p. 69). Nancy overhears a conversation between Fagin and Sikes, and learns that they are planning to get Oliver back. Secretly she goes to Rose, a potential rescuer of Oliver, and warns her. Going to Rose entails entering "other" parts of London, outside her "home". Nancy does it always in a hurry: "she tore along the narrow pavement: elbowing the passengers from side to side; and darting almost under the horses' heads, crossed crowded streets, where clusters of persons were eagerly watching their opportunity to do the like" (Dickens 1994b, p. 367).

Nancy reached "the more wealthy quarter of the town, the streets were deserted; and here her headlong progress excited a still greater curiosity in the stranglers who she hurried past. (…) It was a family hotel on a quiet but handsome street near Hyde Park" (Dickens 1994b, p. 367). Nancy is given the first chance to change her life, as Rose offers her help, but Nancy refuses both assistance and money. Nancy feels she cannot leave Sikes, even though he is a cruel man; she tells Rose: "I must go back, I am drawn back to him through every suffering and ill usage" (Dickens 1994b, p. 375). In this quotation it seems that Nancy recognizes the borders, both physical and metaphorical that she has crossed, and signals that she feels compelled to go back as though she is neither able to, nor deserving of, living on the other side of these boundaries. I believe this represents Dickens's recognition of what today is a fairly common understanding that battered women often feel they deserve their mistreatment.

Nancy and Rose meet one more time to discuss how to save Oliver. Nancy reveals the information to Rose and Mr. Brownlow, on the steps leading to the river at the side of the old London Bridge. This is a dreary and gloomy location for Nancy to visit alone (Miller 1970, p. 41). Unfortunately, she is overheard by Noah Claypole:

> It was a very dark night. The day had been unfavourable, and at that hour and place there were few people stirring. (…) A mist hung over the river, deepening the red glare of the fires that burnt upon the small craft moored off the different wharfs, and rendering darker and more indistinct the murky buildings on the banks. The old smoke-stained storehouses on either side, rose heavy and dull from the dense mass of roofs and gables, and frowned sternly upon water too black to reflect even their lumbering shapes. (Dickens 1994b, pp. 426–427).

This "strange place", as the gentleman calls it, must be frightening indeed if it scared even an older man: "for what purpose can you have brought us to this strange place? Why not have let me speak to you, above there, where it is light, and there is something stirring, instead of bringing us to this dark and dismal hole?" (Dickens 1994b, p. 429). Dickens recognized that these women inhabit two separate worlds: Nancy has to travel to the dangerous sections of London alone, whereas Rose does not travel unescorted.

Nancy tells Rose that she is afraid of something all the time, showing how powerful her imagination is (Hartley 2009, pp. 67–68):

> Horrible thoughts of death, and shrouds with blood upon them, and a fear that has made me burn as if I was on fire, have been upon me all day. I was reading a book to-night, to wile the time away, and the same things came into the print. (…) I saw 'coffin' written in every page of the book in large black letters,- aye, and they carried one close to me, in the streets to-night. (Dickens 1994b, p. 429)

It was imagination that Dickens recognized in the women who lived in Urania Cottage, therefore he provides Nancy with a quality of a vivid imagination. This passage functions also as a premonition as it prepares the readers for a petrifying scene in which Sikes murders Nancy.

When Nancy is offered a second chance—a quiet asylum, in England or in some foreign country—she points to the Thames and cries: "Look at that dark water. How many times do you read of such as I who spring into the tide, and leave no living thing to care for or bewail them"[12] (Dickens 1994b, p. 435). The river then becomes the only refuge Nancy can think of. Ironically, it is there, near the river bank, where the tragedy of Nancy's sacrifice comes to fruition and her life comes to an end. She knew that revealing the secret about Oliver to Rose and Mr

[12] Commenting on that passage J. H. Miller claims that this is the imaginary death of Nancy, even though she actually is beaten to death by Sikes (Miller 1970, p. 41).

Brownlow would be extremely risky to her safety. What is more, this event means crossing both borders: the physical one, meaning travelling to the dangerous area alone, and the metaphorical one which is disloyalty to Fagin's gang. Noah spies on Nancy and later tells Fagin everything; in turn Fagin informs Sikes, Sikes in fury and anger beats Nancy to death. Just before her death, trying to raise, Nancy is holding the white handkerchief she got from Rose, "as high towards Heaven as her feeble strength would allow, breathed one prayer for mercy to her Maker" (Dickens 1994b, p. 445). Going to the river bank was a sacrifice that cost Nancy her life.

The meeting on the bridge also shows how different the two worlds are: the appalling and unhappy world of Nancy, and the beautiful and safe world of Rose and Mr. Brownlow. These worlds turn out to be impossible to join. The London Bridge links and separates these two worlds, too estranged to unite, and therefore impossible to cross. Nancy had a chance to cross the symbolic bridge and start a better life, but because of her deep devotion and love of Sikes she knew she would never be able to do that.

According to Vincent Newey (2004, p. 94), this meeting may also symbolize both the connection and distance between the two women: the high and immaculate Rose, and the low and unclean Nancy. Dickens's performances of the murder scene stimulated and sensationalized Victorian audiences. He continued performing it for over 30 years (Andrews 2006), despite his doctors' warnings that it would be too emotionally tiring for him, and his health would worsen with the constant travel.

In his study *Dickens and women* Michael Slater states that Nancy is a very complex character: she is clever and moody (Slater 1983, p. 221), fearful and courageous.[13] As Slater appositely claimed, the independent Nancy is both devoted and loyal to the criminal, and at the same time able to feel for the small boy (Slater 1983, p. 241).

3.3 Martha: "Take Me Out of These Streets…"

Dickens (1994a), p. 283 Unlike Nancy, Martha neither lived with nor belonged to any group of people. She was alone as a social outcast, disrespected by the people in her village. Seduced and then deserted by a local undertaker Mr. Omer, she became a prostitute. Ham Pegotty calls her "a poor worm as is trod under foot by the entire town. Up street and down street" (Dickens 1994a, p. 281). Like Nancy though Martha is a repentant prostitute who wants to earn respect. Her life almost ends in despair and suicide, as she was on the verge of killing herself in the river,

[13] Two scenes illustrate this point well: Nancy goes to the London Bridge at night and alone, which she found scary and terrifying, but she was not afraid of giving Sikes a medicine to put him to sleep when she secretly runs to Rose.

but fortunately David Copperfield and Mr Pegotty spot her in time to save her and console her, giving her advice on what to do with her life. Mr Peggotty gives her a mission to seek out his lost niece Emily and to rescue her from the brothel into which she has been tricked. Martha is rewarded: she goes with them to Australia where she settles down and marries a farmer.

Martha's first real border is the area of her hometown, Yarmouth, which she leaves and goes to London, crying: "Take me out of these streets, where the whole town knows me from a child! No one knows me there. Everybody knows me here" (Dickens 1994a, p. 283). She feels unhappy and embarrassed, which can be observed in the way she behaves: "gathering her shawl about her, covering her face with it, and weeping aloud, went slowly to the door. She stopped a moment before going out, as if she would have uttered something or turned back; but no word passed her lips. Making the same low, dreary, wretched moaning in her shawl, she went away" (Dickens 1994a, p. 283). The next difficult border was, as in Nancy's case, the dreadful sections of London through which she travels alone: the riverbank near Westminster. The narrator, David, describes them as following:

> The neighborhood was a dreary one at that time; as oppressive, sad, and solitary by night, as any about London; the heavy and unbroken smoke that poured out of their chimneys; nightmare condition; the ground was cumbered with rusty iron monsters of steam-boilers, wheels, cranks, pipes, furnaces, paddles, anchors, diving-bells, windmill-sails, and I know not what strange objects, accumulated by some speculator, and groveling in the dust, underneath which—having sunk into the soil of their own weight in wet weather—they had the appearance of vainly trying to hide themselves (Dickens 1994a, p. 555).

Martha crosses these physical borders more often than Nancy who cannot leave the gang's area. Martha does not have to obey the group's loyalty, because there is no group she belongs to. It adds to her misery, and pushes her to end her life in the Thames.

4 The Metaphorical Borders

These borders are more complex and difficult to define explicitly. There are some borders which both Nancy and Martha have in common, as well as the borders which are individual for each of them. All of them are very deeply rooted in the time and environment these two women lived in, the society's expectations and demands, and the moment in their lives when they had to make crucial decisions. In the following sections I will examine prostitution and performance as forms of the metaphorical boundaries both Nancy and Martha cross becoming social outcasts.

4.1 Prostitution as a Social Border

Prostitution can be seen as the clearest of these metaphorical boundaries, since fallen women were seen as a deformation of the true and high nature of women. Since society offered few economic opportunities for women outside the home, those who had been deprived of an appropriate domestic environment either by accident, drunkenness, poverty, seduction, crime, or elements of societal patriarchy, inevitably turned to prostitution once other alternatives had finished. For this reason, many people have argued that all prostitution is economically coerced (Sullivan 2004). This is what happened to Martha and Nancy. They were fully aware of the society's attitude towards them. As has been mentioned, Nancy's cry to Rose: "How many times do you read of such as I who spring into the tide, and leave no living thing to care for or bewail them!" (Dickens 1994b, p. 435) exemplifies and symbolizes the lack of opportunities for working class women, their subordinate position, no alternatives available, and no possibility for change.

Martha's emotional monologue after David and Mr Peggotty save her from committing a suicide, also suggests a spiritual connection between herself and the river: "I know that I belong to it. I know that it's the natural company of such as I am! (…) I can't keep away from it. I can't forget it. It haunts me day and night. It's the only thing in all the world that I am fit for, or that's fit for me. Oh, the dreadful river!" (Dickens 1994b, p. 557). When David observes Martha near the side of the river, he makes a similar comparison and thinks that she was "a part of the refuse it had cast out, and left to corruption and decay, the girl we had followed strayed down to the river's brink, and stood in the midst of this night-picture, lonely and still, looking at the water." (Dickens 1994b, p. 556) The reading of water plays an important part here: the thoughts of a person standing on the edge of the river stem from the necessity to go forwards and backwards. If the river is "suspended in time", poised between two worlds, then when you pass from land to water you cross some boundaries (Ackroyd 2009).

This brings us closer to the idea of the two women approaching the riverbank, and finding themselves in risky and dangerous situations, that are at the same time crucial to moving the plot of the novels forward. Even though she did know how risky her plan was, Nancy crossed the border between loyalty and disloyalty towards her gang when she met and collaborated with "the good people" (Dickens 1994a, p. 435). Just as Martha stood at the river bank and ostensibly looked across the boundary between her present and her (imagined) future, Nancy also clearly sees her own future in the murky waters of the Thames.

4.2 Nancy's Performance

When we meet her for the first time, Nancy walks near the gang area, laughing. Then Nancy is forced by Fagin to perform the role of Oliver's sister after he was arrested by the police: the boys attempted to rob a gentleman, and they all escaped except for Oliver. She plays this part once more to assist the gang in kidnapping Oliver. To some extent, however, Nancy begins to perform this role in her real life. First, she becomes concerned about the boy, telling Sikes to keep his dog away from Oliver because it may hurt him. Then she starts to realize how appalling the gang's treatment of the boy is, and finally, risking her life, she decides to find the people who had saved Oliver previously, and tell them where the boy is being kept. In making this shift to actually caring for Oliver, Nancy consciously crosses the border between "acting" and real life.

5 Conclusion

The river Thames seen in these two novels is "dreadful" indeed. The areas near the riverbank and the bridges were secluded and dangerous, which is why it was not seen as appropriate for the 19th century women to walk there unescorted. Yet some of them are forced to break with socially accepted models of behavior (Slater 1983, p. 265), and approach the river alone; for both of them going to the riverbank triggers a turning point in their lives.

For Nancy it was in many ways a sacrificial act: she knew how much she risked by meeting with Rose and Mr Brownlow. The river then is a border between both worlds, between her poverty and life of subsistence, and their life of plenty. Nancy approached this border, but despite the warm welcome of "the other side", she felt unable to leave the world she knew. The penalty for even contemplating crossing the other side was Nancy's death.

Martha came to the riverside with the clear purpose of committing suicide there. She saw herself as being dirty and polluted like the Thames, which is why she did not think that she deserved to lead a peaceful and moral life. Martha approached the border of water (death) and land (life), and was only prevented from crossing this boundary through the intervention of others who were steadfastly rooted on the good side. Like Nancy, she was offered help and encouragement, unlike Nancy though Martha did not refuse it. She leaves the river Thames, but crosses an ocean, a boundary to a better life, to settle down in Australia.

Mr Micawber, a comic character from *David Copperfield,* calls London "the Modern Babylon" (Dickens 1994a, p. 138, 435), which highlights Dickens's interpretation of this city as a place in which some types of communication between certain people are impossible, either due to class differences, or social mores. This is a problem depicted in *Oliver Twist* and *David Copperfield.* Approaching and crossing various social borders was a necessary act on the part of these characters, but not always a successful one. If "knowledge of the city comes only with formation of self" (Sicher 2003, p. 14), then studying Dickens's characters as placed inside the city and among its inhabitants unquestionably adds to a deeper understanding of them.

References

Ackroyd, P. 2009. *The Thames.* New York: First Anchor Books Edition.
Andrews, M. 2006. *Charles Dickens and his performing selves: Dickens and the public readings.* Oxford: Oxford University Press.
Arnold, D. 2005. *Rural urbanism: London landscapes in the early nineteenth century,* Manchester: Manchester University Press.
Ayres, B. 1998. *Dissenting women in Dickens's novels. The subversion of domestic ideology.* Westport, London: Greenwood Press.
Barnes, D. S. 2006. *The great stink of Paris and nineteenth-century struggle against Filth and Germs.* Baltimore: The John Hopkins University Press.
Bell, A. D. 1967. *London in the age of Dickens.* Norman: University of Oklahoma Press.
Collins, Ph. 1973. Dickens and London. In *Victorian city. Vol. 2,* ed. J. Dyos, M. Wolff. 537-557. London: Routledge Kegan and Paul.
Czaya, E. 1987. *Rzeki kuli ziemskiej.* Warszawa: Państwowe Wydawnictwo Naukowe.
Depa, Bożena. 2009. The river as a boundary in *Our mutual friend* by Charles Dickens. In *From queen Anne to queen Victoria: readings in 18th and 19th century British literature and culture,* ed. G. Bystydzieńska, E. Harris. 139-48. Warszawa: Wydawnictwo Uniwersytetu Warszawskiego.
Dickens, Ch. 1994a. *David Copperfield.* London: Penguin Books.
Dickens, Ch. 1994b. *Oliver Twist.* London: Penguin Books.
Dickens, Ch. 2007. *Dictionary of the Thames.* Devon: Old House Books.
Dunn, R. J. (ed). 2004. *A Routledge literary sourcebook on Charles Dickens's David Copperfield.* Routledge: New York.
Dyos, H. J. 1982. The slums in Victorian London. In *Exploring the urban past: essays in urban history,* ed. D. Cannadine, D. Reeder, 129-153. Cambridge: Cambridge University Press.
Eliot, T.S.1969. The waste land. In *The complete poems and plays,* ed. V. Eliot. 59–80. London: Faber & Faber.
Freeman, N. 2007. *Conceiving the city: London, literature, and art 1870–1914.* New York: Oxford University Press.
Halliday, S. 2001. *The great stink of London: Sir Joseph Bazalgette and the cleansing of the Victorian capital.* Stroud: Sutton.
Hartley, J. 2009. *Charles Dickens and the house of fallen women.* London: Methuen.
Ingham, P. 1996. *The language of gender and class. Transformation in the Victorian novel.* London, New York: Routledge.
Laurence, D. H. 2004. Victorians unveiled: some thoughts on *Mrs Warren's profession. Shaw: The Annual of Bernard Shaw Studies,* vol. 24: 38-45.
Liggins, E. 2003. Prostitution and social purity in the 1880 s and 1890 s. *Critical survey,* Vol. 15, Issue 3: 39-55.
Litvack, L. 2003. Images of river in *Our mutual friend. Dickens quarterly* March 2003, vol. XX nr 1: 34-55.
Masłowska, E. 1990. Ludowy stereotyp rzeki – zarys struktury. In *Językowy obraz świata,* ed. J. Bartmiński, 197-206. Lublin: Uniwersytet Marii Curie-Skłodowskiej.
Mayhew, H. 2010. *London labour and the London poor: a selected edition.* Oxford: Oxford University Press.
Miller, J. H. 1970. *Charles Dickens: the world of his novels.* London: Oxford University Press.
Niemczyk, E. 2002. *Cztery żywioły w architekturze.* Wrocław: Zakład Narodowy im. Ossolińskich.
Pendergrast, Ch. 1995. *Paris and the nineteenth century.* Oxford: Blackwell.
Piskozub, A. 2001. *Rzeki w dziejach cywilizacji.* Toruń: Wydawnictwo Adam Marszałek.
Poovey, M. 1998. *Uneven developments: the ideological work of gender in mid-Victorian England.* Chicago: Chicago University Press.
Porter, R. 2001. *London: a social history.* Cambridge: Harvard University Press.

Sanders, A. 2000. London. In *Oxford reader's companion to Dickens,* ed. P. Schlicke. 348-365. Oxford: Oxford University Press.

Sanders, A. 2010. *Charles Dickens's London.* London: Robert Hale.

Schor, H. M. 1999. *Dickens and the daughter of the house.* Cambridge: Cambridge University Press.

Schöpp J. C. 2000. Of time and river: Thoreau's "A week on the Concord and Merrimack rivers" as a Passageway into Nature and History. In *Rivers and the American experience,* ed. J. Durczak, 95-108. Lublin: Uniwersytet Marii Curie-Skłodowskiej.

Sicher, E. 2003. *Rereading the city/rereading Dickens: representation, the novel, and urban realism.* New York: AMS Press.

Slater, M. 1983. *Dickens and Women.* London: J. M. Dent & Sons Ltd.

Slater, M. 2009. *Charles Dickens: a life defined by writing.* New Haven: Yale University Press.

Spenser, E. 1989. Prothalamion. In *The Oxford library of English poetry, vol. 1. Spenser to Dryden,* ed. J. Wain, 24-27. London: Guild Publishing.

Sullivan, B. 2004. Prostitution and consent: beyond the liberal dichotomy of 'Free' and 'Forced'. In *Making sense of sexual consent*, ed. M. Cowling, Paul Reynolds, 127-140. Aldershot: Ashgate Publishing Limited.

Tambling, J. 2009. *Going astray: Dickens and London.* Harlow: Pearson Education Limited.

Zapędowska M. 2000. Reading the face of the earth: travelling into nature in Henry David Thoreau's "A week on the Concord and Merrimack rivers" and Seweryn Goszczyński's "Diary of a journey to the Tatras". In *Rivers and the American experience,* ed. J. Durczak, 145-160. Lublin: Uniwersytet Marii Curie-Skłodowskiej.

Valverde, M. 1989. The love of finery: fashion and the fallen women in the nineteenth century social discourse, *Victorian Studies,* January, vol. 32, nr 2: 169-188.

Walkowitz, J. 1980. *Prostitution and Victorian society: woman, class, and the state.* Cambridge: Cambridge University Press.

Westland, E. 2000. Prostitutes and fallen women. In *Oxford reader's companion to Dickens,* ed. P. Schlicke, 475-477. Oxford: Oxford University Press.

Wolfreys, J. 1998. *Writing London: the trace of the urban text from Blake to Dickens*. New York: Palgrave.

Wordsworth, W. 1970. *The prelude or growth of a poet's mind. 1805,* ed. E. de Selincourt, Oxford: Oxford University Press.

"Women with Iron in'em. Women Who Wanted Land and a Home": Female Pioneers Staking Out New Territories in the American Popular Fiction of the 1920s and 1930s

Barbara Leftih

Abstract In 1893 Frederick Jackson Turner declared the Western frontier "closed"—westward expansion came to an end. Nevertheless, the appeal of the West persisted and pioneers and settlers, both male and female, were easily transformed into literary characters. This article analyzes how two popular female writers of the 1920 and 1930's, Rose Wilder Lane and Edna Ferber, depicted the enterprise of staking out new territories and crossing frontiers by their female characters. Ferber's *Cimarron* (1929) revolves around the opening of Oklahoma to non-Indian settlement in 1889. Lane's *Young pioneers* (1933) and *Free land* (1938) concern the decade of the 1880s and portray the struggle of pioneering settlers in Dakota. I will attempt to examine what roles are assigned to the selected female characters or what roles they assign to themselves and how these characters influence the new territories or in reverse how the new territories affect the characters. I would be interested to analyze how staking out new territories and crossing frontiers affects the female characters' gender roles imposed on them by the late nineteenth-century American culture and their racial attitudes when the characters are faced with ethnic or racial difference.

1 Introduction

In 1893 Frederick Jackson Turner declared the Western frontier "closed"—westward expansion came to an end. Nevertheless, the appeal of the West persisted and pioneers and settlers, both male and female, were easily transformed into

B. Leftih (✉)
University of Bielsko-Biała, Bielsko Biała, Poland
e-mail: barbarakaletka@interia.pl

J. Fabiszak et al. (eds.), *Crossroads in Literature and Culture,* Second Language Learning and Teaching, DOI: 10.1007/978-3-642-21994-8_7,

literary characters. In this article I would like to analyze how two popular female writers of the 1920 and 1930s, Rose Wilder Lane and Edna Ferber, depicted the enterprise of staking out new territories and crossing frontiers by their female characters. Ferber's *Cimarron* (1929) revolves around the opening of Oklahoma to non-Indian settlement in 1889. Lane's *Young pioneers* (1933) and *Free land* (1938) also concern the decade of the 1880s and portray the struggle of pioneering settlers in Dakota. I will attempt to examine what roles are assigned to the selected female characters or what roles they assign to themselves and how these characters influence the new territories or in reverse how the new territories affect the characters. I am going to analyze how staking out new territories affects the female characters' gender roles imposed on them by the late nineteenth-century American culture and their attitudes when the characters are faced with ethnic or racial difference.

2 Social and Literary Background

2.1 Emergence of New Genres

The closing of the frontier, declared by Turner, sparked nostalgia for the pioneer past, which became reflected in the emergence of two literary genres, namely the western and the farm novel. The publication of Owen Wister's *The Virginian* in 1902 marked the emergence of the western as a separate genre. The first farm novel, *Main-Travelled Roads*, was published by Hamlin Garland in 1891. The genres, which differed significantly in their portrayal of the frontier experience, reached the apex of their popularity in the 1920 and 1930s. The heightened migration of people from rural areas to industrial centres accelerated in the 1920s and according to the 1920 population census half of the American population lived in the cities. This marked change in demographics contributed to the growing interest in the life on the frontier of yesterdays. In the 1930s, overshadowed by the Great Depression, the fact that farm and pioneer novels foregrounded characters with features such as enterprise, persistence, determination, industriousness, self-reliance, enabling the characters to attain the American dream of prosperity, served an insecurity-assuaging role, convincing the readers that contemporary adversities could be surmounted and were not permanent. Simultaneously, these novels constituted a critique of American consumerism, which reached new heights in the 1920s (Raub 1994, pp. 58–61). Irrespective of the optimistic message they carried in the times of crisis, farm and pioneer novels were "politically reactionary", because they rendered poverty as a romantic struggle against surmountable odds and upheld the social status quo in terms of politics and gender relations (Ehrhardt 2004, p. 118).

2.2 *Portrayal of Pioneer Women in American Culture and Women's Literature of the 1920s and 1930s*

Apart from targeting American infatuation with consumerism and perpetuating the myth of American dream, farm and pioneer novels of the 1920 and 1930s responded to another social phenomenon, namely the changing position of women in American society. As mentioned earlier, many farm and pioneer novels of these two decades revealed conservative tendencies presenting the "'golden age' of gender relationships" in which women were content with their traditional roles (Raub 1994, p. 61). According to Campbell (2003, p. 27), who draws on Annette Stott's article "Prairie Madonnas and pioneer women: Images of emigrant women in the art of the old West", between 1840 and 1890, the heyday of Western expansion, white women used to be portrayed on paintings and prints as "Prairie Madonnas"—passive, morally superior true women accompanying their husbands west. This image upheld the dominant perception of women and division into spheres. However, towards the end of the nineteenth century the submissive Prairie Madonna evolved into an independent Pioneer Woman performing hard work along with men. As Browder (2008, p. 77) claims, the physically strong Pioneer Woman was frequently favourably contrasted with the modern late nineteenth-century women who were perceived as sickly and weak. The above-mentioned images of white women in the nineteenth-century West proliferated in the 1920 and 1930s (Campbell 2003, p. 27). Moreover, Jameson (1987, p. 145) contends that Western women have been stereotypically divided into two categories: good women (Prairie Madonnas or Pioneer Women) and bad women (prostitutes and troublemakers). Modern scholars of Western women (Jameson 1987; Myres 1982) argue that this division, persistent in both histories of the West and popular culture and literature, does not reflect adequately the experience of frontier women.

According to Raub (1994, pp. 62–72), farm and pioneer novels of the 1920 and 1930s written by women novelists frequently portray female characters adhering to traditional roles the society expected them to fulfill. Nevertheless, female characters that go beyond their conventional roles in society do appear in such fiction. Raub suggests that Willa Cather provided models for both types of characters in her novels *O, pioneers!* (1913) and *My Antonia* (1918). The first novel features Alexandra Bergson, a self-reliant woman who is able to withstand numerous adversities and successfully runs a farm in Nebraska; the second novel introduces Antonia Shimerda—a more conventional female character that does not question the traditional role assigned to her by society. In Raub's opinion, female characters portrayed in farm and pioneer novels written by many women novelists of the 1920 and 1930s, such as Ellen Glasgow, Bess Streeter Aldrich, Gladys Hasty Carroll, tend to oscillate between the models established by Carther.

3 Rose Wilder Lane's *Young Pioneers* and *Free Land*

3.1 Young Pioneers

As Ehrhardt (2004, p. 118) states, Rose Wilder Lane was commissioned to write her first farm novel by George Horace Lorimer, the editor of the *Saturday Evening Post*, who argued against the New Deal legislation, as he considered it to be a governmental intrusion into the American economy, and whose ideas were compatible with Lane's libertarian views. Hence, *Let The Hurricane Roar*, later retitled *Young pioneers*, seems to portray American ability to survive independently against the odds. Ehrhardt (2004, pp. 134–135) considers Lane's subsequent farm novel *Free land* to be more contradictory and states that the novel undermines the myth of pioneer success. However, Campbell (2003, p. 28) contends that both Lane's farm novels debunk the optimistic pioneering myth by emphasizing the harshness of frontier life and portraying the frontier as "the site not of limitless opportunity but of inescapable violence over contested territories".

The main heroine of Lane's *Young pioneers* Molly marries young and somehow naturally decides to go west with her husband: "Many settlers had come to the settlement in the Big Woods while David and Molly were growing up. When they were married there was little good land left. Farther west, the country was not yet settled and the land was said to be rich and level, and without forests. So they went west" (Lane 1998, p. 5). The thirst for their own land propels them to leave their home settlement, set off west and finally settle down on a homestead previously occupied by a man who, disillusioned with hard life, decides to leave it.

Molly and her husband David soon realise that staking out and inhabiting new territories is marked with excessive toil and danger. Molly and her husband's homestead lies far from other abodes; therefore, what Molly needs to cope with is loneliness, overwhelming silence and fear. Violence is implied by claim jumpers ready to kill for the land or lonely rider figures that lurk on the horizon and who might turn out to be hostile Indians or white outlaws. The new territory is presented as a vacuum in which the old inhabitants have been pushed to the margin and new organisation and law enforcement have not yet been imposed. Significantly, the dispossession of Native Americans is never discussed in the novel; they have been reduced to distant figures that fuse with the skyline. The main female character of the novel needs to cope with the physical frontier, but not the racial one.

Molly sets off to the West as a conventional young girl of sixteen, reliant on her husband, later a mother-to-be, who allows her husband to make decisions concerning their future. Nevertheless, from the very beginning Molly seems to think that entering a new, unknown territory transfixes people and requires a new set of features. The new territory that Molly is traversing changes her physically, "Her hands and face were brown as an Indian's from the prairie wind and sun. When she unpinned the collar of her basque, it was odd to see the brown face on the milk-white neck" (Lane 1998, p. 10). Giving birth to her child, she resolves not to cry as

Indian women bear their children silently. Molly, while she never foregoes her own racial category, associates the new territory with stamina, resilience, and toughness, of which American Indians seem to be an embodiment. Molly possesses these features, as some of her decisions, even the early ones, suggest, but Lane does not emphasize these qualities in Molly, who is portrayed at the beginning of the book as a rather retiring and timid character. They lie dormant until Molly is left alone to fend for herself. Still, they are not absent as, for example, it is Molly who decides to bear her child in the West as she is afraid that the land might be taken from them if they spend the winter in the East. Molly is more cautious than her husband and she is appalled by David's improvidence and the fact that he incurred a debt against their uncertain future income.

Although Molly allows her husband to take over the reins of their life, as David is the man with a vision, Molly is not a dominated wife relegated to the domestic sphere. Interestingly, Molly and David's secluded life does not seem to have a clear-cut division into spheres. Their roles are shown as complementary as both kinds of work that they perform are indispensable for their survival—David's in the fields and Molly's in the house. Moreover, the scene of birth underlines partnership between David and Molly. It is David who delivers Molly's baby and initially, when Molly is incapacitated, takes care of their son. As Griswold (1988) argues the early conditions of the West, namely the lack of the rigid gender-based allocation of labour and thus lack of the rigid separation between male and female spheres, invalidated some aspects of Eastern domesticity, including female roles and features.

The family idyll is interrupted by a natural disaster that destroys all crops and deprives them of their means of subsistence. David is thus forced to seek work in the East, which transforms Molly into a decision maker and a self-reliant woman. Left alone, "[s]he must decide what to do; all the responsibility was now hers" (Lane 1998, p. 107). When David is away and she is unable to find any affordable accommodation in the town, Molly decides to spend winter on their homestead: "If she had to face loneliness, cold, wolves, outlaws, she'd face them. She'd stay there; she'd be right there when David came back" (Lane 1998, p. 131). Her features that might not have been visible earlier now come to the fore—she steps into her husband's role in order to hold the fort, to guard an outpost in the hostile territory, proving that she is a proper helpmate for her pioneer husband. She challenges nature and is able to withstand or even overcome its perils: she survives blizzards, an encounter with wolves, kills a buffalo for food. Hence, for Lane, staking out new territories involves a heroic struggle with the land and nature, which is presented both as a harrowing and mystical experience. Molly perceives the land as an opponent over which she needs to prevail. When she decides to stay on the homestead, she states, "The gates of hell shall not prevail against me" (Lane 1998, p. 148). Molly tames the physical frontier that she staked out with her husband by her valiant attitude and perseverance.

As Campbell (2003, p. 26) suggests, Lane avoids a stereotypical portrayal of her female character as a Prairie Madonna. It seems that Molly is a character with a contradictory mixture of features, which, as Griswold (1988, p. 15) claims, characterised all Anglo pioneer women who wanted to succeed and settle in the West—she is "both brave and timid, resourceful and dependent, aggressive and retiring". Staking out new territories, which invalidated some aspects of the domestic ideology of the era, releases in Molly features, such as resourcefulness, resilience, persistence, and propels her to action.

3.2 Free Land

Free land, another book by Lane, is far less celebratory than *Young pioneers*, as it concentrates to a greater extent on the socio-economic aspect of staking out new territories and complicates the relationship between the young pioneer couple. The novel introduces a number of female characters, of whom the most important for this analysis seem to be Mary Beaton, another young bride who travels west with her husband, and Nettie Peters. Campbell (2003, p. 29) perceives Mary Beaton as an embodiment of the Prairie Madonna image and Nettie Peters as an exemplification of the subsequent "Pioneer Woman" image. However, it seems that Mary is not an entirely stereotypical submissive Prairie Madonna, as she is frequently impatient with the limits imposed on her by her own body and the role her husband assigns to her. What appears to be more important in Lane's portrayal of her female characters is the awareness that women responded to staking out territories in a variety of ways.

Mary, like Molly at the beginning of *Young pioneers*, is a woman protected and sheltered by her husband. However, contrary to Molly, for whom the travel west is a natural course of action, another stage of westward expansion, Mary is portrayed as a "reluctant pioneer" who seems to regret her decision. At some point in the novel, she states, "'I hate it!' The words burst out with an angry sob, Mary had not known how much she hated this country, the bareness, the emptiness, the winds" (Lane 1984, p. 92). While for Molly the land is an opponent that she respects but needs to conquer, for Mary the frontier is a place of exile from the civilized surroundings of her hometown. Her need to plant something green around the house exemplifies her, and other Western women's, as Norwood (1988, p. 155) avers, desire to transform the Western landscape into "a landscape resembling as much as possible their homes back East".

Mary's inability to cope in the new territory leads to her constant feeling of inadequacy and even to marital tensions. She is dismayed by the fact that her physical limitations do not allow her to become David's helpmate. Contrary to Molly, Mary gives birth to her first child in the East without her husband's presence. Moreover, she views herself as a burden because she is unable to be David's physical equal. For example, she insists on helping David pick turnips, but the task proves to wear her out. When David asks her to rest, she retorts,

> What's killing me is being no use. I can't even trade-in my eggs and butter to help on the groceries. I'm nothing but an expense. It's not right. It's against nature. It makes me feel like we're not married. ... I thought we were starting out like other folks do. I thought I'd do my part. For all the use I am out here, I'd better stayed home and saved you the cost of my keep. (Lane 1984, p. 227)

Mary seems to have expected that staking out new territories would result in complementarity of roles, partnership between her and David; however, she is defeated by the new territory and unable to expand or negotiate her role as Molly does.

Mary realises that her lack of stamina and spunk, as she calls it, limits her choice of roles and prevents her from transcending the role of a domestic wife. She expresses anger at David making decisions without consulting her and being infantilized by her husband: "Some day I hope and pray you'll have sense enough to stop lying to me as if I was a child" (Lane 1984, p. 117). The West does not offer Mary a new status and she remains, in her husband's eyes, a vulnerable woman to be protected. Importantly, at some point, although David is never disloyal to his wife, he is attracted to a different kind of womanhood, exemplified by Nettie who seems to thrive in the West, while Mary wastes. The new kind of womanhood which is alluring, as it implies partnership and female independence, but also disconcerting, as these notions are a revision of domestic ideology and accepted female roles.

Nettie first appears in the novel as a sixteen-year-old girl carrying a rifle and resembling an American Indian: "Her hair, redder than a buckeye, hung in braids beside her cheeks, like an Indian woman's hair, and her thin face was as brown as an Indian's" (Lane 1984, p. 23). Again it seems that for Lane the woman who succeeds in staking out new territories and thriving on the frontier needs to acquire Native American physical and mental characteristics. Ironically, acquiring such characteristics is never translated into a cross-cultural understanding between women. Although Native American men appear in *Free land*, for example in a potentially hostile situation triggered by a theft of human remains perpetrated by a culturally unaware white character, Native American women are non-existent.

Nettie, the Indian-like character, is contrasted with Mary, who embodies domesticity and civilization, whereas Nettie is always portrayed as having a streak of wildness. Nettie, raised in the West and forced by circumstances to move with her family from place to place, has been forced to carry out tasks that in the East would be performed by men. She hunts, repairs the roof, knows how to behave in a blizzard and accepts vigilante violence if a situation requiring it should arise. Her Western upbringing instilled in her the features that Mary lacks but would like to possess—self-reliance, endurance, which as David observes is "in the marrow of her [Nettie's] bones", independence, and fearlessness (Lane 1984, p. 68). Nettie is potentially an excellent helpmate for her future husband, but her decision to postpone marital union in order to acquire education and become a schoolteacher signifies to some extent that she would expect to make her own decisions and become her husband's partner in decision making.

4 Edna Ferber's *Cimarron*

While Lane had her parents' pioneer memories of the West to rely on, Ferber's only encounter with the Southwest was a short sojourn into Oklahoma (Campbell 2003, p. 30). Ferber intended her novel to be a satire of the pioneering myth and pioneer women: "*Cimarron* had been written with a hard and ruthless purpose. … It contains paragraphs and even chapters of satire, and I am afraid, bitterness" (cited in Campbell 2003, p. 30). However, to Ferber's disappointment, the overwhelming majority of her readers did not detect satire or irony in her novel (Campbell 2003, p. 30). The novel introduces the character of Sabra Cravat—a self-ordained civilizer of a Southwestern town. Sabra, her husband Yancey Cravat, an adventurer and newspaper editor, and a young son, Cim, take abode and begin pioneering in Oklahoma. After she moves West, Sabra, steeped in the domestic ideology, which as Griswold (1988, p. 26) explains was "an ideology of social order", attempts to transplant into and impose on the West middle-class cultural standards and values that were adhered to in the other parts of the country. She becomes a reformer and organizer of women's clubs and searches for "female moral authority", as Peggy Pascoe (1993) would phrase it, and entry into the public sphere of activity.

However, before she assumes the role of a female civilizer, Sabra needs to evolve from a passive, submissive wife, "a charming little fool", to a self-reliant woman (Ferber 1930, p. 93). Sabra, even more conspicuously than Molly, experiences the transformative power of the new territory. Almost at the very beginning of her life in the new territory, which blurs to some extent the division into spheres, Sabra realises that men are not always right and that women seem to perform tasks more efficiently than men: "These last 3 weeks had shown her that the male was often mistaken, as a sex, and that Yancey was almost always wrong as an individual" (Ferber 1930, p. 93). Moreover, Sabra gradually takes over newspaper printing, the basis of their income, from her husband who dedicates himself to other pursuits.

After some time spent in the new territory Sabra rebels, for her daughter's sake, against the masculinity and male rules of the town: "It was a man's town. The men enjoyed it. They rode, gambled, swore, fought, fished, hunted, drank. …. The saloon was their club, the brothel their social rendezvous, the town women their sweethearts" (Ferber 1930, pp. 140–141). The first stage of her rebellion takes the shape of female bonding and networking, which develops into female organizing against male interests and vision of the West. Initially, Sabra publishes tips for homemakers in the newspaper and her home becomes a social centre. Later, she starts a women's club, which proliferates into other clubs, as an attempt to tame the territory and transform it into a liveable space for women and children. She seems to be strangely unaffected by the territory, apart from the fact that it triggers her independence, and vehemently adheres to her middle-class standards of life: "In the three and a half years of her residence in Osage Sabra had yielded hardly an inch. … She had set herself certain standards, and those she had maintained in spite of almost overwhelming opposition" (Ferber 1930, p. 159). While the struggle between the sexes is largely absent from Lane's novels, in *Cimarron* it becomes central. Sabra challenges the male conception of the territory that she helps to stake out and combats its social and moral inadequacy. For Sabra staking out a new territory shifts from territorial expansion to widening the sphere of female influence, which is synonymous for her with civilization and implementing middle-class standards.

What appeals most to Sabra, who is initially a "reluctant pioneer", is the fact that the new territory is still in flux and offers a potential for transformation. When she goes home to visit, she realises the difference between "uncivilized" Oklahoma and her hometown: "Their existence [in Wichita—Sabra's hometown] was orderly, calm, accepted. For herself and the other women of Osage there was everything still to do. There lay a city, a country, a whole vast Territory to be swept and garnished by an army of sunbonnets" (Ferber 1930, pp. 160–161). Sabra wholeheartedly yields to the excitement of taming and transforming a new frontier, which is the main source of satire in the novel. Sabra and her aides are likened to "an army of sunbonnets" who, blinded by zeal, believe without any reservations in their civilizing mission. Underwood (1982, p. 9) contends that Ferber, writing her novel in the 1920s, when the significance of female organizing diminished, accedes to the stereotypical image of Western women as self-righteous civilizers.

One of the natural opponents of a Western civilizer would be moral vice, which in *Cimarron* takes the form of Dixie Lee, the proprietor of the town's brothel. Seemingly, Ferber stereotypically divides female characters into good and bad ones; however, as it turns out, the good character is quite ruthless in her single-minded witch hunt and the bad character has redeeming qualities. The task of rooting out prostitution and unlawfulness undertaken by Sabra and other women is sabotaged by Sabra's own husband: "Grimly Sabra ... set about making this new frontier town like the old as speedily as possible. Yancey ... tried to make the new as unlike the old as possible" (Ferber 1930, p. 142). When Sabra, as a newspaper editor, finally manages to take Dixie Lee to court for immoral deportment, Yancey acts as Lee's defence and the case against her is dropped. Hence, ironically, Sabra the civilizer is unable to "tame" her own husband who is an eternal pioneer and thrives on novelty.

Abandoned on several occasions by Yancey, Sabra has to fend for herself and their children—she runs the newspaper on her own, supports herself and the children financially, and does not cease her moral amelioration campaigns. Finally, she rises from her domestic position to become "Oklahoma's first feminist" and the Oklahoma Congresswoman (Ferber 1930, p. 195). In other words, she evolves from her housekeeping to a national housekeeping role. Importantly, Sabra, like Lane's Molly, fulfils her potential after she is deserted by her husband and thus freed from the conventional role of a passive, domestic woman: "It was not until he left her, and the years rolled round without him, that she developed her powers" (Ferber 1930, p. 91).

In order to underscore the satirical tone of the novel, Ferber portrays Sabra as an Indian hater who incites racism on the frontier, which as Myres (1982) suggests, is another stereotypical image of Western women. Sabra is burdened with racial prejudices that verge on the Indian phobia. Lane's heroines are not confronted with the racial frontier as their encounters with American Indians come down to a figure on the horizon (Molly) or a danger of an open confrontation that is handled by men (Mary). The only cultural difference that Molly is faced with is a Swedish couple who occupies a neighboring homestead—Molly easily accepts difference and bonds with the Svensons. Sabra, whose perception of Native Americans was shaped by school teaching, the legacy of pioneer women's fear, family tales, reacts with racial phobia to any element of Indian culture or behaviour.

Sabra's narrow-mindedness leads her to perceive Indians as "dirty and useless two-footed animals" (Ferber 1930, p. 41). She explains the undesirable features of her Native American servant, Ruby, who later becomes her daughter-in-law, not with the girl's character, but with her racial characteristics. Indians in Sabra's mind are always untrustworthy, unpredictable, cunning, uncivilized, and seething with rebellion. Although they seem to acquire the trappings of civilization, they tend to go to the blanket and persevere in their ceremonies and customs. Sabra's civilizing gaze abhors any deviations from the middle-class norm; moreover, the belief in white superiority makes her distrust any Indian attempts at the so-called civilization. Her gaze is exclusionary—she is not interested in incorporating Indians into the world she strives to build in her town. Only at the end of the novel is she able to metaphorically cross the racial frontier between herself and Native Americans when she cannot deny the fact that her son is married to an Indian woman and she is a grandmother of racially-mixed children.

5 Conclusion

The fact that Edna Ferber's *Cimarron* was intended to be a satire unfortunately strips Sabra of her potential as a character. Staking out a new territory transfixes Sabra, releases in her a new set of features, makes her reevaluate gender roles, and turns her into a female organizer. However, what comes to the fore is her hysterical racism, inability to cross racial boundaries and overzealous civilizing mission, the aim of which is to transform the new territory into a middle-class, racially-exclusive, moral space. Rose Wilder Lane portrays her characters in a much more realistic way, emphasizing the fact that women responded to staking out new territories in a variety of ways. In accordance with Frederick Jackson Turner's thesis, the West endows Molly with independence and self-reliance. Significantly, she is aware of the fact that her decision to stake out new territories with her husband requires new features, which she will not be able to find in the domestic ideology. The character assumes a conquering attitude towards the land that she helps to stake out and somehow naturally becomes a success-oriented agent of westward expansion. Moreover, she never considers the fact that she resettles a territory that was cleansed of its previous inhabitants. On the contrary, Mary Beaton is shown as a reluctant pioneer for whom staking out new territories is not a liberating, but rather imprisoning force. As she is unable to cope in the West, she is left with a constant feeling of inadequacy and inferiority. She is contrasted with Nettie Peters whose ability to cope with the new environment resembles Molly's.

References

Browder, L. 2008. *Her best shot: Women and guns in America*. Chapel Hill: University of North Carolina Press.

Campbell, D. 2003. "Written with a hard and ruthless purpose": Rose Wilder Lane, Edna Ferber, and middlebrow regional fiction. In *Middlebrow moderns: Popular American women writers of the 1920s*, ed. L. Botshon and M. Goldsmith, 25-44. Boston: Northeastern University Press.

Ehrhardt, J. 2004. *Writers of conviction: The personal politics of Zona Gale, Dorothy Canfield Fisher, Rose Wilder Lane, and Josephine Herbst*. Columbia: University of Missouri Press.

Ferber, E. 1930. *Cimarron*. London: Pan Books Ltd.

Jameson, E. 1987. Women as workers, women as civilizers: True womanhood in the American West. In *The women's West*, ed. S. Armitage and E. Jameson, 145-164. Norman: University of Oklahoma Press.

Griswold, R. 1988. Anglo women and domestic ideology in the American West in the nineteenth and early twentieth centuries. In *Western women: Their land, their lives*, ed. L. Schlissel et al., 15-33. Alburquerque: University of New Mexico Press.

Lane, R. Wilder. 1984. *Free land*. Lincoln: University of Nebraska Press.

Lane, R. Wilder. *Young Pioneers*. 1998. New York: Harper Trophy.

Myres, S. 1982. *Westering women and the frontier experience 1800-1915*. Alburquerque: University of New Mexico Press.

Norwood, N. 1988. Women's place: continuity and change in response to Western landscapes. In *Western women: Their land, their lives*, ed. L. Schlissel et al., 155-181. Alburquerque: University of New Mexico Press.

Pascoe, P. 1993. *Relations of rescue: The search for female moral authority in the American West, 1874-1939*. New York: Oxford US.

Raub, P. 1994. *Yesterday's stories: Popular women's novels of the twenties and thirties*. Westport: Greenwood Press.

Underwood, J. 1982. The civilizers: Women's organizations and Western American literature. In *Women and western American literature*, ed. H. Winter Stauffer and S. Rosowski, 3-16. New York: Whitston Publishing Company Troy.

Dante Gabriel Rossetti and the Complexities of Gender

Małgorzata Łuczyńska-Hołdys

Abstract This paper proposes to explore the mode in which Dante Gabriel Rossetti's fatal woman figure subverts traditional Victorian gender categories. The analysis is based on three poems, i.e. "Soul's Beauty", "Body's Beauty" and "Astarte Syriaca". Additionally, the paintings "Sybilla Palmifera" and "Lady Lilith", the visual equivalents of "Soul's Beauty" and "Body's Beauty", as well as the picture accompanying "Astarte Syriaca" will also be discussed. Within the 19th century theories of normative masculinity and femininity certain stable features were attached to the conceptions of male and female roles. As a result, 'separate spheres' debate became a standard notion for a discussion of gender issues in the 19th century texts. Thus, femininity was conceived of as emotionality, home, withdrawal from scenes of public life and lack of self-interest, in contrast to public activity, desire for power, and emotional reserve associated with manliness. Rossetti's treatment of these categories calls for a special awareness, as he seems to upset this traditional perception, figuring his *femme fatale* character in possession of features associated with manliness rather than femininity, and yet making her an object of male desire. In this way, his fatal woman is both seen as a male fantasy and a threatening agent. What follows is an equally subversive presentation of men who, instead of maintaining manly reserve and controlling their sexual drives, 'fall into' desire, an act which unequivocally questions their manliness.

M. Łuczyńska-Hołdys (✉)
Warsaw University, Warsaw, Poland
e-mail: m.holdys@uw.edu.pl

J. Fabiszak et al. (eds.), *Crossroads in Literature and Culture,* Second Language Learning and Teaching, DOI: 10.1007/978-3-642-21994-8_8,

1 Introduction: Dante Gabriel Rossetti and 19th Century Conceptions of Gender

This paper proposes to explore the mode in which Dante Gabriel Rossetti's fatal woman figure subverts traditional Victorian gender categories. Within the 19th century theories of normative masculinity and femininity certain stable features were attached to the conceptions of male and female roles. As a result, 'separate spheres' debate became a standard notion for a discussion of gender issues in 19th century texts. Thus, femininity was cast in such terms as emotionality, homely existence, withdrawal from scenes of public life and lack of self-interest and self-gratification, in contrast to public activity, desire for power, and emotional reserve associated with manliness.

Anderson (1993, p. 38) in her book *Tainted Souls and Painted Faces* describes the link between the issue of self-control and agency, understood as active shaping of one's character and life and the issues of gender. Through her analysis of the writings of John Stuart Mill, particularly his seminal *On the Subjection of Women*, she shows that femininity in the Victorian period was seen as divorced from a resilient agency, and rather cast into terms of greater susceptibility, malleability and artificiality. Hence, femininity may be seen as a negative foil on which men exhibit their independence, reserve and control.

Manly reserve and calmness seems to be of paramount importance, truly a touchstone in the Victorian construction of gender. Sussman's (1995, p. 13) comments are revealing here:

> For nineteenth-century men, manhood was conceived as an unstable equilibrium of barely controlled energy that may collapse back into the inchoate flood or fire that limns the innate energy of maleness, into the gender-specific mental pathology that the Victorians saw as male hysteria or male madness. For the Victorians manhood is not an essence, but a plot, a condition whose achievement and whose maintenance forms a narrative over time.

It may not be entirely coincidental that femininity for some Victorians was a plot as well, and that it needed to be struggled for in exactly the same way. We can find in Victorian writings specific warnings against particular innate female vices: vanity, caprice, self-indulgence, indolence. Sarah Stickney Ellis (1839, p. 40) thus addresses women in her *The Women of England, Their Social Duties and Domestic Habits:*

> It is necessary for women to lay aside all her natural caprice, her love of self-indulgence, her vanity, her indolence—in short, her very self—and assuming a new nature, which nothing else than watchfulness and prayer can enable her constantly to maintain, to spend her mental and moral capabilities in devising means for promoting the happiness of others, while her own derives a remote and secondary existence from theirs.

Female innate *very self* is thus opposed to the assumed one; the latter needs to be maintained with effort and prayer lest the woman lapses into her natural vices. However, this effort is crucial to meet the demands of femininity as socially constructed: officially, the only woman worthy of worship was a monument of selflessness, with no existence beyond the loving influence she exuded as daughter, wife, and mother. (Auerbach 1982, p. 185)

Similarly, Sussman (1995, p. 25) defines Victorian manliness as "a hard-won achievement, a continuous process of maintaining a perilous psychic balance characterised by regulation of potentially destructive male energy". Masculine self must control itself, as much as man must actively influence his surrounding circumstances. Unrestrained desires and impulses pose a serious threat to male identity and autonomy. John Stuart Mill writes in *On Liberty:* "One whose desires and impulses are not his own, has no character, no more than a steam-engine has a character" (1869, p. 108)

In the light of the quoted passages, then, it seems that Dante Gabriel Rossetti in his construction of male and female characters subverts the normative Victorian categories of gender. In poems and paintings concerned with a female type that may be understood as a fatal woman figure he represents all-powerful, self-indulgent, indolent, capricious, active and controlling female and a weak, passive, emotional male, yielding to his impulses and not actively striving to overcome or control his desires. Rossetti's fatal women as well as submissive men exhibit sensibilities which cast them as deeply unwomanly/unmanly.

1.1 Three Sonnets: Three Goddesses

Reading the sonnets "Body's Beauty", "Soul's Beauty" and "Astarte Syriaca" the reader notices primary characteristics of all three women to be power, independence and certain indifference, as if they were fully emotion-resilient. The sonnets feature, respectively, Beauty Incarnate, enthroned and enshrined, Lady Lilith, threatening demon-enchantress and Syrian Venus, a powerful goddess. All three can be seen as unpredictable, hence capricious, looking for self-gratification and most definitely active agents of their own, as well as their admirers', fate. Their worshippers, in turn, become awe-struck, mute, devoid of both decisiveness and agency.

1.2 Sybilla Palmifera or Soul's Beauty: Is This Goddess Alive?

In "Soul's Beauty" the reader is presented with a vision of Beauty Incarnate—sitting on the throne, mysterious and inaccessible. The quasi-religious elements are easily noticeable—her dwelling place is a shrine, guarded by all-combining opposites—life and death, terror and mystery, and the awe that she invokes in the speaker is clearly a mixture of aesthetic and sensual admiration. The universality of the experience—encounter with absolute, ideal beauty—is stressed by invoking another, apart from the lyrical I, witness of the scene who is directly addressed in such phrases as 'The sky and sea bent on thee' (l.5), and in the sestet, 'Thy voice and hand shake still—long known to thee' (l.10). Through this device the reader directly participates in the aesthetic experience, feels awe as a result of meeting the lady's gaze (l.3). By direct apostrophe to the implied (male) reader in the sestet, he becomes one of those who pursue the image of the ideal beauty, in whose praise his hand and voice tremble.

Looking at the verbs used in this sonnet one notices that they all conspire to express the power of the enthroned goddess as much as powerlessness on the part of the enchanted male. Thus, her eyes can *draw* the spectator to one law which she dictates (of her worship), her gaze *strikes* awe and makes the viewer 'the allotted bondman of her palm and wreath'. In the praise of her beauty the viewer's hand *shakes still*, oxymoronically enacting his lack of stability and control in the face of the woman. Furthermore, hers is the power to bend sky and sea over the spectator and make them meet—although the syntax of these verses is deliberately obscure—creating a horizon line, the never ending emblem of eternity, a merge of the human and temporal (sea) and the divine and eternal (sky), a place which is never to be reached yet tantalisingly tempts the traveller.[1]

The notion of this everlasting pursuit is another example of unequal power distribution within the sonnet—those who remain under the spell of the goddess's gaze join in the flight, submissive, following her day by day. The adverbs of manner accompanying the description of the flight—'passionately' and 'irretrievably' stress the fact that the viewers/followers are deeply emotionally involved, they have fallen into desire. Having given into their emotional and sensual side, their manly reserve is utterly crashed or melted. The power of the goddess is further rendered in the concept of the vain nature of their flight—the ideal is not ever to be caught. This concept is strengthened by the attributes of the woman revealed to us only in line 10—flying hair and fluttering hem—the synecdoches suggesting instability, temporal presence and impalpable, perhaps capricious nature. The elaborate phonemic device: an alliteration of 'fluttering', 'flying', 'flight' and 'follow' all combine into a mental picture of elusiveness and intangibility; a viewer is compulsively drawn to follow the capricious, fluttering and evasive personification of the ideal. Starting his pursuit, he simultaneously admits the lady's dominance and his own submission.

Rossetti's sonnets, however, are far from unambiguous. Even in the case of "Soul's Beauty" there are elements which subvert one stable interpretation. As Pollock (1990, p. 145) notices, shrines are places traditionally connected with the deceased—in which case Lady Beauty can be seen as dead, but commemorated and thus far less threatening. Secondly, in the octave of the sonnet agency falters due to the overuse of the verb 'draw'. While the lady's eyes draw the spectator to her power, it is the spectator who 'draws the gaze in' 'as simply as his breath'. Since the breath connotes the breath of life, it can denote complete absorption of

[1] The importance of the horizon symbol as a meeting place of heaven and earth is thoroughly elaborated by Stephen Spector's essay entitled "Love, unity and desire in the poetry of Dante Gabriel Rossetti".

the goddess's life. Is the Lady Beauty, the beauty of the soul, too threatening alive? Or is it a projection of the speaker's desire to usurp her power?

The painting "Sybilla Palmifera", which Rossetti meant as a companion piece to "Soul's Beauty", comes as a surprise when compared with the sonnet. First and foremost, it is unusually static, one of those paintings of women which William Rossetti called "female heads with floral adjuncts" (Hough 2007, p. 41). The only allusion to flight and movement can be detected in the butterflies painted on the right, which are not literally present in the poem, yet there is a connection to be made between the flutter of their wings and the fluttering hem of the goddess's dress. The painting primarily impresses through its sensual elements—colours and textures. Thus, we have the silky fabric of the dress and velvety touch of the roses contrasted with stony, cold fireplace and marble-like complexion of the woman. Further, the unnatural paleness of her skin acutely clashes with deep red of her dress and auburn hair. While seeming to show the all-powerful Ideal Beauty, the painting implies in fact something visibly different. If in the sonnet she was the goddess, awesome, sublime and alluringly threatening, whose presence sufficed to make the spectator's voice and hand tremble, in the picture these qualities disappear. The viewers see the unnaturally regular, characteristically withdrawn face of Alexa Wilding, with virtually no forehead at all (Pollock 1990, p. 132). The gaze which was meant to strike awe is completely devoid of life and energy, strikingly absent. This is an aesthetic image of beauty enshrined, dead but commemorated, robbed of power and influence. As Sussman (1995, p. 168) states in his *Victorian Masculinities*, Rossetti's post-Brotherhood portraits of women represent not the models but the erotic power of the male gaze; the female body becomes a screen onto which he projects sexualised male desire. In this way Rossetti's painting pictorially articulates a deep anxiety over gender issues: the threat of female power becomes tamed and domesticated.

1.3 Lady Lilith: The Femme Fatale or an Independent Woman?

"Body's beauty" and "Lady Lilith" figure prominently a more unequivocal personification of the *femme fatale*, at the same time highlighting a tension between threat and desire. The central idea of sonnet 78—"Body's Beauty"—is the enchantment, the spell which Lady Lilith casts on men. In Hebrew mythology she was believed to be Adam's first wife, who refused to submit to Adam and left heaven for the region of the air, next entering the relationship with the devil. Already in her mythological realm Lilith connotes desire for power, independence, refusal of submission, active agency, and decisiveness in pursuing her aims. In short, she is far from what the Victorians deemed feminine. Yet, the way Rossetti renders her in his sonnet, she is both alluringly desirable and extremely threatening. What is more, it becomes evident that her desirability is precisely conditioned by the danger she poses. The sonnet is built around one central concept—Lilith's golden hair, which acts both as a fetish and a direct threat. Becoming the woman's irresistible allure, it can be fatal, as it leads to total submission of the person on whom the spell is cast, and directly to his death by strangling (l.14). It is further linked to the idea of a web, or net, which she weaves, to ensnare her admirers in. The motif of weaving is fully and ironically orchestrated, covertly alluding to the Victorian concept of female education. Weaving and needlework were the most appropriate occupation for young girls, and we can detect this idea in innumerable Victorian literary and cultural works—from Alfred Tennyson's "The Lady of Shalott" to Elisabeth Gaskell's *Ruth.* Lady Lilith, although also involved in weaving, produces a curious embroidery—with dead insect—like men entangled in her golden web of hair. The sonnet enacts this idea verbally, through alliterative play on such words as 'spell', 'snare', 'soft sleep' and 'sweet kisses', where the abundance of hissing sounds pervert the two last positive ideas into ominous concepts, reverberating of serpent—like deception and poison. Under Lady Lilith's spell the observers become tranced, and in this trance they are summoned to come nearer and nearer, until they are ensnared in the golden web and lost forever. The image of an admirer in the trance further emphasises how much devoid of agency Rossetti's lovers are. Pensive and passive, they cannot do anything but watch the goddess pulling them towards her embrace, a promise of love and imminent death. The sonnet is powered by overt misogyny, finding outlet in intense fascination and equally intense fear.

Lilith in both the poem and the picture is presented looking into the mirror. In the text this idea is conveyed by the phrase 'subtly of herself contemplative' (l.6), while in the painting she reposes on the armchair, with the mirror in one of her hands, and a hairbrush in the other. The aesthetic impression of the painting relies on the impact of the woman's hair, long, bushy, auburn web. As Lilith draws men to watch this bright net, so are the viewers drawn to watch the auburn storm of hair in the picture (again, the verb 'to draw' from line 8 of the sonnet is an example of exquisite verbal play). While representations of women with mirrors usually denoted vanity and moral criticism (however, hypocritical) of this vice (Berger 1972, p. 51), we can see how Rossetti's seemingly vain, bodily beauty turns this concept upside down. I would like to suggest that this poem yields very well to the feminist reading 'against the grain', in which Lilith, instead of being the deadly *femme fatale*, can be regained for more positive reading. In this reading her look into the mirror stops connoting mere vanity, and starts to mean self-assurance and independence. She is the woman sure of herself, self-confident in her esteem, who does not rely on other spectators of herself for opinion and judgement. Meeting eyes not with the male spectator of the picture but with herself, yet probably conscious that she is being looked at, her body language seems to convey indifference and lack of interest. She embodies all the calmness and the emotional reserve which was traditionally attributed to truly manly men, thus subverting social constructions of gender. Interestingly, there is another mirror in the picture—in the upper left corner of the painting. This mirror, formally speaking, functions as a window, and it reflects a garden, which is not present in the enclosed room. As McGann (2000, p. 18) suggests, "it is as if the mirror presented a memory of the Edenic garden which Lilith fled". Lilith sits with her back towards this mirror/window, symbolically spurning the paradise in which she would have to admit Adam's dominance and her own submission. Moreover, it further suggests events which happened in garden of paradise—temptation, meeting the serpent, and, consequently, the fall of man. In this reading, however, Lilith *becomes* the serpent, the agent of temptation and fall, threatening the world of male dominance with her hissing sounds, a mass of wavy, serpent-like hair, and her independence.

In line with the feminist reading, Allen (1984) in her article "One Strangling Golden Hair" makes a suggestive link between Lady Lilith and the emerging figure of the New Woman. She mentions periodicals such as *Aetheneum*, *Tinsley's Magazine* and *Saturday Review*, which Rossetti is believed to have read, and which engaged in scathing criticism and ridicule of the New Woman Movement, and claims that Rossetti's Lilith may be seen as embodying the common attitude to the Woman Question in mid-Victorian period. Seen in this way, Lilith's seeming independence may be read as threatening to male viewers. An enlightening comment made by the same scholar regards the fact that mythical Lilith is believed to be a direct threat to children. Interestingly, in the debate surrounding the New Woman Movement one of the strong critical arguments against women's suffrage was that it led to limitation of the childbirth and that independent, educated women were unable to nurse their children (Allen 1984). Thus, Lady Lilith epitomises a threat, not only to male readers themselves, but also to the respected Victorian institution of family. As much as she is feared, however, she is desired with just the same intensity. What is more, probably the desire the gentlemen felt encountering independent women like Lilith was powered and intensified precisely by the threat she embodied. She became the *other,* unknown, unpredictable, fascinating, the exact opposite of domesticated, tamed Victorian wives and mothers.

1.4 Astarte Syriaca's Unstable Gender

Let us move now to the final, third poem-painting combination, "Astarte Syriaca". Similarly as in the case of "Soul's Beauty", the reader needs to brace himself for a breathtaking encounter with a goddess, whose imperial presence dominates both the sonnet and the picture. Astarte Syriaca is the Syrian-Babilonian goddess of love, fertility, and war, identified with the planet Venus (Kopaliński 2001, p. 411). Definitely not confined to the feminine domestic sphere, Astarte is a woman of action and power. The word 'mystery' which opens and closes the sonnet seems to be an echo and a summary of the previous poetic encounters, and points out at ambiguity, elusiveness, and incomprehensibility of the feminine ideal. As the other fatal women, she is definitely beyond ethical judgement. Her atemporal, universal, infinite nature is stressed even further when she is described standing 'betwixt' the sun and the moon. In this way her affinity with the cosmic spheres is spelled out. Additionally, being situated between the archetypal contraries of day and night, heaven and earth, male and female element, gold and silver, light and darkness, she seems to embody both—which is particularly visible in the picture—and in this way she abolishes the traditional dichotomies. In the poem she is both the bliss and the fright, the virgin and the whore, exerting power and uttering the promise, since the Syrian Venus Queen was frequently associated with sacred prostitution (Kopaliński 2001, p. 411), *hieros gamos*. In her girdle we see the "boon of bliss" where "heaven and earth commune" (l.4–5). Terror and mystery, and fascination, the destruction and inspiration, all are mingled in the perspective of the encounter with this ineffable goddess, who, in the sestet of the sonnet, is specifically called 'Love'.

The gender of this 'love', however, seems deliberately unsettling. While in the poem her neck is figured as an "inclining flower-stem" (l.6), connoting fragility, meekness and willingness (double meaning of 'inclining' being both 'leaning' and 'willing'), her lips are "love-freighted" and her eyes "absolute" (l.7). The "love-freighted" lips, metaphorically carrying the promise of "freight" of love can easily be changed into "love-frightened" or, more in tune with the whole poem, "frightening" lips. Her "absolute" eyes exert power, weaning' "the pulse of hearts to the spheres dominant tune," (l.8) spelling out dominance and the demand of submission.

It is sufficient to look at the painting "Astarte Syriaca" to be convinced about ambiguous gender of Rossetti's Syrian Venus. The picture featuring demonic, unsettling, troubling Jane Morris became one of the better known paintings by Rossetti. It was commissioned by Clarence Edmund Fry for the highest amount of money Rossetti ever received for any of his paintings. The whole foreground of the picture is dominated by the figure of Astarte, presented at the same time as a demonic goddess and a personification of desire. Full lips, translucent skin, dark, long, wavy hair became a trademark of Rossetti's women painted in the later period of his life. In the picture, Astarte's gaze focuses with great intensity on the viewer, staring boldly back at the person who dares to admire her as a work of art. This look is unsettling, resonates of resentment, provokes confrontation. Positioned "betwixt the sun and moon" (l.1), she is painted distinctly differently than both Sybilla Palmifera or Lady Lilith. Astarte is not exuding delicate, evanescent femininity, even with a premonition of danger. On the contrary, she is exuding power, dominance and strength, and her body, although her skin is still pale and almost transparent, has lost a lot of its traditional female charm. In fact, her right arm seems to belong to a man rather than a woman. Similar conclusion can be made about her thick although long neck and muscular shoulders. She is literally positioned between the sun and the moon—between a male and female element, becoming—like some of Blake's figures—almost androgynous. Hugeness of the canvas and of the woman suggest both temptation and intimidation, unambiguously emphasising the threatening element of female beauty (Riede 1992a, p. 160). Such treatment may be symptomatic of what Herbert Sussman (1995, p. 78) calls "characteristically early Victorian practice of displacing a threatening male sexuality onto the female Other". The painting did not meet with feminist acclaim: During the Woman's Suffrage demonstration in Manchester in 1913, rioting women stormed the art gallery and threw rocks at Astarte Syriaca (Allen 1984, p. 293). The response of the critics, in turn, has been extremely contradictory, from treating the work as a masterpiece to writing it off as an aesthetic failure.

2 Conclusion

In this paper I hope to have demonstrated that Rossetti's poems and paintings of the *femme fatale* are furrowed with complex issues of gender and difference. Interestingly, they record multiplicity of divergent attitudes. It is as if Rossetti was speaking different voices, at times misogynistic, a moment later subverting and criticising normative approaches to male and female roles. In all the texts, however, he himself constructs an artificial and highly abstract model of femininity. The woman is always scrutinised and stylised from the male perspective, and usually it happens at the cost of obliterating any individuality, even if some of Rossetti's women seem to acquire a life of their own and boldly stand against conventional readings. As Christina Rossetti wrote in one of her poems: "One face looks out from all his canvasses./Not as she is, but as she fills his dream" (2000, p. 1586).

References

Allen, V. M. 1984. One strangling golden hair: D. G. Rossetti's Lady Lilith. *The art bulettin* Vol.66, No.2 285-294.

Anderson, A. 1993. *Tainted souls and painted faces. The rhetoric of fallenness in Victorian culture.* Ithaca and London: Cornell University Press.

Auerbach, N.1982. *The woman and the demon.* Massachusets and London: Harvard University Press.

Berger, J. 1972. *Ways of seeing.* London: BBC and Penguin Books.

Hough, G. 2007. *Last romantics.* London: Methuen.

Kopaliński, W. 2001. *Słownik symboli.* Warszawa: Oficyna Wydawnicza Rytm.

McGann, J. 2000. *Dante Gabriel Rossetti and the game that must be lost.* New Haven and London: Yale University Press.

Mill, J.S. 1869. *On liberty.* London: Longmans, Green, Reader and Dyer.

Pollock, G. 1990. *Vision and difference. Femininity, feminism and the histories of art.* London and New York: Routledge.

Riede, D. 1992 (A) *Dante Gabriel Rossetti revisited.* New York: Twayne Publishers.

Riede, D.(ed) 1992. *Critical essays on Dante Gabriel Rossetti.* New York: G.K. Hall & Co.

Rossetti, D.G. 2008. *The complete writings and pictures of Dante Gabriel Rossetti.* A Hypermedia Archive, ed. J. McGann. http://www.rossettiarchive.org/index.html. Accessed 3.04.2010.

Rossetti, Ch. 2000. "In the artist studio" In *The Norton anthology of English literature. The Victorian age.* Seventh Edition, Volume 2B, ed. M.H.Abrams, 2000. London: Norton and Company, 1586.

Spector, S. 1992. Love, unity and desire in the poetry of Dante Gabriel Rossetti. In *Critical essays on Dante Gabriel Rossetti,* ed. D. Riede, 89-112. New York: G.K. Hall & Co.

Stickney Ellis, S. 1839. *The women of England, their social duties and domestic habits.* New York: D. Appleton.

Sussman, H. 1995. *Victorian masculinities. Manhood and masculine poetics in early Victorian literature and art.* Cambridge: Cambridge University Press.

“Outlaw Emotions”: Carol Ann Duffy’s “Eurydice”, Dramatic Monologue and Victorian Women Poets

Magdalena Pypeć

Abstract The purpose of the paper is twofold: first, to analyse how Carol Ann Duffy appropriates the form of the dramatic monologue in her notable volume *The world’s wife* (1999), putting “Eurydice”, one poem from the collection, under close scrutiny. Second, the analysis of the poem will seek to link “Eurydice” and *The world’s wife* to some Victorian women poets who also used the form to explore questions of gender and for the purpose of revisionist mythmaking, rewriting and “recycling” myths, fairy or folk tales, religious stories and clichés, for instance Adah Isaacs Menken’s “Judith” (1868), Augusta Webster’s “Circe” (1870) and “Medea in Athens” (1870), Amy Levy’s “Xantippe” (1881), and Catherine Dawson’s *Sappho* (1889). Although the paper connects Duffy’s work with her nineteenth century female predecessors, its aim is to discuss differences in their construction of the poetic persona and their experimentation with the genre the tendency of which was usually to question rather than confirm. The paper also seeks to underscore the generic innovations that women poets introduced to the dramatic monologue.

1 Introduction: Women and the Dramatic Monologue

Cynthia Scheinberg, a feminist and a scholar, who has been actively involved in promoting interest in the nineteenth century recently discovered “lost” women writers, thus commented on the critical work concerning Victorian dramatic monologue: “in essence women writers have been locked out of the theoretical

M. Pypeć (✉)
Warsaw University, Warsaw, Poland
e-mail: magdalenapypec@gmail.com

J. Fabiszak et al. (eds.), *Crossroads in Literature and Culture,* Second Language Learning and Teaching, DOI: 10.1007/978-3-642-21994-8_9,

discourse on Victorian dramatic monologue… I remain struck by the lengths to which past and even more recent criticism, even that of explicitly feminist literary critics, has gone in order to separate Victorian women's poetry from the discourse of generic innovation that marks so much work on Victorian dramatic monologue" (1997, pp. 173, 184).[1] Indeed, considering the following seminal criticism on the genre one may have the impression that Victorian dramatic monologue and women's poetry of the period are two separate issues, see for instance Ina Beth Sessions' "Dramatic monologue" (1947), Robert Langbaum's *Poetry of experience. The dramatic monologue in modern literary tradition* (1957), A. Sinfield's *Dramatic monologue* (1977), Dorothy Mermin's *The audience in the poem. Five Victorian poets* (1983), E. A. Howe's *The dramatic monologue* (1996) or W. D. Shaw's *Origin of the monologue: The hidden god* (1999). The aforementioned critics of Victorian poetry either excluded female practitioners of the genre and based their research on two male canonical poets—Alfred Tennyson and Robert Browning—or briefly referred to two women poets who rarely used the form: Christina Rossetti and Elizabeth Barrett Browning. Although Glennis Byron in the most recent study of 2003 records the historical recovery of many Victorian poetesses, her critical work is divided into "Men's and women's voices"; the canonised male tradition and specific female poetic tradition that is presented as different and separate (Byron 2003). To validate such a division Byron refers to arguments presented in Dorothy Mermin's earlier work:

> The women's dramatic monologues are different from the men's. (…) The women seem usually to sympathise with their protagonists, and neither frame them with irony as Browning does nor at least partly objectify them like Tennyson by using characters with an independent literary existence. The women did not find figures in literature or mythology or history. (…) Nor do they show their virtuosity the way Browning does in "My last duchess", for instance: we are not made aware of the poet signalling to us from behind the speaker's back (…) where men's poems have two sharply differentiated figures—in dramatic monologues, the poet and the dramatised speaker—in women's poems the two blur together (1983, pp. 75–6).

However, most of the aforequoted observations do not remain applicable, taking into consideration such notable poems as Elizabeth Barrett Browning's "Bertha in the lane" (1844), "A man's requirements" (1850), Adah Isaacs Menken's "Judith" (1868), Augusta Webster's "Circe", "Medea in Athens" and "The happiest girl in the world" (1870), Amy Levy's "Xantippe" (1881) and "A minor poet" (1884). According to Scheinberg, such an approach based on emphasising the essential difference between male and female poetry places women constantly in a "tradition of their own" and excludes them from being equal agents of literary history:

---

[1] For more of her critical research on Victorian women poets see, for instance Cynthia Scheinberg's "Canonising the Jew: Amy Levy's challenge to Victorian poetic identity," *VS* 39, no. 2 (winter 1996), pp. 173–200, *Women's poetry and religion in Victorian England,* CUP 2008.

> My point (...) is not to suggest that we must never make comparisons to male writers in our work, but simply that when we do make comparison, we should resist discourses that depict certain authors (...) as "owning" certain genres. The problem of relying on such methodology of genre ownership is that women writers who use the genre tend to get cited in terms that cast them as "lacking", "other", or different in their use of the genre, rather than as poets who have equal claim to generic innovation (1997, p. 186).

Reading Victorian dramatic monologues written by women I cannot but admit that there is a number of poems which prove to be very original specimen of the genre rather than passive adoptions of previously created models. In this essay, I will analyse Amy Levy's "Xantippe" (1881) to show the poet's unique approach to the genre. Next, I will lay particular emphasis on generic innovations that Victorian poetesses like her might have bequeathed to later poetic tradition by comparing "Xantippe" with a contemporary dramatic monologue "Eurydice" from Carol Ann Duffy's much acclaimed collection of 1999, *The world's wife.*

2 Victorian Innovations: Amy Levy

Amy Levy was a woman of letters who took an active part in the mainstream of Victorian culture as a contributor of essays, poems and stories to such Victorian journals and magazines as *The Cambridge Review, The Jewish Chronicle, the Spectator, the Gentleman's Magazine, the Pall Mall Gazette* and Oscar Wilde's *Woman's World.* She was the first Jewish woman to attend the Cambridge women's college. Before her suicidal death at the age of twenty seven she wrote three volumes of poetry—*Xantippe and other verse* (1881), *A minor poet and other poems* (1884), *A London plane-tree and other verse* (1889) and three novels—*The romance of a shop* (1888), *Reuben Sachs* (1888) and *Miss Meredith* (1889), (Cunningham 2000). The eponymous speaker in Levy's poem is Socrates's notorious wife whose image as a proverbial scold, a nagging and shrewish wife has been deeply embedded in culture, partly by means of literature. "As Socrates' Xantippe or a worse," Petruchio says about Katharina in *Taming of the shrew* (Shakespeare 2002, Act 1, sc.2). In Levy's monologue Xantippe is dying in her weaving room attended by her young maids and recounts to them the story of her life with the great philosopher, thus giving the reader some glimpses of the grandee from the private perspective of home. In the poem this alternative strictly female space of the weaving room is contrasted with the male sphere of Socrates's leafy arbour where he discusses "high philosophy" with Plato and Alkibiades (Levy 1999, p. l.119). Such a strict division is symbolic of the conventional assumptions about gender and ideological norms assigned to men and women, not only in Xantippe's ancient Greece but also in Victorian England. It seems obvious that by aligning similar social situation of women Levy indirectly calls forth the contemporary context. Such a division has also the implication of the gendered Victorian spheres of public and private, male and female, a division that Xantippe desired to cross all her life. The speaker's earlier memory of her youth takes the reader again to a weaving room where Xantippe is working with her female peers, suffering "the merry mockeries/Of other maidens sitting at the loom," or "sharp voices bidding [her] return/To maiden labour" (Levy 1999, pp. l.33–36). The poetic persona uses the image of "a shining sea" that distracts her from feminine chores to stress her insatiable curiosity about the world and its mysteries and her capacity for wonder (Levy 1999, pp. l.21–30). Xantippe's language reveals her self-conscious status as an outsider, a stranger among other women for she refuses to learn the "lesson of dumb patience" (Levy 1999, p. l.51) and instead is yearning for knowledge and education though frequently reminded that her "woman-mind had gone astray"—"For maidens, mark, such are not woman's thoughts" (Levy 1999, pp. l.42–44). The first encounter with her future spouse did not seem to be very promising, though she was bid to marry the man: "I saw him, all ungainly and uncouth", "I saw his face and marked it, half with awe,/Half with a quick repulsion at the shape" (Levy 1999, pp. l. 58, 60–1). Despite this Xantippe saw in the marriage a chance for fulfilling her dreams and hoped that her wise husband would be a willing teacher. Yet Socrates never allowed Xantippe to access the male world of the leafy arbour and treated her thirst for knowledge with cold contempt and disdain:

> Twas only that the high philosopher,
> Pregnant with noble theories and great thoughts,
> Deigned not to stoop to touch so slight a thing
> As the fine fabric of a woman's brain—
> So subtle as a passionate woman's soul.
> I think, if he had stooped a little, and cared,
> I might have risen nearer to his height,
> And not lain shattered, neither fit for use
> As goodly household vessel... (Levy 1999, pp. l.119–127)

Xantippe does not accuse her late husband of intentional wrongdoing, she merely implies that despite his breakthrough approach to many important issues of the day, he was very conventional and conservative in his own household: "Yet, maidens, mark; I would not that ye thought/I blame my lord departed, for he meant/No evil, so I take it, to his wife" (Levy 1999, pp. l.116–118).

A ubiquitous convention of the dramatic monologue is a certain kind of dramatic irony that had been defined on the basis of Browning's monologues as "the strategy of the double mask":

> In addition to the primary creation of the character or speaker, there exists a secondary creation, a mask which the speaker uses in dealing with the dramatic situation in the poem. The speaker's use of this mask is in the form of rhetorical strategy, designed to conceal some aspects of the speaker's personality and to replace it with another more favourable to the speaker's purpose (Garratt 1973, p. 115).

Thus the irony in the poem seems to be the result of the disparity between what the speaker intends to say and the wider and different understanding of the reader. In her poem Amy Levy displays an innovative approach to dramatic irony, perhaps suggesting that there are other ways by which a poet may achieve a sense of doubleness or discourse splitting, namely through a very ingenious use of the audience. "Poems with auditors are about communication, regarding the individual as part of society and speech in terms of its effect on an audience", thus Dorothy Mermin emphasises the importance of this generic feature in the dramatic monologue (1983, p. 8). At the end of Levy's poem it dawns on Xantippe that her story has been misinterpreted by her female auditors. They do not seem to identify with her aspirations, but rather take her story as a cautionary tale. "Enough, enough. In vain, in vain, in vain!" (Levy 1999, p. l.278) the speaker sums up her endeavours at explanation and again seems to feel as a misfit in the female space of the weaving room. Commenting on the role of audience in the poem Cynthia Scheinberg suggests that:

> the auditors' different judgements of Xantippe affect their various capacities for sympathy. Thus, Levy reminds us how poetic or linguistic power is not merely something that the speaker/poet controls; on the contrary, the speaker is totally dependent on an audience's power to validate that speech—to claim it as representative (1997, p. 182).

Xantippe's auditors do not claim her "predicament" as "representative" and by their dubious sympathy Amy Levy essays to show the difficulty of challenging certain deeply ingrained stereotypes and preconceptions about the world. Over and above, I would like to underscore the fact that in revisiting the iconic story of Xantippe, the poetess proves to be one of the earliest revisionists who employed the technique in the dramatic monologue. "The Poet's function is not, like the novelist's, to devise new stories, but make old stories new," wrote another Victorian revisionist, Augusta Webster in her prose writings published under the misleading title *Housewife's opinions* in 1879 (2000, p. 366). Revisionism has become now one of the recognised feminist strategies defined by Alicia Ostriker as "revisionist mythmaking":

> Whenever a poet employs a figure or story previously accepted and defined by a culture, the poet is using myth, and the potential is always present that the use will be revisionist: that is, the figure or tale will be appropriated for altered ends, the old vessel filled with new wine, initially satisfying the thirst of the individual poet but ultimately making cultural change possible. (…)These poems generically assume the high literary status that myth confers and that women writers have often been denied because they write "personally" or "confessionally". But in them the old stories are changed, changed utterly, by female knowledge or female experience, so that they can no longer stand as foundations of collective male fantasy. Instead (…) they are corrections; they are representations of what women find divine and demonic in themselves; they are retrieved images of what women have collectively and historically suffered; in some cases they are instructions for survival (1982, pp. 72–73).

So far revisionism has been associated with such post-modern women writers as Ann Sexton (*Transformations,* 1971), Angela Carter (*The bloody chamber and other stories,* 1979), Liz Lochhead (*The Grimm sisters,* 1981) and Carol Ann Duffy with her much acclaimed *The world's wife* (1999) which has brought the poet to the notice of ordinary book-buying public.

3 Contemporary Dramatic Monologue: Carol Ann Duffy

The world's wife is a collection of thirty dramatic monologues spoken by the wives of famous men, be they mythic, biblical, fictional or real, women who have been so far defined by their male partners while themselves have remained in the shadows. "I think the poems are looking for the missing truth, rather than accepting the way we've been taught", Duffy said about her volume in an interview for the *Guardian* of 25 September 1999 (Viner 1999). In revising cultural archetypes Levy and Duffy seem to have well understood the power of the tale or myth in western culture that has been thoroughly analysed by Roland Barthes or Claude Lévi-Strauss.[2]

In "Eurydice" the poet is concerned to debunk the archetypal myth of love between Orpheus and his wife.[3] Like in Levy's poem the audience in "Eurydice" is also singled out as female, yet the speaker stresses the female intimacy and the bond between them by her warm and straightforward address with an implication that she has got something juicy to reveal:

Girls, forget what you've read.
It happened like this—(Duffy 1999).

Duffy also deprives her dramatic monologue of the immediate setting in time and place by placing Eurydice and her female audience in "nowhen":

Girls, I was dead and down
in the Underworld, a shade,
a shadow of my former self, nowhen (Duffy 1999).

In fact "nowhen" seems to be a key word not only for "Eurydice", but also for the whole collection of her poems. For the problems and predicaments voiced by her personae are universal female problems, regardless of time and place, as much ancient as contemporary. In Levy's poem the contemporary reader might find it difficult to identify with Xantippe's impossibility of fulfilment through learning and treat it merely as a historical curiosity of the time. To give Eurydice's experience the air of universality, Duffy creates in her poem an interesting plural discourse, for it reads as if spoken by various "Eurydices", differentiated by age, education and time. To achieve such an effect she uses language drawn from various registers like slang, colloquialisms and language of literature and literary theory; spoken expressions like "ye Gods" or "bollocks" are placed along with "octaves, sestets, (…) elegies, limericks, villanelles" (Duffy 1999). At times the speaker appears to be a psychologically harassed young woman who managed to ditch a stalker boyfriend: "What did I have to do, I said,/to make him see we were through?" (Duffy 1999) The image of Orpheus called "Big O" with "a familiar

---

[2] See for instance Roland Barthes, *Mythologies* (Hill & Wang, 1972), Edmund Leach, *Lévi-Strauss* (Chicago: University of Chicago Press, 1989).

[3] See Robert Graves, *The Greek myths* (London and New York: Penguin, 1993).

knock-knock-knock at Death's door" parodies Bob Dylan's song "Knockin' on Heaven's Door", thus the speaker may sound as a wife of some self-obsessed rock star with a "lyre" as a guitar. Moreover, Orpheus is also presented as a "legendary" arrogant, over-sensitive poet who relies too heavily on his wife. "I'd done all the typing myself,/I should know", she confesses to the "girls" (Duffy 1999). This Eurydice intends to start writing herself and to find her own poetic voice:

> And given my time all over again,
> rest assured that I'd rather speak for myself
> than be Dearest, Beloved, Dark Lady, White Goddess,
> etc., etc. (Duffy 1999)

Duffy seems to place Eurydice (with Big O's appropriation of her as a muse and typist) in the tradition of such talented women like Milton's daughters or Dorothy Wordsworth who aided their male partners or relatives in their poetic work. "I like to use simple words, but in a complicated way so that you can see the lies and truths within the poem", Duffy commented on her deliberate use of cliché in an interview given in 1988 (McAllister 1988, p. 75). This specific use of language can be well observed in "Eyrydice", for there is also a double entendre to the witty tongue-in-cheek speech of the eponymous speaker. In the poem Duffy explores the nature of language and the female writing identity. Eurydice felt imprisoned in the masculine language of her husband, the poet—"trapped in his images, metaphors, similes, /... histories, myths" whereas the underworld is "a place where language stopped, /... where words had to come to an end" (Duffy 1999). To be released from the psychological imprisonment of her oppressive husband, Eurydice must also escape from being theorised, enclosed and encoded by a male discourse which cannot account for feminine experiences and perspective. Such an approach to language calls forth the theories of Luce Irigaray and Hélène Cixous, two feminist critics and philosophers who saw language as a masculine construct that cannot accommodate female experience since it is a result of damaging hierarchical relations within western culture in the spheres of philosophy, mythology, psychoanalyses.[4] In her much lauded essay "The laugh of the Medusa" (1975) Cixous called for the creation of "écriture féminine", a female liberating practice of writing, devoid of "the phallocentric tradition". If Eurydice intends to express herself in language, she would have to find "small gaps" in the masculine language, "the very possibility of change, the space that can serve as springboard for subversive thought, the precursory movement of a transformation of social and cultural structures" (1989, p. 1093).

---

[4] See Elaine Marks and Isabelle de Courtivron (eds), *New French feminism: An anthology* (Brighton: Harvester Press, 1981) and Margaret Whitford (ed.), *The Irigaray reader* (Oxford: Blackwell Publishers, 1991).

4 Silence, Freedom, Strength

"I look everywhere for grandmothers, and see none"—Elizabeth Barrett Browning's famous comment of 1845 has been often quoted to stress the difficult situation of female writers in Victorian England due to the lack of female poetic precursors with whom they could identify (Kelley, Lewis 1984, p. 14, vol. X). Yet many of her female contemporaries were appropriating the voice of Sappho and Erinna and created their own poetic identity through "forms of imaginative projection"—dramatic monologues (Flint 1997, p. 156). The genre where "as a rule I does not mean I" offered the female poets a form of protection from the Victorian readers and literary critics who were studying poets "as if they were putting themselves through a process of vivisection for the public to see how they were getting on inside" (Webster 2000, pp. 370–1). Dramatic monologue allowed women whose cultural and social place in the nineteenth century was that of meek domesticity and silence particular freedom to cross the borders of the self and inhabit multiplicity of voices disregarding their class, gender, time and even topographical boarders (see for instance "The Indian with his dead child" by Felicia Hemans, "The runaway slave at Pilgrim's Point" by Elizabeth Barrett Browning, "Rukhmambi" by Catherine Dawson"). Some Victorian poetesses used the genre as a means of social protest ventriloquising feelings of those who were less privileged then themselves, for instance "the fallen woman"—"Magdalen" by Amy Levy or "The castaway" by Augusta Webster. "Levy's dramatic monologues focus on representing speakers who stand in opposition to certain dominant cultural beliefs, and as a result find no authorising audience to validate their poetic speech", Cynthia Scheinberg comments on the poet's "minority poetry", Levy's preoccupation with giving voice to the silenced, disempowered and oppressed (2002, p. 212). Looking at the bulk and variety of dramatic monologues written by Victorian women one may venture an opinion that they seem to challenge the frequent attempt to pigeon-hole their poetry as subjective, personal, confessional, emotive, limited, and unable to grasp a wider vision (see for example Mary Ann Stodart's "Female writers: Thoughts on their proper sphere", 1842). According to Kate Flint the genre allowed Victorian poetesses to "relish in the hidden power which another's voice can give them", they could assume a position of knowledge and power inaccessible in life (1997, p. 166). In time they also discovered that silence did not have to be solely associated with the appropriate female virtue, but also with the condition necessary to produce art. "Man, on his way to Silence, stops to hear and see", observed Alice Meynell in her late nineteenth century poem "To Silence" (Meynell 2005). Likewise Carol Ann Duffy's Eurydice recognises this quality of Silence as a prerequisite to any artistic creation. In Duffy's poem Silence becomes a desirable state that is planned and scrupulously executed by a budding poet—Eurydice herself. She intends to be in the Underworld, "near the wise, drowned silence of the dead" (Duffy 1999).

5 Conclusion

In summary, the two dramatic monologues separated by more than a hundred years unquestionably display a very innovative conceptualisation of the genre. Both women poets employ the strategy of revisionism in the dramatic monologue to convey a different message from a well-known tale. Although there is now an established tradition of revisionist mythmaking among post-modern women writers, I would like to italicise the fact that Victorian poetesses like Amy Levy and Augusta Webster much earlier recycled stories with deep resonances in western culture in order to change and undermine female archetypes. Amy Levy frames her poetic persona with irony as a result of an innovative use of the female audience who are unable to identify with her and miss the larger point Xantippe attempts to make. Thus the poet stresses Xantippe's isolation even in the feminine space of the weaving room. Xantippe's desire to cultivate her intellect beyond weaving room accomplishments perceived by all around her as a violation of the order of Nature implies the nineteenth century context of Victorian England and women's limited educational possibilities. On the contrary, Duffy does not situate her speaker in any time specific setting to universalise her female experience. She does not seem to be interested in any specific historical moment, but rather with the creation of the strong female identity. Moreover, the poetess draws on language from a wide variety of contexts, which creates the impression of a collective female persona. Unlike in Levy's poem, the references to female auditors seem to emphasise the bond and solidarity between women due to their similar experiences. In "Eurydice" Duffy also explores the female writing identity and its relation to language. Furthermore, both poetic personae present the two grandees, Socrates and Orpheus with a witty irreverence, for instance previously mentioned Socrates's far from prepossessing appearance, Orpheus described as "the kind of a man/who… sulked for a night and day/because she remarked on his weakness for abstract nouns"(Duffy 1999). To bring the paper to a close, I would like to make note of one more feature that the two dramatic monologues have in common, both poems celebrate what has been dubbed by a feminist critic as "outlaw emotions" that seek to challenge the traditional gendering of feeling. In Duffy's poem Eurydice seems to be a strong, practical, single-minded woman whereas her husband is weak, sensitive, sulky, and almost "womanish". Levy's Xantippe feels that she intellectually belongs to Socrates's philosophical discourse in the male sphere of "leafy arbour":

> Outlaw emotions are distinguished by their incompatibility with the dominant perceptions and values, and some, though certainly not all, of these outlaw emotions are potentially or actually feminist emotions… Outlaw emotions stand in a dialectical relation to critical social theory: at least some are necessary to developing a critical perspective on the world, but they also presuppose at least the beginnings of such a perspective (Jaggar 1996, pp. 180–181).

References

Barthes, Roland. *Mythologies.*Hill & Wang, 1972.

Byron, Glennis. 2003. *Dramatic monologue.* London and New York: Routledge.

Cixous, Hélène. 1989. The laugh of the Medusa. In *The critical tradition. classic texts and contemporary trends,* ed. David H. Richter, 1090-1102. New York: St. Martin's Press.

Cunningham,Valentine (ed.). 2000. *The Victorians. An anthology of poetry & Poetics.* Oxford: Blackwell.

Duffy, Carol Ann. 1999. *The world's wife.* London: Picador.

Flint, Kate. 1997. "... As a rule, I does not mean I". Personal identity and the Victorian woman poet. In *Rewriting the self. Histories from the Renaissance to the present,* ed. Roy Porter, 156-166. London and New York: Routledge.

Garratt, Robert F. 1973. Browning's dramatic monologue: The strategy of the double mask. *Victorian Poetry* 11: 115-125.

Graves, Robert. *The Greek myths.*London and New York: Penguin, 1993.

Jaggar Alison, M. 1996. Love and knowledge: Emotion in feminist epistemology. In *Women, knowledge and reality: Explorations in feminist philosophy,* ed. Garry and Marilyn Pearsall, 166-182. London and New York: Routledge.

Kelley, P. Lewis, S. (eds.) 1984. *The Browning correspondence.* Winfield KS: Wedgestone Press.

Leach, Edmund. *Lévi-Strauss.* Chicago:University of Chicago Press, 1989.

Levy, Amy. 1999. Amy Levy (1861-89). In *The Broadview anthology of Victorian poetry and poetic theory,* ed. Thomas J. Collins and Vivienne J. Rundle, 871-876.Toronto: Broadview Press.

Marks, Elaine, Isabelle de Courtivron (eds), *New French feminism: An anthology.*Brighton: Harvester Press, 1981.

McAllister, Andrew. 1988. Carol Ann Duffy Interview. *Bête Noire* 6: 69-77.

Mermin, Dorothy. 1983. *The Audience in the poem. Five Victorian poets.* New Brunswick, NJ: Rutgers University Press.

Meynell, Alice. 2005. *The poems of Alice Meynell.* USA: Kessinger Publishing Co.

Ostriker, Alicia. Autumn 1982. The thieves of language: Women poets and revisionist mythmaking. *Signs* 1: 68-90.

Scheinberg, Cynthia. 1996. "Canonising the Jew: Amy Levy's challenge to Victorian poetic identity," *VS* 39, no. 2. *Women's poetry and religion in Victorian England,* CUP 2008, 173-200.

Scheinberg, Cynthia. 1997. Recasting 'sympathy and judgement': Amy Levy, women poets, and the Victorian dramatic monologue. *Victorian Poetry* 35/2: 173-192.

Scheinberg, Cynthia. 2002. *Women's poetry and religion in Victorian England.* Cambridge: CUP.

Shakespeare, William. 2002. *Taming of the shrew.* Oxford: OUP.

Viner, Katharine. 25 September 1999. Meter Maid. *The Guardian weekend,* 20-6. http://www.guardian.co.uk/books/1999/sep/25/costabookaward.features.print. Accessed 11February 2010.

Webster, Augusta. 2000. Poets and personal pronouns. In *Portraits and other poems. Augusta Webster,* ed. Ch. Stuphin, 366-372. Toronto: Broadview Press.

Whitford, Margaret (ed.), *The Irigaray reader.* Oxford: Blackwell Publishers, 1991.

"The One Great Drawback to the Life of Women is That They Cannot Act in Politics"? Political Women in Anthony Trollope's Fiction

Agnieszka Setecka

Abstract Although the nineteenth century in England was dominated by the figure of the female monarch, Queen Victoria, women had virtually no political rights. The voices critical of their political impairment, notably John Stuart Mill's *The subjection of women*, did not bring any significant change, so that women were granted political rights on equal terms with men as late as 1928. However, women's political rights and their involvement in the world of politics did become an important issue, which found its way into fiction. This paper is concerned with Anthony Trollope's *The Prime Minister* and the way he presented women who, dissatisfied with the limitations of their prescribed domestic life, got involved into politics. Although women could participate in politics only through their husbands or fathers, Trollope nevertheless represented them as powerful, if invisible, agents, grey eminences of the political life in England. Not only was the social world dominated by women a shadow of the male political world, social coteries being an equivalent of political parties, but it also influenced the political world. However, although Trollope did grant women a relatively large scope for action in his novels, he nevertheless presented them as limited by the demands of the ideology of separate spheres, especially that their too direct and indiscreet interference into the political world might meet with condemnation. Therefore, far from abolishing the boundary between the private and the public altogether, Trollope nevertheless revealed how blurry and porous it was.

The quotation in the title comes from Trollope's *Phineas Finn* (1999b, p. II 26).

A. Setecka (✉)
Adam Mickiewicz University, Poznań, Poland
e-mail: saga@ifa.amu.edu.pl

J. Fabiszak et al. (eds.), *Crossroads in Literature and Culture*, Second Language Learning and Teaching, DOI: 10.1007/978-3-642-21994-8_10,

1 Introduction

The Victorian period begins with the year of the first Reform Act,[1] and the issues connected with reforming the electoral system and the growth of democracy remain major concerns later in the century. In the 1850s and especially the 1860s the reform debates resurge with the Liberals proposing the increase in the number of voters by lowering financial qualifications, which, finally, led to introducing the second Reform Act in 1867. Significantly, the first Reform Act of 1832, although it doubled the number of voters, absolutely excluded women from participation in the political life of the country as it referred to "male persons only" (Wingerden 1999, p. 8), and the final shape of the second Reform Act disappointed the hopes of the growing host of the supporters of female suffrage. However, the 1860s became the time particularly important in the history of women's movement, as it is the period when it consolidates, which is manifest in the founding of the Langham Place Circle and of *The English Woman's Journal* (Wingerden 1999, p. 3), as well as the time when the debates concerning women's involvement in politics become particularly heated, to grow in intensity later in the period. The debates, the high point of which must have been the Ladies' Petition[2] and John Stuart Mill's address to the House of Commons in 1867, much like other major issues of the period, neither began nor ended in the Parliament, but extended to the press and the novel, Trollope's Palliser novels being a most prominent example.

In Trollope's six Palliser novels, the first of which, *Can you forgive her?*, was published in 1864 and the last, *The Duke's children*, in 1880, politics constitutes the major concern although, as Trollope puts it in his *Autobiography*, "[i]f [he] write[s] politics for [his] own sake, [he] must put in love and intrigue, social incidents, with perhaps a dash of sport, for the benefit of [his] readers" (Trollope 1999a, p. 317). This paper will focus on women's special relation to politics and the available ways in which they could participate in the political life of the country in spite of their inferior position, as represented in Trollope's *The Prime Minister* (1876). Trollope's fiction abounds in strong-minded women who have their own political views and who take deep interest in their husbands' political careers. Paradoxically, however, they are mostly opposed to enfranchising women and do not desire the vote for themselves. Although they would probably disagree with the first of arguments against women's suffrage quoted by John Stuart Mill, namely that "[p]olitics are not women's business, and would distract them from their proper duties", they would not oppose the remaining three, which say that "Women do not desire the suffrage, but would rather be without it: Women are

[1] Chew and Altick (1967, p. 185) claim that "[t]he precisian might limit the Victorian period to the years between the Queen'a accession in 1837 and her death in 1901, but a new era really began with the passage of the Reform Bill in 1832 and closed at the end of the Boer War in 1902.".

[2] The Petition was a plea that the Parliament grants the vote to all people fulfilling the financial qualifications, regardless of their sex. It contained 1,499 signatures (Wingerden 1999, p. 2).

sufficiently represented by the representation of their male relatives and connexions: Women have power enough already" (Mill 2003). Even the last statement, which met with laughter in the House of Commons, would not seem in any way preposterous for Trollope's female characters, who did not underestimate the power of their feminine influence.

2 "Influence" and de Certeau's "Tactics"

The idea of women's influence or "the true queenly power", as Ruskin described it (2002, p. 69), was often evoked in Victorian writing, and it referred to women's moral authority. According to Ruskin, it both depended on and justified the separation of the spheres, which ensured that women are "protected from all danger and temptation" (Ruskin 2002, p. 77). If men were naturally predisposed for "battle" and "rough work in open world" (Ruskin 2002, p. 77), the woman was to rule, since "her intellect is not for invention or creation, but for sweet ordering, arrangement and decision" (Ruskin 2002, p. 77). The two spheres, although separate, were supposed to be complementary, so that it is unjust, Ruskin (2002, p. 70) insists, to write of "the 'mission' and of the 'rights' of Woman, as if these could ever be separate from the mission and the rights of Man—as if she and her lord were creatures of independent kind, and irreconcilable claim". Thus, the ideology of separate but complementary spheres, of which Ruskin is so ardent an advocate, does not so much totally preclude women from participation in the political life of the country, as it limits the scope of their action, since they can only affect political life through their husbands. Indeed, as an anonymous (1850, p. 60) author claims in an article to *Eliza Cook's Journal*, "The influence of women on the mind of the country is still great, even in matters of government and justice. They exercise an unseen power and with invisible reins guide the opinions of men", which would probably reflect the sentiments of a great majority of Victorians.

However, if the idea of female influence is so dear to Victorian writers, Michel de Certeau's "tactics" seems a more adequate term to describe the involvement of Trollope's women into politics. Significantly, although many of his female characters evidently do not limit themselves to moral guidance only, they do not either demand the vote or lament the ideology of separate spheres which seems to incapacitate them politically. Rather than rebel against the established order and place themselves outside it, which could jeopardize their social position and respectability, they learn to act from within, employing what de Certeau called "tactics", or "an art of the weak" (Certeau 1984, p. 37). As de Certeau (1984, p. 37) writes, "the space of tactics is the space of the other. Thus it must play on and with a terrain imposed on it and organized by the law of a foreign power" so that it works "within enemy territory". Through the use of tactical practices, among which de Certeau enumerates "clever tricks, knowing how to get away with things, 'hunter's cunning,' maneuvers, polymorphic simulations, joyful discoveries, poetic as well as warlike" (Certeau 1984, p. xix), women can "turn to their own ends forces alien to them" (Certeau 1984, p. xix). Trollope's "political women" operate within the world as defined by men but they find the ways of operating that subvert the very principles on which this world is based without openly contesting it. The result is an ambiguity in Trollope's representation of female characters, at once weak and powerful, complying with the ideology of separate spheres and at the same time subverting it from within, blurring the boundary between the public and the private realms.

3 Glencora Palliser: A Political Woman

As she is presented in *The Prime Minister*, the Duchess of Omnium, known in earlier Palliser novels as Glencora Palliser, embodies the ambiguous position of a Victorian political woman: possessed of no direct political power, she is nevertheless one of the wealthiest women in England, which, coupled with her rank, makes her an unexceptionally influential character. In *The Prime Minister* she is presented at the point when her husband is asked to form a new coalition cabinet and becomes "the leading man in the greatest kingdom in the world" (I, 50).[3] Glencora wants to have a share in his political success and offers her support. However, not only does she go beyond the mere moral guidance, but she also seems to play the dominant part in the attempts to strengthen his authority. Most importantly, she superficially conforms to the norms and limitations of Victorian ideology, or, in other words, she tactically turns the Victorian ideas about women's role to her own advantage, paradoxically, by choosing to understand them very literally. She thus exploits the position she holds in the social world, traditionally a female domain, as well as the ideals most dearly held by Victorians: unity between the spouses and a wife's devotion to her husband, whose interest she should serve. Thus, although her scrupulous husband accuses her of vulgarity (I, 177), her behaviour does not seem to go much beyond the scope considered proper for women.

3.1 The Husband and Wife [are] One Subject in Law

After her initial resistance to her marriage described in the first novel of the series, *Can you forgive her?*, Glencora Palliser learns to love her rather reserved and prosaic husband, and becomes a model of a devoted wife. Her devotion is manifest in the interest she takes in politics and in her desire to foster Plantagenet Palliser's political career. Although, as the narrator indicates, "politics were altogether unnecessary to her" because she had "a wider and a pleasanter influence than

[3] All references to *The Prime Minister* will include the information about the volume (the Roman number) and to the page (the Arabic number).

could belong to any woman as wife of a Prime Minister" as well as "a celebrity of her own, quite independent of [her husband's] position, and which could not be enhanced by any glory or any power added to him" (I, 50), she devotes all her efforts and her fortune to ensure her husband's success as the Prime Minister. Glencora identifies her husband's interest with her own, so that when he goes to the Queen to be commissioned, "her heart throbbed with excitement" (I, 50) and when he finally resigns from office after unsuccessful attempts to form a coalition government, she cannot hide her disappointment and considers it to be her own failure, so that she complains that "[a]ll her grand aspirations were at an end" and that "[a]ll her triumphs are over" (II, 343).

Such identification, or unity, between spouses, was one of the major tenets of Victorian ideology, and it was reflected in the law of coverture. It held that "the husband and wife become one subject in law" and that "the very being or legal existence of the woman is in many respects suspended during the marriage, or at least is incorporated and consolidated into that of her husband" (Broom and Hadley 1869, p. 543). Not only did it lead to women's financial dependence, as it transferred the rights to a woman's property to her husband after marriage unless special provision was made for her, but it was also a frequently quoted reason against female suffrage, since women were considered to be sufficiently represented by their husbands. In the novel, Plantagenet Palliser takes the idea of the unity between husband and wife for granted, when he identifies Glencora's "ambition of seeing [her] husband a great man" with her ambition to be "a great man's wife" (I, 170), while failing to notice any subtle difference between the two. He also feels personally offended when his wife's name appears in a negative context in *People's Banner*, a radical newspaper exposing what its editor saw as political scandals, after Glencora meddled with the elections at Silverbridge, traditionally in the hands of the House of Omnium. At the same time however, although he claims to be "anxious… that [Glencora] should share everything with [him],—even in politics", he assumes his own dominance by insisting that "in all things there must at last be one voice that shall be the ruling voice", by which he means his own voice (I, 303).

Glencora's identification with her husband might create a misleading impression that she accepts her subordinate status and her disadvantageous legal and political position. Indeed, rather than rebel against them, Glencora tactically turns them to her own advantage. If she and her husband are one, his power and authority might be represented by herself. As Victorian lawyers explain, "[a] woman might indeed act as agent for her husband, for that implies no separation from, but is rather a representation of, her lord" (Broom and Hadley 1869, p. 544). Therefore, acting with the best intentions, and disregarding her husband's scruples, she interferes with the elections for a member for Silverbridge, by showing her preference for one of the two candidates, which amounted to revealing "what

---

[3] All references to *The Prime Minister* will include the information about the volume (the Roman number) and to the page (the Arabic number).

people call the Castle interest" (II, 102). Thus, Glencora exercises the only form of power available to women especially in the times when aristocratic influence in the boroughs was still very strong. As Reynolds (1998, p. 129) writes, "the return of the landowner's candidate remained a sound indication of the influence of the aristocratic landowner. Despite the changes to franchise carried in 1832 and 1867, and an increasingly vociferous debate on the nature and scope of 'legitimate influence', before the radical extension of the county franchise in 1884–5, the aristocracy continued to play a significant part in the return of members to the lower house". Significantly, aristocratic women could promote the interests of their families and friends by "follow[ing], without intervening in, the election of a particular candidate … tak[ing] a position of ostentatious non-interference… [and] a direct intervention in the constituency" (Reynolds 1998, p. 140).

Interestingly, although in the case of Silverbridge elections Glencora acts against the wishes of her husband, who believed that "[t]he influence which owners of property may have in boroughs is decreasing every day, and there arises the question whether a conscientious man will any longer use such influence" (I, 194–195), she considers her action justified, as she claims to be acting for his benefit. Following the "little syllogism… as to the Duke ruling the borough, the Duke's wife ruling the Duke, and therefore the Duke's wife ruling the borough" (I, 196), Glencora can thus exercise her power without actually going against the Victorian laws banishing women from the sphere of politics. Although the Duke sees Gelncora's behaviour as one that might destroy the unity between himself and his wife, so that he begs her not to separate herself from him (I, 302) as he believes that she "intend[s] to put [herself] in opposition to [him]" (I, 301), Glencora interprets her political intervention as an action resulting from the identification between her own and her husband's interests.

3.2 *She had Made Herself a Prime Ministress*

Glencora's identification with her husband results in her desire to match his political and her social success. Although, as a woman, she is banned by education and custom from holding any official function, she is nevertheless well prepared to take role of the political hostess, a figure extensively described by Reynolds (1998, p. 156–77).[4] As a political hostess, she has the "particular speciality" of "entertaining political connections, exercising patronage, guarding political confidences, and offering advice" (Reynolds 1998, p. 156), and she sets down to her job very seriously. Indeed, when she first hears of her husband's nomination as Prime Minister, she claims that she is ready to "slave" for him (I, 48), which, were it not told in earnest, might be read as a satirical reference to the relations between spouses. In the novel, her work is represented as analogous to Plantagenet's work in the political sphere.

[4] Reynolds distinguishes between a political hostess and an aristocratic political wife, a distinction that is not of great relevance for the purposes of this paper.

> It may be a question whether on the whole the Duchess did not work harder than he did. She did not at first dare to expound to him those grand ideas which she had conceived in regard to magnificence and hospitality. She said nothing of any extraordinary expenditure of money. But she set herself to work after her own fashion, making to him suggestions as to dinners and evening receptions, to which he objected only on the score of time. "You must eat your dinner somewhere," she said, "and you need only come in just before we sit down, and go into your own room if you please without coming upstairs at all. I can at any rate do that part of it for you." And she did do that part of it with marvellous energy all through the month of May,—so that by the end of the month, within six weeks of the time at which she first heard of the Coalition Ministry, all the world had begun to talk of the Prime Minister's dinners, and of the receptions given by the Prime Minister's wife (I, 72).

Thus, Glencora fulfils her promise to "entertain after a fashion that had never been known even among the nobility of England" (I, 51), and the success of her parties seems to exceed the less her husband's less spectacular triumphs.

Plantagenet at first considers Glencora's understanding of her role as a wife of the leading politician to be childish, but he is finally forced to admit that her understanding of the inextricable connection between the social and the political was much superior to his own. He recognizes in Glencora (rather than in himself) the social skills and qualities that would ensure the success of the coalition, which he tries to maintain:

> And now, gradually,—very slowly indeed at first, but still with a sure step,—there was creeping upon [Plantagenet Palliser] the idea that his power of cohesion was sought for, and perhaps found, not in his political capacity, but in his rank and wealth. It might, in fact, be the case that it was his wife the Duchess,—that Lady Glencora of whose wild impulses and general impracticability he had always been in dread,—that she with her dinner parties and receptions, with her crowded saloons, her music, her picnics, and social temptations, was Prime Minister rather than he himself. It might be that this had been understood by the coalesced parties,—by everybody, in fact, except himself. It had, perhaps, been found that in the state of things then existing, a ministry could be best kept together, not by parliamentary capacity, but by social arrangements, such as his Duchess, and his Duchess alone, could carry out. She and she only would have the spirit and the money and the sort of cleverness required. In such a state of things he of course, as her husband, must be the nominal Prime Minister (I, 161).

Glencora herself is also fully aware of her crucial role, believing that "[s]he had made herself a Prime Ministress by the manner in which she opened her saloons, her banqueting halls, and her gardens" (I, 264). *The Prime Minister* thus makes conspicuous the inextricable connections between the political and the social spheres, which cannot exist independently. Not only do the numerous dinners and parties serve to secure the support for her husband to make it possible for him to carry out his political work, but, as Plantagenet finally comes to realise, they constitute the essence of political activity. Most importantly, the social world both reflects political alliances, with ladies forming their own "parties" or "cabinets" analogous to the ones functioning in the Parliament, and crucially contributes to the shaping of alliances or even determines them. This is why Glencora attempts to "ha[ve] everybody" (I, 97) to her huge parties, so that "[i]n London there should not be a Member of Parliament whom she would not herself know and influence by her flattery and grace" (I, 51). Indeed, one of the characters' ironic remark that "the country can be ruled by flowers and looking-glasses" (I, 100) is truer than it originally seems.

Glencora's understanding of the complementarily of the political (male) world and the social (female) sphere, or, in fact, conflating the political and the social, contributes significantly to the blurring of the line between them, but it also exposes her husband's deficiencies as a politician. Plantagenet Palliser is described in the novel as an upright and conscientious man, unable to bear even the smallest stain on his reputation, or, as Trollope writes in his autobiography, he is "a perfect gentleman" (1999a, p. 361). Glencora seems to realise that it is only "sufficient audacity, a thick skin, and power to bear for a few years the evil looks and cold shoulders of his comrades" that can ensure political success while "[t]o the man who will once shrink at the idea of being looked at askance for treachery, or hated for his ill condition, the career is impossible" (II, 364). Plantagenet's oversensitivity as regards his honour, or his tender skin (II, 21), as Glencora calls it, seems to make him an effeminate character and prevent him from holding long in office. It is Glencora, "ambitious both on her own behalf and that of her lord" (I, 46), who has to make up for his weakness. Not only does she claim that she "should have been a man, [her] skin is so thick" (II, 21) but she also declares her readiness to do "all the dirty work":

> They should have made me Prime Minister, and have let him be Chancellor of the Exchequer. I begin to see the ways of Government now. I could have done all the dirty work. I could have given away garters and ribbons, and made my bargains while giving them. I could select sleek, easy bishops who wouldn't be troublesome. I could give pensions or withhold them, and make the stupid men peers. I could have the big noblemen at my feet, praying to be Lieutenants of Counties. I could dole out secretaryships and lordships, and never a one without getting something in return. I could brazen out a job and let the 'People's Banners' and the Slides make their worst of it. And I think I could make myself popular with my party, and do the high-flowing patriotic talk for the benefit of the Provinces… a Prime Minister should never go beyond generalities about commerce, agriculture, peace, and general philanthropy. Of course he should have the gift of the gab, and that Plantagenet hasn't got. He never wants to say anything unless he has got something to say (II, 154–155).

Glencora is evidently fully aware of the harshness and corruptibility of the public world, from which women are excluded, as Ruskin claimed, to protect their purity. It is men who were to "encounter all peril and trial" to emerge "wounded, or subdued; often misled; and *always* hardened" (Ruskin 2002, p. 77, original emphasis). Glencora, however, professes herself to be a more competent person to face the dangerous and morally ambivalent world than her husband. Not only does she thus blur the boundary between the male and the female domains, but she also actually reverses the male and female roles, as they were understood in the Victorian period.

In her great ambition on behalf of her husband as well as herself, her lack of any scruples in achieving her goals and in her "masculine" thick skin, Glencora might remind of Lady Macbeth, to whom she compares herself several times in the novel. Like Lady Macbeth, who notoriously prays to be unsexed and to have "the access and passage to remorse" stopped so that "no compunctious visiting of nature [might] shake [her] fell purpose" (Shakespeare 1992, p. 37), Glencora takes on the qualities considered to belong "naturally" to men and urges her weaker husband to action.

Glencora is obviously properly toned down to fit a realistic novel rather than a great tragedy, and she seems to be talking about her being like Lady Macbeth in jest; she says "I feel myself to be a Lady Macbeth, prepared for the murder of any Duncan or any Daubeny [a former Prime Minister and a conservative] who may stand in my lord's way. In the meantime, like Lady Macbeth herself, we must attend to the banqueting. Her lord appeared and misbehaved himself; my lord won't show himself at all,—which I think is worse" (Shakespeare 1994, p. I, 96). Yet, the comparison leaves the reader with the feeling, that, like Lady Macbeth, she entered a forbidden territory of the male world, from which she cannot return unstained. Glencora is fully aware of a kind of moral degradation that any political career might involve, and yet, to keep her husband in power, she was ready to undergo ridicule and accusations of vulgarity. Paradoxically, even through her unfeminine behaviour she only proved her truly feminine devotion to her husband.

4 Conclusion

In the light of Lady Glencora Palliser's political activity, the quotation that was used in the title of this paper—"The one great drawback to the life of women is that they cannot act in politics" (Trollope 1999b, p. II, 26), a sentence spoken by Madame Max Goesler (later Mrs Finn), seems ironic and, indeed, it was not spoken in earnest. The very same character, when asked in *The Prime Minister*, whether she would "go in for [women's] 'rights'" answers: "Not by Act of Parliament, or by platform meeting. I have a great idea of a woman's rights; but that is the way to throw them away (I, 103–104) and she jokingly suggests that women might achieve political goals without spending her days getting bored in the House of Commons. Although the idea of influence might justify women's exclusion from the political life, Trollope's political women redefine it and tactically use to their own advantage. Therefore, far from abolishing the boundary between the private and the public altogether, Trollope nevertheless revealed how blurry and porous it was.

References

Anonymous. 1850. The vocation of women. *Eliza Cook's Journal*, 3, 56: 59-61 http://archive.org/details/elizcooksjournal03cookrich Accessed 27 May 2008.

Broom and Hadley. 1869. *Commentaries on the laws of England*. 4 vols. London: William Maxwell &Son, Henry Sweet, and Stevens and Sons. http://books.google.pl/books?id=thitAAAAMAAJ&dq=Broom+hadley Accessed 12 February 2010.

Certeau, Michel de. 1984. *The practice of everyday life*. (Translated by Steven Rendall). Berkeley, Los Angeles and London: University of California Press.

Chew, S. C. and Altick, R. D. (ed.) 1967. *A literary history of England. The nineteenth century and after*. London: Routledge and Kegan Paul.

Mill, J. S. 2003. *The admission of women to the electoral suffrage*. http://www.pinn.net/~sunshine/book-sum/js_mill2.html. Accessed 16 January 2010.

Reynolds, K.D. 1998. *Aristocratic women and political society in Victorian Britain*. Oxford and New York: Oxford University Press.

Ruskin, J. 2002. *Sesame and lilies*. (edited and an introduction by Deborah Epstein Nord). New Haven and London: Yale University Press.

Shakespeare, W. 1994. *Macbeth*. London: Penguin Books.

Trollope, A. 1999a. *An autobiography*. (Edited by Michaels Sadler and Frederick Page, introduction and notes by E.D. Edwards). Oxford: Oxford University Press.

Trollope, A. 1999b. *Phineas Finn. The Irish Member*. 2 vols. (Edited with and introduction and notes by Jacques Berthoud). Oxford: Oxford University Press.

Trollope, A. 1999c. *The Prime Minister*. 2 vols. (Edited with notes by Jennifer Uglow, instroduction by John McCormick). Oxford: Oxford University Press.

Wingerden, S A. van. 1999. *The women's suffrage movement in Britain, 1866–1928*. Basingstoke and New York: Palgrave Macmillan.

Transgressing Boundaries to Metamorphose: *The Outlander* by Gil Adamson

Ewa Urbaniak-Rybicka

Abstract Gil Adamson's debut novel *The outlander* (2007) is set in 1903 and presents a nineteen-year-old Mrs. Mary Boulton, a widow by her own hand, who starts her rebirth with an act of violence. Adamson's heroine incites change and fashions her life as a process of transformation at the cost of transgressing visible as well as invisible boundaries to self-realisation. Fleeing from the external dangers, the widow confronts the overwhelming darkness of her own mind, which provides this remake of the western and a Canadian romance with a Gothic dimension. Haunted by the nightmares of the past, the enormity of her crime, the dead within her as well as the phantom of her former married Other, Mary Boulton learns survival skills, communes with nature, sides with other misfits she encounters on her way, and saves herself from hanging. Enlightened as well as strengthened by her traumatic experiences, pursued while pursuing, abandoned but herself abandoning, the widow stakes out a new territory with a promise of life inside her. She travels towards ultimate liberation, acceptance, and independence to the North, the romanticised place for renegotiating her identity and final erasure of constraints. In the peculiar detective story the widow's revengeful pursuers are joined by the reader who from the mosaic of fragments and memories reconstructs the story of the villainess' life. Evading easy categorisation and liberating herself from the gender stereotypes the murderess echoes the subversive heroines of such Canadian writers as Margaret Atwood and Aritha van Herk.

E. Urbaniak-Rybicka (✉)
Adam Mickiewicz University, Kalisz, Poland
e-mail: ewaur@poczta.onet.pl

J. Fabiszak et al. (eds.), *Crossroads in Literature and Culture,* Second Language Learning and Teaching, DOI: 10.1007/978-3-642-21994-8_11,

1 Introduction: A Violent Rebirth

The heroine of Bharati Mukharjee's novel *Jasmine* remarks that "there are no harmless, compassionate ways to remake oneself. We murder who we are so we can rebirth ourselves in the images of dreams" (Mukherjee 1989, p. 29). Similarly, in Gil Adamson's awarded and critically acclaimed debut novel *The outlander* (2007), set in 1903, the process of metamorphosis starts with destruction. The novel presents nineteen-year-old Mrs. Mary Boulton, a widow by her own hand, who begins her self-creation with an act of violence.[1] The murder of her husband is a metaphorical annihilation of her imprisoned self and a gateway to rebirth. Adamson's heroine incites change and fashions her life as a process of transformation at the cost of transgressing visible as well as invisible boundaries to self-realisation. Thus, the widow escapes through the Canadian wilderness not only from her brothers-in–law, who seek revenge, but also from her previous identity. During the journey, functioning in the novel as a path towards redefinition, the heroine metamorphoses from a passive victim into an active, independent woman responsible for her own existence and orchestrating her future. Fleeing from external dangers, the widow confronts the overwhelming darkness of her own mind, which provides a Gothic dimension to the narrative. Haunted by the nightmares of the past, the enormity of her crime, the dead within her as well as the phantom of her former married Other, Mary Boulton, a contemporary incarnation of the pioneer woman, learns survival skills, communes with nature, sides with other misfits she encounters on her way, and saves herself from being hanged. Enlightened as well as strengthened by her traumatic experiences, pursued while pursuing, abandoned but herself abandoning, the widow stakes out a new territory with a promise of life inside her. She travels towards ultimate liberation, acceptance, and independence in the North, the romanticised place for renegotiating her identity and final erasure of constraints. In this unusual detective story the widow's revengeful pursuers are joined by the reader, who from a mosaic of fragments and memories must reconstruct the story of Mary's life. Evading easy categorisation and liberating herself from gender stereotypes, the murderess echoes the subversive heroines of such Canadian writers as Margaret Laurence, Margaret Atwood and Aritha van Herk.

[1] The novel has received several literary awards, such as the 2007 Drummer General's Award, the 2007 Amazon.ca/Books in Canada First Novel Award, the 2007 International Association of Crime Writers' Dashiell Hammamett Prize, and the 2007 ReLit Award. It was nominated for the 2009 Canada Reads Award, and was short-listed for the Commonwealth Writers' Prize, Best Book, Canada and Caribbean and the Trillium Book. Additionally it was short-listed for the Commonwealth Writers' Prize and long-listed for the International IMAC Dublin Literary Award.

2 The Canadian Western: Rewriting the Genre

One of the generic intertexts for *The outlander* is definitely the western. This genre probably came into existence in the fiction of James Fenimore Cooper (1789–1851), who combined tales of American settlement in the wilderness with the archetypal pattern of adventure stories (Cawelti 1976, pp. 192–259). The western does not have a set pattern of action. However, its recurring and essential ingredients are a challenge to the hero, an ultimate confrontation with an antagonist, the motif of revenge, as well as chase and pursuit. What characterises the genre and crucially defines it is the landscape, with its influence on the hero. Consequently, the western landscape provides space for the confrontation between civilisation and wilderness, life in the clearing and life in the wild, law and chaos, or constraints and total freedom/lawlessness. Historically, the western portrays a moment of harmony between the forces of civilisation and wilderness, while the hero is positioned between two contrasting ways of life. Cawelti (1976, p. 194) sees the most significant aspect of the western in its representation of the relationship between the hero and the clashing forces of civilisation and wilderness. The prototype of the western hero is Nathaniel Bumpo, the romanticised man of the wilderness from James Fenimore Cooper's Leatherstocking Tales. His extraordinariness is in the paradoxical combination of the civilised and the natural man, the skills which make him suitable for facing and taming the wild. Natty Bumpo was for Cooper the embodiment of the natural virtues that civilisation had lost. Furthermore, the western often included the conflict between domesticity and its values, such as marriage, family, social respectability, and security, and the ideal masculine way of life, which is liberated from constraints. The western emerged from setting and plot patterns through which the conflicts of society and individual autonomy, civilisation and unrestrained violence, could be settled.

In the 19th century American imagination the West was pregnant with romantic connotations offering ultimate freedom from the complex institutions, corruption and limitations associated with the East. In the further developments of the genre, the western setting functioned also as a place of moral regeneration due to the strong influence of nature. For Frederic Jackson Turner (1861–1932), the essential aspect of the West became social flexibility as well as the equality of opportunity, which led to the metamorphosis of men into free individuals. On the other hand, Zane Grey (1872–1939) and W.S. Hart (1864–1946) emphasised the West as a testing ground for a character, disclosing the truth about him. The outcome of the confrontation with wild nature and violent men is an affirmation of traditional American values—monogamous love, the settled family, the basic and stereotypical separation of gender roles, and the importance of religion. The classic western of the post-World War Two period presents a hero similar to Cooper's literary characters, as he is both involved in and separated from society. He often takes law into his own hands. Since the middle of the 1960s, the basic formula of the western has not been possible to be discerned, though such trends as emphasis on graphic violence and treating sex in a more open manner can be noticed.

In Canadian literature, due to the country's different course of literary development in the 19th century, a Canadian western genre did not exist, and there were no Canadian counterparts for the male-centred markers of the American frontier as traditionally represented in the genre (Davidson 1994, pp. 98, 99). The literary form was not particularly relevant to the Canadian experience. Therefore, the romance with its emphasis on action and adventure rather than the western with the focus on character and setting, was the predominant form of prose fiction about the Canadian West until the First World War (Ricou 1973, p. 14). The appropriation and deployment of American western iconography would have been, in Davidson's opinion, re-inscribing Canada's paracolonial status. According to Davidson, "Canada (…) stands to the United States in the world of global politics as 'woman' stands to 'man' in the plots of the conventional western—on the sidelines, occasionally counselling restraint or offering advice and moral support, but definitely not packing the big guns" (Davidson 1994, p. 99). Furthermore, the Canadian West was settled later and in a different fashion than the American West. It was not the settlers but the North West Mounted Police who were first in the Canadian West and worked to prevent any acts of violence resulting from the clashes between Indians and the pioneers. The American Wild West became, as Daniel Francis claims, the Mild West (2003, pp. 29–51). The Canadian Mountie, unlike the traditional hero of American westerns, did not resort to violence to solve a problem, but rather avoided violence while settling the conflict between civilisation and savagery (Harrison 1977, p. 162). Additionally, the Canadian settlers came to the West mostly by steam train, therefore such characteristic western determiners as wagon trains and Indian wars, cowboys and rustlers, shoot-outs and stampedes were out of place there. Due to the above differences, Canadian western writers have challenged the American western with a realistically portrayed Canadian West. What is more, because they have equalled the western with the American and the male, their re-writing of the genre has often resulted in a subversion and parody of the classic form.

Since the Canadian western was not a copy of the American one, the form itself was not perceived as exclusively male, either. Nevertheless, though some women, such as, for example, Nellie McClung (1873–1951), and later Martha Ostenso (1900–1963), Laura G. Salverson (1898–1970), Margaret Laurence (1926–1987), Gabrielle Roy (1909–1983) and Ethel Wilson (1888–1980), appropriated the western space in their novels, as Aritha van Herk claimed in her essay "Women writers and the prairie: spies in an indifferent landscape," first published in 1984, for most women writers the traditional image of the West was masculine. Therefore, in the West and the western, "men are men and life a stern test of man's real attributes. The fabric of this living, breathing landscape has been masculinized in art, descriptive passages of a land instinctively female perceived by a jaundiced male eye" (van Herk 1992, p. 141). Women were traditionally, like in other masculine genres, to fit stereotypical roles.[2]

3 Gendered Space: The Pioneer Woman in the Wild

By placing her character in the Alberta wilderness and sending her on a journey through the Rocky Mountains Adamson definitely puts her heroine within the framework of the western, though the novel is very eclectic, including also elements of the gothic, the picaresque, and the historical novel (Bond 2009). Adamson's western, however, is feminist, and a pretext for challenging, testing and reshaping the protagonist's identity through violence and quest. It also examines the social context of women's lives at the turn of the 20th century. In a very romanticised setting, the healing power of nature and solitude as well as the survival test in the mountains result in Mary's moral regeneration and independence. The process is completed through still another final action—a journey to the North, even emptier and wilder than the mythical West, a symbolic equivalent of the West on Canadian land and in the Canadian imagination with the potential for freedom, autonomy, and independence.[3] Thus, experiencing not only the clash of wilderness and civilisation as it happens in the western, Adamson's heroine also faces her internal frontier, which makes her—to a certain extent—a contemporary embodiment of the pioneer woman. The pioneer woman as a character type was created by Catherine Parr Traill (1802–1899) in her non-fiction books *The backwoods of Canada* (1836), *The Canadian settler's guide* (1860) and her fiction *Canadian Crusoes* (1852) (Thompson 1991, pp. 112–157).[4] This character type can be adapted to various definitions of the frontier, and it refers especially to the frontier as a state of mind rather than a specific place. According to Thompson, the

---

[2] Van Herk believes that this masculine view masquerades

> a world view for women and men, [and is] a selective realm where women are portrayed as mothers, saints or whores, but never people in their own right. Women as victims, cripples, betrayers, servants, objects, bitches. Female characters consistently maimed or killed for their rebellion. Or ignored for their acquiescence (1992, p. 83).

In her own novels, Aritha van Herk joined other Canadian women writers in appropriating the West and the western as women's space, and infiltrated the "kingdom of the male virgin" with her spies in the "indifferent landscape" (van Herk 1992, pp. 139–151) in such novels as *The tent peg* (1981), *No fixed address* (1986) and *Places far from Ellesmere* (1990).

[3] Compare, for example, Hulan 2002.

[4] Adamson's heroine is not the only example of the pioneer woman in Canadian literature, of course. In her study Thompson discusses several literary characters who adhere to the pattern, and also enumerates the 20th century writers (prior to the year 1991) who, according to her, use the character type in their fiction (Thompson 1991, p. 113). She gives only a few examples and does not mention the remaining books by van Herk which contain this type of heroine, such as *No fixed address* (1986), for example, or several heroines from Margaret Atwood's fiction, which, I would argue, are also modelled on the pioneer woman archetype.

characteristics of the pioneer woman are unchangeable even though the frontier environment alters. The pioneer woman, then, is courageous, resourceful, and pragmatic, accepts adverse circumstances with equanimity, and possesses strength to act decisively when she faces discomfort and danger. She should act promptly to improve her situation. Through her pioneering efforts, the woman discovers her intrinsic value. The pioneer woman reveals within herself a concealed or unexpressed potential for independence, a liberation from fear, restraint as well as social criticism, and becomes proud of accomplishing repulsive or difficult new tasks. Pursuing her new activities, the character discovers hidden internal strength and courage.

Though Adamson's heroine does not adhere ideally to the model, especially during her mostly solitary life in a cabin in the Alberta wilderness, she shares certain recognisable features with the literary prototype. Therefore, Mary's frontier landscape is "hostile, disorienting, and confusing" (Thompson 1991, p. 113), and not only external objects and physical perils constitute difficulties for her, but it is the heroine herself who creates internal problems. The story of Mary Boulton's life is disclosed in *The outlander* in little pieces of memories scattered non-chronologically throughout the novel. She had a solitary and miserable childhood in a house infused with illness and then mourning. Her father, after marrying a rich, though sickly, woman, treated his job, first as a pastor and then a lawyer, only as a pleasant distraction. Forever bedridden and ill, Mary's fading mother did not devote any time or attention to her daughter. What Mary remembers about her mother is only her death: "Everything is remembered by its moment of intensity. Dying was hers" (Adamson 2009, p. 96). For Mary's widowed father, life consisted of cessation and decay, death being the inevitable end of it. During his life as a widower he was permanently drunk, trying to erase his grief. As a young adult, when analysing her childhood and early youth Mary becomes certain that "she was invisible to both her parents. For her father, there had only ever been one her, and once she was gone, so was his connection to everything else" (Adamson 2009, p. 246). Furthermore, Mary's family deprived her of education, and throughout her life Mary tried with various ingenious tactics to hide her virtual illiteracy. The widow's grandmother believed that education was perilous for girls—damaging their brains and causing reproductive malfunction. Later, the widow realises that her tragic fate could be attributed not only to her father's despair and her grandmother's inability to deal with him and her, but also to her ignorance, to "knowing how little she herself knew" (Adamson 2009, p. 49).

Ignorant, innocent, and utterly obedient, the widow married the first candidate her family found suitable for her. Paraded at the marriage market as her grandmother's "project of the moment" (Adamson 2009, p. 137), Mary was to be prevented from becoming an old maid and ruining her life. The young girl, devoid of control over her own life, perceived herself as "a story, a tale still in the telling," curious about its end (Adamson 2009, p 137). Although she disliked being forced to attend parties, part of the scheme to find a husband, the elemental terror of spinsterhood that her grandmother had experienced soon spread to her. Still, at parties she remained mostly a recluse, trying to make her self invisible. It was this potential for loneliness in Mary and her apparent desire for a solitary life that must have attracted John, her future husband and victim, to her during one such party:

> There was a girl who could stand the quiet and isolation, a girl who didn't need social life, didn't want it. How much better she would be in a lonely log cabin than would these happy playful girls who ran about the lawn holding hands (Adamson 2009, p. 138).

She became John's "best and only choice" since "like all gamblers, and to his peril, he trusted such moments of intuition" (Adamson 2009, p. 139). John Boulton presented himself to his future wife as a businessman and landowner whose father was a magistrate, and the courtship began. Still, nothing was known about the man, except for what he wanted to be known. Again, Mary was letting the current carry her away from her childhood and family. The marriage was to be for her a desired metamorphosis: "If she left, she might be free to change, to be something and someone else" (Adamson 2009, p. 139). She was attracted physically to her fiancée and felt unaccustomed passion, though doubts nagged her, as well. In her uncertainty, Mary was reassured that she would grow to love her husband, like other generations of women before her. Consequently, at the age of nineteen Mary was given away to a complete stranger who was thirty-five years old. Leaving her old home "drunk with the promise of transformation" (Adamson 2009, p. 140), she followed John into an unknown future.

John's promises of an affluent and comfortable life were not kept. Subsequently, Mary, a pampered girl, was offered a life in the solitude of a cabin, with an unfaithful and a gambling husband who was constantly displeased with her service as a housewife, and who resorted to sexual abuse. On the journey to the new home Mary wondered what her life in a "well-appointed house 'with hundreds of acres of land' would look like" (Adamson 2009, p. 127). John told her that in the new place "there's nothing but room" (Adamson 2009, p. 140). This turned out to be very true, because when they reached the destination, instead of a new home Mary saw only a small, square foundation. For two months the newlyweds lived in a tent in the trees, without a roof over their heads or privacy. Mary was totally unprepared and untrained for this type of existence:

> When she had found herself finally in the cabin with her new husband and she had unpacked her trousseau, the dresses with silk-covered buttons had lain in her hands like artifacts from another world. (…) It had been obvious what she must do. She had packed away her former self and begun sewing clothes, rough simple things to fit her new life (Adamson 2009, p. 35).

Furthermore, Mary had different expectations of marriage than John, who would disappear for weeks, and who treated her like a servant and his property. She was forced to acquire a new identity—that of the pioneer woman—and she slowly arrived at the truth about her husband.

At first the protagonist believed that her married life was not successful because there was something in her that caused her husband's dissatisfaction. Later, she noticed that John led a double life, the echoes of which started reaching her. The illuminations, "a process of seeing things again," came gradually, as "painful little kink[s] in the flow that forced all thought to adjust to the truth. These were the seeds of her despair and madness" (Adamson 2009, p. 170). The accumulation of her husband's wrongdoings—unfaithfulness, gambling, sexual abuse, stealing her things to give them away to his lovers, neglecting her emotionally, and leaving her without assistance or support in the most dramatic moments of her life—led, in due course, to Mary's violent rebellion and the murder. On this path towards disaster, her initial happiness was fading to give place to bitterness:

> She'd been properly happy on her wedding day, gaily waving goodbye on the train, embracing her husband when he finally came to her in bed. Happy as expected. Then: the happy duped wife, the happy inept housekeeper left constantly alone, with winter roaming outside the cabin and the voices roaming inside her. The happy mother of a sick and dying baby (Adamson 2009, p. 210).

The traumatic event of her baby's death and her helplessness amid the wilderness, to which Mary ironically referred as a "small devastation" (Adamson 2009, p. 36), started the chain of tragic events. When her husband returned after a week to the cabin to check on his wife and the fading child, the baby had already died and his wife was in the midst of post-natal depression. What pushed Mary over the brink and triggered her rebellion was her humiliation when John's pregnant lover visited the cabin and when, after the child's burial, she was left alone again. Mary could remember only in flashes what happened later. There was a sudden realisation that she had been horribly impoverished: "Everything had been taken from her—her father, her birthplace, any money she had ever possessed, her engagement ring, her only child and now her husband," and this made her view her life in a new way: "The girl had looked around at nothing and had seen everything" (Adamson 2009, p. 217). A series of actions and decisions changed her life forever. First, she buried her wedding ring. Then she started sewing her widow clothes, waited for her husband, shot him, left him bleeding and dying on the floor of the cabin, resumed sewing and then began her run for a new life and a new identity.

4 The Flight Towards Self-Liberation

Mary's journey through the Alberta wilderness and the Rocky Mountains functions in *The outlander* as a metaphorical passage to the Conrad-like "heart of darkness"—into the widow's haunted mind and the horrors of her former life. It becomes also an important step towards self-liberation. It is in the mountains, "a monument in her path, promising freedom and camouflage" (Adamson 2009, p. 58), where the widow faces the internal frontier. This personal frontier is the result of the social context of women's lives at the turn of the century, and a result of Mary's mentality which has been shaped by her mourning father and her grandmother's inability to raise her. In her desire to overcome her internal constraints and her desire for freedom, she surprises the environment with her previously unrecognised layers of power, both physical and spiritual, and her amazing yearning for personal autonomy. Throughout the novel she acts upon the frontier to bring about change. The process of self-understanding and metamorphosis starts in the mountains, when floods of memories cause Mary to ponder over her past. She disintegrates, and then puts her shattered self back together. The external landscape merges with the internal one:

> It was with grinding certainty that her mental lapses came. Sometimes accompanied by noises – a booming in her ears (...) but also voices, strange and distorted. Terrible things were imparted to her in non-words, in senseless howling, or the sound of a cricket chirruping. Or a clatter, like a spade thrust into a fan. She would press her hands over her ears – pointless. Because the noise came from within – press her palms there as if to keep the horrors from leaking out of her into the room. First the sounds, then the visions. And every time, she suffered a sense of fatedness, of punishment (...) Memories, gestures, her husband's indifferent voice. She suffered the echo of her baby's fussing, his shallow breathing. She snapped her hands round herself, rolled to her side, and closed her welling eyes. Wait, wait and they will pass, and perhaps nothing will rise in you (Adamson 2009, pp. 10, 157).

The wilderness offering her freedom, autonomy and adventure lets her realise that the greatest dangers come from the inside, and that the only darkness to be anxious about is of her own mind:

> The thing to be feared always came from within: exhaustion, unsound thoughts, ignorance, starvation. This was the locus of fear for her, a worm in the heart, where hope rotted in its dark whorls, where unwanted visions leaped – the darkness of her own mind. And yet here she was alone in the wilderness strangely content (Adamson 2009, p. 65).

Gradually, the widow starts seeing the world around her in a simpler and clearer way than before. Her former life and her previous identity as a submissive wife seem to her like a story about a different girl whom she has left behind. Looking back at her life, the widow becomes conscious of the fact that ignorance, humility, obedience, dependence and fear are her own private wilderness.

While fleeing from her former life and self, the widow meets several men, also outlanders, who contribute to her gradual metamorphosis and autonomy. One of them is the Ridgerunner—William Moreland. This chronic thief, just like the widow, is on the run from society and later also from himself. Resembling American transcendentalists, the man has spent 13 years away from society in solitary communion with nature, rejecting civilisation, suffering no loneliness, and enjoying absolute freedom in the mountains. The Ridgerunner teaches Mary some basic skills of survival in the wilderness and offers her the warmth, tenderness and sensuous pleasures which her egocentric husband refused to give her. The widow, still emotionally famished, quickly falls in love with the recluse and dreams of their mutual future. Then, however: "The shadow of something fretful passe[s] over his face and then [is] gone" (Adamson 2009, p. 105). Promising at first that he will never abandon the widow, William Moreland suddenly leaves, unable to sacrifice his freedom for the unexpected union.

After the Ridgerunner's disappearance Mary is unable to live on her own and looks for the care of another man. She starts living with the Reverend Mr. Angus Lorne Bonnycastle in the mining town of Frank, the only woman in a community of men. She becomes a pragmatic and an efficient housewife, a hunter and a barber at Charles MacEchern's trading post, and also advises the Reverend on alternative ways of appealing to the mining congregation. In her simple and quiet life with Bonny, devoid of any sexual subtext, she finds refuge from her stormy past and is slowly able to silence the raging emotions and haunting visions of her mind. She constantly remains in the process of healing from her disappointment with William Moreland: "How foolish it was to allow a man in, how terrible his power once you did" (Adamson 2009, p. 288). Surrounded by Bonny's security and care, she finds temporary happiness, though the enormity of her murder does not stop overwhelming her: "Like a woman rising from the damp sheets after a fever, the widow looked about her at a new life. All that was left now was her crime" (Adamson 2009, pp. 246, 247). Emotional independence from Bonny comes violently to the widow when the Reverend dies in a landslide. Modelled on the historical geological transformation that took place in the town of Frank in April 1903, the event described in Adamson's novel devastates the place and annihilates its numerous inhabitants. Mary helps in nursing the injured and dressing the dead. In despair again, "Twenty years old and she had already reached the border of her heart's endurance twice" (Adamson 2009, p. 318), she finally frees herself from the ghosts of the past and appreciates the miracle of her own survival amid the ruins.

It is at that moment when she becomes able to make conscious and independent decisions about her life. However, there awaits still another confrontation with her past when her brothers-in-law finally capture her, having recognised her in a picture in the coverage of the Frank landslide. Mary struggles against the brothers, for whom she is "an unfinished task" (Adamson 2009, p. 347). Due to their physical similarity to their murdered brother the widow feels as if she was facing her dead husband John once again:

> his jaw, his face, twinned, staring up at her in astonishment, and she, yearning to murder him again. His mouth falling open in surprise. A pink mist floating on the air, its dull taste on her lips, droplets on her hands, chin, forehead. His soundless open mouth as he fell. His mouth, saying, "You can have another," saying, "They come and go, like calves" (Adamson 2009, p. 332).[5]

Finally captured, tied up, furious but not afraid, Mary awaits a trial and possibly hanging for murder. With her survival beyond her control, the widow becomes a local spectacle, a murderess in Indian clothes, a curiosity. The desire for freedom does not die within her as she scrutinises the prison, "considering the objects in her

[5] The sentences ringing in the widow's ears were spoken by her husband after the death of her baby to cheer her up.

room as some kind of disassembled key to her escape" (Adamson 2009, p. 356). Realising she is pregnant with Ridgerunner's child and learning that one of the options given to her may be postponing the execution for the child to be born and then handing it over to its closest relatives, John's brothers, the widow discovers new strength and ingenuity. With a stolen knife, managing to dismantle the window bars through desperate efforts, she escapes once again, determined to keep the baby. Each step leads towards freedom: "She feels the rising exquisite gladness, a pride bursting inside her, for with any luck, her life was now her own, no one else's. The great fortune of it!" (Adamson 2009, p. 377). She sets the dead within her free, liberating herself from the haunting memories and guilt, to cling to the living and the promise of life within her. She discards her love for the Ridgerunner and with an air of certainty, determination, and newly-discovered inner power she sets out for the North to embrace its possibilities and test the potential for self-realisation.

5 Conclusion: From Dependence to Responsibility

Adamson's heroine of this Canadian feminist western novel with Gothic and picaresque undertones, like a literary prototype of the pioneer woman, does not remain idle in the face of hardships and difficulties. Although trained for a different life, in the Alberta wilderness she acquires survival skills that enable her to escape, after a period of submissiveness and mental breakdown, into the arms of the promising freedom of the Rocky Mountains, where she starts a process of healing and transformation. Her metamorphosis begins with an act of overwhelming violence. During her journey she confronts not only the physical but, more importantly, the inner frontier of her mind, and gradually tames her personal fears. Finding but always quickly losing the care, tenderness and security offered by men, the widow learns that she has to be responsible for her own existence and shape her own fate, rather than rely on others. Resembling other contemporary incarnations of literary pioneer women, such as the heroines of Margaret Lawrence, Aritha van Herk and Margaret Atwood, throughout the novel Mary Boulton sheds her former selves, discards self-imposed restraints, and grows into independence. The widow's never satisfied desire "refuses to believe that the world is finished and complete, (…) [It] stretches romanticism, defies history, transforms detail, perplexes tradition, welcomes freedom" (van Herk 1992, p. 84). This desire makes her transgress internal and external boundaries, and liberates her from the restricting labels of the western female character. Therefore, the disappearing murderess from Gil Adamson's *The outlander* can be measured by her striving for what seems inaccessible and her longing to explore human as well as the world's possibilities.

References

Adamson, G. 2009 (2007). *The outlander*. London: Bloomsbury Publishing.

Bond, L. 2009. Mixing periods and genres. *Canadian Literature/Littérature canadienne: a quarterly of criticism and review (Univ. of British Columbia, Vancouver)* (201): 131–133.

Cawelti, J., G. 1976. *Adventure, mystery, and romance. Formula stories as art and popular culture*. Chicago and London: University of Chicago Press.

Davidson, A. E. 1994. *Coyote country: fictions of the Canadian West*. Durham and London: Duke University Press.

Francis, D. 2003. *National dreams. Myth, memory and Canadian history*. Vancouver: Arsenal Pulp Press.

Harrison, D. 1977. *Unnamed country: the struggle for a Canadian prairie fiction*. Edmonton: University of Alberta Press.

Hulan, R. 2002. *Northern experience and the myths of Canadian culture*. Montreal: McGill-Queen's University Press.

Mukherjee, B. 1989. *Jasmine*. New York: Grove Weidernfield

Ricou, L. 1973. *Vertical man/horizonal world. Man and landscape in Canadian prairie fiction*. University of British Columbia Press.

Thompson, E. 1991. *The pioneer woman: A Canadian character type*. Montreal: McGill Queen's University Press

van Herk, A. 1992. Desire in fiction: De-siring realism. In *The frozen tongue*. 79–85. Sydney: Dangaroo Press

van Herk, A. 1992. Women writers and the prairie: Spies in an indifferent landscape, In *The frozen tongue*. 139–151. Sydney: Dangaroo Press

Forbidden Territories: the Sexual and the Social in Allan Hollinghurst's Novel *The Line of Beauty*

Przemysław Uściński

Abstract The following paper discusses the fourth novel by a British writer Alan Hollinghurst, entitled *The line of beauty* and published in 2004. The novel is set in London at a time of the political reign of Margaret Thatcher. The author of this paper tries to focus on the representations of sexuality and of the British society in the era of Thatcherism that the novel offers. These aspects are analyzed in the light of the texts by Eve Kosofsky Sedgwick and Judith Butler, among others, which are often referred to as catalysts for many recent developments within Literary Theory and Cultural Studies. The paper thus tries to examine some of the main themes of the novel, such as sexuality, the private/public dichotomy, social class and style by applying some recent theories on these subjects.

1 Introduction

Allan Hollinghurst's fourth novel, *The line of beauty*, published in 2004, received the *Man Booker Prize* the same year and gained a wide readership both in Britain and abroad, thus being the first such a big success for the author both in terms of a commercial triumph and critical acclaim. It was also made into a *BBC* mini-series in 2006. In the following paper I wish to analyze those aspects of the novel that are connected with its representation of sexuality as a shaping presence in the main protagonist's social life.

The action of the novel is set in London in the mid and late 80s, the time of the political reign of Margaret Thatcher. To be more specific, it opens in 1983 with the

P. Uściński (✉)
Warsaw University, Warsaw, Poland
e-mail: przemek.u@hotmail.com

J. Fabiszak et al. (eds.), *Crossroads in Literature and Culture,* Second Language Learning and Teaching, DOI: 10.1007/978-3-642-21994-8_12,

reelection of Margaret Thatcher and the installation of 101 new Tory members of Parliament. The main protagonist of the novel, Nick Guest, is a 21 year-old former Oxford student who is about to start writing his doctorate on style in the 19th century novel, principally Henry James's work. Nick lives in the attic of the house owned by the rich and powerful parents of his school-friend, Toby. Toby's father, a Tory MP, Gerard Fedden, and Toby's mother, Rachel Fedden, are both from aristocratic and apparently very influential families (the list of quests at Rachel's brother's mansion had in the past included King Edward VII and Henry James). Nick himself, however, a son of an antique furniture dealer, is a shy, talented and introvert student from a lower middle-class family who aspires to a better social position. When invited to live with Toby's family, Nick tries by means of language, behaviour, attitude and mimicry to become an indiscernible part of the social class he is fascinated with and aspires to. At the same time, Nick is a closeted gay man and so he also hides the fact of his homosexuality whenever he is with people he suspects might frown upon it, including his close friends. As a consequence, he is almost constantly being entangled in a double strategy, or twofold tactics of sorts, of keeping up appearances. The reader thus becomes trapped in Nick's mind, contemplating his solipsist, if not paranoid, deliberations on what others might possibly make of his words and behaviour.

Nick seems to operate throughout the novel within the two worlds, or two different territories; the first one is the "public" sphere of his friendly yet somehow formal relationships with the Feddens and the outside world in general. This is the territory where constant pretending and strategic planning are a necessity. The other world, a more "private" world, comprises dates with other gay men, the underground gay life; the personal sphere of romantic relationships and sexual experiences, a territory within which Nick can be open about himself. The biggest effort on Nick's part is directed towards keeping the two territories separate and to guard their boundaries. It appears that the political climate of Thatcherism, especially its moral conservatism, is in the center of Hollinghurst's attention, providing a context for the novel's attempts at political satire.

2 The Closet and its Discontents

It might be considered a truism to say that all people live their lives within two considerably differing territories—the public and the private. Yet the fact that one of the basic components of one's personality, identity and life is regarded as socially unacceptable, as is (or, perhaps, was) the case with gay people, produces a remarkable effect whereby the two spheres become enormously at odds with each other. This orthodox separation between the public life and the intimate relations, primarily with reference to gay people, has, at least within English-speaking cultures, been most insistently described by way of the metaphors of "being in the closet" and "coming out of the closet". This self-guarded secrecy that constitutes the "closeting" of gay life has recently been scrutinized by numerous academic endeavors. Eve Kosofsky Sedgwick, in her *Epistemology of the closet*—a work often proclaimed to be one of the founding texts of the so-called "queer theory"—observed that

> the deadly elasticity of heterosexist presumption means that (...) every encounter with a new classful of students, to say nothing of a new boss, social worker, loan officer, landlord, doctor, erects new closets whose fraught and characteristic laws of optics and physics exact from at least gay people new surveys, new calculations, new draughts and requisitions of secrecy or disclosure (Kosofsky Sedgwick 1990, p. 68).

In this passage she seems to suggest that one is never fully "out of the closet" since, however open and straightforwardly honest one tries to be, every possibly significant encounter with a new person brings one back to the point of departure, i.e. the question whether to reveal the truth about one's "private" life or not. Consequently, epistemology of the closet is put into motion every time there is a new acquaintance by a self-perpetuating circle of uncertainty and reassessments of the previously adopted tactics. The impossible rationality of the closet is the irrationality of the two spheres, the private and the public, caught in-between a variety of irresolvable dialectic tensions. In *The line of beauty*, accordingly, there is, on the one hand, a pressure produced by Nick's ideal of the mutual exclusion of the two worlds and his boundary-drawing efforts to keep his "private" life away from his "official" life that he would preferably let others see as a record of more admirable achievements. On the other hand, his unspoken yearning to melt and unify the two worlds, i.e. to "come out of the closet", and be a consistent and openly honest person produces pressure in the opposite direction.

According to Zygmunt Bauman, whose discourse often focuses on uncertainty as a socially meaningful phenomenon, certainty—or the sense of certainty—can be defined as

> knowing the difference between reasonable and silly, trustworthy and treacherous, useful and useless, proper and improper, profitable and harmful (Bauman 1999, p. 17).

Certainty is a result of being able to trust one's own competence and reason and is thus tantamount to an ability to plan and act rationally. In his novel Hollinghurst seems, through the employment of constant deliberations on the part of the main protagonist, to show not only that the management of the secrecy of one's sexuality is a strategy that needs permanent operation but also that the very rationality of such a strategy is an impossibility. Hence, the sense of certainty is perhaps never to be accomplished. This is being proven throughout the narrative not without a touch of irony and a comic effect. During Toby's birthday party, after which Nick joins the group of friends smoking marihuana in Toby's room, he gets stoned and under the soothing influence of the drug discloses the truth about his sexual preferences and sexual life:

> Pot was a kind of truth drug for him—with a twist. He had an urge to tell, and show himself to them as a functional sexual being, but as he did so he seemed to hear how odd and unseen his life was, and added easy touches to it, that made it more shapely and normal (Hollinghurst 2004, p. 90).

What for Nick seemed a dangerous life-changing event was received by his friends with a light air of acceptance and positive irrelevance of the fact just revealed; his laboriously crafted strategy of pretending to be straight seemed to Nick, at that moment, violently contradicted, and for some time he was not comfortable with other people knowing his secret. He intuitively distrusted the positive reaction on the part of his friends.

Nick's inability to rationally decide whether it is better to reveal or to conceal his sexuality stems from the intangible nature of heterosexist presumption that assumes that all people are, or should be, heterosexual—its "deadly elasticity" meaning it is both an imaginary and real, "virtual" and "actual" assumption and amounting also to the simple fact that it may be Nick's imagined presumption more than his interlocutor's actual opinion. It also stems from the cultural relegation of (homo) sexuality to the sphere of private life, and thus to the sphere of presumably no social importance (or, a relegation of gay identity to homosexuality, i.e. to the intimate sphere of sexuality). Therefore, Nick's "confession" during the party has merely a private character and may simply be read as irrelevant since he continues to deny his sexuality before others. His sexuality becomes an open secret, silently acknowledged inside the family. The "open secret" metaphor, which in itself is a contradiction in terms, has also been frequently employed to describe the policing of the private/public boundary characteristic of the gay closet (Sinfield 2005, pp. 45–49). Homosexuality is therefore implicitly acknowledged but never explicitly accepted. Rachel never addressed the issue directly, but she offered Nick that he should bring his "special friend" home, thus implying she accepts his "friendships" but only as far as, outside his room, they remain simply "friends" (Hollinghurst 2004, p. 127). The paradox of this specific etiquette of the "open secret" is that gay subjectivity is simultaneously implied and bracketed, silenced, kept at a distance (Kosofsky Sedgwick (1990, p. 3) refers to it as a "learned ignorance" and a "speech act of silence"). Homosexuality is thus referred to by other characters in the novel usually by means of prolepsis. This term, known also as *praeteritio*, *preterition*, *cataphasis*, *antiphrasis,* or *parasiopesis*, denotes a rhetorical figure of speech wherein the speaker or writer invokes a subject only to deny that it should be even invoked. Rachel thus refers to homosexuality as an "unsafe and vulgar activity" which should better not be discussed (Hollinghurst 2004, p. 174), and this phrase reappears several times later on in the novel, as a label for Nick's secret activities.

As the novel progresses, more and more people become aware of Nick's sexual orientation, yet it is almost never discussed. The acceptance on the part of family and friends encourages Nick to be relatively more open. Throughout the second Chapter Nick goes to parties, meets new gay people, becomes more self-confident, but also engages in risky sexual practices and extensive use of drugs. *The line of beauty* seems to suggest that Nick becomes increasingly cynical and his coldly calculated attitude, by means of which he hopes to win it all, makes him seem desperate and entangled within the web of his own machinations. In this sense, *The line of beauty* seems to be a parodic opposite of the *Bildungsroman*: there is no chance for the main protagonist, being an anti-hero of sorts, to learn and become a mature adult since he is forced to focus on working on the illusion of his self rather than on his true self. The social consensus of the "open secret" teaches him that double standards are simply a profitable necessity.

Nick Guest's American cousin could, to some extent, be Fitzgerald's (2004) Nick Carraway, who is as much of an outsider almost admitted to the inner circle of his rich idols as Guest is. Indeed, *The Great Gatsby* seems to have been an inspiration as legitimate as Henry James' novels definitely were for crafting *The line of beauty*; both authors focused on social illusions of wealth, aesthetics and manners. Also Nick's surname—*Guest*—in quite an obvious way underscores the fact of him being, at the same time, inside and outside of social networks and ranks he aspires to. Being only a guest at Fadden's he does not really belong there, and he is, in a last stance, homeless:

> 'I know – we are homeless' Nick said.
> 'Homeless love' said Leo, shrugging and than cautiously nodding, as if weighing up a title for a song (Hollinghurst 2004, p. 117).

Nick's homelessness can be read in a number of ways. It denotes a lack of his own place but also the lack of a sense of belonging, either within his own family, within the Feddens family but also within the society at large. Nick posits himself in opposition to society and its values, not by means of critical engagement since, "officially", he tries to fit in, but rather through his antisocial behaviour (using drags, engaging in reckless sexual intercourse with strangers) in his "private" life. His homelessness is thus tantamount to his life as "open secret", to his Machiavellic cynicism and uncaring manner of life but also, perhaps first and foremost, to a lack of emotional engagement due to his sense of being an outcast, even if temporarily and provisionally accepted.

The whole story reaches its climax in a massive scandal when the press is informed that Nick organized gay sex-parties in the house owned by Gerard Fedden, already a prominent Tory MP at that time (ibid. 461–501). The scandal is thus an effect of breaching the line between the private and the public, of the open secret being no longer any secret, but something publicly acknowledged (it was mentioned in the press), something fully and unequivocally known. As long as there is some secrecy there is a controllable, albeit growing, circle of trust and some space for maneuver, denial, recantation. The closet, however, offers merely a fragile and dubious shelter against violence or discrimination. Being too much "out", being unexpectedly "outed" by the press against his will, Nick becomes defenseless. The Feddens decide, to Nick's surprise, that he needs to immediately leave their house since he betrayed their trust by allowing the secret they had been sharing together to be revealed (ibid. 482). What is satirized here is the hypocritical ideology of Thatcherism repulsively sweeping under the rug the emerging AIDS crisis. The double standard mocked by Hollinghurst is that the Feddens and their friends comfortably accepted Nick's homosexuality in the privacy of the house, yet they were appalled by the fact after it had been publicly revealed.

3 The Double-Curved Line

It seems the author wished *The line of beauty* to be as much about gay life as about style. Nick considers himself a true aesthete, always delighted when he spots a Guardi, Rembrandt, or Cezanne hanging in these aristocrats' living rooms. Style seems to be a decisive social factor in the novel as it is possessions, clothes, manners and ways of speaking that the narrative focuses on. For Nick the sense of taste seems to be a ticket not only to a Ph.D. degree but also, rather naively, to a higher social rank. Still, however, as Judith Butler reminds us: "Certainly, one can practice styles, but the styles that become available to you are not entirely a matter of choice. Moreover, neither grammar nor style are politically neutral" (Butler 1999, p. 19). Nick's tactics include mimicking and melting into the styles already existing and not questioning them. Whether he speaks or thinks, and however eloquent he appears to himself, his language remains caught in the socially prevalent mode of reasoning. Consequently, his tools for possible political action and ideological disobedience are limited. Thus, in order to escape the heterosexist presumption, he chooses, instead of openly opposing it, to live his "personal" life within the confinements of secrecy. Ironically, even though Nick is surrounded by influential politicians, he remains utterly powerless.

The title of the novel refers to William Hogarth's concept of the double-curved, serpentine, s-like line, which Hollinghurst describes as the snakelike flicker of an instinct, of two compulsions held in one unfolding moment (Hollinghurst 2004, p. 200), and which Hogarth originally defined in his influential treatise on aesthetics, *The analysis of beauty* (Hogarth 1997, pp. 48–59). The 18th century British painter, to whom Henry Fielding referred as a "Comic History-painter" (Bindman 1998, p. 107), has been applauded for sharp precision in depicting social nuances through a cannonade of his vivid "Characters and Caricatures". Like Henry Fielding (1993) in *Tom Jones*, Hollinghurst refers to Hogarth several times throughout his novel, notably when Nick employs the concept to describe his lover's body: "He [Nick–P. U.] didn't think Hogarth had illustrated this best example of it, the dip and swell—he had chosen the harps and branches, bones rather than the flesh. Really, it was time for a new analysis of beauty" (Hollinghurst 2004, p. 200).

Had Hogarth lived after Freud and Foucault, he probably would make his *Harlot's progress* include some sexually-charged nakedness alongside caricaturized social costumes. Hollinghurst, in this double, opposable move of the line of beauty, combines subtle yet comic social satire with erotic undertones and overt, if not bold, descriptions of gay sex in order to impressively highlight the risky tension between the "public" and the "private", which perhaps still trembles potently in the context of gay experience.

4 Conclusions

It seems that Hollinghurst's novel by taking us into the social and political reality of very recent times is asking how much has changed through the course of the last two decades. "The closet is the defining structure for gay oppression in this century," Eve Kosofsky Sedgwick wrote in 1990, thus seeing "the closet" as historically contingent and not a universal condition of gay life (Kosofsky Sedgwick1990, p. 71). The closet, as a theoretical concept, grasps the self-perpetuating structure of gay oppression which powerfully reproduces itself exactly because it is, at the same time, an often "reasonable" and often absolutely necessary strategy of survival within the framework of the very oppression it constitutes. It is, therefore, a sociologically lethal strategy; a strategic invisibility that, every time and in a long run, reinforces the "spell" of heterosexist presumption and the private/public boundary. Moreover, it forces gay people to constantly make impossible choices between secrecy and disclosure, the consequences of which remain simply unpredictable. Whether the 21st century will see an end to the closeting of gay life is a difficult question the importance of which *The line of beauty* seems to be highlighting in its own, subtle and slightly satirical, way. "Hollinghurst went to the same college as Oscar Wilde. But he inhabits another world", Stephen Moss (2004), a *Guardian* reviewer, optimistically stated.[1] It seems that homosexuality has indeed found its place in a more open debate, both academic and political, thus increasingly putting into question the legitimacy of the gay closet. Writers such as Hollinghurst now openly interrogate the social and political implications of homosexuality with psychological depth and frankness so appreciated by the Booker Prize committee.

References

Bauman, Z. 1999. *In search of politics.* California: Stanford University Press.
Bindman, D. 1998. *Hogarth.* New York: Thames and Hudson.
Butler, J. 1999. *Gender trouble. Feminism and the subversion of identity.* New York and London: Routledge.
Fielding, H. 1993. *Tom Jones. A foundling.* Vol. I and II. Ware: Wordsworth Classics.
Fitzgerald, F. S. 2004. *The Great Gatsby.* New York: Scribner.
Hogarth, W. 1997. *The analysis of beauty.* New Heaven and London: Yale University Press.
Hollinghurst, A. 2004. *The line of beauty.* London: Picador.
Kosofsky Sedgwick, E. 1990. *Epistemology of the closet.* Los Angeles and Berkeley: California University Press.
Moss, Stephen: 'I don't make moral judgments'—an interview with Alan Hollinghurst, Guardian UK, http://www.guardian.co.uk/books/2004/oct/21/bookerprize2004.bookerprize (April 14, 2010).
Sinfield, A. 2005. *Cultural politics—Queer reading.* London and New York: Routledge.

[1] http://www.guardian.co.uk/books/2004/oct/21/bookerprize2004.bookerprize (April 14, 2010).

What Haunts Hundreds Hall? Transgression in Sarah Waters' *The Little Stranger*

Barbara Braid

Abstract The latest novel by Sarah Waters, published in 2009, has received critical acclaim for its realistic representation of post-war Britain, its affiliation with great gothic classics, such as *The Turn of the Screw* or The Fall of the House of Usher, and above all, its ambiguous ending. However, it is a bit of a 'shorthand' to call the novel gothic, as has been admitted by the author herself in a video interview taken by Rebecca Lovell in May 2009. Instead, Waters calls the book "a haunted house novel." The unanswered question of the novel is, what is it that haunts Hundreds Hall? The paper makes an attempt to provide one possible answer to this question, focusing on the idea of transgression. Of course in any gothic work of fiction the occurrence of a supernatural element signals a crossing of the borderline between the possible and the impossible. A fantastic entity questions the stability of this frontier. In Waters' novel, however, the transgression of the real signals another crossing of boundaries. The unreliable narrator of the novel is the crux of its interpretation. The enigmatic events in the house coincide with Dr Faraday's growing attachment to the Ayreses family. He is an educated man who, on the one hand, no longer fits into the working class of his parents, but still, due to his origins, feels insecure in the upper class to which the inhabitants of Hundreds Hall belong. His constant roaming of the class boundaries is the key to the haunting mystery. The paper is going to analyse different images and symbols of transgression in the novel to show the connection between the issue of class and the gothic elements in the book and to provide a possible interpretation of its equivocal ending.

B. Braid (✉)
Szczecin University, Szczecin, Poland
e-mail: barbara.braid@gmail.com

J. Fabiszak et al. (eds.), *Crossroads in Literature and Culture,* Second Language Learning and Teaching, DOI: 10.1007/978-3-642-21994-8_13,

1 Introduction

The latest novel by Sarah Waters, published in May 2009, received critical acclaim for its realistic representation of post-war Britain and its affiliation but yet a clever reworking of great gothic classics. It provides an illustration of the social changes in society of the 1940s, with a new Labour Party government and changing perspectives on the issue of class. Still, above all, for many readers it is a novel of suspense—a spooky ghost story. The unanswered question of the novel is, what is it that haunts Hundreds Hall? The analysis presented below attempts to provide one possible answer to this question. It is based on the observation that the motif of transgression pervades in the novel, and even though at first glance the issue of class might not seem to explain the supernatural events in the novel, it actually holds the key to the mystery of the story.

2 Transgression and the Haunting

In his book on transgression, Jenks (2003) draws our attention to the fact that it is one of the most potent features of postmodernism. In contrast with pre-modern epochs, which promoted convention and therefore often oppression, modernism and even more clearly postmodernism express "the desire to transcend limits—limits that are physical, racial, aesthetic, sexual, national, legal and moral" (Jenks 2003, p. 8). Transgression is about crossing the boundaries, breaking laws and rules, revealing taboos, "violating and infringing" (Jenks 2003, p. 2). Still, paradoxically, by an attempt to break the law, it also attracts attention to it, to the Either/Or logic of Western civilisation and the binary oppositions of the sacred/the profane, open/taboo, sane/mad, real/unreal, normal/abnormal, good/evil, etc. that it is supposed to transcend. Therefore, "transgression is a deeply reflexive act of denial and affirmation" (Jenks 2003, p. 2).

Even though it has been said that transgression is promoted by postmodernism, the theme of crossing the limits with all its consequences can be found in literature since the Original Sin. Gothic literature especially has been characterised by its fascination with breaking the rules, from the laws of physics, through sexual taboos, to criminality. Of course, gothic fiction includes transgression not only in the form of supernatural protagonists it might present, but maybe more importantly, in the law that might be broken or threatened in the course of action, by attempting or committing murder, rape, imprisonment, usurpation of power, etc. It revels in revealing and crossing social, moral or sexual taboos or any other kind of excess. In the eighteenth century gothic romance, transgression is not only an "excess and an ambivalence" (Botting 1996, p. 8), but also a warning of its dangers for the community and social order. Paradoxically, it can be said that the pre-Romantic gothic novel both glorified the transgression and reasserted respected social values (Botting 1996, p. 9). On the other hand, the nineteenth century gothic focused less on a transgression of social laws, but rather on an expression of internalised guilt, anxiety and despair; therefore, the transgression transformed from a social one to an individual one. Stress was put on the transgression of the limits of one's psyche and surrounding reality (Botting 1996, pp. 10–11). To sum up, if we follow the definition of gothic fiction proposed by Anne Williams (1995) in her seminal work *Art of darkness*, which presents a gothic novel as a an "escape fantasy" of a "family romance," then the genre itself "calls attention to the importance of boundaries: the literal and figurative (...), the principles of the elaborate cultural system Lacan called 'the Law of the Father'" (Williams 1995, p. 12). The lines and walls created by the Law of the Father both provoke the possibility of transgression and suggest the proper punishment for those rebels who cross them, who "go too far" (Williams 1995, p. 12).

Obviously, in any gothic work of fiction the occurrence of a supernatural element signals a crossing of the boundary between the possible and the impossible. A fantastic entity questions the stability of this frontier. The idea of haunting, or coming back from the dead involves crossing the threshold between the world of the living and the dead. It has been interestingly presented in vampire folklore: a vampire cannot enter a house unless invited, and a threshold of the world of the dead that it refuses to cross is then symbolised by any kind of borderline—a river or a stream—that it cannot cross in this world (Fader 2000, pp. 88–89). A ghost, too, crosses the boundaries of the afterworld and the here-and-now of the living; it is an ambiguous being, both here and not here, hard to define or capture. "The spectre is an absent presence, a liminal being that inhabits and gives shape to many of the figurations of trauma that characterise the Gothic" (Smith 2007, p. 147). What is characteristic in the interpretation of haunting in gothic fiction is the perception of spectres as "internalized images of other people" (Castle 1995, p. 125), ourselves or our (collective) past. A ghost crosses not only the boundaries of the real and the unreal, but also between the self and the other, or the self and the hidden, subconscious self. "Ghosts are (...) projections of our innermost anxieties and (...) blurring of physical and psychological realities" (Smith 2007, p. 148). What we cannot consciously face or come to terms with in our past comes back to haunt us. As "the energies that haunt us are very much a part of our own selves [and] (...) it is hard to establish who is being haunted and who is doing the haunting" (Cavallaro 2002, p. 61), it is of utmost importance for a postmodern writer or/and reader of the gothic (and not only) to examine what haunts us; in Derrida's (2006, p. 221) words, we need to "let [the ghosts] speak or (...) give them back speech, even if it is in oneself, in the other, in the other in oneself: they are always *there*, spectres".

3 The Fall of the House of Ayres

Sarah Waters' latest novel is haunted by an unnamed spirit. The characters and their relationships are awkward, but the haunting itself is awkward, too. It lacks the smoothness of Stephen King or Clive Barker, where a reader can easily slip into the convention like a well-fitting glove. Here the ghost is doubted and almost unnoticed at first; it turns from a poltergeist, through a presence of a child, into a materialisation of a murderous power.

The novel is narrated by a middle-aged general practitioner, Dr Faraday, who starts telling the story of Hundreds Hall, a Georgian mansion somewhere in Warwickshire, with a description of an afternoon from his childhood when he first sees the interior of the house while accompanying his mother, a nursery maid, and falls in love with it. Almost 30 years later, he finds himself in Hundreds again, on a house call to a maid who turns out to be spooked out by the old house. As in any classic ghost story, the 14 year-old servant and the family dog are the first to feel and be affected by the mysterious presence. Dr Faraday befriends the Ayreses—the owners—and observes, in turn, how the son, Roderick, goes mad after experiencing the objects in his room move of their own accord and a fire which starts mysteriously in his bedroom; how the mother, Mrs Ayreses, commits suicide under the influence of what she believes is her baby daughter's ghost; and finally, how the daughter of the family, Caroline, dies in a suspicious accident after she breaks an engagement with the doctor. At the end of the novel, Faraday and the house stand alone, deserted, the true lovers from the beginning to the end of the story.

Being a postmodern novel, *The little stranger* invites different images and aspects of transgression. The very first scenes of the novel present a misdemeanour committed by the narrator in his childhood. When he first sees the house, he is so taken with its beauty that he decides to take a piece of it with him. Untypically for himself, he commits an act of vandalism by breaking off a piece of stucco. One might wonder if that seemingly insignificant act of destruction is not what sets the house onto the road to total dereliction and ruin (Mantel 2009). The event is the first symbolic transgression in a series of escalating incidents. The decay of the house goes hand in hand with the destruction of the Ayres family. It is at first done by the selling of the land by the family impoverished in the war. Another symbol of transgression ensues: the wall dividing Hundreds Hall and its park from the town is torn down, and in its place council houses for working class families are built. What had been a borderline between lower and upper classes is no longer there; the existing class divisions are not valid anymore. While there is financial trouble looming over the hall, a more sinister transgression takes place inside the house: soon its inhabitants are hit by madness, suicide and possibly murder, typical plagues haunting gothic fiction.

Is *The little stranger* a gothic text as well? As it has been admitted by the author herself in a video interview taken by Rebecca Lovell in May 2009, it is a bit of a "shorthand" to call the novel a ghost story. Instead, Waters calls the book a haunted house novel: "there is something haunting Hundreds Hall" (Lovell 2009). The idea is, according to Waters, for the readers to make their own mind about what it is that constitutes the supernatural element. Still, the reviewers notice several references to classic gothic literature: *The turn of the screw* (Mantel 2009), *The yellow wallpaper* (O'Brien 2009) or The fall of the house of Usher (Charles 2009). It shares with them a building suspense, ghostly messages, and ambiguity of those unusual occurrences, which, as Cavallaro puts it, "is a basic principle of the rhetoric of haunting" (2002, p. 65). Like many gothic texts, this novel also makes use of the technique of an unreliable narrator, but one with a twist; in opposition to Henry James' governess, Dr Faraday is not too ready to accept a supernatural phenomenon, but actually he struggles against all odds to find a rational explanation to more and more irregular happenings at Hundreds Hall. Just as an alert reader might grow suspicious of the determination with which the governess tries to convince them of the existence of ghosts, they might also accuse the narrator of *The little stranger* of being blind to the obvious signs of a haunting. His scepticism and tortured rationality (Mantel 2009) are exactly what make him an unreliable narrator in this gothic novel.

4 Crossing the Line of Class

Dr Faraday's constant roaming of the class boundaries holds the key to the haunting mystery. The enigmatic events in the house coincide with his growing attachment to the inhabitants of Hundreds Hall. He is an educated man who, while no longer fitting into the working class of his parents, still, due to his origins, feels insecure amongst the upper class to which the Ayreses belong. His attitude to his class is at best ambivalent. On the one hand, he feels a descendant of the working class and loyal to the ethos of honest labour that his parents taught him. In one of the first scenes of the novel, he is indignant when listening to Caroline and Roderick poking fun at the servants they used to have in the house; he comments: "[p]erhaps it was the peasant blood in me, rising" (Waters 2009, p. 27). In his thoughts, he criticizes them for their superiority over the kind of people among whom Faraday's mother was also included:

> But Hundreds Hall had been made and maintained, I thought, by the very people they were laughing at now. After two hundred years, those people had begun to withdraw their labour, their belief in the house; and the house was collapsing, like a pyramid of cards. Meanwhile, here the family sat, still playing gaily at gentry life, with the chipped stucco on their walls, and their Turkey carpets worn to the weave, and their riveted china (Waters 2009, p. 27)

He makes the exodus of the servants from Hundreds Hall, and therefore the class changes after 1945, responsible for this decay. Even though he feels flattered by the familiarity and friendship the Ayreses offer him, their disrespect towards their former servants strikes a painful cord. What is surprising about this scene, contrary to his expectations and perhaps a stereotype of higher classes, it is not the Ayreses who make the class division between themselves and Dr Faraday visible, but it is the doctor himself; by mentioning his mother who had been a nursery maid in Hundreds Hall, he makes the difference between their class origin evident, and creates a possible obstacle for the budding friendship between himself and the Ayreses. However, even though there might be a hint of superiority in his behaviour towards the Ayreses whom he sees for a split second as spoilt and decadent, most of the time he desperately tries to fit in at Hundreds. He feels privileged and proud when he is invited to the house for a party and a Christmas dinner; when the neighbours gossip about him and Caroline Ayres coming together for a district hospital dance, he thinks boastfully: "[w]hy the hell shouldn't I dance with the squire's daughter, if the squire's daughter wanted to dance with me?" (Waters 2009, p. 265). There is something wonderful, to his mind, in the fact that within the span of time of one generation the same blood which used to be the Ayreses' servants turns into their friend, confidante, and later even a kin-to-be. It is metaphorically represented by the scene in which, sitting at an old kitchen table, he remembers one afternoon in his childhood when he visited Hundreds Hall with his mother: "I certainly never guessed then that I'd be back here one day, quite like this. I dare say my mother never guessed it, either. A pity she didn't live to see it" (Waters 2009, p. 432). Becoming a doctor is certainly a sign of social mobility, but it does not make him proud; still, potentially becoming a part of Hundreds Hall is certainly something that would make his parents' sacrifice worthwhile.

The ambivalence of his class identity is clearly visible in his relationship to his cousin Tom Pritchett, who is one of the builders constructing the council houses in the vicinity of Hundreds Hall. When Faraday and Caroline come to visit the building site, an awkward situation occurs: even though Dr Faraday knows Tom since childhood, there is an invisible wall between them; Faraday notices that "it was queer for him to call me 'Doctor', but out of the question, too, for him either to use my Christian name or to address me as 'sir'" (Waters 2009, p. 248). Faraday is neither working nor upper class; his origins put him on the equal level with Tom, but due to his education Faraday might have more in common with Caroline Ayres than his cousin. This meeting prompts the doctor to tell Caroline about his parents and how they worked themselves virtually to death so that his natural aptitude could earn him qualifications and a highly skilled profession. However, it also earned him a feeling of guilt for being ashamed of his parents and his working class origins; he notices bitterly: "[t]hey'd have been better off, really, if I'd been like my cousins—like Tom Pritchett back there. Maybe I'd have been better off, too" (Waters 2009, p. 250). Thus, being at the same time proud of his achievements and his social mobility, Faraday seems to be also tormented by it, culpable of rejecting his roots and at internal conflict about his social identity. He becomes yet another image of transgression—a crossing of class division, both a working class and an educated class member, but also neither of them. He is like a ghost, crossing the borderlines not as much of the world of the living and the dead, but of the working and upper class, "an anomalous, shadowy presence" in Hundreds Hall (Thomas 2009).

As Sarah Waters herself mentioned in her interview for *the Guardian*, the 1940s in Britain seemed an exciting time for the working class, with the Labour government and such new institutions as the NHS. At the same time, however, the landed gentry felt that the world they had known was disappearing; they felt in peril (Lovell 2009). This feeling of a threatening menace, of a known world slipping away is also what the Ayreses experience. It seems that struggling with the ghostly presence in Hundreds Hall is a metaphor of fighting inevitable changes that cannot be stopped, but which the Ayreses cannot accept and do not adjust to.

5 Our Ghosts, Ourselves

It has been said already that the ghosts and spectres in the gothic fiction can be a representation of our consciousness haunting us. A similar view is expressed in the novel by Faraday's colleague, Dr Seeley, who suggests a different explanation of things happening at Hundreds Hall—madness; it is a convenient explanation used by Faraday in all cases. He claims that the mysterious phenomena can be explained by a subliminal mind, a shadow self, "a Caliban, a Hyde, (...) a little stranger" (Waters 2009, p. 380). There is something Freudian in this suggestion; "the self is ghosted by the subconscious" (Smith 2007, p. 148); the hidden desires, frustrations and fears create an energy which becomes a poltergeist hunting the inhabitants of the house. Seeley traditionally points at women at the house who could be responsible for unleashing the telekinetic powers: Mrs Ayres, who as menopausal is in "a queer time, psychically" (Waters 2009, p. 380) and the teenage servant Betty. It brings to mind the Victorian discourse of female madness, where women at important turning points for the development of their biology—puberty, pregnancy and childbirth, menopause—are at the greatest danger of plunging into madness (Showalter 1987, p. 55). In this discourse, however, it is a spinster who is the most dangerous as the unregulated one, who fails to fulfil the holy duties of a woman—to be a wife and a mother (King 2005, p. 22). This way of perceiving single women is echoed in Seeley's and Faraday's discussion, where it is suggested that Caroline might be the root of the *miasma* haunting the family (Waters 2009, p. 381).

Dr Seeley claims that the haunting might be caused by an energy "motivated by all the nasty impulses and hungers the conscious mind had hoped to keep hidden away" (Waters 2009, p. 380). Looking at the main characters living in the house, we can see that most of them have feelings and frustrations kept locked up in their subconscious that could have been the source of the poltergeist. Roderick is obviously physically and mentally crippled by the war; he became a bitter, cynical shadow of his old self. He struggles with preserving Hundreds Hall as its owner and squire, but he fails miserably. His feeling of responsibility for his family is later reflected in his haunting; when it culminates with a fire in his bedroom, Roderick readily agrees to be locked up in a mental hospital in order to protect his mother and sister from the spectral presence, a germ that might infect others (Waters 2009, p. 155). Interestingly, Roderick rightly notices that the ghostly presence has its roots in himself:

> Last night I heard noises. I thought there was something at the door, something scratching, wanting to get in. Then I realised that the noise was inside me, that the thing that was scratching was inside me, trying to get out. It's waiting, you see (Waters 2009, p. 225).

Initially, Roderick hopes that removing himself from Hundreds Hall is going to protect his family; and so his feeling of responsibility is used against him.

As it has been said before, an interesting part of the haunting in Hundreds Hall is the fact that it is a different phenomenon depending on the haunted. Therefore it seems that the spectre comes from within the members of the Ayres family, as a reflection of their deep emotions and anxieties. After Roderick ends up in an asylum, Mrs Ayres is tormented by what she gradually recognises as the presence of her deceased baby daughter Susan, her oldest, and by her own admittance, most beloved child (Waters 2009, pp. 219–220). Mysterious sounds fill the house, and cryptic scribbles appear on the walls, first just the letter 'S,' then the name Sucky. In a spooky climax of this haunting, Mrs Ayres gets locked up in her daughter's old nursery room, with a speaking tube emitting strange noises. Finally, Mrs Ayres admits to seeing and hearing the ghost of her daughter who encourages her mother to join her and thus the mother commits suicide to be with her baby. It may be said that her pain after the loss of her daughter as well as the guilt that she had not loved her other children as much as Susan might have materialised in a form of a ghost.

Caroline, on the other hand, does not seem to be affected so much by the ghosts. The haunting does not concern her, she is an observer and also hears and sees strange sounds and signs. However, until the very end of the novel, she does not seem to be the intended recipient of the bizarre phenomena. Dr Faraday and Caroline manage to strike a relationship, an awkward and fragile one in the spooky and sinister circumstances, but after her mother's death Caroline breaks off the engagement and decides to sell Hundreds Hall and leave the country. In his anger Faraday forms an accusation:

> A year ago, what did you have? A house you claimed drained all your time. An ageing mother, a sick brother. What was your future? And yet, look at you now. You're free, Caroline. You'll have money, I suppose, once Hundreds is sold. It seems to me, you know, that you've done really rather nicely (Waters 2009, p. 465).

Yet, it soon turns out that Caroline is not a perpetrator, but also a victim of whatever haunts Hundreds Hall. Just before she is ready to leave, Caroline falls off the stairs at night after what seems to be an encounter with "the little stranger." Later in court Betty testifies that Caroline called out "You," "as if she had seen someone she knew, (…) but as though she was afraid of them" (Waters 2009, p. 483). Betty's interpretation is that the ghost inhabiting Hundreds Hall wants to stop Caroline from leaving the house. Still, as much as the interpretation of the Ayreses being haunted by their own selves rings true for Roderick and Mrs Ayres, in Caroline's case there does not seem to be any hidden desire, frustration or guilt that might have fed the spectre. It seems rather that Caroline has been punished for her rejection of mainstream sexuality and marriage.

6 Who is the Little Stranger?

While one can agree with the diagnosis posed by Dr Seeley, it does not seem right in the case of all three members of the Ayreses family. It also might be a significant limitation to presume that the source of the haunting comes from a woman. There is also another character in the story who feels strongly about the family and the house—strongly enough to unleash subliminal powers to achieve their aims. Yet, as a first person narrator Dr Faraday escapes our judgement, at first at least. If we look into details of the haunting, however, we will see that there is a lot of evidence that Caroline was not the only one who could gain something from the elimination of the Ayreses. Faraday himself admits that he admired the house since their first "meeting;" when he took a plaster acorn from Hundreds' walls, it was an act of possession: "I was like a man, I suppose, wanting a lock of hair from the head of a girl he had suddenly and blindingly become enamoured of" (Waters 2009, p. 3). The need to possess the house is the desire to take over what might symbolically signify higher classes and the old social order. On a metaphorical level, his influence over the Ayreses and their destruction might indicate the destruction of the pre-war class system, taking over the higher class by the working class. Of course, his menace to the house and the family is not a conscious one, so we need to find evidence for it in the events of the novel.

Using the knowledge of the weaknesses that trouble the Ayreses he gets from becoming their doctor and friend, Faraday seems to be using them against the family, to get rid of those who might stand in his way of possessing the house and becoming its inhabitant. His relationship with Caroline may therefore serve a purpose: to marry into the family. The threatening and potentially dangerous presence of the doctor is at first noticed by Roderick:

> Get your hands off me! Don't you f—g well tell me how to behave! That's all you ever do. And when you're not doling out your doctor's advice you're making a grab at me, with your filthy doctor's fingers. And when you're not grabbing you're watching, watching, with your filthy doctor's eyes. Who the f—g hell are you, anyway? Why the hell are you here? How did you manage to get such a footing in this house? You're not a part of this family! You're no one! (Waters 2009, p. 197)

It might not be a coincidence, therefore, that the same night Roderick utters these words the fire which could have killed him breaks out in his bedroom. Finally, when Caroline who was supposed to be the key to Hundreds Hall refuses to marry Dr Faraday, she has to be punished too. The ghostly presence she recognized at the top of the stairs could have been a spirit or a manifestation of the doctor's self. Interestingly, each time a haunting happens Faraday is not in the house; most often he is shown as asleep. Significantly, that last night he is driving to Hundreds Hall to talk to Caroline once again, but decides against it and falls asleep in his car, not far from the hall:

> And in the slumber I seemed to leave the car, and to press on to Hundreds: I saw myself doing it, with all the hectic, unnatural clarity with which I'd been recalling the dash to the

> hospital a little while before. I saw myself cross the silvered landscape and pass like smoke through the Hundreds gate. I saw myself start along the Hundreds drive.
> But there I grew panicked and confused—for the drive was changed, was queer and wrong, was impossibly lengthy and tangled with, at the end of it, nothing but darkness (Waters 2009, p. 473).

Is it possible that his other self, his Hyde, went to Hundreds Hall to punish his fiancée for rejecting him? The final piece of evidence is visible in the ending of the novel. Hundreds Hall is not sold after all and Faraday having still kept his old key, keeps visiting the place from time to time. He hopes to see "the little stranger" there and solve the mystery of Caroline's death. Sometimes he thinks he can see something in the corner of his eye, just to realize that: "[what he is] looking at is only a cracked window-pane, and that the face gazing distortedly from it, baffled and longing, is [his] own" (Waters 2009, p. 499). It seems possible that what Dr Faraday fails to realise is that this face, his own face, is actually the true little stranger he is looking for.

7 Conclusion

The image of transgression is thus a crucial one for the understanding of the novel and its central conflicts. Dr Faraday's transgression of the old, pre-war class divisions makes it possible for him to possess his "beloved," Hundreds Hall. The transgression of bodily constraints, an unconscious creation of "the little stranger" leads to more transgressions which follow the gothic convention of a family romance: madness, suicide and murder. The opening and the closing scenes are the frame which in itself leads us back to the idea of transgression too: the boy who steals a stucco acorn from the wall of the mansion is still haunting Hundreds Hall, his face reflected in a cracked window pane; the little stranger who finally managed to seize the house for himself.

References

Botting, F. 1996. *Gothic*. London and New York: Routledge.

Castle, T. 1995. *The Female thermometer. Eighteenth-century culture and the invention of the uncanny*. Oxford: Oxford University Press.

Cavallaro, D. 2002. *The Gothic vision. Three centuries of horror, terror and fear*. London and New York: Continuum.

Charles, R. 2009. "This old house." Rev. of *The little stranger*, by S. Waters. *Washington Post.* <http://www.washingtonpost.com/wp-dyn/content/article/2009/05/19/AR2009051903162.html>. Accessed 17 April 2010.

Derrida, J. 2006. *Spectres of Marx. The state of the debt, the work of mourning and the new international.* Trans. Peggy Kamuf. London and New York: Routledge.

Fader, S. 2000. "Invite me in!"*Parabola* 25: 88-89.

Jenks, C. 2003. *Transgression*. London and New York: Routledge.

King, J. 2005. *The Victorian woman question in contemporary feminist fiction*. Basingstoke: Palgrave Macmillan.

Lovell, R. “Sarah Waters on *The Little Stranger*.” 2009. *The Guardian*. <http://www.guardian.co.uk/books/video/2009/may/20/hay-festival-sarah-waters>. Accessed 6 April 2010.

Mantel, H. 2009. “Haunted by shame.” Rev. of *The Little Stranger*, by S. Waters. *The Guardian*. <http://www.guardian.co.uk/books/2009/may/23/little-stranger-sarah-waters>. Accessed 30 Dec 2009.

O’Brien, S. 2009. “Sarah Water’s little stranger.” Rev. of *The Little Stranger*, by S. Waters. *The Times literary supplement*. <http://entertainment.timesonline.co.uk/tol/arts_and_entertainment/the_tls/article6387134.ece>. Accessed 31 Dec 2009.

Showalter, E. 1987. *The female malady. Women, madness and English culture, 1830-1980*. London: Virago Press.

Smith, A. 2007. “Haunting.” Spooner, Catherine and Emma McEvoy, eds. *Routledge companion to Gothic*. London and New York: Routledge.

Thomas, S. 2009. “House calls.” Rev. of *The Little Stranger*, by S. Waters. *New York Times*. <http://www.nytimes.com/2009/05/31/books/review/Thomas-t.html?_r=1>. Accessed 17 April 2010.

Waters, S. 2009. *The little stranger*. London: Virago.

Williams, A. 1995. *Art of darkness. A poetics of Gothic*. Chicago and London: The University of Chicago Press.

Crossing the "Gender Frontier": Cross-dressing and Male Impersonation in Sarah Waters' *Tipping the Velvet* (1998)

Malwina Degórska

Abstract Culturally accepted gender roles are what we are born into, therefore from early childhood we are being taught how to find our way within specific gender boundaries. Those who decide to transgress these set limits consciously cross what Kathleen M. Brown (1993) called a "gender frontier"—the borderline where two culturally specific systems of knowledge about gender and its nature meet. During the Victorian Era, there were two main ways in which a woman could cross this frontier: first, and commonly accepted one, was to become a male impersonator and perform on stage; second, and a widely ostracized one, was to cross-dress in real life and pass as a man. Even though both ways stemmed from the idea of one sex dressing in the clothes of another, the repercussions of the undertaking of the latter included long imprisonment and years of hard labour. In her debut neo—Victorian novel *Tipping the velvet* (1998), Sarah Waters presents the story of a protagonist who managed to cross the gender frontier twice, both as a male impersonator and as a cross-dresser. The focus of this paper is therefore on the means and processes of the heroine's transgression of socially accepted gender roles in 19th century London as presented by Waters in her novel.

1 Introduction

In her article entitled "Brave new worlds: Women's and gender history" (Brown 1993) historian Kathleen M. Brown coined the term gender frontier defining it as the borderline where two culturally specific systems of knowledge about gender

M. Degórska (✉)
Szczecin University, Szczecin, Poland
e-mail: mdegorska@gmail.com

J. Fabiszak et al. (eds.), *Crossroads in Literature and Culture*, Second Language Learning and Teaching, DOI: 10.1007/978-3-642-21994-8_14,

and its nature meet. This clash of gender systems proved that "cultural differences in gender divisions of labour, sexual practices and other signifiers of gender identity such as clothing or hair significantly influenced how European and indigenous peoples perceived each other" (Brown 1993c, p. 318). The theory, even though intended for historical study of gender, may be well applied to modern-day research on gender variance as presented in literature. But before approaching the subject of this paper, two clarifications are necessary at the outset.

Firstly, in order to use Kathleen Brown's idea of the gender frontier for purposes of literary analysis one must first define both gender and sex and explain possible implications of transgressing the former one. In presenting my point of view I shall start with sex, as it is the active agent underlying human behaviour, traits, and decisions. Sex is also the reproductive category in which we are all placed at birth; it defines both our biology and our belonging to the well established and socially accepted binary male/female division. Contrary to sex which is a biological notion, gender is a social construct based on assigned behaviour and traits, and just as the dichotomy of being either male or female defines our sex, the socially accepted binary nature of gender, being feminine or masculine, defines our role within society and culture.

Secondly, anyone who decides to transgress gender boundaries, crosses Brown's gender frontier, enters the group of the transgender and by doing so stakes out new territories somewhere in between safe notions of male and female binary. Therefore, the term transgender incorporates those who undergo medical procedures to change their sex because of their gender identity, as well as those who cross-dress in private or onstage without the need of sex change. Examples of cross-dressing which are of my interest in relation to Sarah Waters' novel include both the 19th century male/female impersonators and modern-day drag queens/kings respectively. Therefore, in the light of the transgender phenomenon, Victorian society's notions of gender roles and the use of the concept of queer identity, this article will focus on cross-dressing and male impersonation as presented in *Tipping the velvet* (1998).

2 Queer Identity as Performative

Recent theorizing about queer identity has relied on an identitarian paradigm torn between two opposing views as its foundations. Essentialism, which grew out of 1970 feminist theories on sexuality, presumes that sexual attraction lies in person's nature and that homo/heterosexual dichotomy has remained unchanged through history and across cultures (Blackburn 1996). The opposing view is the so-called social constructionist stance, which draws heavily from Judith Butler's idea of the performative nature of gender. It states that gay/straight binaries reveal not only one's sexuality and gender but also politics and psychology. Their meaning is constant only for a time because of the changeable nature of social beliefs (Foucault 1978). The role of performativity in creation of queer identity, proposed by Butler in *Gender trouble* (Butler 1990), presents gender as a construct which is performed and not integral to our nature, where performing of gender means producing a convincing image of a woman or a man through repeated performative acts of "corporeal signs and other discursive means" (Butler 1990c, p. 136). In this respect, social constructionism perceives gender as the result and not the cause of one's identity. Such understanding of gender combined with the existence of myriad of queer identities, challenges the essentialist claim of sexual orientation as being fundamental to human nature.

The performative nature of gender becomes obvious with cross-dressing because gender-specific clothes mark stable categories that allow the outside world to read both our sex and gender. Any significant disruption in such sex-gender coding misinforms the viewer and deconstructs safe boundaries our culture has been taught to operate within. Accordingly, the primary cultural symptom of cross-dressing was causing anxiety by openly threatening the stability of social hierarchy (Garber 1992, p. 223), which later culminated in social ostracism aimed at those who dared threaten the collective idea of stable gender categories.

For this reason, in the mid 19th century, the only socially accepted female-to-male cross-dressing practice, as an act of crossing the gender frontier, was undertaken in theatre in a form of male impersonation since it:

> appealed to two audiences simultaneously: male and female theatergoers in the Gallery—working-class people who loved to laugh at their social 'betters'—and the aristocratic male patrons in the Boxes and the Orchestra Stalls, who could well afford (in every sense) a little harmless fun at their own expense (Stetz 2004, p. 82).

In order to reinforce the safety of their gender-bending acts, male impersonators of that time played only certain, assigned by their male directors/managers roles, such as: the Champagne Charlie—an epitome of the Victorian Dandy, the soldier or the schoolboy (Rodger 2002, pp. 105–133). These parts did not allow for a variety of possible interpretations as they were not empowered enough to break the Victorian image of a man. Consequently, they were in complete congruity with the audience's expectations and assumptions about the music hall and gender roles performed there. In comparison, female-to-male cross-dressing outside the theatre walls was considered as manifestation of transvestism—an act of conscious and deliberate crossing of the gender frontier. Such transgression resulted in threatening gender hierarchy and turned the transvestite into a queer. Thus for the sake of cultural gender stability, female-to-male transvestism was "forbidden on the street" (Wilson 2006, p. 294). Consequently, breaking that rule was a violation of not only the social but also judiciary and church laws since the Victorians closely followed the Biblical imperative of no woman being able to dress as a man, and vice versa, as it would constitute an "abomination unto the Lord your God" (Couric 2003, Deut. 22:5). A female transvestite was either fined (Munby 1861) or thrown to prison (Zimmerman 2000, p. 279) for not obeying this rule.

There were two additional factors which added to the negative perception of female transvestites in the 19th century England. The first one was the 18th century's fascination with cross-dressed women based on historical accounts of those females who committed crimes when posing as men. The second was the rise of popular fiction propagating the image of an emancipated, self-sufficient woman (Craft-Fairchild 1998, p. 200), such as Daniel Defoe's Moll Flanders (1722), juxtaposing this image with one of a 19th century vision of a mother and wife as the angel in the house. Thus, the Victorian fear of a transvestite woman was a consequence of the assumption that "the crossing of the barrier between the sexes obviously eased the transgression of other norms, and vice versa" (Dekker and Van de Pol 1997, p. 101).

3 Crossing the Gender Frontier: Male Impersonation

Tipping the velvet (1998) is a neo-Victorian novel that relies on the past to provide engendered setting in which cross-dressed queer identities are constructed and performed irrespectively of the normative gender hierarchy of the Victorian Era. In the novel, Waters used the motif of cross-dressing as a medium of the protagonist's journey to find her real gender identity by allowing the character to smoothly slide on a feminine/masculine continuum. This would not have been possible without the author complicating the protagonist's life by a series of peculiar events that led to the formation of her gender-bending identity.

Firstly, a turn from a seemingly heterosexual small-town girl to lesbian male-impersonator in a big city opened the protagonist's eyes on the existence of Sapphists and performative queer identities. Secondly, a denigrating change from respected male impersonator to bisexual transvestite, reinforced her belief that her queer sexual orientation will remain unchanged. The protagonist's subsequent experience of becoming a sex-toy/slave in a Sapphic circle allowed her to understand that living as an oppressed subject was not her fate. Consequently, she realised that the transformation into a Victorian madonna[1] an epitome of a perfect wife and mother, was against her nature. Finally, her becoming a self-confident New Woman socialist transvestite completed the process of her gender-bending identity formation. Interestingly, through all the twists and turns, during a long journey from innocence to experience, the protagonist consciously crossed the gender frontier by cross-dressing only twice: firstly as a male impersonator and then as a transvestite.

An interlude to Nancy's initial crossing of the gender frontier was skilfully presented in the novel in a passage rich in tropes. On meeting Kitty in person for the first time, it was small-town Nancy who behaved like a young boy trying to court a more experienced girl. The protagonist was constantly blushing, stammering, being unable to produce a coherent sentence and marvelling in silence at everything around her idol from dirty cups to jars with paint. Even though during

---

[1] For more information on Victorian division of women into *magdalens* and *madonnas* consult Jannette King's *The Victorian women question in contemporary feminist fiction* (2005).

the meeting the performer was wearing her female garment, she also suddenly changed her manner into a gentleman -like one and "all at once she was the gallant boy of the footlights again. She straightened her back, made me a little bow, and raised my knuckles to her lips" (Waters 1998, p. 14), leaving Nancy completely perplexed. In this particular passage both women bended their gender without being cross-dressed. As we learn later in the novel, Kitty and Nan have strong masculine sides and in the fragment discussed above, it was this aspect of their identities which took over the socially accepted gender roles. Their behaviour in the dressing-room, in Butlerian terms, was therefore an example of crossing the gender frontier by performing the opposite gender identity.

In order to cross the gender frontier in a professional manner Nancy had to master her new stage skills by the means of observing and imitating males. She recalled that with Kitty, they "spent hours, as he [Walter, the manager] had advised in shops and market squares and stations studying the men" (Waters 1998, p. 86) and as Nancy admitted "we learned together the constable's amble, the coster's weary swagger, the smart clip of the off-duty soldier. And as we did so we seemed to learn the ways and manners of the whole unruly city" (Waters 1998, p. 86). Since in the 19th century London the role of street spectators was granted to men (Epstein Nord 1995, p. 1), by the simple act of learning males by observation, Nancy once again subconsciously bended her gender and began to cross the gender frontier again. Nevertheless, by learning the subjects she was later to impersonate she was simultaneously (un)consciously gathering experience required not only for the stage but also for her future life outside the theatre and developing her gender-identity based on (fe)male gaze. Furthermore, according to Simone de Beauvoir one was not born, but rather became a woman and being a woman, in opposition to being female, meant much more than having adequate reproductive organs (de Beauvoir 1989, p. 267). The philosopher also noted that womanhood itself is learned and that this education differs every time and place it occurs. It must be pointed out here that if becoming a woman might be learned, then when one reverses de Beauvoir's theory, by sheer observation and without male reproductive organs Nancy would be capable of becoming a man. She would not become a male, as the lack of appropriate genitalia would pose an obvious obstacle, but she would be definitely capable of learning how to be masculine.

The protagonist's hinted masculinity or rather was skilfully signalled by Waters in the character's name. Since Hughes points to the fact that from 1824 to the late 1888 Miss Nancy and Nancy were used as pejorative terms to refer to male homosexuals and effeminate boys (2006: 236) the protagonist's Christian name should be read as a sort of an aptronym or foreshadowing of Nancy's future crossing of the gender frontier. In this respect, her name change from the familial Nancy Astley to Nan King on stage only emphasised the dual nature of her gender. Indeed, since her first name was shortened to Nan and her surname changed to King, the girl's new pseudonym implied that, as an impersonator, she was of male gender and, according to Victorian standards, of dominating role in the performing couple. Consequently, the pseudonym of the second actress, and Nancy's lover—Kitty

Butler, pointed to a feminine female, as opposed to masculine Nan, of a lower status and/or passive role in the relationship.

While gender-bending during male-impersonating acts was not perceived as improper (Stetz 2004), the Victorians though of actresses as of a 'text' read by mass culture, turning them into commodities available to those who could pay, just like the prostitute (Powell 2004, p. 358). Since both an actress and a prostitute were the opposite of the Victorian image of heterosexual womanhood, because their sexual preferences and gender were debatable, they were treated as queer moral outcasts. Unfortunately, only Kitty was experienced enough to detect the danger of being reduced to queers by those who paid to watch and scrutinise them. It was also Kitty who wanted to keep up appearances, retain femininity and conceal her lesbian desire beyond the stage while Nan felt better in male's clothes and in Kitty's companionship when outside theatre walls. All in all, Kitty's efforts were futile as some weeks after being called a Tom by a drunk spectator, her inner shame for both her sexual and gender identity drove her into marrying her manager and leaving Nan nothing but her own gender-bending identity and theatrical costumes. In consequence, instead of becoming a heterosexual epitome of Victorian angel in the house, a status she so desperately tried to achieve, Kitty remained queer —she simply turned from a lesbian cross-dresser into a bisexual woman living in a seemingly proper Victorian heterosexual relationship.

4 Crossing the Gender Frontier: Transvestism

By taking the majority of her stage attire from the theatre Nan symbolically disposed of her past and consciously crossed the gender frontier for the second time by earning her living cross-dressed as a rent-boy. By selling her body while pretending to be a male she became an exemplary queer subject who rejected family and kinship—the attributes of the mainstream society (Halberstam 2005, p. 10). Naturally, her turn into gender-ambiguous and sexually charged street life was caused mainly by economic reasons, since she knew no craft apart from preparing shellfish and acting. Thus both coming back home and on stage were impossible her situation. The family would not accept her lesbianism, whereas the theatre was over for her without Kitty. Despite that, her ongoing affection for male clothes proved to be as useful in the street as it used to be on stage since it allowed her to assume a secret identity other than Nancy Astley or Nan King. This unnamed other self not only facilitated rejection of inner self-hatred, but also grounded her temporary unwillingness to live as a male full-time (Waters 1998, p. 193). Crossing the gender frontier was now to be confronted not with the theatrical audience but real people.

It should be noted that Nan soon liked her new cross-dressed self and had to rent two rooms at the same time to live and change in. The two rooms then might be read as a metaphor of the protagonist's split gender identity and her true-self which was sliding somewhere in between the feminine and masculine binary. At that point of the novel the character was not sure of her real self anymore, as her drag resembled an addiction, and crossing the gender frontier to and fro became a habit. Consequently, she confused those who observed her as they were unable to tell whether she was a girl who rented one of the rooms in order to "pull on a pair of trousers" or whether she was a lad who "arrived to change out his frock" (Waters 1998, p. 195).

In a broader perspective then, Nan's intensified transvestism and split gender identity were markers of the transitory phase in a long process of identity shaping by means of performing gender. In the course of her life, being feminine in reality and masculine on stage and later shifting into being both a male and a female in real life but distinguishing between two spheres: working- the street and personal—the room, confirmed that her cross-dressing and gender-bending were limited to time and space in which she currently found herself living and working.

However, spatiality was not the only limiting factor. Nan's own image of her emerging 'real' self made her cross-dressing skills perfect. When she was still a music hall performer, she could not disappoint her audience by showing any sign of her girly traits or behaviour as she was to epitomise an effeminate boy singing of and to the ladies. It would have been a personal and professional failure if her ambiguous gender was perceived as feminine, and she was reduced to a girl dressed as a boy singing of and for girls. Likewise, while being a rent-boy and working in drag in the streets of London, she could not disappoint both her clients and herself by letting her sex be recognised by failing to perform her chosen gender.

Nan's cross-dressing reached another level when she was taken from the street by a rich widow Diana Lethaby. Before introducing her to a male-free cross-dressed world of wealthy Victorian Sapphists, Diana changed the girl's name from Nan King into a more aristocratically sounding Neville King. Similarly to the previous name change, Waters hid a message in Nan's new name and also this time it was to foreshadow development of the protagonist's gender identity. Originating from French De Neuve ville, Neville has stood for "of the new town" (Arthur 2006, p. 206) and in queer reading of Nan's story, the new town was London presenting Nan with secrets she never thought existed. It was London filled with rich Sapphists and cross-dressed women who could afford a life with same-sex lover without thinking about social conventions.

For Nancy, crossing the gender frontier for the second time paid off as it confirmed not only her acting skills but also reinforced her masculine gender identity. At entering the Sapphic club in male's attire and being taken for a man, Nancy's pride of her impersonating capabilities revived once more. Seeing all the ladies gathered together in one room cherishing and admiring the newcomer, Neville was both amazed and pleased by their reaction and the setting (s)he was in:

> There were about thirty of them, I think—all women; all seated at tables, bearing drinks and books and papers…They were dressed, not strangely, but somehow distinctly. They wore skirts-but the kind of skirts a tailor might design if he were set, for a dare, to sew a bustle for a gent. Many seemed clad in walking-suits or riding-habits. Many wore pince-nez, or carried monocles or ribbons. There were one or two rather startling coiffures; and

> there were more neckties than I had ever before seen brought together at an exclusively female ensemble (Waters 1998, p. 272).

Even though (s)he was nothing more than a senior member's cross-dressed toy, his/her status was well grounded as (s)he was the most masculine of all young transvestites gathered in the club. Living for a year at Diana's side under her new name, Nan/Neville was so convincing in her disguise that several ladies enquired about her as him in order to find out whether he would be an appropriate candidate for a husband of their daughters.

After a year spent at Diana's house the affair ended and once again Nancy was alone and homeless. By a strange twist of fate she managed to find Florence, a girl she was once attracted to. In order to win her back Nan decided to undergo an impossible metamorphosis from despicable Victorian magdalene to a cherished madonna, the epitome of the angel in the house. The shift from Neville to Nancy proved to be more difficult than the protagonist anticipated because for the first time in her life Nan's fluid gender transgression seemed completely impossible. The change from masculinity to femininity was rather problematic and she felt that girl's clothes did not fit her anymore admitting that "it was as if wearing gentleman's suits had magically unfitted me for girlishness, for ever-as if my jaw had grown firmer, my brows heavier, my hips slimmer and my hands extra large" (Waters 1998, p. 381). All in all, Nancy's/Nan's/Neville's final transition was finally consolidated only when Florence accepted her as she was: a female by birth, a man by gender. Only then did Nancy decide not to think about Victorian social conventions and to live openly as a transvestite in a Sapphic relationship with a socialist suffragist.

5 Conclusion

From the queer theoretical stance the neo-Victorian *Tipping the velvet* (1998) works to complicate one's notion of both sexual and gender identity. The complex construction of two non-normative characters, a couple of queer music hall actresses Nan King and Kitty Butler, whose gender characterization is not fixed and whose object of desire changes accordingly to their current needs, reveals the limits of traditional perception of gender and sexuality per se and challenges our assumptions and understanding of their meaning. On the one hand, the subversive potential of gender as performance creates an illusion of rigid identity categories; on the other, Nan's gender-bending practices in real life confirm the delusive nature of such preconceptions and the imperfect interpretation of sexual and gender identity.

The peculiar crossing of the gender frontier as presented in Waters' novel might also easily be applied to any contemporary lesbian cross-dresser's life; starting with a difficult 'coming out', a family incapable of understanding the feelings of a young lesbian, a search for a more supportive community, through finding first same-sex partner and discovering same-sex love, experiencing first rejection, struggling with life on one's own, being tangled in a relationship one does not truly want, to finally finding a lifetime partner. The life of Nan King proves that no matter what epoch, being a cross-dresser means being, in one way or another, queer. In other words, every woman in history who lived cross-dressed was divided between the homoerotic and gender ambiguity staking out new territories, like all other women who decided to cross the gender frontier once and for all.

References

Arthur, W. 2006 (1856). *An etymological dictionary of family and christian names*. Whitefish: Kessinger Publishing, LLC.

de Beauvoir, S. 1989 (1949). *The second sex*. H.M. Parshley (tran.), New York: Vintage Books.

Blackburn, S. 1996. *The Oxford dictionary of philosophy*. Oxford University Press.

Brown, K. 1993. "Brave new worlds: Women's and gender history". *William and Mary Quarterly 50/2*: 311-328.

Butler, J. 1990. *Gender trouble: Feminism and the subversion of identity*. New York and London: Routledge.

Couric, R.A (ed). 2003. *King James 2000 Bible*. Amazon Kindle Edition.

Craft-Fairchild, C. 1998. Cross-dressing and the novel: Women warriors and domestic femininity. *Eighteenth-Century Fiction 10/2*: 171–202.

Dekker, M. R. and Lotte C. Van de Pol. 1997. *The tradition of female transvestism in early modern Europe*. New York: St. Martin's Press.

Epstein Nord, D. 1995. *Walking the Victorian streets. Women, representation and the city*. Ithaca: Cornell UP.

Foucault, M. 1978 (1990). *The history of sexuality: Vol. 1. An introduction* (R. Hurley, Trans.). New York: Vintage Books.

Garber, M. 1992. *Vested interests: Cross-dressing & cultural anxiety*. New York and London: Routledge.

Halberstam, J. 2005. *In a queer time and place: Transgender bodies, subcultural lives*. New York: New York University Press.

Hughes, G. 2006. *An Encyclopedia of wwearing: the social history of oaths, profanity, foul language, and ethnic slurs in the English-speaking world*. New York: M.E. Sharpe.

King, J. 2005. *The Victorian women question in contemporary feminist fiction*. New York: Palgrave Macmillan.

Munby, A. 1861–1867. *Diaries of Arthur J Munby. Volumes 29 to 35, 1861-1867*. London: Adam Matthew Publications Ltd.

Powell, K. 2004. Gendering the Victorian theatre. In *The Cambridge History of British Theatre, vol. 2*, ed. Joseph Donohue, 352-368. UK: Cambridge University Press.

Rodger. G. 2002.'He isn't a marrying man': Gender and sexuality in the repertoire of male impersonators, 1870-1930. In *Queer episodes in music and modern identity*, eds. Sophie Fuller and Lloyd Whitesell, 105-33. Urbana: University of Illinois Press.

Stetz, Margaret D. 2004. *Gender and the London theatre 1880-1920*. Buckinghamshire: Rivendell Press.

Waters, S. 1998. *Tipping the velvet*. London: Virago Press.

Wilson, Cheryl A. 2006. From the drawing room to the stage: Performing sexuality in Sarah Waters's Tipping the velvet. *Women's Studies: An Interdisciplinary Journal 35/3*: 285-305.

Zimmerman, B. (ed). 2000. *Lesbian histories and cultures: An encyclopedia*. New York: Garland Publishing Ltd.

Part III
Cross-Language Ventures

“Black Women are the Most Fascinating Creations in the World” or How (not) to Translate Alice Walker’s *The Color Purple*

Marcin Jurkowicz

Abstract The paper examines Michał Kłobukowski’s Polish translation (originally published in 1992) of Alice Walker’s *The color purple* (originally published in 1983) focusing on the rendering of African American Vernacular English used by the main characters from the book. The translator used a common strategy of translating a non-standard variety of English (i.e. African American Vernacular English) into an assortment of Polish non-standard forms. However, this leads to a number of problems. For the Polish reader, the main characters’ language is bound to evoke images of either the Polish peasants and the Polish countryside, or uneducated and vulgar people. This is quite unlike the American text for the source reader. Furthermore, the emphasis on lack of education or possible inferiority seems to be contrary to Alice Walker’s intentions. The translator’s strategy leads to bizarre associations for the Polish reader, which clash with the images created by the plot of the book. A close analysis of Alice Walker’s non-fiction writings shows that she holds all African American women in high esteem, which is rather incompatible with intentionally depicting them as vulgar and uneducated creatures. Furthermore, she uses African American Vernacular English not to suggest lack of education or create a feeling of inferiority, but as a medium of African American culture. African American Vernacular English is used in order to emphasise the characters’ commitment to African American culture and their ethnic identity as opposed to white people. Once the issues described above have been discussed, drawing on feminist translation theory, the paper will go on to suggest a number of strategies to minimise the suggested translation problems and reduce the clashes of imagery.

M. Jurkowicz (✉)
Warsaw University, Warsaw, Poland
e-mail: marcin.jurkowicz@onet.eu

J. Fabiszak et al. (eds.), *Crossroads in Literature and Culture,* Second Language Learning and Teaching, DOI: 10.1007/978-3-642-21994-8_15,

1 Introduction

The purpose of this paper is to discuss the Polish translation of Alice Walker's *The color purple*[1] done by Michał Kłobukowski. Only one aspect of this translation will be focused on in this paper—that of the translator's strategy regarding the rendering of African American Vernacular English used extensively by numerous African American characters in the original text. In order to achieve this aim, the cultural and political background of both the book and its writer will be briefly investigated. Next, selected parts of the texts will be compared and discussed, which will present the strategy adopted by the translator and highlight possible problems it is likely to cause. Finally, drawing on the feminist translation theory, the paper will offer suggestions as to how these problems can be minimised.

The feminist translation theory offers tools which can help to solve the problems created by Michał Kłobukowski's translation of *The color purple*. As Luise von Flotow explains, feminist translation appeared in Quebec in the 1970 and 1980s as a way of criticising and drawing attention to patriarchal language. At that time, a number of feminist writers began producing experimental work which tried to attack or deconstruct the conventional language use, which they regarded as misogynist. The writers tried to explore women's experiences, which in patriarchal language could not be properly expressed, and find a new linguistic way of expressing them. They also believed that the only way to express these experiences is to deconstruct the conservative patriarchal language (von Flotow1991, pp. 72–73).

Feminist translators adopted a number of characteristic strategies, such as *supplementing*, or *prefacing* and *footnoting*, to name but a few. *Supplementing* can be described as "voluntarist action" on the translated text, or a process of interfering with it, by means of which the translator, who is supposed to be conscious of his or her political and cultural role as a mediator between languages and cultures, compensates for the inevitable differences between the source and target language (von Flotow 1991, pp. 4–75). As for *prefacing* and *footnoting*, they allow the feminist translator to communicate to the reader numerous meanings, and language and culture subtleties that are otherwise bound to be lost in translation. Consequently, they allow active creation of meaning for the target reader, explaining various aspects of the process of translation and language use (von Flotow 1991, pp. 76–78). It seems that the above feminist translation strategies can be helpful in translating Alice Walker's *The color purple*.

[1] The translation was first published in 1992.

2 Historical and Cultural Background

Alice Walker[2] was born in 1944 and grew up in Georgia in the USA. During her childhood and adolescence, she could experience the rich African American culture of the South together with the everyday use of African American Vernacular English. As most African Americans at that time, she was also exposed to oppression and remnants of slavery, which would be engraved in her memory for ever. She became better known as a writer in the 1970s, when she started writing and publishing various texts, such as essays, novels, short stories or even poetry. Throughout the following decades, she has devoted herself to intellectual, political, cultural and environmental issues, working at colleges and universities, teaching, advocating feminism, and continuing to write. Her novel *The color purple* was published in 1983.

Most of Alice Walker's work seems to be devoted to an offshoot of feminism that she calls *womanism*. She presents this philosophy in a collection of her own essays *In search of our mothers' gardens* (1983). Alice Walker understands *womanism* as a form of feminism which cherishes and esteems women's strength, their emotional life, culture, and everything that is specifically female about women. Generally speaking, womanism advocates a view that there is a need for a tolerant attitude both between sexes and among races. Despite the universalist nature of the womanist philosophy, Alice Walker seems to favour African American women. She says for example, "For me, black women are the most fascinating creations in the world." Walker (1983, p. 251) and "it is the black woman's words that have the most meaning for us, her daughters, because she, like us, has experienced life not only as a black person, but as a woman" (Walker 1983, p. 275). Thus, Alice Walker emphasises the fact that African American women have been discriminated against for two reasons: because they are black, and because they are women. Furthermore, she declares that Black women are called (...) "the mule of the world" because we have been handed the burdens that everyone else (...) refused to carry. (Walker 1983, p. 237) and she also stresses that "black women among all women have been oppressed almost beyond recognition—oppressed by everyone" (Walker 1983, p. 149). What Alice Walker tries to convey is that African American women have been subject to discrimination not only from the white population, but also from members of their own race, i.e. African American men. She argues that "The black movement had given me an ideology that helped explain his [i.e. her father's] colorism (...) [and] his sexism. I was relieved to know his sexist behavior was not something uniquely his own, but, rather, an imitation of the behavior of the society around us" (Walker 1983, p. 339). Thus, Alice Walker presents her views on the reality in which African American women have lived. Their lives have been full of discrimination

[2] Giving a full, detailed biography of Alice Walker is outside the scope of this paper. Only a general background is provided, which will facilitate the understanding of certain translation problems discussed below.

from the white population, but also from within the African American population itself, and African American women have been the greatest victims. This theme is present in a number of Alice Walker's publications. It is also prominent in *The color purple*.

3 Analysis of the Polish Translation

The main character of *The color purple*—Celie—is a good case in point. She has to bear humiliation from her father and husband, often being raped, beaten and treated in a manner that is an affront to common decency. In fact, similar wrongs are often experienced by other African American female characters in the book, which are brought on them by various other African American, but also white characters. In *The color purple* which is an epistolary novel the reader is exposed to letters, which serve as a means to learn about the characters and the plot of the book. Most of the letters are written by Celie, and some of them also by her sister Nettie.

English. The following passage is a good illustration of the language used by Celie throughout the book:

> I don't argue. I git the coffee and light her cigarette. She wearing a long white gown and her thin black hand stretching out of it to hold the white cigarette looks just right. Something bout it, maybe the little tender veins I see and the big ones I try not to, make me scared. I feel like something pushing me forward. If I don't watch out I'll have hold of her hand, tasting her fingers in my mouth.
>
> Can I sit in here and eat with you? I ast. (Walker 2004, p. 49)

The language in which Celie writes her letters is African American Vernacular English. It is the language of a great majority of the African American population. In contemporary America, its use depends largely on social factors (Dillard 1972). African American speakers who endeavour to climb the social ladder of the dominant white culture are bound to prefer white Standard English. It is white Standard English that many people often associate with correctness and good style. Furthermore, it is often regarded as the language of the educated with most, if not all, other varieties projecting an image of vulgarity of behaviour or lack of education on the part of the speakers. It is probably such views that at least partly must have influenced the work of the translator, who rendered the passage in the following way:

> Nie chcę się spierać. Przynoszę jej kawę i przypalam papirosa. Sięga po niego. Jej chuda czarna ręka galanto wygląda przy długiem szlafroku, białem tak jak i papiros. Trochu się boję patrzeę na tę rękę, może bez te cieniuchne żyłki co je na niej widzę i te duże żyły których wolę nie zauważaę. Czuję jakby coś mnie popychało w jej stronę. Muszę się pilnowaę bo jeszcze złapię za tę rękę i zacznę palce ssaę.
>
> Mogę tu posiedzieć i trochu z tobą pojjeść, pytam się? (Walker 2003, p. 53)

As the translated passage demonstrates, Michał Kłobukowski chose to translate African American Vernacular English into an assortment of Polish non-standard forms. As has been pointed above, this strategy may result from the common, rather colonial, view on language varieties in general. However, it is probably also a result of the popular, linguistically-oriented translation procedure of automatically replacing a variety of the source language used in the source text with a variety of the target language in the target text, unfortunately often with no regard to the cultural aspects of doing so.

Whatever the reasons, the translator's strategy causes a number of problems. Being an assortment of various non-standard forms, Celie's speech in the target text is likely to produce various associations that are alien to the source text. Most notably, exposed to Celie's Polish non-standard forms, the Polish reader will activate general mental images of the Polish countryside and peasantry together with their customs. Indeed, for some readers, the target text may even evoke associations with Polish iconic cultural artefacts, such as *The peasants* by W. Reymont, both the book and the film based on it. Obviously, such associations are totally different from the images of the American South, slavery or African Americans in general that will be evoked by the average reader of the source text. Inevitably, this will lead to a cultural clash. While reading the target text and constructing a mental image of the world that is described there, the Polish reader will, like it or not, impose target culture images alien to the reality of the source text. Moreover, other possible mental images that the Polish non-standard forms in the target text may activate are those connected with vulgarity or inferiority. It so happens that people usually associate non-standard speech with lack of education. Consequently, for the average Polish reader, the characters from the book will probably appear to be uneducated, uncouth, pitiful or even primitive. Thus, the non-standard variety constructed by the translator is likely to create a rather negative attitude towards the characters on the part of the reader.

At this point it is worth noticing an important difference between the source and the target text. The non-standard variety used by Michał Kłobukowski is artificial. It is merely an assortment of various non-standard words or expressions taken from a number of real rural or urban varieties, slang or words deliberately distorted by the translator. Taken in its entirety, it is probably not used by anyone in Poland. This would be immediately noticed by a linguist but for the average reader, who is linguistically untrained, this is not so obvious. The non-standard forms introduced by the translator, so different form Standard Polish, will immediately activate mental images of some sort of Polish countryside and peasants somewhere in Poland or some undefined uneducated people. They do not give Celie any specific group identity apart from vague associations with the Polish countryside or uneducated people as a whole. Moreover, any kind of group identity that this or any other real Polish non-standard variety could give will produce a cultural clash with the reality described in the book and therefore should not be recommended. In fact, it seems that the only purpose of Michał Kłobukowski's artificial non-standard variety is to meet the requirements of his own linguistically-oriented translation strategy that the translator apparently chose to pursue. In contrast, African

American Vernacular English is a real variety of English used by a real group of people, who have a certain culture and history. Consequently, what is really important is why Alice Walker chose to use this form of speech in the first place. In other words, what any translator translating this novel should be interested in is what the function of African American Vernacular English in the source text is.

4 Function of African American Vernacular English

It is African American women, the main characters of the novel, who Alice Walker identifies with. Surely, then, she has no intention of additionally downgrading and victimising them by granting them *an inferior way of speaking*. Yet, she does use African American Vernacular English, which means it must have a completely different function. Therefore, it is worth looking at what Alice Walker thinks about certain aspects of writing and culture in general. She writes, e.g., "it is language more than anything else that reveals and validates one's existence, and if the language we actually speak is denied us, then it is inevitable that the form we are permitted to assume historically will be one of caricature, reflecting someone else's literary or social fantasy" (Walker 1988, p. 58). And a few pages further on, she warns that "if we kill off the sound of our ancestors, the major portion of us, all that is past, that is history, that is human being is lost, and we become historically and spiritually thin, a mere shadow of who we were, on the earth" (Walker 1988, p. 62). Moreover, she stresses that "our language is suppressed because it reveals our culture, culture at variance with what the dominant white, well-to-do culture perceives itself to be. To permit our language to be heard (…) is to expose the depth of the conflict between us and our oppressors" (Walker 1988, p. 63). Alice Walker also explains that "Celie speaks in the voice and uses the language of my step-mother (…) Celie is created out of language. In *The color purple* you see Celie because you see her voice. To suppress her voice [i.e. to replace it with Standard English] is to complete the murder of her" (Walker 1988, pp. 63–64). Finally, Alice Walker argues that it is not true that "the 'exposure' of turn-of-the-century Southern black folk language degrades black people, for the language is an intrinsic part of who we are and what has, for good or evil, happened to us" (Walker 1988, p. 65).

The quotes presented above show that Alice Walker views African American Vernacular English as being on a par with Standard English. It is a positive force which gives the African American characters their identity. It is not degrading, nor is its function to signal vulgarity of behaviour or lack of education. Indeed, by using African American Vernacular English, Alice Walker emphasises the characters' commitment to African American culture and their ethnic identity as opposed to white people. When Darlene, one of the characters in the book, complains about Celie's speech and suggests *improving* it (Walker 2004, pp. 194–195), the main character's reaction is that of resistance. Celie will not change her speech as it is familiar, and gives her both security and identity.

5 Suggested Translation Strategy

Given all this, the question remains what to do with Celie's speech in the target text. Would giving her Standard Polish speech be a viable strategy? As has been mentioned above, the original book consists of two kinds of letters—some written by Celie, and some by her sister—Nettie. This division is important because unlike Celie, who uses African American Vernacular English, Nettie writes her letters in Standard English. In fact, Nettie's speech is evolving during the course of the novel. Her chronologically initial letters do contain some rare non-standard features. However, very quickly, as a result of education, which she obtained when she left the USA and started missionary work in Africa, her speech becomes perfect Standard English. To render these initial non-standard characteristics of Nettie's speech, the translator uses the same strategy as he used with Celie's speech, whereas Nettie's Standard English is translated into Standard Polish.

While Celie's speech is that of an insider, a member of the African American community embedded in African American culture, Nettie, because of her standard speech, is portrayed as an outsider using the language of an educated white person. Consequently, the difference plays an important role in the novel and should also be taken into account in the Polish translation. However, to maintain this difference of speech, Polish translators cannot simply resort to rural or urban non-standard varieties of Polish since this will lead to the cultural clash discussed above. This is exactly where Michał Kłobukowski's translation fails. However, given the obvious differences between Polish and American culture, no Polish translation will ever be able to recreate the same kind of associations as those of the average American reader. Therefore, regardless of the adopted translation strategy, there will always be some loss in the process of rendering this novel into Polish. Unfortunately, by using Polish rural and urban non-standard features, Michał Kłobukowski loses much more than he actually preserves.

As has been explained above, for Alice Walker, African American Vernacular English is equal to Standard English. Its function is positive as it gives ethnic identity to the main characters expressing their commitment to African American culture as opposed to outside white culture. Alice Walker certainly identifies with the speakers of African American Vernacular English. Moreover, the whole story is narrated by an African American girl and African American people are its main characters. Consequently, it seems natural that a translation into Polish should also be made from an African American person's point of view with African American Vernacular English translated into Standard Polish. Thus, Alice Walker's intentions will be fulfilled as the average Polish reader identifies with Standard Polish in the same manner as Alice Walker identifies with African American Vernacular English. However, there is still the important difference between Celie's and Nettie's speech to cope with.

African American Vernacular English is deeply rooted in African culture and its origins date back to the early times of slavery when slaves were brought over from Africa to the USA. The language has mainly an oral tradition with a wide variety of characteristic speech behaviours (Kochman 1972), such as rapping,[3] jiving[4] or signifying,[5] to name but a few. Celie's speech is full of idioms, often humorous and in a conversational style. Therefore, while trying to render African American Vernacular English into Polish, emphasis should be put on informality, dynamism, humour and the rhythm of Standard Polish.[6] The passage given above could therefore be translated in the following way:

> Nie sprzeczam się. Podaję kawę i przypalam jej papierosa. Ma na sobie długi, biały szlafrok. Jej chuda czarna dłoń sięga po tego białego papierosa. Wygląda w porządku. Jednak coś w tej dłoni mnie przeraża. Może te małe, delikatne żyłki, które widzę. A może te duże, na które próbuję nie patrzeć. Czuję, jak coś pcha mnie w jej kierunku. Muszę uważać. Inaczej ją chwycę i palce ugryzę.
> Mogę tu usiąść i zjeść z tobą, pytam?

In contrast, Nettie's educated speech is very formal, humourless and often boring for the reader, as through the newly acquired educated style she is virtually unable to express real emotions. Furthermore, she seems to have difficulty in using natural speech, e.g. when she is trying to describe her feelings towards Samuel, she uses rather cliché expressions reminiscent of the style that can be found in some cheap love and romance stories, e.g.: "Especially when I tell you what a total joy it was. I was transported by ecstasy in Samuel's arms" (Walker 2004, p. 215). Nettie's speech is not rich in idioms and is often similar in style to a dry academic thesis. Besides, since her speech is that of a white person, Nettie is presented as an outsider and she is certainly not who Alice Walker identifies with. Resorting to derogative terms discussed briefly at the beginning of this paper, if one really felt compelled to label anyone's speech as *inferior*, given Alice Walker's views and the whole cultural context of the novel, it would not be Celie's, but Nettie's speech. Consequently, while trying to translate Nettie's educated speech into Polish, one could make it extremely formal, excessively educated and dull with long sentences and a boring style, which may require some mocking on the part of the translator. Thus, the formal difference between Celie's and Nettie's speech would be maintained in the target text. Below, there is a passage illustrating how Nettie's speech could be translated into Polish (I include Michał Kłobukowski's translation for comparison purposes):

> Celie, it seemed as good a time as any to put my arms around him. Which I did. And words long buried in my heart crept to my lips. I stroked his dear head and face and I called him darling and dear. And I'm afraid dear, dear Celie, that concern and passion soon ran away with us. (Walker 2004, pp. 214–215)

---

[3] Rapping is a fluent and lively manner of talking; a kind of self-performance.

[4] Jiving is a speech behaviour used in interaction with white people; it is behaving in certain situations in a manner that white people expect African American people to do.

[5] Signifying is a verbal game to create excitement and interest within the group, often initiated out of boredom.

[6] This is in agreement with what Krzysztof Knauer writes in his essay (Knauer 1996).

Michał Kłobukowski's translation:

Poczułam, Celio, że to najlepsza chwila, żeby go objąć. Tak też zrobiłam. Słowa, które od dawna tkwiły ukryte w moim sercu, same zaczęły mi się cisnąć na usta. Głaskałam te drogą mi twarz i głowę, mówiłam mu "kochany" i "najdroższy". I muszę ci wyznać, moja droga Celio, że troska i namiętność wkrótce zawiodły nas, dokąd chciały. (Walker 2003, p. 215)

My suggestion:

Wydawało mi się Celio, że był to odpowiedni moment, aby go objąć, co też uczyniłam. Natomiast słowa od dawna skrywane głęboko w moim sercu zaczęły skradać się ku mym wargom. Pogłaskałam również jakże drogą mi głowę oraz twarz i powiedziałam "kochanie", "najdroższy". I obawiam się, moja droga Celio, że niedługo potem poniosły nas troska oraz namiętność.

6 Feminist Translation Strategy

As has been discussed at the beginning of this paper, the traditional translation strategy used by Michał Kłobukowski does not seem to work, creating more problems than it really solves. In contrast, the new translation strategy presented above does not seem to result in a cultural clash while maintaining the functional differences between the characters' speech. Moreover, it seems to produce a target text that is closer to Alice Walker's intentions as opposed to the translator's intentions. Although English and Polish do not have the same collection of non-standard varieties and associations related to each of them, it is still possible in the target text to construct a similar functional relationship to that designed by Alice Walker in the source text. In so doing, the translator embraces what is in essence a feminist translation strategy called *supplementing* (von Flotow 1991, pp. 74–76). The strategy compensates for the differences between the two languages and cultures by supplementing what lacks in the Polish language, thus making the target text truly meaningful to the Polish reader.

Since *The color purple* is deeply embedded in South American and African American culture, certain aspects of it may not be immediately comprehensible to the average Polish reader. Therefore, in order to facilitate comprehension of the novel and make it more meaningful to the reader, it may be advisable to provide extra information explaining aspects of the culture, novel and the translator's strategies. This can be done by means of a preface and footnotes, which is again in accordance with the feminist translation theory and its two prominent strategies called *prefacing* and *footnoting*[7] (von Flotow 1991, pp. 76–78). In *The color*

[7] A preface and footnotes are not necessarily something that is only present in feminist translations. Other translators who do not call themselves 'feminist' also sometimes use them. However, the feminist translation theory puts great emphasis on these two strategies and formalises them.

purple, both the preface and footnotes would make it possible for the reader to gain insights into e.g. the intricacies of Alice Walker's feminist views and intentions regarding her novel, the use and status of African American Vernacular English, as well as the strategy of *supplementing* chosen to translate the source text. Apart from facilitating comprehension of the novel, both *prefacing* and *footnoting* would also have the additional benefit of making the translator's choices explicit to the reader. In the analysed translation done by Michał Kłobukowski, there are no prefaces or footnotes. Thus, the Polish reader is left completely uninformed about the general cultural context of the novel and the language used in it, and may only wonder why the characters speak in the manner presented by the translator.

7 Conclusion

To sum up, there can be no one perfect translation of Alice Walker's *The color purple*, or any other text, especially one that is so deeply embedded in the source culture. Each strategy adopted by any translator is bound to lead to some loss of, or difference from, the original quality of the source text. However, the question is always which qualities are vital, and which ones can be sacrificed. It seems that in this case, Michał Kłobukowski's strategy sacrifices more than it actually tries to preserve. However, using the feminist translation strategies of *supplementing*, *prefacing*, and *footnoting* makes it possible to avoid a cultural clash, preserve the functional relationships that are present in the source text, and translate it more according to Alice Walker's intentions regarding her novel. Thus, the feminist translation strategies allow minimising translation losses, and preserving more than they actually sacrifice.

References

Dillard, J. L. 1972. *Black English: its history and usage in the United States*. New York: Random House.

Knauer, K. 1996. Między kompetencją lingwistyczną a kulturową. Tłumaczenie odmian angielskiego na podstawie przekładu powieści Alice Walker *The color purple. In Polityka a przekład*, ed. P. Fast, 85-93. Katowice: Wydawnictwo Śląsk.

Kochman, T (ed.) 1972. *Rapin' and stylin' out: communication in urban Black America*. Urbana: University of Illinois Press.

von Flotow, L. 1991. Feminist translation: Contexts, practices and theories. *TTR: traduction, terminologie, rédaction, vol. 4, no. 2*, 1991:69-84.

Walker, A. 1983. *In search of our mothers' gardens*. San Diego, New York, London: Harcourt Brace Jovanovich.

Walker, A. 1988. *Living by the word*. London: The Women's Press.

Walker, A. 2003. *Kolor purpury*. Warszawa: MUZA SA.

Walker, A. 2004. *The color purple*. London: Phoenix.

Part IV
Old World, New World and the Space Between

Female pioneers, Good Indians, and Settler Nostalgia: Colonial Ambivalence in Sally Armstrong's *The Nine Lives of Charlotte Taylor*

Anna Branach-Kallas

Abstract The paper focuses on the Canadian novel *The nine lives of Charlotte Taylor*, published in 2007 by renowned Amnesty International reporter Sally Armstrong. Acclaimed immediately as a national bestseller, the novel depicts, according to the author herself, "Canadians' fascination with their roots and their pride in an unwavering pio cneering ethics". Seen in this Eurocentric light, Charlotte Taylor (a real life pioneer of the late 18th c. Canada) becomes "everyone's ancestor". My analysis focuses on the dynamics of gender, race, and land acquisition, as depicted in the novel. The paper demonstrates how Armstrong rewrites the pioneer experience from a woman's perspective; I approach the protagonist as a boundary crosser, emphasizing women's liminal position in the early stages of colonization in Canada and their ambivalent relationship to settling the land. I also show, however, that Armstrong fails to recognize the complex dynamics of race relationships in early Canada. While the feminist intention of the novel is therefore (at least partly) realized, the figures of First Nations and their relationship with the white pioneers remain highly stereotypical.

1 Introduction: Nova Scotia as a Contact Zone

The nine lives of Charlotte Taylor, published in 2007 by renowned reporter Sally Armstrong, was acclaimed in Canada as a national bestseller. The novel, based on historical sources, the oral legacy of Charlotte Taylor's descendants, and Mary Lynn Smith's website *The life and times of Charlotte Taylor*, fictionalizes the story

A. Branach-Kallas (✉)
Nicolaus Copernicus University, Toruń, Poland
e-mail: annabranachkallas@yahoo.com

J. Fabiszak et al. (eds.), *Crossroads in Literature and Culture*, Second Language Learning and Teaching, DOI: 10.1007/978-3-642-21994-8_16,

of one of the first women settlers in Nova Scotia, who married thrice, mothered ten children, and had a life-long liaison with a Mi'kmaq man, of which, however, little is known (Armstrong 2008b, p. xiv). It is set at the turn of the 19th century (1775–1841)—a particularly turbulent period in the Canadian pre-Confederation history in present-day northern New Brunswick. It records the gradual arrival of British settlers, the later struggle for land between these so-called Pre-Loyalists, the American rebels, and the Loyalists migrating to the British Colony after the American Revolution. Armstrong also creates interesting figures of Acadians, who were perceived as a threat to the English settlers and expelled from the Canadian Maritime Provinces between 1755 and 1763. 14,000 were deported either to Louisiana or to France; some of them were hidden in the Indian camps, while others returned in the 1770s to find their land settled by the British. In 1784, the division of the province of Nova Scotia and the creation of New Brunswick caused further complications regarding the grants of originally registered lands, which exacerbated the conflicts between the old settlers and the new ones. The white pioneers' struggles for land resulted in the gradual dispossession of the indigenes, represented in the novel by the powerful Salmon People, who, ultimately, decimated by epidemics, hunger, and genocide, are forced into reservation land.[1] The social space depicted in the novel can be therefore referred to after Mary Louise Pratt as a contact zone, "the space of imperial encounters, the space in which peoples geographically and historically separated come into contact with each other and establish ongoing relations, usually involving conditions of coercion, racial inequality, and intractable conflict" (Pratt 2008, p. 8; Armstrong 2010; The life 2010).

2 Women Pioneers in the New World: The Ambivalences of Colonization

In *The nine lives of Charlotte Taylor*, the eponymous heroine arrives to Canada at the age of twenty. She is the daughter of an aristocratic family, who eloped to the West Indies with her Black lover, the family's butler. When Pad Willisams dies, she is rescued from Jamaica, where she would probably have become a debased concubine, by Commodore George Walker, another historical figure, Late Commander of the Royal Family Privateers, Justice of Peace for the County of Halifax in the Province of Nova Scotia, and proprietor of the trading post of Nepisiguit. Walker, who is acquainted with her father, takes Charlotte to Nova Scotia, yet he intends to send her back to England on the first ship available before the beginning

[1] The novel thus illustrates the radical change in the policy of the British imperial government in Canada, who, in the 1830s moved away from the earlier trade and military alliances with the indigenous population "towards a policy that effectively reconstituted First Nations people as wards of the state" and initiated an intense civilizing mission (Henderson 2003, p. 91).

of fierce winter. He insists that Nova Scotia is not the right place for a woman of Charlotte's background and that she should make a proper marriage in Sussex, live in a comfortable house, and raise children. "Could any woman want more?" (Armstrong 2008a, p. 90), asks Walker. The female protagonist, however, pregnant and aware of the fact that her father would never accept a mixed-race child, is determined to remain in "the middle of nowhere" (Armstrong 2008a, p. 41), "A place far away from proper civilized places [...] A place where a person could shed a past" (Armstrong 2008a, p. 54). Canada clearly offers more independence than England: the hardships of a pioneer's life contrast vividly with the rigid gender expectations of the Motherland. Despite its difficulties, for Charlotte, as for many other early female pioneers in Canada (see Henderson 2003, p. 153), the isolation and challenges of colonial life provided liberation from the confined existence of women in the Old World, and an opportunity for every-day decision making that was absolutely unavailable to genteel women in England at the turn of the 19th century. In *The nine lives of Charlotte Taylor,* the eponymous protagonist therefore escapes to the Indian camp, where she gives birth to her daughter, acquires some knowledge of aboriginal metaphysics and means of survival in the harsh climate, and is befriended by indigenous women as well as Wioche, Chief Julian's appointed traveler. In spite of the couple's obvious mutual attraction, the relationship does not trespass the boundaries of friendship. Charlotte feels comfortable in the Indian encampment, where "no one pretended to know her mind better than she did herself" (Armstrong 2008a, p. 125). With time, however, and somehow to the reader's surprise, she realizes that she cannot remain with the Mi'kmaq any longer.

Notwithstanding the settler woman's relative autonomy, such agency was enjoyed only by married women. Single women in the colonies were perceived as a threat to the status quo, since it was assumed that women's sexuality had to be protected in marriage, and, when unguarded by patriarchal law, single women could lose their chastity and thus undermine white prestige (Stoler 1997, p. 355). At the end of the 18th century, women, moreover, were still deprived of political and economic rights. Political and social liberties were assigned to men, and Englishness was conflated with masculinity (Hall 1997, p. 174). Women's role in the colony was to act as "custodians of family welfare and respectability and as dedicated and willing subordinates to, and supporters of, colonial men"; if they fell outside these traditional roles, their liminal, uncertain position was unacceptable (Stoler 1997, p. 355). Moreover, the male-dominated structures of the trading posts in 18th century Nova Scotia were not really adjustable to female presence: the later arrival of white women to the colonies, according to Ann Laura Stoler, "put new demands on the white communities to tighten their ranks, clarify their boundaries, and mark out their social spaces" (Stoler 1997, p. 351). In Armstrong's novel, the beautiful, young heroine, is clearly suspicious for she has already given birth to an illegitimate (as it is conjectured), mixed-race child that embodies the most serious challenge to colonial morality, the sexual affront of *métissage* (Stoler 2002, p. 110). She has, moreover, displayed more than once signs of an "undisciplined constitution" (Armstrong 2008a, p. 101) and must be therefore brought under strict control. Having refused the advances of the much older Walker, Charlotte marries the younger and handsome John Blake, whom she, however, hardly knows: "a husband was the only solution and this husband the best husband available" (Armstrong 2008a, p. 161). It must be noted that Armstrong's critique of her heroine's entanglement with the colonial etiquette and sexual norms is curiously implicit.

John Blake is represented in the novel as a typical colonial man, a soldier who served in the British navy, then captained cargo (slaves?) across the Atlantic, and fought the historic Battle on the Plains of Abraham under General Wolf, which marked the end of the French Empire in North America. This fierce, violent, blood-thirsty male protagonist seems to suffer from a sense of anxiety that Ann McClintock characterises as boundary loss in a New World full of terrifying ambiguities (McClintock 1995, pp. 22–36). This acute paranoia, combining fantasies of conquest with the dread of engulfment, is disavowed "by reinscribing ritual excess of boundary, accompanied, all too often, by an excess of military violence" (McClintock 1995, p. 24). Blake, who marries Charlotte in the period of the American Revolution, is determined to annihilate the First Nations tribes, who consort with the American rebels. Interestingly, his hatred is not so much directed against the Americans as against the aboriginal population, whom he (rightly) sees as the main obstacle to the imperial possession of land. By constructing the First Nations peoples as excessively violent, Blake seems to project on the aboriginals his own fantasies of blood-shed: "It is not until every savage has his scalp cut from his skull that we may rest here. Scalp them and leave them to die, as they would us. *That* must be British policy. It is *my* policy" (Armstrong 2008a, p. 164). When they settle on his land plot on the Miramichi River, Blake leaves Charlotte several times, alone in the wilderness, with small children, to fight the rebels and the Indians, and fabricates a sophisticated plot that leads to the murder or imprisonment of many First Nations chiefs. Blake's pathological distrust of the natives is particularly painful to Charlotte, who sees the Mi'kmaq as the most hospitable people and locates her true home in the camp of the Salmon tribe (Armstrong 2008a, p. 212). Furthermore, her husband combines his excessive violence towards his racial inferiors with an excess of gender hierarchy, demanding absolute obedience from his wife (Armstrong 2008a, p. 172). A few hours after their wedding, he is enraged when Charlotte returns from the Salmon camp, where she said her farewells to her hosts, because he suspects her of having betrayed his military plans to his enemies. He also deeply resents her ambivalent relationship with Wioche. Charlotte must therefore carefully conceal the visits of her Indian friend, who keeps protecting her (Armstrong 2008a, p. 191). Blake clearly conflates the body of his wife with his land, the possession of which he is uncertain about. Nira Yuval-Davis has demonstrated that in imperialistic and nationalistic discourses there often appears a close association between "collective territory, collective identity and womanhood" (Yuval-Davis 1997, p. 45). Women's bodies are thus constituted as signifiers of ethnic or national differences, and the purity of a woman's body represents the integrity of the group boundaries (Anthias and Yuval-Davis 1996, p. 115). Charlotte's contacts with Wioche are as unacceptable to Blake as the Indian's presence on his land: trespassing social space is thus identified with intimate, bodily transgression, which reveals Blake's intense anxiety about white male and imperial power in the New World. Although Armstrong emphasises Charlotte's erotic attraction to her first husband, she also implies that her heroine realises that Blake's violence is another version of her father's authority that she has tried to escape. In this sense, Charlotte is constructed as utterly powerless in the novel.

Nevertheless, as both Jennifer Henderson and Kate Pickles emphasise, the settler woman's position in Canada was deeply ambivalent. Restricted to the home, the settler woman enjoyed within that home an independence her European sisters could only dream of. Thus, in *The nine lives of Charlotte Taylor*, during Blake's absence, it is Charlotte who is responsible for the survival of her growing family: she is in charge of field work, of planning and organising their lives before and during the harsh winters, and she uses Indian herbal remedies to cure their bouts of sickness. When the settlement is attacked, she does not hesitate to defend her children and her neighbours from the American rebels and their Indian allies. According to Pickles, "During colonization, women and bodies mattered and were bound up in creating and perpetuating an often hidden, complex, contradictory, and fraught history. Women occupied the spaces of colonial encounter between Aboriginals and newcomers as both colonizers and the colonized, transgressing restrictive boundaries and making history" (Pickles 2005, p. 1). Henderson even argues that the settler women "occupied the site of the *norm*, not a position outside of culture and external to the machinations of power" (Henderson 2003, p. 4; italics in the original). Situated in this perspective, Armstrong's eponymous heroine also seems involved in the colonial project.

3 Good Indians: From Indigenization to Tender Violence

In contrast to her first husband's violent explosions of hypermasculinity, however, Charlotte uses the complex strategy that Terry Goldie defines in *Fear and temptation: The image of the indigene in Canadian, Australian, and New Zealand literatures* as indigenization, the (impossible) process of becoming indigenous and erasing the "separation of belonging:" "The white Canadian looks at the Indian. The Indian is Other and therefore alien. But the Indian is indigenous and therefore cannot be alien. So the Canadian must be alien. But how can the Canadian be alien within Canada?" (Goldie 1989, p. 12). Charlotte first attempts to become indigenous by sharing the Salmon People's lives in their encampment. She eats the food they eat, sleeps in the wigwam with Marie and her family, learns the codes specific to this culture. Interestingly, when she leaves the camp to become a settler by Blake's side, she feels poignantly "unsettled" (Armstrong 2008a, pp. 186, 226). Her non-belonging to the land is reaffirmed formally after Blake's death and the creation of New Brunswick, when she learns that she has no right to her land unless she puts her name in the new land register of the province. Since she distrusts the white men around her, fearing they might steal her property, she asks Wioche, her Indian friend, to take her to Frederick Town in the middle of winter to put a claim on her land. Paradoxically, therefore, it is the indigenous figure, dispossessed of land by the white settlers, that helps Charlotte to secure the claim to "her" land and thus to become *settled.* Furthermore, Charlotte and Wioche's sexual union also functions as a strategy of indigenization. The aboriginal figure is thus constructed stereotypically as "an emanation of land" (Goldie 1989, p. 73); by possessing the man sexually, Charlotte possesses the land.

According to Goldie, sex, together with violence, nature, orality, and mysticism, functions as a stereotype by means of which indigenous peoples are represented in the white literatures of settler-invader nations. Wioche in *The nine lives of Charlotte Taylor* is indeed a highly stereotypical figure; he is not really characterized, rather marginal in the action of the novel, always watching the female protagonist from a distance. Thus, the novel can be clearly situated in the tradition of imperial white discourse, which "implacably reduces the non-white subject to being a function of the white subject, not allowing her/him space or autonomy, permitting neither the recognition of similarities nor the acceptance of differences except as a means of knowing for the white self" (Dyer 1997, p. 13; see also Said 1978; JanMohamed 1986). Wioche's closeness to the natural world allows him to be transformed into a "noble savage" by the elevated power of nature (see Goldie 1989, p. 31). Whereas traditionally, the male indigene is associated with violence and the female indigene with sexuality (Goldie 1989, p. 65), Armstrong reverses this conventional model by attributing the stereotypical features of the female indigene, such as sexual attraction, beauty, purity, devotion, and protection (Goldie 1989, p. 68), to Wioche. Consequently, while violence as the most frequent literary representation of the dehumanized indigene, though projected by Blake, is clearly undermined in the novel, sexuality still functions in *The nine lives of Charlotte Taylor* as a crucial characteristic. Moreover, Armstrong's love-plot shares much with late 18th century imperial narratives (such as John Stedman's *Narrative of a five years' expedition against the revolted Negroes of Surinam* [1796]) that, according to Mary Louise Pratt, invoke love and devotion as an alternative to colonial domination. The exploitation and debasement of the local populations are thus eradicated and replaced by the wilful submission of the colonised. Nonetheless, transracial love plots are imaginings "in which European supremacy is guaranteed by affective and social bonding" (Pratt 2008, p. 95).

Significantly, in *The nine lives of Charlotte Taylor*, as in the 18th century imperial love plots, the lovers are separated and the heroine is reabsorbed by the colonial world (see Pratt 2008, p. 95). Although "From the moment she laid eyes on [Wioche] at the Mi'kmaq camp on the hill over Alston Point, she was bewitched by his gentle strength, his clear pride in his people, the way his kindness to her lit the dark intensity in his eyes" (Armstrong 2008a, p. 273), once she comes back home, Charlotte quickly marries the tender and affectionate William Wishart and dutifully bears his children. The process of indigenization in Charlotte's case is therefore an ambivalent process of individuation: the indigene acts as an agent of liberation from an authoritarian white husband (Blake), yet not from the dominant patriarchal white culture. Contrary to the reader's expectations, Charlotte (together with Armstrong) is not ready for a radical questioning of the status quo, the imagining of a life outside the patriarchal family and open miscegenation. If she transgresses, she does it once, in-between two respectable marriages, in the liminal space of wilderness between the white settlement on the Miramichi River and Frederick Town. Consequently, although more independent than other settler women around her, Charlotte ultimately contributes to the colonial ideal of white femininity. When her second husband dies, she remarries immediately for equally pragmatic reasons. Finally, having survived three husbands and given birth to ten children, surrounded by seventy grand-children, Charlotte becomes a genuine "matriarch" (Armstrong 2008a, p. 340), combining the paradoxical functions of "an emblem of sexual vulnerability" and an agent of normalization (Henderson 2003, p. 43). Although she keeps visiting the Indian camps, deeply moved by the terrible fate of the First Nations people, she does it as a social superior, teaching them the skills necessary to survive in an increasingly colonised world. In spite of her idealization of First Nations' society, Charlotte seems to be guilty of what Laura Wexler has called the "tender violence" (in Henderson 2003, p. 98) of literary figures who, though they seem to defend the slave or "savage" characters, in fact reinforce the norms threatening to the racial others. It is after all the presence of white women in the Crown colonies that "legitimated the violent suppression of native insurgency" and the elaboration of strict "measures of protection and containment" (Henderson 2003, p. 104).

4 Charlotte: The Spirit of White Enterprise

Charlotte's struggle for land throughout the novel and her ultimate position as a wealthy landowner also situate her as an agent of colonial power. In this sense, Charlotte clearly embodies the spirit of white enterprise: industry, courage, will to power, energy (Dyer 1997, p. 31). Depicting the accumulation of property by this apparently powerless woman, her fight for the organization of schools in the district, and her demand for recognition of her rights as a woman in North America (Armstrong 2008a, pp. 281–282), the novel shows that "What justifies the White displacement of Native people is enterprise—that passion for organization, civic improvement, and development—invested in the soil" (Coleman 2006, p. 57). In spite of her intimacy with vulnerable racial figures, her black and aboriginal lovers, Charlotte Taylor contributes to the expansion of imperialist power in North America. Although she maintains her quasi-mystical, and apparently sexless, relationship with Wioche till the end of her life, spending idyllic afternoons with him in his canoe (oblivious to the gossip spread by her neighbours and the discontent of her husbands and sons), Charlotte experiences little guilt in relation to the First Nations. Having secured her land claim with Wioche's help, she dares ask him how to find more farmland, feeling just "*a little* ashamed: why should he collaborate so directly in her taming of the land?" (Armstrong 2008a, p. 322; my italics) Although at times she confesses that she is fully aware of her complicity in displacing the indigenes (Armstrong 2008a, pp. 356, 359, 382), she accepts it sadly as a necessity on the path of colonial progress and does not hide her satisfaction as a successful pioneer and proprietor of beautiful land. By approaching the question of the female settler's responsibility with ambivalence, *The nine lives of Charlotte Taylor* shows not so much the extent of the Europeans' culpability in the New World "but, rather, how far the criteria by which guilt and innocence are determined have been historically constituted" (Stam 1997, p. 193). Charlotte's view of Canada as a refuge offering peace to people of all countries, religions, and political beliefs, as well as her insistence at the end of the novel that her "own hearth could be described as Acadian, British and Micmac" (Armstrong 2008a, p. 375) are attempts to conceal—under the mask of hybridity and tolerance—complex histories of dispossession and the settler's intense determination "to tame this wild land" (Armstrong 2008a, p. 375) at the cost of its "savage" inhabitants.

5 Conclusion: Settler Nostalgia

Sally Armstrong*'s The nine lives of Charlotte Taylor* is filled with settler nostalgia. Charlotte Taylor's life is presented in an absorbing, romanticized narrative; the figure of the central protagonist is beautiful, strong, compelling; her friendships with Acadian and Mi'kmaq characters as well as the tragedies of their peoples are depicted in a moving way. Commenting on the novel, the author herself declares that it reveals "our (i.e. the Canadians') fascination with our roots; our pride in an unwavering pioneering ethics; the bold, dauntless and often tragic beginning of Canada" (Armstrong 2008b, p. 4). Seen in this Eurocentric light, Charlotte Taylor becomes "everyone's ancestor" (Armstrong 2008b, p. 4). Armstrong thus posits British settler experience as a model for the immigration experience of other immigrant groups, which denies the unequal distribution of privilege in the history of immigration to Canada, shaped by imperial politics. Furthermore, the novel traces disturbing historical parallels between the dispossession of Canadian First Nations, the Acadians, and the British pre-Loyalists threatened by the arrival *en masse* of the Loyalists from the south and the administrative changes in the colony.[2] The indigenous peoples' loss of native land is thus equalled with the expulsion of the Acadians, and the frustrations of British pioneers resulting from confusion in colonial bureaucracy, while the genocide of the First Nations, the

[2] This tendency to generalize diaspora experience and loss of homeland is not uncommon in contemporary Canadian fiction. Examples of sentimental novels, filled with settler nostalgia, in which the relationship between the settler communities and Canadian First Nations are depicted in a perplexingly ambivalent way include Jane Urquhart's celebrated *Away* and Alistair MacLeod's *No great mischief*. See Smith 2008; Goldie. In a review, Sally Armstrong's *The nine lives of Charlotte Taylor* has been praised for its "enlightened historical perspective" (Quirk 2000, p. 158).

ethnic cleansing of French colonists, and the conflicts between the white settlers at the turn of the 19th century are all conflated within a vicious circle of violence in the New World contact zone. Although it is easy to imagine that such might have been the real Charlotte Taylor's perspective, the lack of explicit political commentary in a text written by a renowned reporter and human rights activist, who has covered conflicts from Bosnia and Somalia to Rwanda and Afghanistan, is rather disconcerting. Consequently, while the novel is successful in portraying the ambivalences of the woman's settler condition, the figures of First Nations and their relationships with the white pioneers remain highly stereotypical. The political issues relating to land acquisition are depicted in *The nine lives of Charlotte Taylor* in a misleading way, covering the problem of indigenous territory appropriation with layers of settler nostalgia, which does not so much obliterate the issue of violence, but rather mystifies the reader as to who were its principal victims.

References

Anthias, F. and N. Yuval-Davis. 1996. *Racialised boundaries. Race, nation, gender, colour and class and the anti-racist struggle*. London and New York: Routledge.

Armstrong, S. 2008a. *The nine lives of Charlotte Taylor*. Toronto: Vintage Canada.

Armstrong, S. 2008b. The nine lives of Charlotte Taylor: Special features. In *The nine lives of Charlotte Taylor*, 1-7. Toronto: Vintage Canada.

Coleman, D. 2006. *White civility: The literary project of English Canada*. Toronto-Buffalo-London : University of Toronto Press.

Dyer, R. *White*. 1997. London and New York: Routledge.

Goldie, T. 1989. *Fear and temptation: The image of the Indigene in Canadian, Australian, and New Zealand Literatures*. Montreal: McGill – Queen's University Press.

Hall, S. 1997. The local and the global: Globalization and ethnicity. In *Dangerous liaisons: Gender, nation, and postcolonial perspectives*, eds. A. McClintock, A. Mufti, and E. Shohat, 173-187. Minneapolis, London: University of Minnesota Press.

Henderson, J. 2003. *Settler feminism and race making in Canada*. Toronto: University of Toronto Press.

JanMohamed, A. R. 1986. The economy of Manichean allegory: The function of racial difference in colonialist literature. In *"Race," writing and difference*, ed. H.L. Gates, Jr., 78-106. Chicago and London : The University of Chicago Press.

McClintock, A. 1995. *Imperial leather: Race, gender and sexuality in the colonial contest*. New York and London: Routledge.

Pickles, K. 2005. *Contact zones: Aboriginal and settler women in Canada's colonial past*. Vancouver: University of British Columbia Press.

Pratt, M. L. 2008. *Imperial eyes: Travel writing and transculturation*. London and New York: Routledge.

Quirk, L. 2000. Voracious for life. *Canadian literature 200*: 157-158.

Said, E. 1978. *Orientalism*. New York: Vintage Books.

Smith, A. 2008. The culture of whiteness: Immigration, race, and Ethnic Mythology in Alistair MacLeod's No great mischief. In *Migrance comparée/Comparing migration*, eds. M. Carrière and C. Khordoc, 339-351. Bern-Berlin-Bruxelles-Frankfurt am Main-New York-Oxford, Wien: Peter Lang.

Stam, R. 1997. Multiculturalism and neoconservatives. In *Dangerous liaisons: Gender, nation, and postcolonial perspectives*, eds. A. McClintock, A. Mufti, and E. Shohat, 188-203. Minneapolis, London: University of Minnesota Press.

Stoler, A. L. 2002. *Carnal knowledge and imperial power: Race and the intimate in colonial rule*. Berkley-Los Angeles-London: University of California Press.

Stoler, A.L. 1997. Making empire respectable: The politics of race and sexual morality in Twentieth-Century colonial cultures. In *Dangerous liaisons: Gender, nation, and postcolonial perspectives*, eds. A. McClintock, A. Mufti, and E. Shohat, 344-373. Minneapolis, London: University of Minnesota Press.

Yuval-Davis, N. 1997. *Gender and nation*. London: SAGE Publications.

Sally Armstrong www.speakers.ca/armstrong_sally.aspx. Accessed on 6 April 2010.

The life and times of Charlotte Taylor by Mary Lynn Smith www3.bc.sympatico.ca/charlotte_taylor. Accessed on 30 March 30 2010.

Crossing the Frontiers of Death: A Journey Through War Memories and Continents in Anne Michaels' *Fugitive Pieces*

Dagmara Drewniak

Abstract The aim of this article is to analyze the journey of a Holocaust survivor from Poland to Canada via Greece in Anne Michaels' famous novel *Fugitive pieces* published in Canada in 1996. The voyage becomes a multilayered quest through war memories of the main protagonist, a seven year old Jakob Beer who has to overcome the specters of the war such as witnessing his parents' death and vanishing of his sister. Not only is the book a discussion on how to deal with the haunting ghosts of the past but it also transforms into a poetical treatise on the burden the Holocaust survivors carry and the possibilities of liberating oneself from it. The analysis of the novel is intertwined with the recent theoretical studies on memory and its value in shaping one's personality and identity. Therefore, Marianne Hirsch's book on postmemory titled *Family frames. Photography narrative and postmemory* (1997) and Anne Whitehead's *Memory* (Routledge, London, 2009), for instance, serve as tools to render Jakob Beer's identity comprised to a large extent his memories of pre- and post-Holocaust childhood. The influence of the trauma of the war extermination is tremendous and it has an impact on all spheres of Jakob's life, even as an adult. As a result of that the article will address the issue of capability of crossing the frontiers of death and starting a peaceful life, devoid of these ghosts in a different country or even possibly on a different continent.

D. Drewniak (✉)
Adam Mickiewicz University, Poznań, Poland
e-mail: dagmarad@amu.edu.pl

J. Fabiszak et al. (eds.), *Crossroads in Literature and Culture,* Second Language Learning and Teaching, DOI: 10.1007/978-3-642-21994-8_17,

1 Introduction: *Fugitive Pieces* as the Holocaust Narrative

The aim of this article is to present and analyze various aspects of crossing frontiers during the journey of a Holocaust survivor from Poland to Canada via Greece in Anne Michaels' famous novel *Fugitive pieces*, published in Canada in 1996. The critical acclaim the book has received and its bestseller status place the novel among the most important voices referring to WWII and the Holocaust not only in Canada but worldwide. The travel experience presented in the book becomes a multi-layered quest through war, memories, and the post-war existence of the main protagonist, seven-year-old Jakob Beer, who has to overcome the specters of the war such as witnessing his parents' death and the disappearance of his sister. Not only is the book a discussion of how to deal with the ghosts of the past, but it is also a poetical treatise on the burden that Holocaust survivors carry and the possibilities of liberating themselves from it. The multiplicity of interpretations pertaining to the problem of frontier-crossing is a striking phenomenon of the story.

The present study will undertake only a few of these interpretations, and will refer to crossing the frontier of death understood as the physical escape of Jakob from the Nazi slaughter, followed by his constant flight and weaving between life and death until his safe arrival in Greece via Biskupin. Another layer of the motif is connected to Jakob's travel from Poland to Greece, then to Canada and back to Greece, staking new territories and trying to find consolation for the trauma he has experienced. The final question, though the topic is still far from exhausted, is how memory crosses the boundaries of generations and affects survivors as well as their children with the frequently unbearable burden of remembering and simultaneous acceptance of their lives.

2 Trauma: Escape and Memory

The focal and opening event in the narration is the traumatic murder of Jakob's parents, which he witnessed while hiding in a cupboard when the Nazis entered their house somewhere in occupied Poland. Even though he was only seven years old at the time, when he heard the screams of his parents and realized they were dead, he instinctively decided to escape as far from the place as possible. Atrocious as this experience was, he was troubled even more by the fact that he did not hear the cry of his beautiful, fifteen-year-old sister, Bella. Her spiritual presence remains strongly with him throughout his life, and he dreams of receiving a confirmation of her safe escape or proof of her death. Jakob felt an urgent need to run away from the house, from the site of the slaughter, abandoning the dead bodies, to find a way out of the death zone. This represents the first dimension of crossing the frontier of death, of the direct Nazi threat. The boy was afraid they would come back if they realized they had spared somebody. At the same time he acquired an overwhelming certainty that his parents' spirits, through crossing the boundaries of life and death, were accompanying him in his quest into the unknown: "[t]hen—as if she'd pushed the hair from my forehead, as if I'd heard her voice—I knew suddenly my mother was inside me. Moving along sinews, under my skin the way she used to move through the house at night, putting things away" (Michaels 1998, p. 8). Leaving the house, and drastically leaving his childhood behind, Jakob starts his journey into the darkness of the uncertain. He walks at night, eating anything edible that he finds, hiding during the day in the forest, among roots, in dry leaves, and in mud, not to be seen by anybody. Feeling pain, cold and fear, he gradually escapes the death which has marked his life forever. Not only does he cross the frontier of death and fear, but also a figurative boundary between his innocent childhood and the awareness of adulthood. This is indicated through a reference to other deaths and mass killings: "From the other bank, I watched darkness turn to purple-orange light above the town, the colour of flesh transforming to spirit. They flew up. The dead passed above me, weird haloes and arcs smothering the stars" (Michaels 1998, p. 7).

The escape undertaken by Jakob, while physically exhausting and difficult for a small boy, is also a tremendous psychological challenge, as he does not know his destination. His running at night is an attempt to flee from death rather than an escape to a certain delivery. It consists in constantly crossing the frontiers between life and death, day and night, staying still and going forward, and breathing the air and immersion in mud. The first phase of his journey ends when he arrives in Biskupin, where he sees a geologist working in the excavations of a famous prehistoric settlement, now fully reconstructed: "I leaped from the streets of Biskupin; from underground into air. I limped towards him, stiff as a golem, clay tight behind my knees. I stopped a few yards from where he was digging—later he told me it was as if I'd hit a glass door, an inarguable surface of pure air—'and your mud mask cracked with tears and I knew you were human, just a child. Crying with the abandonment of your age'" (Michaels 1998, p. 12). Jakob's appearance in Biskupin, frequently called the Polish Pompeii, and numerous references to archeology that are scattered throughout the whole book, is compared symbolically by Eaglestone (2008, p. 120) to the fate of Jews in Europe during the Second World War. The first and the only words that Jakob utters, upon seeing a man in Biskupin, are "dirty Jew" (Michaels 1998, p. 13), repeated over and over again. The settlement of Biskupin, a wooden complex, "preserved by the action of the silt and water, becomes a motif for the murdered communities of European Jewry, the survival of some of the Jews, the processes of archeology-like memory" (Eaglestone 2008, p. 120). Although this moment becomes the moment of Jacob's salvation as Athos, the Greek archeologist, smuggles him to Greece and raises him, this is also the point when Jakob starts being haunted by memories of what he has gone through. Not fully conscious of what has happened, the boy's memories are stored, layer by layer, like the layers of mud, clay, ash and soil covering the once destroyed civilizations of Pompeii or Biskupin. Therefore, the second experience of crossing the frontier is never finished, and remains a focal element shaping Jakob's life.

3 Ben and Jakob: Annihilation, Hell, and Rebirth

The motif of a drowned settlement of Biskupin is accompanied by another, significant meaning. 'The drowned city' is a title of two main parts of *Fugitive pieces* and it may also be a metaphor of the annihilation of European Jews in the Nazi ravages. Such a drowning in silence can be a reference to the fact that the whole Europe remained silent on this mass killing. This drowning in silence is also visible in the narration overtaken by Ben who, as a child of the Holocaust survivors, born four years after his parents' escape from the horrors of the camp, never really heard a story of his family and according to Eaglestone: "Ben's family in Toronto [is] drowned in silence about the Holocaust" (Eaglestone 2008, p. 122). Ben admits that "[t]here was no energy of a narrative in my family, not even the fervour of an elegy" (Michaels 1998, p. 204) and, as he continues: "my parents and I waded through a deep silence, of not hearing and not speaking" (Michaels 1998, p. 204). These two bitter statements prove Ben's lack of knowledge about his roots and the past of his parents. He waits vigorously for some revelation, some stories disclosing the war years. Since he was not even born at that time, he was spared suffering and death perhaps but he was also devoid of memories. Ben is only offered some insight into his mother's family stories but he still lacks a substantial part. Moreover, his mother is only ready to reveal some of her past when the father is not at home, which Ben sums up: "The code of silence became more complex" (Michaels 1998, p. 223). In this way Ben is spiritually linked with Jakob's life not only by their Jewishness but also through the fact that Jakob's life becomes a key to overcome the void of Ben's silenced past.

Professor Rachel Falconer, a critic interested in the vision of hell and descent to the underworld, sees *Fugitive pieces*, and thus Jakob's journey, as a descent narrative (Falconer 2007, pp. 91–2). The idea of such a narrative is closely connected to the fact that Jakob has an overwhelming feeling of the presence of his family's ghosts.[1] Despite the fact that Jakob's communication with Athos is hindered by lack of knowledge of languages, Athos teaches the boy English and Greek. The war, which also haunts the beauty of the Greek island of Zakynthos, where they live, or Athens, where they stay for some time, forces them finally to leave for Canada. But during their time in Greece Jakob learns a lot, and the sum of his wisdom does not only comprise languages, but also Athos's professional interests in geology, archeology, and poetry, as well as the calm and stable love that Athos is able to offer. All of these do not, however, erase Jakob's memories, and Jakob is painfully aware of this as he consciously tries to suppress such memories with new languages he learns, with the care and histories Athos confides in Jakob and yet the erasure of the past is unavailing. He admits his failure despite

[1] Apart from naming the book a 'descent narrative', Falconer also locates the novel within 'hauntological' narratives and thus compares it to W. G. Sebald' *Austerlitz*, where memories and haunting by the war images play important roles as well. Falconer (2007, p. 91–92) acknowledges her inspiration with Jacques Derrida's use of the term.

many undertakings: "Years later I would try a different avalanche of facts: train schedules, camp records, statistics, methods of execution. But at night, my mother, my father, Bella, Mones, simply rose, shook the earth from their clothes, and waited" (Michaels 1998, p. 93). Athos's role in comforting Jakob is significant. Not only is he a caring guardian and mentor, who comforts Jakob each time he wakes at night by saying: "Jakob, I long to steal your memories from you while you're sleeping, to siphon off your dreams" (Michaels 1998, p. 92), but, according to Falconer, he becomes Virgil leading Jakob in a descent to find consolation (Falconer 2007, p. 105). Apart from the obvious references to Dante's *Divine comedy* and Virgil's guidance through Hell, Purgatory and Paradise (led by Beatrice), and the well-known myth of Orpheus visiting the underworld, there are other mythological connotations that Falconer's reading of *Fugitive pieces* offers. One of them is the figure of Isis from the myth about Osiris, revisited by Milton. Falconer claims:

> The myth of Isis, allegorized as a truth-seeker, (...) who wandered the world, gathering up the limbs of the mangled body of Osiris (...), provides perhaps the closest mythic analogue to Jakob Beer's story. Like Isis, Jakob gathers up fragmentary images of his sister Bella, and Ben [Jakob's successor] in turn collects and publishes Jakob's memoirs (Hammill 2007, p. 105).

The motif is closely connected to the third understanding of crossing frontiers proposed at the beginning of the present study. Crossing the boundaries of countries, first from occupied Poland to Greece in a car, hidden under Athos's coat, then from war-stricken Greece to Canada, to return to Greece at the end to look for Athos's heritage, corresponds to a descent into the underworld and is an attempt to quench Jakob's torment, to come to terms with his traumatic memories, "to retrieve the fugitive pieces of the past," to use Falconer's term (2007, p. 105). This quest goes through various stages, from immersion in the muddy night landscape of Poland, through his sunny but fear-filled stay on Zakynthos when Jakob remained inside the house due to his fear of being found and recognized, to an exploration of the Canadian landscape. The Polish phase represents the primal urge to escape from death, then Athos's storytelling serves as therapy, while in Canada they try to find consolation through "escapes to ideal landscapes; lakes and primeval forests so long gone they could never be taken away from us" (Michaels 1998, p. 102). Canada is, on the one hand, a safe hiding place because it is physically far away from the land of terror, but on the other hand it offers new dimensions to the torment by the specters of memory. Once it is the language Jakob hears accidentally in the street, and another time it is the love he feels for a woman which suddenly awakens memories of his sister Bella. English seems to be a safe zone for him, and Jakob realizes this: "when I began to write down the events of my childhood in a language foreign to their happening, it was my revelation. English could protect me; an alphabet without memory" (Michaels 1998, p. 101). The erasure of the past and soothing the traumatic memories becomes an incorrigible dream even in Canada. It was partially due to the fact that the area they lived in bordered the Jewish district with a vivid representation of this ethnic group: "As if determined by historical accuracy, the Greek neighbourhood bordered the Jewish. When I first discovered the Jewish market, I felt a jolt of grief. Casually, out of the mouth of the cheese-seller and the baker came the ardent tongue of my childhood. Consonants and vowel: fear and love intertwined" (Michaels 1998, p. 101). The physical boundaries of countries and landscapes are crossed, new territories are staked, the descent into hell perhaps comes to an end, but there is still a long way to travel before the nearly impossible aim is achieved—namely to go through purgatory and finally arrive in paradise.

4 From Burden to Gift: Redemption Through Love

While the English language and Canadian settlement far from the war and Europe may be seen as a chance to forget, this narrative proves that forgetfulness after the trauma of the Holocaust is unattainable. It has been widely accepted that the Holocaust has changed the perception of memory, and has initiated a series of new explorations of this field (Whitehead 2009). This assertion leads, however, to a quandary about "the very possibility of representation and remembrance" (Whitehead 2009, p. 84). It is only a starting point to further debates whether memory is single, unitary or collective, to what extent prose or poetry can be a representation of the traumas hidden in memory. Many scholars (Falconer 2007, p. 90; Hammill 2007, pp. 122–123; Rzepa 2004, p. 86) claim, however, that memories of trauma are inescapable and can be converted from a burden into a gift only on two conditions: that the redemption is achieved through love, and that the experience is transformed into a narrative. It is Athos who tells Jakob, "Write to save yourself (…) and someday you'll write because you've been saved" (Michaels 1998, p. 165). Jakob writes his memoir, composes poetry, translates the works of other writers as well as completes Athos's *magnum opus* on Nazi archeology. His own memoir is completed by his spiritual heir, Ben, who travels to Greece in order to follow Jakob's path, and being himself a child of a Holocaust survivor, continues Jakob's story. The story is then realized by a second generation affected by the trauma, and as Rzepa states: "the memory of death seeps through generations, enters the pores of living" (Rzepa 2004, p. 86). The process of crossing boundaries has arrived thus at the fourth phase of moving into the next generation. This exemplifies Marianne Hirsch's theory of postmemory, which calls for "the condition of exile" and "diasporic experience," as well as an idea borrowed from Nadine Fresco of "absent memory" (Hirsch 1997, pp. 242–243), to comprehensively depict memory haunting and thus determining the lives of the children of Holocaust survivors, who carry on the task of remembering.[2]

[2] The term 'absent memory' used extensively by Marianne Hirsch in her book *Family frames: photography, narrative and postmemory* published in Cambridge by Harvard University Press in 1997 is borrowed from a text by Nadine Fresco "La Diaspora des Cendres" which is not directly quoted in the present paper as it is only available in French. The article was originally published in *Nouvelle revue de psychoanalyse,* 1981, pp. 205–220.

Ben, who has already been mentioned in this article, proves Fresco's and Hirsch's ideas with his whole existence. Having been devoid of hardly any memories of his parents' past, he has to fill in the gap of the unknown himself. Throughout his whole life, Ben tries to get to know his "absent memory" (Hirsch 1997, p. 243). In her study, Hirsch discusses such instances as Ben mentioning a "generation whose parents never spoke of their abandoned world or of their wartime experiences and who thus had almost no access to the repressed stories that shaped them" (1997, p. 243). This impossibility of talking about the war experiences is a frequent motif in various studies on the Holocaust survivors. It is a subject of a lecture given by Rabbi Sacha Pecaric in the Center for Dialogue and Prayer in Auschwitz in 1999 where he explains different meanings of the word 'silence'. He differentiates between the two understandings: on the one hand, silence, *sztika*, means simply being quiet, not speaking and on the other silence, *dumija*, is when a human being is speechless and is unable to speak in the face of unspeakable atrocities (Pecaric 1999). A similar phenomenon is observed by Cathy Caruth who has been an interviewer and scholar employed in gathering recordings of the Holocaust survivors at the Furtnoff Video Archive. She claims that "the imperative to tell the story of the Holocaust is inhabited by the impossibility of telling, and therefore, silence about the truth commonly prevails" (Caruth 1995, p. 64). The decision not to talk about the past is not an innocent one as it always has serious consequences. These who are silent "become victims of a distorted memory" (Caruth 1995, p. 64). This memory influences their own lives as they live in the constant shadow of the evil from the past as well as their children.

Such a situation refers clearly to Ben whose lack of knowledge haunts him and the fragments of information he received from his mother make him realize the enormity of unspeakable atrocities as he says: "Loss (…) drained everything from my father (…) what my father had experienced was that much less bearable" (Michaels 1998, p. 223). His father refused to talk about his life during the war and probably therefore Ben willingly undertakes a search for Jakob's past to reconcile with his own parents though Jakob's life. Ben's travel to Greece reveals a lot and the discovery of Jakob's diaries helps Ben come to terms with this debilitating silence inherited from his parents. They probably wanted Ben to avoid thinking about their suffering and humiliation as this would stigmatize him forever. This silencing could have also been incapacitating especially for Ben's father. It turns out that the war, though never mentioned at home, left an imprint on Ben's mind. It is Naomi, Ben's wife, who "says a child doesn't have to inherit fear" (Michaels 1998, p. 280) but Ben responds firmly: "But who can separate fear from the body? My parents' past in *mine molecularly*" (Michaels 1998, p. 280, emphasis mine). In spite of the silenced words, stories, and facts, the emotions and fears are passed down to the next generation. Children can feel the heritage and instead of memories they have to fill in the void of their post memory in order to cope with their lives.

The condition mentioned above, namely that of redemption through love, is also visible in Jakob's life. It is falling in love with Michaela which probably saves him from spiritual annihilation. It is not an easy experience, but a truly liberating one. Despite the fact that Michaela causes Jakob to constantly recall his sister

Bella: "I watch Michaela bake a pie. She smiles and tells me that her mother used to roll the pastry this way. Unknowingly, her hands carry my memories. I remember my mother teaching Bella in the kitchen" (Michaels 1998, p. 192), it is her love and her courage which cause Jakob to cross the frontiers of death. Although their dream of having a child named Bella/Bela never comes true and such a cycle of life cannot be completed, Michaela's love accomplishes the task of symbolic salvation: "Michaela takes me to one of the meccas of her childhood, a birch forest growing out of sand. This is where I become irrevocably unmoored. The river floods. I slip free the knot and float, suspended in the present" (Michaels 1998, p. 188). After spending a day in nature, Jakob and Michaela fall asleep in a tent. The rain comes and in the morning the landscape around is beautiful yet not so welcoming. In spite of the weather it is this place which makes Jakob find consolation. It is not clearly stated why but one may risk a claim that the birch forest in a rainfall seems similar to Jakob's childhood memories of his homeland and its scenery: "When we wake, there is a pool of water by our feet. It is not on Idhra or on Zakyntos but among Michaela's birches that I feel for the first time safe above ground, earthed in storm" (Michaels 1998, p. 189).

5 Conclusion: Ways of Memory

Fugitive pieces, along with numerous interpretative approaches and a complicated structure, is definitely a novel which reflects upon psychoanalysis as it closely draws on memories and suppressed pain. Falconer uses Freud's theory "in which remembering consciousness digs backward in time to uncover an encrypted source of psychic pain" (2007, p. 91). This obsessive going back to the past is present in Jakob's as well as Ben's lives. In both cases they return to the painful experience of the Holocaust. Jakob does it in order to find out whether or not his sister, Bella, was killed in the slaughter and Ben in order to cross the frontier of silence about the past. Falconer claims that the leading motive for Jakob's coming back is a confrontation of "the sources of [his] pain" (2007, p. 93). Such a psychoanalytical rendition of the novel helps to combine a few, important elements: childhood, trauma, distant past, and repressed or non-existent memories.

The condition of exile and trauma, the absence of parents and sister, their ghosts haunting Jakob, as well as Jakob's haunting of Bella by not letting her go, make the novel a multi-layered discussion on the forms of memory, the ways of crossing the boundaries of life and death and chances of ascension from hell. The proposed frontiers which are crossed in the book reveal only a fragment[3] of the possible

[3] Meredith Criglington (2006) proposes another reading of the novel according to Walter Benjamin's idea of *Bilddenken*. or "thinking-in-images," being a frame for a connection between philosophy and its historical images for the generation born after WW II. He explains this reading technique as the one "in which philosophical concepts are made visible through concrete historical images" (2006, p. 88).

interpretations, and show the fusion of history and memory, but also death and love, as the powers holding a strong grip over humanity.

References

Caruth, C. 1995. *Trauma. Explorations in memory.* Baltimore: The Johns Hopkins University Press.
Criglington, M. 2006. Urban undressing: Walter Benjamin's 'Thinking-in-images' and Anne Michaels' erotic archeology of memory. *Canadian literature*. 86-102.
Eaglestone, R. 2008. *The Holocaust and the postmodern.* Oxford: OUP.
Falconer, R. 2007. *Hell in contemporary literature. Western descent narratives since 1945.* Edinburgh: Edinburgh University Press.
Hammill, F. 2007. *Canadian literature.* Edinburgh: Edinburgh University Press.
Hirsch, M. 1997. *Family frames: Photography, narrative and postmemory*. Cambridge: Harvard University Press.
Michaels, A. 1998. *Fugitive pieces.* London: Bloomsbury.
Pecaric, S. 1999. Niezdolność mówienia.
http://www.cdim.pl/index.php?option=com_content&view=article&id=145 %3Asacha-pecaric-niezdolno-mowienia&catid=53 %3Ateologia-w-obliczu-auschwitz&Itemid=10&lang=pl. Accessed 8.07.2010.
Rzepa, A. 2004. Cultural and personal haunting in *Fall on your knees* and *Fugitive pieces.* In *Individual and community: Canada in the 20th century. 2nd International Conference of Central European Canadianists. Proceedings.*. Brno: Masaryk University, 83-87.
Whitehead, A. 2009. *Memory.* London-New York: Routledge.

India Re-loaded: Vikas Swarup's *Slumdog Millionaire* as a Postcolonial Novel

Ryszard W. Wolny

Abstract Once "the jewel in the crown" of the British Empire, India for some time ceased to be the focus of interest for the British writers. Largely due to the success of its film version, Vikas Swarup's *Slumdog Millionaire* (Black Swan, London, 2005, originally called *Q and A*) again drew the attention of the Western world to the problems postcolonial India has to face: poverty, crime, sex abuse, exploitation of children, police brutality and many more. In this paper, however, we are not going to compare the two versions of the story, i.e. the novel and the film, but primarily concentrate on the textual commentaries in the context of postcolonial theory and literature. Of particular interest for us there will be the problem of language since one of the most fundamental questions in postcoloniality is in what language the postcolonial writer should write, and even though it is evident that Swarup—an experienced Indian diplomat—uses English for his novel, we know it is certainly not a complete answer. Therefore, Kamau Brathwaite's theory of a "nation language" will be discussed in an attempt to formulate the paper's main thesis as to the national, postcolonial character of Vikas Swarup's *Slumdog Millionaire*.

1 Introduction: Postcolonial Erosion

Once "the jewel in the crown" of the British Empire, India for some time ceased to be the focus of interest for British writers. After E. M. Forster's *A passage to India* (1924), there was just one significant—though fourfold—attempt by the British

R. W. Wolny (✉)
University of Opole, Opole, Poland
e-mail: rwolny@uni.opole.pl; ryszardwolny@onet.eu

J. Fabiszak et al. (eds.), *Crossroads in Literature and Culture*, Second Language Learning and Teaching, DOI: 10.1007/978-3-642-21994-8_18,

novelist to tackle the problem of once important British colony, the *Raj Quartet* by Paul Scott. In June 1964, Scott began to write *The jewel in the crown*, the first novel of what was to become the *Raj Quartet*. It was published in 1966 and it did not stir a furore either among the readers or critics. However, it has been considered the most significant in the whole sequence—the other being *The day of the scorpion* (1968), *The towers of silence* (1971) and *A division of the spoils* (1974). With their outstanding output, Salman Rushdie and S. V. Naipaul (though Trinidadian by birth, he was Indian by origin) made great contributions to the Anglo-Indian literary heritage of the second half of the 20th century, particularly with their *Midnight's children* (1981) and *India: A wounded civilization* (1977), respectively. But, then, there was a silence.

Largely due to the success of its film version, Vikas Swarup's *Slumdog Millionaire* (2005, originally called *Q and A*) again drew attention of the Western world to the problems postcolonial India has to face: poverty, crime, sex abuse, exploitation of children, police brutality and many more. In this paper, however, we are not going to compare the two versions of the story, and our objective is to show that *Slumdog Millionaire* is an entirely new type of postcolonial novel—so much different from what has commonly been acknowledged as postcolonial, that is, that type of a novel which signifies its opposition to, in the main, (European) imperialism and Eurocentrism. Thus, of particular interest for us there will be the problem of language since one of the most fundamental questions in postcoloniality is in what language the postcolonial writer should write, and the issue of language will be discussed in an attempt to formulate the paper's main thesis as to the changed, re-shaped postcolonial character of Vikas Swarup's *Slumdog Millionaire.*

In the second decade of the twenty first century, we may unhesitatingly speak, paraphrasing Terrence Hawkes's renowned statement of the "erosion of the assumptions and presuppositions central to the study of literature" (in General Editor's Preface to the first and second edition of Bill Ashcroft's et al. *The empire writes back* (1989), of the erosion of the term "postcolonial" and various presuppositions in regards to "postcolonial literatures," alongside a clearly visible and ongoing departure from Afrocentrism in postcolonial theory.

We may generally agree with the assumption formulated by Ashcroft et al. that "the literatures of African countries, India, Malaysia, Malta, New Zealand, Pakistan, Singapore, South Pacific Island countries, and Sri Lanka are all post-colonial literatures" and that "the term 'post-colonial' … [covers] all the culture affected by the imperial process from the moment of colonization to the present" (2010, p. 2). Of course, we may also accept as true various commonsensical notions of the term that relate it to the time following the establishment of independence in a colony: "occurring or existing after the end of colonial rule" (Shorter Oxford Dictionary 2007, p. 2297). But multifarious debates in recent times (e.g. Howe 2002; Young 2001 and 2003, Edwards 2008 or Mullaney 2010) bring to fore the problem of a rapidly changing world, where the old labels of the Third World do not fit any longer to such emerging *postcolonial* nuclear powers as India or Pakistan, with the former also dominating the world stock exchanges with their booming mining and steel industries. What, then, does India have in common with, say, Sudan or Kenya? The British colonial rule—may be the shortest possible answer. So have Australia and New Zealand. So has the U.S. No doubts, the term "postcolonial" needs a thorough re-working and redefinition in the face of economic, military and cultural change in the early 2010s.

2 The Question of Language

Vikas Swarup's *Slumdog Millionaire* (2005) does not deal with India's political or military problems at all, nor—directly—with British imperial heritage, but instead it tells a story of a simple boy from the lowest sphere of Indian society who wins a TV quiz only to be immediately arrested for an apparent fraud. However odd it may sound to the ear of a Westerner, he is not surprised at all by this fact since he knows what it means to live in Dharavi, Asia's biggest slum:

> Arrests in Dharavi are as common as pickpockets on the local train. Not a day goes by without some hapless resident being taken away to the police station. There are some who have to be physically dragged off by the constables, screaming and kicking all the while. And there are those who go quietly. Who expect, perhaps even wait for, the police. For them, the arrival of the jeep with the flashing red light is actually a relief. (Swarup 2005, p. 11)

He is also aware of his low social standing:

> When your whole existence is 'illegal', when you live on the brink of penury in an urban wasteland when you jostle for every inch of space and have to queue even for a shit, arrest has a certain inevitability about it. You are conditioned to believe that one day there will be a warrant with your name on it, that eventually a jeep with a flashing red light will come for you. (Swarup 2005, p. 12)

The main character, Ram Mohammad Thomas, a former tiffinboy from Mumbai, can be viewed as exemplary of India's misplaced children, and the novel as a scathing social commentary on today's India, with all its (post)colonial and multicultural heritage. His full name, consisting of three names denoting the three main religions of India: Hindu, Islam and Christianity, makes him an emblematic figure of multicultural India, the fact also noticed by the quiz presenter: "Mohammad Thomas, now that's a very interesting name. It expresses the richness and diversity of India" (Swarup 2005, p. 46). Ram's social awareness does not stop here: he knows perfectly well that he has already trespassed a foreign territory—the territory of the rich and famous, where the classically Marxian class struggle occurs:

> They will wag a finger at me and remind me of what the elders in Dharavi say about never crossing the dividing line that separates the rich from the poor. After all, what business did a penniless waiter have to be participating in a brain quiz? The brain is not an organ we are authorized to use. We are supposed to use only our hands and legs. (Swarup 2005, p.12)

Therefore, if the fundamental—and at the same time—the founding question of the book is how a poor boy can win the top prize in a TV quiz, which demanded the knowledge of not only popular culture (popular film stars) but also high culture (Beethoven, Shakespeare), our founding problem in this paper is to disclose these elements of the novel that make it postcolonial in the 21st-century sense. Providing that one of the measures of "postcoloniality" is the dependence, relationship or a link—stronger or weaker—between the former colony (the periphery) and the former empire (the centre), *Slumdog Millionaire* demonstrates such a relationship between India and England in a variety of ways, the most significant being the language and popular Western culture. What seems to be an issue here is the fact that Swarup decisively rejects all postcolonial, mainly African and Caribbean, experiments with the English language.

The language of the novel is Standard English, only with occasional lapses into the vernacular dialect of Mumbai (formerly Bombay). Characteristic will be the use of local nouns and adjectives alongside some corrupt English phrases as in the dialogue between the main hero and a prostitute:

> "No, wait. I want to ask you some more questions."
> "*Arrey*, have you come here to fuck or to talk? You are like that *firang* reporter who came here with his tape recorder and camera. Said he was not interested in me and was only doing his research. But the moment I opened my *choli* he forgot all about his research. The only sounds on his tape recorder will be his moaning and groaning. Let me see now whether you're the same." (Swarup 2005, p. 300)

Swarup does not attempt to construct what could have been an Indian equivalent of Edward Kamau "nation language" (1984), nor does he reject English altogether like a Gikuyu writer from Kenya, Ngugi wa Thiong'o (1981), who—after a successful career writing in English—decided to return to his native language in order to save the culture of his people. Swarup does not follow the steps of the Nigerian novelist Ken Saro-Wiwa (1995), whose *Sozaboy* (1985) was written in rotten English—a local, though corrupt, form of English, nor does he take the advice of his more distinguished compatriot, Salman Rushdie, who advocated changes to English when it is used by Indian Anglophone writers:

> One of the changes has to do with attitudes towards the use of English. Many have referred to the argument about the appropriateness of this language to Indian themes. And I hope all of us share the opinion that we can't simply use the language the way the British did; that it needs remaking for our own purposes. Those of us who do use English do so in spite of our ambiguity towards it, or perhaps because of that, perhaps because we can find in that linguistic struggle a reflection of other struggles taking place in the real world, struggles between the cultures within ourselves and the influences at work upon our societies. To conquer English may be to complete the process of making ourselves free. (Rushdie 1992, p. 17)

Notwithstanding the words of the great Indian-born master, the author of the highly acclaimed novel *Midnight's children* (1981) about the Indian independence, Swarup made his own choice for Indian postcolonial novel, and his choice was Standard English.

3 Embracing the Postcolonial Heritage

It is noteworthy that the early beginnings of postcolonial criticism are popularly associated with Indian independence (1947) and "as part of a general leftist reorientation of the 'Third-World' struggles (above all in Algeria) from the 1950s onwards" (Selden et al. 2005, p. 221). Although born in Martinique (the Caribbean), Frantz Fanon, one of the founders of postcolonial theory and an Algerian by his personal choice, situated language at the centre of colonization, servitude and marginalization. In *Black skins and white masks* (1952), he declared that "a man who has a language consequently possesses the world expressed and implied by that language" (1986, p. 18), to conclude with the statement in which he argued that "[m]astery of language affords remarkable power" (1986, p. 18). Putting language together with power does not seem to be a particularly original idea—just to recall Michel Foucault's famous *bon mot* to the effect that knowledge is power—but, situated within the Indian context, it acquires a fresh meaning. The English language—the language of the former colonizers—would thus enable the Indians—the former colonized—to possess "the world expressed and implied by that language," (Fanon 1986, p. 18), that is, as the novel teaches us, the world of money, power, glamour, superstars, luxurious houses and cars, swimming pools, expensive drinks and beautiful women—the world inaccessible to 95 per cent of poverty-stricken Indians, living on a few dollars a day. *Slumdog Millionaire* is a product of Western-turn-Eastern popular culture, the heritage of the founding myth of Hollywood-turn-Bollywood: to become rich and famous alongside a myth of a quick and easy career: from a shoeshine boy/slum boy—to a millionaire/billionaire (in rupees).

What is, however, different here is that it is not exactly the U.S. that was the former colonial power, but it was the U.S. which was Britain's former colony that now re-colonised culturally, materially, ideologically its colonial sister of long-standing. The U.S. plays in India—as much as in other Asian countries—the role of the former empire or, should we rather say, the very contemporary and modern empire—the empire of popular culture and popular myth. The dominance it practices over India is, therefore, a dominance of a different kind than most former empires use to practise: it is no longer a military, political and economical dependence—India is after all also a nuclear power, particularly in Indian sub-continent–, but, principally, it is a cultural dependence, particularly in the sphere of collective consciousness and language, the heritage of their common colonizer.

If language is power, then language is by all means also a site of power struggle—to use a Marxist formula—a site of class struggle or, as Fanon would have it, a site of a struggle for decolonisation/de-neocolonisation. India, then, would be a country in which the conflict occurs more generally on the level of cultures (also religions), one of which is devalued and the other takes dominance over it. So is with language: the English language dominates worldwide, repressing other languages, also local or regional dialects, which leads—in the best of solutions—to hybridization of local or national languages.

Such a hybrid would be—what Edward Kamau Brathwaite, the poet and literary critic from Barbados, postulates—a nation language, a regional form of English to articulate a specific identity that resists Standard English and would be different from a mere dialect since—he argues—dialect carries with it pejorative connotations as an inferior form of the European language. In his critical work, *The Development of the nation language in Anglophone Caribbean poetry* (1984), he pointed out the differences between the official English language as an imposed imperial language and the actual variety spoken in the Caribbean by the Caribbeans. Distinguishing it from Creole English, he defines nation language as "the kind of English spoken by the people who were brought to the Caribbean, not the official English now, but the language of slaves and labourers, the servants who were brought in" (1984, p. 6). The 20th-century India was by no means the Caribbean, and its language problem was of a different kind: it was still a continuation of nation's power struggle. As Arvind Krishna Mehrotra (2009) has it in Introduction to his *Concise history of Indian literature in English*

> The Official Language bill was brought before [Indian] parliament in 1963 and passed the same year, extending the constitutional recognition of English beyond the fifteen-year period. It reassured non-Hindi speakers that English would continue to be used in their states so long as they did not themselves want a change, nor would a knowledge of Hindi be compulsorily required of anyone seeking employment in the central government. (...) In the debate on the bill, speaking with a passion that came to be expected of him when the subject was language, Nehru [the first Indian prime minister] explained why he was pressing for retaining English in India. The British invasion, he said, had 'administered a shock' to people, but the shock had a positive side. Our languages, which like our lives had become 'static', were made 'more dynamic' through their contact with English. English would 'serve as a vitaliser to our languages' in the future, as it had in the past. Succeeding events, both linguistic and literary, have proved Nehru right. (2009, pp. 17–18)

Ram Mohammad Thomas, obviously a fictitious character, was a very lowly paid Indian-born servant who acquired his Standard English not though formal education—which he had none—but because he was brought up by a British priest whom he considered for a long time his biological father (which is a hilarious thing since the priest was white and the boy dark-skinned). Nonetheless, the world presented in the novel is the world of sexual abuse, child slavery, notorious theft, racism, poverty, homelessness shown as the heritage of the British imperial system and the English as the language of oppression and dominance. We are to notice here a long way the postcolonial novel travelled from the early African stage, through the Caribbean to the Indian one, and the various ways in which vernacular authors wanted either to completely get rid of English as a vehicle of communication with their readers or to change it, mix it or hybridize it, to finally accept it as a global language of popular or mass culture. The English, as shown in the novel, is a means of communication, uniting the divided nation of the rich and poor, the Christians and the Muslims and Hindus, the actors and the viewers, and it is largely due to it that the protagonist—a social outcast—is at all allowed to enter for the quiz and ...win it.

Planned as a picaresque adventure novel—the main hero is a foundling born on Christmas Day—*Slumdog Millionaire*, probably not quite coincidentally, appeals to the reader with the vivid descriptions of how majority of Indians live and what they have to go through in their lives dominated, on the one hand, by poverty, but, on the other, by television and the Bollywood myths of being rich and famous. After half a century of their independence, they are still very much dependent on the imperial language and the ideologies of—paradoxically—another former British colony and the whole of the Western civilization alongside its popular and highly electronicized culture. Yet, as it seems evident, the people of India are not at all unhappy with the colonial and neo-colonial heritage, and their discourse is not a reaction to the discourse of colonialism, as it was in the early African stage of postcolonial literature. Further, what they produce does not signify their clear position against imperialism and Eurocentrism. On the contrary, it may even be conjectured that, as Vikas Swarup's *Slumdog Millionaire* tries to argue, their world is becoming a copy, though imperfect but instead with a very local flavour and flare, of the Western world as seen on their TV screens or in cinemas. Genrewise, the novel itself is also a copy or, shall we say, a 21st-century pop culture variant of picaresque novel, a more distant offspring of Henry Fielding and other English forefathers.

4 Conclusion

We may state that *Slumdog Millionaire* is a new type of the postcolonial novel: no longer does it attempt to rid the country of the heavy burden of colonialism; on the contrary, it just presents it, partially in an entertaining, light form, as an unending chain of lucky—but unlikely—coincidences that take "a stupid orphan boy" (2005, p. 67) to the level of a millionaire, giving him at the same time the status of a TV star and popular idol among the low caste; and partially in a decisively gruesome tone, when it speaks of India's pervading crimes and ever-present illiteracy. At the same time, *Slumdog Millionaire* attempts to shift the focus of attention of the postcolonial novel from Afrocentrism and the problematic of, broadly speaking, "decolonisation" of dependent nations to something which just emerges in the New Millennium: India as a regional superpower, India as a prosperous capitalist, multicultural society, India in which even a slumboy can make a career providing he knows Standard English, tries hard enough and the luck is on his side.

References

Ashcroft, Bill, Gareth Griffiths and Hellen Tiffin. [1989]. 2010. *The empire writes back. Theory and practice in post-colonial literatures*. Second Edition. London and New York: Routledge.

Brathwaite, Edward Kamau. 1984. *The development of the nation language in Anglophone Caribbean poetry*. Boston: Beacon Press.

Edwards, Justin D. 2008. *Postcolonial criticism. A reader's guide to essential criticism*. Houndmills: Palgrave Macmillan.

Fanon, Frantz. [1952]. 1986. *Black skins, white masks* (trans. C. I. Markmann). London: Pluto Press.
Howe, Stephen. 2002. *Empire: A very short introduction*. Oxford: Oxford University Press.
Mehrotra, Arvind Krishna. 2009. *A concise history of Indian literature in English*. London: Palgrave Macmillan.
Mullaney, Julie. 2010. *Postcolonial literatures in context*. London: Continuum International Publishing Group.
Ngugi wa Thiong'o. 1981. *Decolonising the mind: The politics of language in African literature*. Portsmouth, NH: Heinemann.
Rushdie, Salman. 1992. *Imaginary homelands*. New York: Granta.
Saro-Wiwa, Ken. 1995. *Sozaboy*. London: Longman.
Scott, Paul. [1966]. 2010. *The jewel in the crown*. London: The Folio Society.
Selden, Raman, Peter Widdowson and Peter Brooker. 2005. *A reader's guide to contemporary literary theory. 5th Edition*. Harlow: Pearson Education Limited.
Shorter English dictionary. 2007. Sixth Edition, Vol. 2. Oxford: Clarendon Press.
Swarup, Vikas. [2005]. 2006. *Slumdog millionaire*. London: Black Swan.
Young, Robert. 2001. *Postcolonialism: An historical introduction*. Oxford: Blackwell Publishing.
Young, Robert. 2003. *Postcolonialism: A very short introduction*. Oxford: Oxford University press.

Narration and Representation of Race in Matthew Kneale's English Passengers

Maciej Sulmicki

Abstract Matthew Kneale's multiple-perspective novel, *English passengers*, written around the centenary of Queen Victoria's death, attempts to portray the realities of the underprivileged races in pre—and early Victorian times. In order for the image to be as impartial as possible, both the repressed and the oppressors are given time on the podium and allowed to present their opinions without apparent interference on the side of the author. Yet, not all the speakers are equally credible. The problematic nature of colonization is shown, with negative consequences arising also from seemingly good intentions. The article presents the differences in modes of presentation of the English, Manx and Tasmanian narrators and characters—the differences in perception of the world and language, as well as their dependence on the circumstances. The interweaving of various identities—the colonizers and the colonized and the subjective nature of the 'other' are also explored. As a whole, the article explores the impact of the narrative structure on the credibility of the narrators and narratives and the way they are perceived by the reader.

1 Introduction

The cover of the Penguin paperback edition of Matthew Kneale's *English passengers* gives an impression of the variety of forms of presentation which awaits us inside. Thus, we have a *wood engraving* of the globe placed in an *inked* stand. On the globe in turn floats a black-and-white *copperplate engraving* of a ship, stand

M. Sulmicki (✉)
Warsaw University, Warsaw, Poland
e-mail: m.sulmicki@gmail.com

J. Fabiszak et al. (eds.), *Crossroads in Literature and Culture,* Second Language Learning and Teaching, DOI: 10.1007/978-3-642-21994-8_19,

three *water-color* painted gentlemen and a *printed and colored* pig. But this is not all—we receive a full impression of the novel as a whole, not just the story-proper, if we look at the back cover, where the same gentlemen who towered over Australia a moment ago now balance on the edge of an oval containing the bar code—the modern point of view underlies the whole nineteenth-century narrative. At the same time we should notice the orientation of the globe on the front cover. Both England and Tasmania, the main subjects and settings of the novel, are at the extremes, in fact invisible. The focus is on neither of the nations, but rather the space in between—the interaction between cultures, individuals and languages. This reflects the structure of the novel, a large part of which takes place on a ship from the British Isles to the Antipodes. The vessel functions as an exterritorial limbo aboard which the Manx and English attempt to find a common tongue.

Considering the multiplicity of heritages represented by the characters in the novel and their largely incompatible points of view, the technique of presenting the story through a series of interweaving independent (unconscious of each other) monologues seems an apt choice. In effect, rather than one narrator, there are twenty-two intradiegetic and two extradiegetic sources of information. The structure of the novel allows both the colonizers (the English), the colonized (Tasmanian Aborigines) and the in-between (the Manx, the half-English half-Tasmanian character Peevay) to speak, with each narrative undermining the others and through adding another facet to the story forcing the readers to revise their understanding of the events and characters described in earlier narrations. The readers are subject to what Dominick LaCapra described as the "retrospective (…) effects of later knowledge" (as in Shiller 1997, p. 543). A good example of this mechanism is when we learn that the missionary Robson led the Aborigines to an island colony not just in the belief of acting for their own good and wanting to convert them… but also because there was a prize of five pounds a head. Robson in fact never speaks for himself, so it is up to the reader to decide which interpretation to believe. The novel abounds with such discordances between the various speakers. Although each of the numerous narratives presents its own point of view, they can be roughly grouped according to ethnic outlooks into three types: the English, the Manx and the Aborigine.

2 The English

Robson is not the only English character presented in a negative light. Both the main English narrators (the titular English passengers): Reverend Geoffrey Wilson and Dr Thomas Potter, serving to represent, respectively, the religious and pseudoscientific aspects of colonization, are largely caricatural portraits. The former organizes the journey to Tasmania in order to prove his theory of 'divine refrigeration' claiming that Van Diemen's Land (Tasmania) is in fact the biblical Eden. The latter is writing a treatise on the nature of human races, intent on proving the superiority of the Anglo-Saxons. Nota bene, Potter's own personality, his petty quibbles with Reverend Wilson and the cutting comments in the Manx narrative demonstrate the opposite to be true, at least in the case of the main English characters of the novel.

> One must nevertheless admit that the Saxons, as represented by Wilson, Potter et al., are surprisingly adaptable in their mental obstinacy. New data, contrary to their theses, rather than disprove them is quickly adapted into a modified version of the same claims. Such is the case when Dr Potter is dissatisfied with the Dutch in Cape Colony as not fitting to his description of the Saxon Type. Rather than revise the bases of his theory, he invents a new category of 'Belgic Celts' whose 'natural' status is that of 'assistants to British' (Kneale 2001, p. 168).

Potter's approach also exemplifies the divisiveness of colonization (Ross 2006, p. 267)—different races are considered to be nearly different species, largely irreconcilable. This approach is well illustrated through the metaphor of the Rabbit-proof Fence popularized through a novel by the Australian Aborigine author Doris Pilkington (born Nugi Garimara).The fence was constructed in the early 20th century in order to prevent the spread of the rabbit population into Western Australia. Measuring around 2,000 miles, the pest-exclusion fence served to segregate land where rabbits were allowed from the settlers' pastoral areas where they were not; government-employed inspectors were responsible for ensuring that no rabbits trespassed onto the other side. Similarly the Aborigines appearing on English settlers' pastures in Tasmania (sometimes treating the sheep as traditional game) in *English passengers* are repeatedly killed. Accordingly the Governor of Van Diemen's Land lays out the plans of dividing the island into two parts—one for the colonizers and the (mountainous) other for the natives.

Kneale's novel emphasizes the role of the British in the demise of native Tasmanians and therefore fits largely into the 'black arm-band history' view of Antipodean history. The term was coined by Geoffrey Blainey as part of the ongoing 'History Wars', consisting of a debate as to the nature of the nineteenth century Australian and Tasmanian colonialism—whether it was far more humane than the colonization of, for example, America, or if it bordered on genocide. If the former interpretation is adopted, most of the faults of the nineteenth century are considered to be the effect of the British rule rather than the actions of the settlers themselves. Although Kneale seems to opt for the black arm-band view in terms of the influence of colonization on the indigenous peoples,[1] many of his decision-making characters seem oblivious to the effects their actions will have. They appear in fact to wish the Aborigines well and do what they consider to be the best for them, or at least what they have convinced themselves to be the best. The veracity of the Australian settlers' and governors' good intentions is however, somewhat undermined by the contemporary state of affairs. As Miles Fairburn points out, "unlike Britain which has granted independence to its indigenous minorities—the Irish, the Welsh and the Scots—Australia and New Zealand still

[1] As Wallhead puts it, "Kneale's objective is to tell the "truth" about the Aborigines' fate, a story that has been silenced or glossed over in the history books" (2003, p. 24).

hold their indigenous peoples—Aborigines and Maoris—in what looks like subjugation" (2004, p. 240).

Not only the English, but also the major figures among the colonized (the female Aborigine tribe leader, the Manx captain) have their minds set on conflict (be it through warfare or smuggling) rather than integration (from their point of view—with the English)—the one exception being the hybrid Peevay. Still, even he decides in the end that completely forsaking violence in the attempts at fighting for his clan's rights was a mistake. The conclusion seems to be that he would have met his death either way, but it would have been less painful psychically, to have died in battle, retaining cultural integrality. Peevay's problem from the start, however, is that he is neither fully Aborigine nor fully English and cannot fully forsake either part of his identity and heritage. His tragic dilemma symbolizes the point of no return which the arrival of European settlers was to the Tasmanian Aborigines.

3 The Aborigines

The profoundness of the perceived disparity between Europeans and autochthonous inhabitants of Oceania may result not only from physical differences (such as skin color), but also geographic localization—on the opposite side of the world from Europe to the Antipodes. As the Roman philosopher Macrobius wrote in the 5th century, long before Australia was discovered, of the other side of the world, "the southern is inhabited by men who plant their feet in the opposite direction to yours and have nothing to do with your people; the other, the northern, is inhabited by the Romans" (Macrobius 1952, p. 200). The main difference in the English attitude and approach was that the other side of the world was also to be settled and ruled by the people of the empire. But the perception of the inhabitants of the Antipodes as profoundly different creatures with whom conversation or negotiations are impossible remained. The Aborigines are seen by the English as at best a people to be subjugated and indoctrinated in the basics English civilization[2] or kept in isolation.

The lack of any serious search for a common language on the part of the English colonists has been pointed out as early as the late 1930s by H.G. Wells, who likened the British settlers to Spanish conquistadors who

> slaughtered [the natives], they robbed them, they enslaved them, and baptized them; but they made small note of the customs and motives that changed and vanished under their assault. They were as destructive and reckless as the early British settlers in Tasmania, who shot at sight the Palaeolithic men who still lingered there and put out poisoned meat for them to find (Wells 1949, p. 783).

[2] Mrs Catherine Price, Wife of the Storekeeper in the Wybalenna Aboriginal Settlement phrases the goal of dealing with the Aborigines as hoping "with time [to] transform them into something like a happy band of English villagers" (Kneale 2001).

Wells, however, retains the view that the Tasmanians were not fully human, as is visible in the above use of 'Palaeolithic' and much more overtly in *The war of the worlds,* wherein the narrator speaks of the "ruthless and utter destruction our own species has wrought, not only upon animals [...] but also upon its own *inferior races*. The Tasmanians, in spite of *their human likeness*, were entirely swept out of existence [...] by European immigrants, in the space of 50 years" (2008, pp. 10–11, my emphasis). This underlying presumption even in narratives condemning the colonists' behavior throws some light on the problems with interracial communication and the unease the Governor of Van Diemen's Land feels when receiving Peevay's pleas and postulations. Serious consideration thereof would require the treatment of Aborigines on equal terms as the white colonists. Instead, even though Peevay learns and uses the language of the English, his words hardly seem to be understood.[3]

The fact that a large part of the culture of the Tasmanian natives was eradicated justifies Kneale's creating his own version of the language of the 19th century Aborigines. A language, as he points out, quite different from that of the Aborigines still living today. Peevay's language is said to have as a goal portraying "someone [...] who is from a culture completely remote from that of the white men but has been educated by them." (Kneale 2001, p. vi) Although he is able to construct a letter or speech in mainstream English, when narrating his part of the story, Peevay uses a specific vocabulary, and sometimes atypical syntax and capitalization.[4] Other Aborigine characters speak in the novel only through being quoted by Peevay—he is therefore the only translator of the indigenous language into English. Hence it is in large part his idiolect which we hear, though there is no reason not to believe that he is translating the words of other representatives of his race as faithfully as he can. As has been mentioned, however, this is not an easy task.

The narratives of characters speaking from the non-mainstream points of view serve to defamiliarize certain notions which may seem obvious to white Europeans, such as their skin color being the default one. When Peevay first sees colonists he describes the "three somethings" (Kneale 2001, p. 54) as clumsy and colored like raw meat[5]—hence unnatural. Similarly English settlers, when forming a posse officially supposed to "persuade" the autochtonous people to move to the mountainous part of the island, speak of the Aborigines as "crows" to be bagged. This mental construct is shaken when the colonists see Peevay's mother, Walyeric,

---

[3] Eventually Peevay decides to fight the white man – his rejection of adopting European culture is expressed symbolically in the decision to kill 'Father' – the English convict who raped his mother. (Kneale 2001) The mock name given to him by the English, Cromwell, turns out to be more fitting than intended.

[4] One should not forget, however, that the Englishmen's own discourse is also far from homogeneous. Among the forms of expression used by English characters in the novel are diary entries, a newspaper article condemning the theft of an Aborigine corpse, prayers, lists, memoirs, legal documents and an excerpt from a tourist guide.

[5] Elsewhere Peevay states with relief that "at least my skin was human colour" (Kneale 2001).

armed with a gun. Although their reaction—letting the natives pass—may be attributed only to fear, it seems that the firearm allowed the Europeans to see the Aborigines as humans rather than animals.

Perhaps one of the factors which determine this dichotomy between Europeans and native Tasmanians is a difference of perception resulting not only from being brought up in a different culture and style of life, but also using a completely different language. As the Sapir-Whorf hypothesis claims, this would determine a disparate point of view. Kneale attempts to illustrate this through an invented Anglo-Aborigine idiolect. The various races can be in fact said to be speaking in tongues, with each one referring to its own reality or at least viewpoint. Problems with understanding each other due to these profound differences were still visible in the second half of the 20th century when the art historian Karol Estreicher in a fairly objective chapter as far as the art of Australian natives is concerned wrote of their "astonishing conservatism, visible in the fact that they do not willingly allow themselves to be civilized and once they are, they are not happy" (1983, p. 297, my own translation).

An interesting example of the endogenic differences of perception is the 2009 movie *Contact*, which describes the history of the remains of a tribe of Australian Aborigines meeting white man for the first time in the 1960s, when their hunting ground is cleared for the testing of military rockets. The women, later placed in a camp for Aborigines and taught English, speak of running from the dragon with fiery eyes when describing the white Australians looking for them in a car, and of escaping to a place where one of their gods would protect them, successfully invoking rain on the way there. They speak in English, however, and laugh at some of the thoughts they describe. A similar change is visible in Kneale's novel. While at first the parts of the novel narrated by Peevay may be somewhat difficult to understand, as the speaker learns new English words and spends more time with Europeans, the speech comes closer to standard British English and many of the concepts at first approached with suspicion and curiosity become commonplace. There remains, however, a definite disparity between the language of the English and Aborigine characters, as does a profound difference between their opinions and viewpoints. As Ross puts it, the bilingualism allows for "a heteroglossia not channeled by the single conceptual groove of "standard" English" (2006, p. 249).

4 The Manx

Also the Manx characters, although speaking English, do not fully fit into the official forms of discourse, as best exemplified by an Anglo-Manx glossary being appended at the end of the novel. The author precedes this appendix with a short commentary on the nature of the language and what it allows to be known about its speakers, i.e. there being "a wealth of words concerned with beatings, inheritance and small amounts of money," but first and foremost presents it as showing that the inhabitants of the Isle of Man "delighted in playing games with language"

(Kneale 2001, pp. 459–460). As Kneale stated in an interview, despite the last natural Manx speaker having died over 30 years ago, the attempts to revive Manx Gaelic are a noteworthy effort. Largely due to the fact that the death of a language entails the death of not only words and forms of expression but also certain notions and a unique point of view they serve to express.[6]

One could venture to say that it is the Manx captain who comes closest to an objective truth in the novel. Nota bene, his name—Illiam Quilliam Kewley—and particularly the initials, IQ, suggest a preference for the character on the author's part. Considering Kneale's Manx roots, this seems a natural sympathy.

The Manx seem more straightforward than the English, for example through not trying to hide from the reader the falsehoods they need to declare when smuggling.[7] It is interesting to compare their overtness with a letter by George Baines, an employee of the New World Land Company, disclosing an instance of mass murder of the Aborigines on the part of early colonists. Despite this being a criminal offense, the white men go unpunished, as in all such cases in the then short history of Van Diemen's Land. In this particular case this is justified with care for the well-being of the settlement and its mother-company. "Shares will lose their value," warns the company chairman. The epistle ends with the at first disgusted settler stating that now, that he has "written it all, every part. It is for that very reason [...] that this letter will never be sent" (Kneale 2001, p. 76).

Another difference in the Manx and English attitudes appears to be an acceptance on the part of the inhabitants of the smaller island of the world as it is rather than attempting to impose their thoughts and ways thereupon. A strong contrast is visible between the narratives of Dr Potter and particularly Reverend Wilson and the opening of the novel by Captain Kewley. While the Reverend goes to the Antipodes in search of proof of the world having begun at a specified date (6,000 years ago), Captain Kewley wonders about the subjective nature of beginnings altogether.

Peevay and the Aborigines in general serve to defamiliarize through showing various aspects of the English way of life from the point of view of those who have never before come into contact with many of the objects, rituals and words which seem commonplace to the average English-speaking European. The Manx, in turn, and in particular Captain IQ Kewley, fulfill the role of the jester, deriding the theories and behavior of the English from the point of view of one seemingly free

---

[6] A significantly simplified example of varying viewpoints is that of Manx toasts, which as Captain Illiam Quilliam Kewley explains, always referred to herring:

> 'Bois da dooine as baase da eease,' we called out, which means in English 'Life to men and death to fish.' and is about herring, as are all Manx toasts. Then it was 'Death to the head that never wore hair,' and 'Here's death to our best friend.' Meaning herring, of course. (Kneale 2001, p. 31)

[7] On the other hand, the addressee (intended audience) of Captain Kewley's narration is never specified, so one cannot in fact be sure whether he is speaking more overtly or just keeping a secret diary.

from their rules and mental constraints such as schematic modes of thought. As Ross points out, "Kewley's personal idiom, like Peevay's, contrasts pungently with the cut-and-dried correctness of most of the other monologuists' idiolects" (2006, p. 263).

5 Conclusion

As has been shown, Matthew Kneale put much effort into making his Neo-Victorian novel polyphonic, not only by admitting a plurality of voices, but also through making use of a variety of forms of expression other than the traditional first—or third-person narration. The types of narration in each case form an integral whole with the substance of what is being said, thus presenting not only events, but also the narrators and their mindsets, dependent largely on who and where they were born. Thus the structure of the novel manages to "preserve the past while always underscoring that we can never know it prior to its transformation into narrative" (Shiller 1997, p. 547).

The lack of a general narrator means that the only elements of the novel-proper which are not elements of the characters' discourse are the chapter numbers followed by information on the name of the speaker and the date. However, these elements as well as the characters' monologues are portioned and arranged in sequence, which along with the epilogue and notes on language and dialect show that the novel is a novel rather than a series of transcriptions and cutouts (Wallhead 2003, p. 27). This in turn suggests reading it as one text rather than a compilation of many independent narratives.

Although *English passengers* can be read as an accusal against the tyrannic nature of British colonialism in Tasmania and in this sense be seen as rather monophonic,[8] it retains many of the characteristics of the polyphonic novel described by Bachtin. Due to the forsaking of a supervising narrator, Kneale has gone even further in creating the impression of the reader dealing, to use Bachtin's terminology, rather than with one "author-artist", with several "authors-thinkers" forming "a plurality of equally important consciousnesses and their worlds" (Bachtin 2009, pp. 153–4) own translation). Accordingly, "any attempt at presenting the world as finished and closed in the monologic sense must inevitably end with failure" (2009, p. 298).

[8] On the other hand, it is possible that this interpretation is a superstructure imposed by the reader, in this case myself, committing the "cardinal error" Bachtin described as understanding the polyphonic novel as "one word, one voice, one accent" (2009, p. 180, own translation). Nevertheless, it is difficult not to read the novel with a sense of the titular English passengers being the symbolic negative characters (cf. Ross 2006, p. 267) who were rightly punished and see this as the author's intended main message to which all the seemingly disparate voices contributed.

It is the tragic results of the monologic perception of the world by the English passengers, governors and settlers which lead to the demise of the Tasmanian natives. *English passengers* fits well into the contemporary attempts at presenting the world as able to be perceived in a plurality of equally valid ways—the only reservation being the Hippocratic one—that they will not entail harm to others. One is inclined to wonder, however, if the presentation of the English as the oppressors (even if often seemingly unmindful of their negative role) of the definitely more positively portrayed Manx and Tasmanians, does not lead to a semi-conscious simplification of the novel's warning against racism, voiced most overtly in the epilogue. Namely, that if one is not English, there is little risk of doing wrong. Or that when one belongs to the oppressed, any deed is sure to be justified.

References

Bachtin, M. 2009. Problemy twórczości Dostojewskiego. In *Ja—Inny: wokół Bachtina. Antologia,* ed. D. Ulicka. Kraków: Universitas, 148–298.

Contact. Dir. Martin Butler and Bentley Dean. Perf. Grant Judson, Juwali Nixon. Contact Films, 2009. Film.

Conversation with Matthew Kneale. Bold Type vol. 3.12. April 2000. http://www.randomhouse.com/boldtype/0400/kneale/interview.html Accessed 19 March 2010

Estreicher, K. 1984. *Historia sztuki w zarysie*. Warsaw: Państwowe Wydawnictwo Naukowe.

Fairburn, M. 2004. The state of Victorian studies in Australia and New Zealand. In *The Victorians since 1901. Histories, representations and revisions,* ed. Taylor, Wolff. Manchester: Manchester University Press, 230–243.

Kneale, M. 2001. *English passengers*. London: Penguin.

Macrobius A.T. 1952. Commentary on the Dream of Scipio. Translated, with an introduction and notes by William Harris Stahl. *Records of Civilization, Sources and Studies*, New York: Columbia University Press, 48.

Pilkington, D. [Nugi Garimara] 2002. *Rabbit-proof fence.* New York: Hyperion.

Ross, M. 2006. *Race riots: comedy and ethnicity in modern British fiction.* Montreal: McGill-Queen's University Press.

Shiller, D. 1997. The redemptive past in the neo-victorian novel. In *Studies in the novel.* Denton: Winter 1997.Vol. 29, Iss. 4:538–571. Literature Online. http://lion.chadwyck.co.uk. Accessed 10 May 2010.

Wallhead, C. 2003. To Voice or not to Voice the Tasmanian Aborigines: Novels by Matthew Kneale and Richard Flanagan. *Revista Estudios Ingleses* 16:6–34.

Wells, H.G. 1949. *The outline of history, being a plain history of life and mankind.* Garden City, New York: Garden City Books.

Wells, H.G. 2008. [1898] *The war of the worlds*. Rockville: Arc Manor.

A Dominican-American Experience of Not Quite Successful Assimilation: Junot Diaz's *The Brief Wondrous Life of Oscar Wao*

Brygida Gasztold

Abstract The aim of this article is to examine Junot Diaz's novel *The brief wondrous life of Oscar Wao* (2007) through the post-colonial lens. Portraying the complicated story of the de Leon's family, Diaz draws a compelling link between the tragic history of the Dominican Republic and equally disturbing issues related to the experience of immigration, in the context of contemporary, multiethnic American society. As a member of the Dominican-American Diaspora, the author explores the anxiety of assimilation and the quest for self-identity. By the means of a polymorphic narrative and unconventional language, the author shows his interest in the persistence of historical memory and the politics of storytelling. Oscar, the titular protagonist, is an adolescent who challenges common immigrant markers such as language, race, class, and masculinity in an attempt to forge his own identity. Re-discovering and re-telling his family history, on the one hand, helps him to make sense of the past, but, on the other hand, encumbers his efforts to come to terms with his immigrant present.

1 Introduction: *The Brief and Wondrous Life of Oscar Wao* as the Immigrant Narrative

The highly acclaimed first novel of the Dominican born, North-American author Junot Diaz is distinguished by an important authorial voice which stretches the boundaries of post-colonial and immigrant discourses. Skilfully connecting past and present, exploring themes of exile and nationality, authenticity and

B. Gasztold (✉)
Technical University of Koszalin, Koszalin, Poland
e-mail: brygida.gasztold@tu.koszalin.pl

J. Fabiszak et al. (eds.), *Crossroads in Literature and Culture,* Second Language Learning and Teaching, DOI: 10.1007/978-3-642-21994-8_20,

globalization, Diaz challenges the well established notions of a normative Anglophone literary model creating a unique immigrant voice which inhabits a "pluriverse," rather than a "universe," and where old linguistic, cultural, and social boundaries need to be reframed. The painful process of shaping the protagonist's identity is closely connected to the ambiguous factors, which are deeply embedded in the dispute about racial, class, and gender issues. Diaz's literary comment adds to the discussion about the idea of ethnicity in the contemporary American society by positing that one's ethnic affiliation is determined as much by politics as by biographical data, language, race, and gender. In other words, the political status of a social group determines whether it is classed as ethnic or not. This essay examines Diaz's novel *The brief wondrous life of Oscar Wao* as a means to investigate normative roles versus their social practices in an attempt to describe the trauma, which the protagonist's family's history and their experience of immigration inflict on the adolescent, male, mind.

2 Language as a Medium of Identity Construction

What firstly draws the reader's attention to Diaz's novel is his use of the English language as a medium through which a new identity is constructed. Academic English is mixed with the streetwise variety of Spanglish, slang, numerous references to pop culture, especially to science fiction, comic books and computer games. The historical footnotes, written in a non-academic style, familiarize the readers with the tragic history of the Dominican Republic. It is a discourse which mixes high culture with low, sophisticated with vulgar, important with trivial, and is one which promotes a variety but privileges none. "Oscar went home morose to his pre-Korean-sweatshop-era cartoons—to the Herculoids and Space Ghost. What's wrong with you? His mother asked.[…] When Oscar whimpered, Girls, Moms de Leon nearly exploded. Tu ta llorando por una muchacha? She hauled Oscar to his feet by his ear" (Diaz 2007, p. 14). By choosing not to italicize Spanish words, the author breaks grammatical standards but, more importantly, he refuses to marginalize those Americans who do not speak English. Integrating Spanish words into an English text, he not only legitimize the hybrid language itself but also the experiences of those does who speak it.

Literary code-switching between Spanish and English has been used, to some extent, by most Chicano and Chicana writers since the 1970s (Martin 2005, p. 405). In 1967 Rodolfo Gonzales published an epic poem "I am Joaquin" written both in English and Spanish. Addressing the urban Chicano, the poet interweaves a retelling of the Mexican history with the current American issues. He represents a *barrio* youth punished for speaking Spanish while facing the challenges of discrimination on the way to assimilation. In the 1980s the social concept of assimilation was supplanted by diversity, encouraging ethnic writers to voice their own concerns in their heritage languages. The language code-switching, which is under the author's control, is not an indication that the speaker lacks the necessary skills to communicate fluently in another language, but adds extra layers of meaning to the text by means of a conscious, artistic device. Alfred Arteaga calls it a heterotext, which aptly reflects the hybrid nature of the *mestizos*: Mexican identity, which is the mixture of Indian and Spanish blood, coming to terms with the Chicano, people of Mexican heritage born in the US, and the US Americans (Arteaga 1997, pp. 24–43). Gloria Anzaldua in *Borderlands/La frontera: the new mestiza* (1987) argues for the acceptance of multilingual narrative as a legitimate mode of an artistic expression: "The switching of 'codes' … reflects my language, a new language—the language of the Borderlands" (Anzaldua 1987, p. 8). Similarly, a Chicana poet, Lorna Dee Cervantes uses English and Spanish to render a linguistic split which mirrors the multiple perspective of an immigrant identity. Spanish, just like Native American and Chinese, within an English discourse, enables the author to draw upon the resources of both languages thus creating a useful tool to express the dynamics of literature written by multilingual authors.

In the country where Diaz's native tongue is not dominant, his unique use of language is a key marker of his own identity as well as a manifestation of a narrative which comes into a dialogue with the past tragedies of the country of his birth, and the present demands of his adopted home. It is the mother tongue which binds the characters to the past, offering the comfort of the familiar, but, at the same time, denying participation in the new linguistic environment, at least, at the beginning of their new lives in the U.S. That is why Diaz's characters must constantly negotiate the terms, in which they identify with the dominant culture, and their own. Western thought, which is based on binary oppositions, is at a loss to explain this multi-faceted experience, hence, a new dialogical context is necessary to accommodate this vision. Junot Diaz's language represents the colonizer and the colonized, the past and the present, the feelings of displacement, otherness, discrimination and racial stereotypes.

Yet, the hybrid dialect, which makes use of English ridden with Spanish expressions, questions its own communication values. The kind of patois, which Diaz, uses is common to bicultural immigrants like Oscar and the author himself, but what about the non-Spanish-speaking readers? Are they supposed to miss parts of the text and enjoy the book nonetheless? Is this a novel for a target (English and Spanish speaking) audience only? In an interview Diaz comments on these questions: "I want people to research, to ask each other, to question. But also I want there to be an element of incomprehension. What's language without incomprehension?" adding, "Isn't it about time that folks started getting used to the fact that the United States comprises large Spanish-speaking segments?" (O'Rourke 2008, p. 4) Diaz's fiction does not only require the reader's active participation during the act of reading (English-Spanish dictionary and the internet to look for the meaning of the words no standard dictionary contains, in my case) but it also legitimizes the presence of an ethnic, post-colonial, immigrant, multicultural voice stressing the Other on the American literary map. The voice which refuses to resign from linguistic authenticity for the sake of acceptance by the mainstream culture. "Elite artists have no problems with employing elite language without translation—Greek aphorisms, say, or entire passages in French," argues

Diaz, "Elites never punish themselves for being esoteric" (Somerset 2008, p. 1). Sophia A. McClennen in *The dialectics of exile: Nation, time, language, and space in Hispanic literatures* adds her argument: "The experimental language ... appears ... as a concrete effect of being outcast by authoritarian regimes that tried to impose their own rule over language. ... Exiled writers want to find a home in language (and see it as their only means of political intervention) but at the same time distrust language because they know it is an instrument of power" (qtd. in Rueda 2007, p. 509).

The mighty power of words constitutes a major thread in the novel, which is the Curse and the Doom of the New World, a.k.a. fuku, which afflicts Oscar's family: the fall and deprivation of the de Leon's family, their forced immigration to the U.S., broken family ties, unhappy marriages, and unfulfilled loves. Diaz employs the idea of a curse, which alludes to magic realism, as another dominating narrative that orders the chaotic events by introducing a sense of meaning and purpose. As an organizing principle, a curse is a lense through which the narrative flows, offering a plausible explanation when rational reasons do not suffice, *nota bene*, the tragic story of Belicia'a parents, her lost siblings, and her own miraculous survival. In Diaz's words: "Curses were also a really great way to warp between huge, amorphous History and the personal history of this family. The Greeks, the early Hebrews, they understood that. The way you could connect the vast history of the gods and of the nation to the person is through curse. A curse takes that vast vapor of history and condenses it into a form that we can imbibe" (Boldtype 2008). For people living under the regime, blaming their misfortunes on a curse is a safety valve which helps to transform losers into victims, shifting the blame onto something mythical, something you cannot argue with, and which needs no further explanation. Muffling the mind by arresting further enquiry, a curse may be an excuse which frees one from personal responsibility, at the same time, enhancing a miserable life by lifting it up from the mundane. Finally, a curse, also gives hope that one day it will be lifted, and a new, happy life will start. That is why Oscar relentlessly pursues what everybody is talking about—love, and what he finds at the end is his own demise. His tragic fate may have been the "zafa," another powerful word, just like Junot Diaz wonders "if this book ain't a zafa of sorts. [His] very own counterspell" (2007, p. 7).

3 Diverse Narrative Voices, Which Tell the Story of the Leon's family

The story of three generations of Dominican immigrants is, in accord with the post-modern theory, as fragmented as are its many narrative voices: Yunior, who tells Oscar's story, with contributions from Oscar's sister Lola, mother Belicia and grandmother La Inca. The author oscillates between different perspectives offered by the narrative voices: from the present-day to the historical one, creating a space in which the characters are not fixed, but fluctuate and reveal themselves depending on who is telling their story: the third person narrator, the second or the first person. The main voice, Yunior, is not an omniscient narrator, and he must rely on the second hand information, especially when relating Belicia's Dominican youth under the Trujillo dictatorship. Thereby, Diaz's narrative creates a world which feeds on gaps and silences, which, according to Diaz, lie at the core of the immigrant experience: "The fundamental byproduct of trauma is silence. Immigration put a gag on so many families" (Diaz Johnny 2007). The figure of silence dominates the tragic history of de Leon's family during the dictatorship of Rafael Trujillo, when people disappeared without a trace, as well as, the first years of settling in an alien, both linguistically and culturally, American society. The lack of words which results in the inability to articulate one's emotions haunts Oscar's family long after they gain command of the English language, the fact that stands in the way of their mutual relations. Diaz constructs a narrative which resists univocal reading by virtue of employing unreliable voices, which struggle to be objective. Neither the multiple narrators nor the readers can assume full authority over the text and remain in parts ignorant. Escaping authorial authority, Diaz foregrounds the dangers of uniformity in terms of language, culture, race and policy, which foster the feelings of discrimination and marginalization. His dialogue with both his native and adopted homeland is the source of his creation of his present "Dominican-American" self.

The disjointed and polyphonic narrative, which zigzags from present to past, using creolized English as a medium of communication, eludes easy generalizations and highlights another issue which the novel interrogates—dictatorship. The lives of Oscar's family are intricately linked with the history of the Rafael Leonidas Trujillo's Dominican Republic: the private becomes historical. The ample footnotes and the multiple narrators act as native informants familiarizing the readers with the tragic history of the Dominican Republic. The political events are entangled with the characters' lives making a vague concept of global history palpable to the readers. It is a matter of historical fact that Trujillo was supported by the U.S., which means that one cannot tell the history of the Dominican Republic without mentioning the United States, whereas, hardly any American school textbook mentions the American invasion of the Dominican Republic. The post-colonial discourse begins with the uneven distribution of power and goes down deep into the fabric of the society, contesting its original structure in order to establish its own rhetoric. Immigrant voices, like Yunior's and Oscar's, not only must face the constraints of their own ethnicity but also address the challenges of the dominant culture.

Dictatorship is about power and control, on the one hand, and subjection and silence on the other. Dictators assume all authority and refuse to acknowledge any other voice: "In dictatorships, only one person is really allowed to speak" (O'Rourke 2008, p. 2). Violence, which is an ineluctable companion of colonial power, is reflected both on the state and individual level. The atrocities, which accompanied the bloody *régime* on the island, do not stop with the death of the dictator. History leaves its mark on the lives of people and the pattern of violence is repeated on the personal level. Hence, the human relations among the de Leon's family are characterized by constant tug—of—war in which none of the members assumes the full authority over the others, conversely, none can enjoy tranquillity. The colonizer's legacy: the breach of trust and the inability to communicate permeates further generations of Dominicans. The author uses a figure of "running" to convey the characters' feelings of entrapment, which urge them to escape both from the particular location and circumstances: Lola is a track and field star, Bela, her mother, "had always wanted through her Lost Childhood: to escape" (Diaz 2007, p. 80). Put another way, the feeling of transience is inherent in immigrant experience, as rooting takes longer than one generation. Lola reflects on the fate of her family: "you can never run away. Not ever. The only way out is in" (Diaz 2007, p. 209). As much as traumatic experiences divide the peoples' lives across the political axes in the old country, once shared by the Dominican immigrants, they may trigger a formation of a vital characteristics of that community: the collective immigrant memory persists and haunts the characters.

4 Humor and Satire as Rhetoric Tools, Which Challenge Authority

Humor and satire are rhetoric tools, which Diaz uses to undermine the authority of the ruling power. When Oscar is confronted by his rival—a Dominican policeman, he shouts in self-defence that he is an American citizen, to which the policeman replies: "I'm an American citizen too. I was naturalized in the city of Buffalo, in the state of New York. I bought mine in Miami [another policeman added]. Not me, [the third one] lamented. I only have my residency" (Diaz 2007, p. 295). This scene would not be humorous if it was not for the fact that it is all happening in the present-day Dominican Republic. The sarcasm and humor, which help the family survive the three decades of the Trujillo dictatorship, also prove effective in their American life, allowing them to express their subaltern's perspective. As satire is based on the specific power of words to construct a distorted reality, it suspends judgment, giving the characters respite from their daily hardships. Verbally, humor and satire, mediate the distance between the character's expectations and the failing reality, if not in hope to remedy it, then, to undermine the colonial and post-colonial authority. By making fun of racial stereotypes, satire offers a vent to accumulated negative emotions, in other words, satire is an effective tool in harnessing aggression. A transgressive power of humor, wit, self-irony, and satire, in Diaz's novel, functions as a strategy of the characters' survival, as well as, a means to expose racial and gender hierarchies dominant in the American culture.

5 The Importance of Whiteness

Another issue, which is of paramount importance to Diaz's characters, is the notion of whiteness. Dark skin color stands for the racial Other in the country where white is considered normative. White supremacy is an infamous legacy of the colonizer's culture imprinted on the post-colonial countries like the Dominican Republic: "The family claims the first sign was that Abelard's third and final daughter,..., was born black. And not just any kind of black. But black black-kongoblack, shangoblack, kaliblack, zapoteblack, rekhablack—and no amount of fancy Dominican racial legerdemain was going to obscure the fact. That's the kind of culture I belong to: people took their child's black complexion as an ill omen" (Diaz 2007, p. 248). One cannot escape the observation that in Diaz's novel skin color does not help the characters to define themselves as much as it differentiates them from the privileged others, which results in the feelings of discrimination and ostracism: dark-skinned Bela lands in a Dominican school where the majority of pupils are rich and white-skinned, and the school's handsomest boy is the whitest: "all those pale eyes gnawing at her duskiness like locusts" (Diaz 2007, p. 83), Oscar comments on his time in an American school: "more kids of color—but some things (like white supremacy and people of color self-hate) never change...Every day he watched the 'cool' kids torture the crap out of the fat, the ugly, the smart, the poor, the dark, the black, the unpopular, the African, the Indian, the Arab, the immigrant, the strange, the feminine, the gay" (Diaz 2007, p. 264). Even Lola internalizes and reiterates the white, negative rhetoric saying that "we colored folks talk plenty of shit about loving our children but we really don't" (Diaz 2007, p. 35). Diaz's characters adopt the self-depreciating colonizer's rhetoric which reinforces the old colonial stereotypes and transplants the same scheme from the Dominican to American society. Junot Diaz sees it clearly: "The problem isn't in labeling writers by their color or their ethnic group; the problem is that one group organizes things so that everyone else gets these labels but not it" (O'Rourke 2008, p. 4).

6 The World of Phantasy as Oscar's Refuge

Although Oscar is fully immersed in American mass culture and his dreams are white, he cannot escape marginalization and discrimination common to people of color. To be "[a] smart bookish boy of color in a contemporary U.S. ghetto" is "[l]ike having bat wings growing out of your chest" (Diaz 2007, p. 34): Oscar does not comport with the stereotypical image of an Latino immigrant, which results in his social ostracism and the feeling of alienation. However, he neither tries to battle the system nor comes to terms with it, but, instead, finds a refuge in the world of his imagination, devouring fantasy, science-fiction and comic books featuring invincible heroes and offering happy endings: "In his dreams he was either saving [girls] from aliens or he was returning to the neighborhood rich and famous—It's him. The Dominican Stephen King [...] In these apocalyptic dreams he was always some kind of platano Doc Savage, a super genius who combined world-class martial artistry with deadly firearms proficiency. Not bad for a nigger" (Diaz 2007, p. 27). The withdrawal from the real world into the realm of the imaginary restores his inner balance allowing him to experience a different set of values, where evil is punished and good rewarded, however, painful the moments of awakening are. Yet, there is another reason why he chooses science-fiction genres over the others: the immigrant journey from the underdeveloped Third World Dominican Republic to the modern superpower like the U.S.A. is like traveling through a time zone and requires an imaginative leap similar to the one evoked while reading about imagined worlds. The science fiction conventions lend themselves well to describe what an average Caribbean immigrant felt like, in the time when there was no satellite TV or the internet. The metaphors of the science fiction genre do not only bridge two seemingly incompatible worlds but they offer "the missing emotional cognitive disjunction" (Danticat 2008, p. 4): the eternal other can easily be reclaimed to denote a racial or linguistic other, which is common in immigrant discourse.

7 Oscar as a Denial of the Stereotype for Latino Masculinity

Masculinity, like race and class, is a social construct, which is bonded to its social and historical context. In Latin American context, John Riofrio claims, the "discussions of masculinity have long been dominated by the notion of machismo, a manliness that overpowers and in fact seems to spill over, an excess of masculinity" (2008, p. 24). Notably, he argues that the discussion of masculinity "moves us beyond the simple binary of patriarchal privilege and dominance to a more nuanced understanding of masculinity as embodying both oppression and the oppressed" (Diaz 2007, p. 26). In the Dominican context, Oscar represents the whole generation of young people who grow up fatherless because the economic situation on the island drives their fathers to the U.S., to seek employment opportunities that allow them to provide for their families. No doubt, female protagonists are also affected by growing up in a fragmented family, but young boys on the brink of adolescence who are deprived of masculine role models, must rely on the remembered vestiges of their fathers' presence. That is why their vision of masculinity is mostly affected by the peer dynamic, at the center of which, there is a demonstrative contempt for feminine qualities. In fact "femininity functions as a symbolic border" (Diaz 2007, p. 29), the trespassing of which undermines a boy's own sense of manhood, on the one hand, and threatens his position in the masculine social hierarchy, on the other. As much as Oscar's story is endemic to any immigrant fate, he is the contradiction of what Diaz refers to as a stereotyped Latin hyper maleness. An overweight boy, not aggressive, a book worm who cries easily, "too dorky, too shy, too weird" (Diaz 2007, p. 17), a "sci-fi reading nerd"

(Diaz 2007, p.19) with no knack for music or dance, he "couldn't play sports or dominoes" (Diaz 2007, p. 20)," and never had much luck with the females, how very un-Dominican of him (Diaz 2007, p. 11),—a distant image of both a sleek Latino lover and a sporty, popular American youth. Although assimilation provides Oscar with new tools to construct his identity, his refusal to internalize the normative roles leaves him both distanced from his ethnic roots, and excludes his emotional identification with the American mainstream society. Diaz challenges traditional masculine markers of a lewd and lustful Latino, perpetuated by multiple cultural and social representations, on the one hand, and the Caribbean tourist industry, on the other. Oscar's story demonstrates how the deviation from the norm is pronounced as weakness and castigated.

Diaz's narrative interrogates a central platitude of the traditional Dominican male, which is based on sexual conquests: "One thing you can count on in Santo Domingo is sex" (Diaz 2007, p. 206). Riofrio (2008) explains the importance of male sexuality in Latin American context, claiming that it "is the only tangible cure for the humiliation of poverty and the [protagonist's] only means of establishing the virility he craves both personally, for himself, and externally in the desire for his absent father" (Diaz 2007, p. 27). The stereotypical Latin American character's interest in sexuality should not be confused with his responsibility for family: in Oscar's family men sooner or later disappear, leaving women to carry the burden of bringing up the children. Encumbered by the expectations to conform to the Latin alpha male mould, which haunts the immigrants from the Dominican Republic to the United States for they begin their American lives in the Dominican Diaspora, Oscar is obsessed with the idea of sex, recklessly trying to find a date, as he "heard from a reliable source that no Dominican male has ever died a virgin. […] O, it's against the laws of nature for a dominicano to die without fucking at least once" (Diaz 2007, p. 174). His quest for a date has at its roots not so much his own desire for love and intimacy, but the assurance of his own masculinity: Oscar seeks endorsement among his peers to dispel his own doubts, which result from his fatherless childhood and diasporic anxieties.

However, much as Oscar tries to defy the normative roles, he subconsciously strives to live up to the expectations embedded in his mind by the Caribbean macho culture he inherits. However, when he finds role models among the American mass culture idols, he realizes that he is not representative of them, either. Neither a stereotypical Dominican boy, nor an average American youth, Oscar escapes definite labels and challenges the readers' expectations. He straddles two different cultures and is reluctant to make a choice, yet, the story seems to ask why should he? As his past and present worlds do not meet, he relentlessly shifts between them trying to work out his own space within. As his attempts to meet the expectations of his Dominican family, on the one hand, and his American school friends, on the other, are unsuccessful, he questions the viability of the two constructs. Diaz creates a protagonist whose dynamic is impossible to grasp as he moves inside and outside of the postcolonial discourse, nullifying all the preconceived notions.

8 Immigrant Adolescents: The Question of Oscar's Identity

Oscar's restlessness also derives from the fact that he is going through a difficult period of adolescence, and, just like his mother Bela at the same period of her youth, exhibits "inchoate longings of nearly every adolescent escapist" (Diaz 2007, p. 80). The formative time in any teenager's life acquires another dimension in a life of an immigrant youth. Although Oscar does not personally experience immigration, he is born in the U.S., his story "demonstrates how the children of immigrants are affected by their parents' torn relationships to both homeland and spouse" (Friedman 2009, p. 85). Oscar's father is an absent figure in the boy's life; similarly, his mother was brought up by a female relative: women are the breadwinners in Oscar's family. The old world fosters stereotypical roles for men and women in a patriarchal family: the "macho" Latino and the subservient and domestic female. Simultaneously, Diaz's narrators observe the men stray or leave their spouses, commit acts of adultery, and later come back home, where the women forgive them. This family reunions are based on the shifting grounds as the story repeats itself, regardless of the geographical location. A lack of a father figure for a young boy deprives him of the feeling of safety, and, at the same time, questions the paternal role model. On top of that, Oscar must negotiate the validity of the historical patriarchal patterns against American modernity.

After graduation, Oscar comes back to Paterson to teach at his old college. His success at education asserts his assimilation, yet, the fact that he comes back home to live in the Dominican Diaspora, questions its validity. Put another way, the fact that Oscar chooses the familiarity of his ethnicity over the challenges of the American mainstream culture, indicates that his assimilation is not complete. Consequently, his anguish and pain: "Some mornings he would wake up and not be able to get out of bed" (Diaz 2007, p. 268), results from his failure to construct a stable and coherent identity. Immigrant's experience requires re-inventing oneself in a new environment and, *en route*, confronting one's presence to accommodate one's past, hoping that the final result will yield a frame of reference strong enough for the character to lean on. As long as Oscar's return to the homeland of his ancestors might suggest his own renewed interest in his ethnic roots, the tragic events which ensue challenge that claim: his journey back home proves fatal for him. What it illuminates, though, is the fact that the old world is no longer a safety zone, nor is the new one. The idea of "home" is deconstructed in the way in which neither place acquires a definite meaning, which, in turn, interrogates one's sense of belonging.

Diaz portrays not only the history of the Dominican-American immigrants, but also the working class experience, which seems to be endemic to this Diaspora. Although the story of the de Leon's family goes back to the times of the Dominican affluence, the country's stormy history deprives its inhabitants not only of their wealth, but also denies them education and takes away their dignity: La Inca constantly reminds Beli "of her Family's Glorious Past. [...] Remember, your father was a doctor, *a doctor*, and your mother was a nurse, *a nurse*" (81), your parents "owned the biggest house in La Vega" (Diaz 2007, p. 82), and Beli "was the Third and Final Daughter of one of the Cibao's finest families" (Diaz 2007, p. 82). However glorious the remembered past may be, the present American reality for Diaz's characters is filled with hours of hard work, a constant lack of money, and economic uncertainty about the future. Diaz's portrayal of immigrant life is permeated with physical labor because the Dominican immigrants usually perform the most basic, domestic chores, which are underpaid: "the cold, the backbreaking drudgery of the factorias, the loneliness of Diaspora" (Diaz 2007, p. 164). Amidst the hardships of everyday existence, there is no room for hopes about the American Dream, thus, none of Diaz's characters expresses a desire to reach this, seemingly common, immigrant goal. Instead, Diaz's narrative portrays immigrants who are content with minor representations of success, like home ownership or becoming a teacher.

9 Conclusion: Ways to Challenge Post-Colonial Discourse

Junot Diaz's novel is an attempt to describe the immigrant condition in a discontinuous discourse, which makes use of cross-cultural references, depicts the hybridity of the contemporary American society, and denies sole authority to any of the fictional voices. The narrative takes the shape of an archipelago, "a textual Caribbean," explains Diaz, "shattered and yet somehow holding together" (O'Rourke 2008, p. 1). Such a form resists the singularity of perspective, which is often used to establish the book's authority. The author revises the notion of a traditional post-colonial discourse, challenging its popular concepts of language, dictatorship, whiteness, class, and masculinity. Instead, he creates a discourse, which encompasses a narrative immigrant voice, which comes into dialogue with the literature of its host nation. Diaz's novel negotiates the terms between the two, which results in a multi-dimensional text that resists a definite reading. De Leon's family story, with the Dominican dictatorship at its background, is about piecing together what can be retrieved and drawing inferences from the silences. The author interrogates the readers' expectations consciously manipulating various literary genres: he makes use of science-fiction, parodies history and biography. His idiosyncratic language, full of references to comic books, fantasy and computer games, recognizes younger generations as the target of his fiction. Part of the narrative's appeal stems from the fact that, apart from the issues connected with ethnic identity, it is also a story about adolescence, which is an inescapable aspect of human existence, regardless of whether one is an immigrant, or not. What readers can learn by reading about Oscar Wao is that adolescence, just like being an immigrant, is "the inextinguishable longing for elsewheres" (Diaz 2007, p. 77).

References

Anzaldua, G. 1987. *Borderlands/La Frontera: the new mestiza*. San Francisco: Aunt Lute Books.
Arteaga, A. 1997. *Chicano poetics: Heterotexts and hybridities*. Cambridge: Cambridge University Press.
Boldtype. An interview with Junot Diaz. In *Monthly reviews of books worth reading*. http://www.boldtype.com/152726. Accessed 6 March 2008.
Danticat, E. 2008. Junot Diaz by Edwige Danticat. *Bomb magazine*. http://www.bombsite.com./issues/101/articles/2948. Accessed 3 June 2008.
Diaz, Junot. 2007. *The brief wondrous life of Oscar Wao*. New York: Riverhead Books.
Diaz, Johnny. 2007. Eleven years later...Call it a curse or writer's block. But after the acclaim of his first novel, Junot Diaz struggled to write the follow- up. An interview with Junot Diaz.
Friedman, N. J. 2009. Adultery and the immigrant narrative. *Melus* Vol.34, No 3:72–91.
Martin, H. E. 2005. Code-switching in US ethnic literature: multiple perspectives presented through multiple languages. *Changing English*. Vol. 12, No 3:403–415.
O'Rourke, M. 2008. Questions for Junot Diaz. *Slate magazine*. http://www.slate.com/id/2177644/paganeum/all. Accessed 6 March 2008.
Riofrio, J. 2008. Situating Latin American masculinity: Immigration, empathy and emasculation in Junot Diaz's *Drown*. *Atenea*. Vol. 28. No. 1. 23–36.
Rueda, M. H. 2007. Review of *The dialectics of exile: nation, time, language, and space in Hispanic literature*, by Sophia A. McClennen. *Comparative literature studies*, Vol.44, No.4: 507–510.
Somerset, G. 2008. Sauron of the Caribbean. Review of *The brief wondrous life of Oscar Wao*, by Junot Diaz. http://www.listener.co.nz/issue/3542/artbooks/10747/. Accessed 3 June 2008.

Part V
Private Territories, Social Spaces

Louis MacNeice's *Autumn Journal* as a Place of Asylum in Space and an Archive in Time

Teresa Bruś

Abstract This chapter sets out to apply Philip Lejeune's idea that a diary is both a "place of asylum in space" and an "archive in time" (Katherine Durnin, Versa Press, Manoa, 2009, p. 334) to a reading of Louis MacNeice's *Autumn journal* written from August 1938 until the New Year and published with a preamble in 1939. Lejeune interestingly marks that the diary is "virtually unfinishable from the beginning, because there is always a time lived beyond the writing, making it necessary to write anew" (Katherine Durnin, Versa Press, Manoa, 2009, p. 188). MacNeice did return to his journal 15 years later in 1953 with *Autumn sequel*. It consists of 26 cantos that contain a number of characters drawn from his personal friends and also "hinged to the autumn of 1953" Louis (Collected poems, Faber and Faber, London, 1979, p. 329). Both *Autumn journal* and *Autumn sequel* are defined by the poet as occasional poems. *Autumn sequel,* however, was not received well by critics. The unfavourable reviews stressed the fact that it "look[ed] back over its shoulder beyond the fire-blackened pit of the war" and that it was a mere "ironic gloss on its predecessor" (in Stallworthy (Louis MacNeice, Faber and Faber, London, 1979, p. 410)). Most critics writing about this powerful personal document seem to agree that *Autumn journal* caught the spirit of the 30s in Britain as Eliot's *The waste land* did for the 20s, and that in its "honesty" and range it can be compared to *The prelude* (Jon Stallworthy Louis MacNeice: His own changing self, The Web, p. 248). This comparison is also brought to attention by John Powell Ward. In *The English line: Poetry of the unpoetic from Wordsworth to Larkin* he admits that "MacNeice's time does not allow him Wordsworth's tenacious moral grip against his guilts" (Ward John Powell The English line: Poetry of the unpoetic from Wordsworth to Larkin. Macmillan, Hountsmills, 1991, p. 167). *Autumn journal*'s "unflinchingly personal" meanings and MacNeice's position of engagement with history, of acceptance of political

T. Bruś (✉)
Wrocław University, Wrocław, Poland
e-mail: tbrus@poczta.onet.pl

J. Fabiszak et al. (eds.), *Crossroads in Literature and Culture,* Second Language Learning and Teaching, DOI: 10.1007/978-3-642-21994-8_21,

nature of man not only at times of profound public crisis, make it "one of the few unremittingly good long poems in English" in the twentieth century (Bergonzi Bernard Reading the thirties. Texts and contexts. Macmillan, London, 1978, p. 109).

Days are where we live…
Philip Larkin
…because a diary is like lacework, a net of tighter or looser links that contain more empty space than solid parts
Philippe Lejeune

1 Introduction: Diarizing Synthesized by Poetry

This partial diary, the type Lejeune defines as devoted to a single phase and organized around a particular era of experience, foreshadows a clearer, more explicit closure than a real diary. In the words of Virginia Woolf, the all-purpose diary, a "captious hold-all", plays on evanescent oscillations and diverse rituals of closure. *Autumn journal's* personal expressive function is determined by the diarist's apprehension over the impending war. MacNeice says his journal is "hinged to that season" (1979, p. 329); it closes as "The New Year comes with bombs". (1979, p. 153).[1] The banal title of the journal does not immediately signal that "alarm that came in the autumn" (MacNeice 1996, p. 174) or its setting in the incommunicable reality which Stephen Spender aptly diagnosed as a "mysterious and unimaginable" interval of history (1978, p. 22). The significance of the journal lies not only in its symptomatic character as an example of *l'ecriture de l'exceptionel*. *Autumn journal* is a poet's diary written under the pressure of the moment in a precise, rhythmical scheme. Because the diarist's activity of urgent accumulation of experiences is framed as a lyric poem,[2] we receive its composite nature and standards "which are not merely personal" (MacNeice 1979, p. 101) as "contaminated." The poet surrenders himself to poetry and his experiences are elaborated and synthesized in a poetic structure.[3] Though written *as* a journal, the poem "is deceptively coherent artistically, revealing to a marked degree the art

[1] In his autobiography *The strings are false* he writes of the moment: "The alarm came in the autumn. Perched on the very joists I came down very quickly, began shrieking down for solidarity" (1996, p. 174).

[2] In the Note to *Autumn journal* MacNeice says it is "something half-way between the lyric and the didactic poem"(1979, p. 101).

[3] I am using W. B. Yeats's idea which MacNeice does not assimilate. Yeats wrote his journal in the form of loose notes kept separate. He says he "must keep one note from leading on to another. To do that is to surrender oneself to literature". The process involves first "a casual thought" and then it becomes his life (Yeats 1972, p. 139). It seems MacNeisian way is informed by the distrust of the fragmentariness and pretentions of the modernist kind. His journal is socially constituted and yes, it looks onto a sustained narrative.

that conceals art" (O'Neill 1992, p. 187). For this reason it is to be carefully considered as shaped not so much by introspection but by auto creation, as a diary which is most of all an aesthetic creation with ambivalences sewn into the conflicting identities of its subject.

The choice of verse form in which to contain the record of difficult times bears a strong connection with MacNeice's personal preference for impure poetry which is musical and rhythmical in content. "Unsuited to controversy," verse "convey[s] a certain skepticism about its conclusions" (Auden 1989, p. 26). Verse is responsible for the poem's seemingly unified communication pattern. The design of *Autumn journal* in a verse form is not only a way to sustain this skeptical and anxious mood but also to integrate the personal and the public coordinates. And most importantly, as Peter McDonald argues, the verse pattern in the journal reveals the poet's most careful orchestration of the "responsibility towards society and responsibility towards form" (1991, p. 94). MacNeice himself explains that the artificiality of form, including verse is a necessity.[4] For him verse is an instrument of precision thanks to which every word acquires a nuanced signification. He says that "The writer who despises form must still formalize even in selecting material....The chunk-of-life may be all right for the journalist to write and the man in the tube to read but for serious writers it is a missed opportunity" (MacNeice 1987, pp. 142–143).

Lejeune, dismissing such hybrid forms, says that real diaries "have nothing to do with invention" (2009, p. 204). Other critics like Michael O'Neill and Gareth Reeves defend "genuine" journals in which "the present in the writing is the same as the time of writing" and when the diarist is "constantly ignorant of the outcome of the events and experiences" he is recording. They feel though that MacNeice, by not drawing on hindsight, saved his journal's claims to authenticity (1992, p. 186).

2 Between the Diary and the Journal

To anticipate possible confusion, I want to make some preliminary distinctions between the concept of the diary and the journal. Until recently, these forms have received inadequate theoretical attention. Lejeune argues that misunderstandings concerning the objectives of diary-writing led to its misrepresentation. The diary form has been received as unhealthy, hypocritical, cowardly, worthless, artificial, sterile, shriveling, and feminine (2009, pp. 148–150). Critical ambivalence and even negligence have also resulted from diarists' lack of pretense to diaries' artistic structure, their excessive length and their uncertain status as a literary genre. Misjudged as a book of confessions or a self-portrait, the diary continues to

[4] In MacNeice's words: "Verse is a precision instrument and owes its precision very largely to the many and subtle differences which an ordinary word can acquire from its place in a rhythmical scheme" (1987, p. 143).

invite irrational responses; Lejeune suggests that in the diary "everything reinforces people's biases" (2009, p. 154).

Because the terms *diary* and *journal* have been used interchangeably by major critics including the apostle of the diary, Lejeune himself,[5] there also persists a sense of confusion concerning the formal boundaries of both. In *On diary*, published in 2009, Lejeune continues to blur the differences between the diary and journal to direct his analysis to more critical distinctions between autobiography and diary, between a private document and a public text, especially in his preferred realm of manuscripts. Lejeune, exhausted by autobiography, turns to what he feels is a "superior form of truth" (2009, p. 182). He claims that the diary, "hooked on truth" (2009, p. 182) is nothing like an autobiography; it possesses a distinctive value system and its own principles. Yet the diary can always evolve into an autobiographical construction when, for example, it is self-published. Thus Lejeune presupposes that the diary is revisionary and that it is always in process.

According to the historical definition in the *OED*, *journal* is the older of the two terms, with its original ecclesiastical meaning first recorded in 1355–1356. From 1728, *journal* in the sense of newspaper established the term in the public sphere as a document, as more privileged, and as more orderly. In the *OED*, the journal entry takes three times as much space as the entry for *diary*. The term *diary* appears 200 years later than the journal and its development seems ambiguous. In *The encyclopedia of life writing*, Rachel Cottam suggests not history but etymology as a clarifying context in which to understand the nature of the differences between the journal and the diary. "Both these words," she says "are derived from roots meaning 'day'" with the reservation that "only rarely is a diary written daily" (2001, p. 268). Elizabeth Podnieks accepts the ambiguous terms, and uses what various writers say about these two forms to justify continuing to use *journal* and *diary* interchangeably. She quotes Arthur Ponsonby, who argues that "the word *journal* should be reserved for the purely objective historical or scientific records, and the word *diary* for the personal memoranda, notes and expressions of opinion" (in Podnieks 2000, p. 13). In Podnieks's survey, Felicity Nussbaum uses the terms interchangeably, despite her claims that "diaries are considered to be the less elaborate form." Judy Symours explains that "although strictly speaking 'diary' can be used as a generic term to cover both a diary record of engagements and more intimate writing, while 'journal' tends to refer more specifically to a personal chronicle, writers themselves do not always keep such nice distinctions" (in Podnieks 2000, pp. 13–14).

MacNeice himself responds to the strong impulse within to gather his varied selves which are about to enter the public sphere. Grappling with questions of ethical responsibility, he turns to ancient Greeks who recognized that "… While we chat, envious time threatens to give us/the/Slip;" and himself sets to "… gather

[5] I suspect Lejeune blurs the terminological distinctions to allow the etymological variety to coexist. Also Lejeune uses the term *journal personnel* as a much broader and precise one than e.g. *journal in time* which is usually not addressed to a wide audience and which in view of the development of modern media has lost some of its potent secrecy.

the day, never an inch trusting futurity" (MacNeice 1979, p. 550).[6] The strong unifying impulse motivates the process of obtaining self-knowledge and self-possession at the exceptional historic moment.

3 Keeping and Reading a Diary

Recourse to the journal can be motivated by many forces in the writer's life. Lejeune insists with the openness of a theorist and a practitioner of the form, that during the life of a diarist, the journal is first of all an *occupation*, a practice, a "*to-and-fro* movement" which for brief moments "sculpts life as it happens and takes up the challenge of time" (2009, p. 173). Keeping a journal is a way to structure days, providing an organizing principle of behaviour (2009, p. 153). It is a "way of living before it is a way of writing" (2009, p. 153). Unlike Maurice Blanchot, Lejeune sees the practice as sacred. Blanchot's proposition, based on a much narrower choice of literary examples, defines the journal as a substitute for literary creation. Blanchot posits that journal writing is a "cowardly way of writing by avoiding writing," a trap for any writer (in Lejeune 2009, p. 160).[7] These two contrasting perspectives concerning diarists' very personal motivations touch on the question of secrecy.[8] Oscillating between private and public, the journal re-enacts the privatization of the self, its sole authority. As readers, though, we rarely have access to manuscripts and even when we do, we often deal with altered versions, with finished texts, and as a result we are removed from the conditions of its original intentions. Once the author dies, when his journal becomes a published text ratifying experience and identity, the immediacy of the moment of writing is removed.

But for readers, the diary's attraction lies not in questioning the realm of the manuscript but in "the feeling of touching time." Lejeune says that one of the paradoxes of the journal is that it is dated. The date "immunizes it against aging, and it is even an advantage, a daunting face-off with time" (2009, p. 209). Yielding to time, the diary takes up its challenge (Lejeune 2009, p. 173). In a selective and discontinuous "series of dated traces" of the moment, diarist eventually communicates to the curious reader a series of surprises. Because the

6 Derek Mahon, an admirer of MacNeice, resonates the same idea in "How to Live." He says: "the days are more fun than the years/Which pass us by while we discuss them. Act with zest/One day at a time, and never mind the rest" (1990, p. 76).

7 In "The Essential Solitude", Blanchot sees the practice of journal writing as a pause in creation, as a means to "recall himself to himself"; it ties him to the inessential everyday, and such a perspective, he claims, leads not to remembering but to forgetfulness. As a "parapet walk it is viable," but it really belongs to the person who is not capable of "belonging to time through the community created by work." In the journal the I "sinks into the neutrality of a faceless 'he'" (1992, pp. 828–829).

8 Nussbaum's phrase (1988, p. 135).

diary "taps into a new type of relationship between author and reader, with a more active role for the reader" (Lejeune 2009, p. 209), readers experience it as a personal text with provisions for the future. This peculiar relationship becomes more apparent if we realize that the journal is not only a fragmentary record of events but that its dated entries create a growing archive. The diary invites the reader to a place where traces of time are deposited, both a physical place of writing and metaphorically a social place addressed not to a select body of readers, but anyone interested and able to read. In this broad context, the reader confronting traces of what was acts as a guest and a potential historian.[9]

Because of the essential porousness of the diary, reading it is bound to be incomplete, fragmentary and possibly conflict-ridden. To illustrate the discontinuity of the diary's structure, Lejeune relies on metaphors of "lacework, a net of tighter or looser links that contain more empty space than solid parts" (2009, p. 153), a "spider web," or the art of improvisation. Likewise, in *Autumn journal* the poet hopes to survive his despair by "spinning out his reams/Of colourless thread" (1979, p. 104). Because the reader reads any journal as a finished book, not as a manuscript, and because he is not present as the web is spun, he must read it impersonally and as a continuum, an experience Lejeune identifies with violation. There is always a gap between the reader's perception, says Lejeune, and the reality of what happened; there is a gap between how a diary is written (that is, without a knowledge of the ending) and how it is read (with knowledge of ending). There is yet a different gap when a writer reads his journal after it is published. For example, in the Note to his *Autumn journal* written in 1939, MacNeice signals that shift in reception when he says that "in a journal or a personal letter a man writes what he feels at the moment" (1979, p. 101). Faithful to the nature of the poem, as "neither final nor balanced," he resists alterations of "any passages relating to public events in the light of what happened after the time of writing" (1979, p. 101).[10] Yet he feels that he needs to write this short, revisionary prefacing Note as kind of response to T. S. Eliot's direct criticism and as a short comment on his political "overstatements". The Note does not explain his personal lapses of judgements, but presents them as confessions of the moment. A listener to himself, the poet denies authority to the public.

4 Autumnal Archiving

"MacNeice kept nothing" (Stallworthy 2007). No manuscripts of *Autumn journal* survive, there are no traces or drafts available and so we cannot get a sense of its unpolished former selves and its scattered fragments. We cannot gloss its

---

[9] I am drawing here on Ricoeur's idea of an archive as a place and on his understanding of the role of reader as a historian connecting ideas with places (2000, p. 221).

[10] Stallworthy says, "There is no reason to doubt the truth of this statement" (1995, p. 227).

intentions.[11] Instead, following Lejeune's suggestions, we can study it focusing on the problems of its anxious rhythm.[12] *Autumn journal*'s archival accretion of experience and its peculiar vertigo of repetitions create a strong alliance with its time in the face of what the poet calls the "zero hour": "We have come to a place in space where shortly/All of us may be forced to camp in time" (1979, p. 148). Like Lejeune, MacNeice hopes that the confusion of the moment will be recorded and communicated, and the traces of time will prove instrumental both for him and a future reader. Writing a journal is his way of not stopping to write in the face of fear.[13]

Before I turn to some examples of the journal's "camping in time," I will bring the whole text closer to our attention by quoting a statement from MacNeice himself, who in a response to the publisher's request, defined the nature of his poetic journal, before it was finished (on November 22, 1938).[14]

Autumn journal[15]

- A long poem of from 2,000–3,000 lines written from August to December 1938. Not strictly a journal but giving the tenor of my intellectual and emotional experiences during that period.
- It is about nearly everything which from first-hand experience I consider significant.

---

[11] Robyn Marsack supposes that MacNeice did not think them worthy of preservation but also suspects that many 1930s poems were not "heavily worked over" (1985, p. 42). But we find MacNeice praising Yeats for preserving his manuscripts as mines for researches where the poet's "stiching and unstiching" can be studied. MacNeice, for example, is known to have carefully compared versions of Yeats's manuscripts (1987, p. 217).

[12] Lejeune suggests three approaches to the diary study. His preferred method is the study of published and unpublished diaries in archives, then analysis of the diary as a form of behaviour, and study of a diary as text, which should, he says, focus primarily on problems of rhythm (2009, p. 165).

[13] Shared consciousness of the inevitability of war and need to preserve life invites strong comparisons with other diarizing practices in the 1930s. In a prose journal written under the title *Journals* 1939–1983 Stephen Spender believes that "The best thing is to write anything, anything at all that comes into your head, until gradually there is a calm" (1985, p. 161). Like MacNeice, Spender produces the journal feeling that his life is "very unsatisfactory" (1985, p. 175) and like many of his contemporaries he keeps a 1939 diary. Spender mentions discussing the practice with Virginia Woolf who at the time thought "it was the only thing she could do" (1985, p. 198). Spender also reports a conversation with T.S. Eliot who told him that "Just writing every day is a way of keeping the engine running, and then something good may come out" (1985, p. 199).

[14] Stallworthy suggests that *Autumn journal* has everything "that, in 1938, he asked of a poem: it would admit the impurities of the world, the flux of experience, in a documentary form that, for all its seeming spontaneity, would be directed into patterns on a page—as indexes on film—by the invisible imagination" (1995, p. 228).

[15] He sent Eliot the complete manuscript before 2 February 1939 and Eliot's reaction to it was positive. Eliot seemed to like the imagery which he felt was "all imagery of things lived through" (in Stallworthy 1995, p. 237). The text of the statement is reprinted in Stallworthy (1995, pp. 232–233).

- It is written in sections averaging about 80 lines in length. This division gives it a *dramatic* quality, as different parts of myself (e.g. the anarchist, the defeatist, the sensual man, the philosopher, the would-be good citizen) can be given their say in turn.
- It contains reportage, metaphysics, ethics, lyrical emotion, autobiography, nightmare.
- There is a constant interrelation of abstract and concrete. Generalisations are balanced by pictures.
- Places presented include Hampshire, Spain, Birmingham, Ireland, and—especially London.
- It is written throughout in an elastic kind of quatrain. This form (1) gives the whole poem a formal unity but (2) saves it from monotony by allowing it a great range of appropriate variations.
- The writing is direct; anyone could understand it.

I think it is my best work to date; it is both a panorama and a confession of faith. (in Stallworthy 1995, pp. 232–233)

The 24 sections of the journal have a distinctive internal morphology of beginnings and endings; they come one after another evoking a dramatic continuity of calendared, seasonal time. It is not the pattern of his life, but the pattern imposed from outside, defined by the anguish and uncertainty. Its many sections stand at odds with one another; a whole sequence is made up of what Lejeune calls "microorganisms caught up in a discontinuous whole: between two entries, a blank space" (2009, p. 178). The poet records "summer is ending" (1979, p. 101); "August going out" (1979, p. 101); "September has come" (1979, p. 106); "Today was a beautiful day" (1979, p. 108); "October comes with rain whipping" (1979, p. 117); "I see the November sun at nine o'clock" (1979, p. 134); "December the 19th" (1979, p. 145). Despite forebodings about the impending war, subsequent endings open up methodically and repetitively to moments of hope. There are signs of promise and a sense of purpose: "to begin/The task begun so often" (1979, p. 104); "And while a man has voice/He may recover music" (1979, p. 139); and finally "there will be sunlight later/And the equation will come out at last" (1979, p. 153). These beginnings and endings contain repeated allusions to the life experiences of the diarist's "various and conflicting" selves, caught up by the movements they shape while being shaped by them. They articulate a desire for a cure. From the pattern of endings, he hopes to move through the work of recollecting and writing to enduring.[16] Beyond the personal,

[16] Despite its anachronistic appeal, Freud's "Beyond the pleasure principle" (1920) comes to mind. The whole poem lends itself to a reading informed by Freudian claims to pleasure gain resulting from excessive impulse to obtain the mastery of a situation by means of repetition. Painful anticipations coming from the external world give rise to such a principle. Freud emphasizes its active component which he contrasts in this essay with mere "passivity of experience" (1990, p. 646). He also remarks that the repetition compulsion which constitutes psychic life, goes beyond the pleasure principle, that it seems "more primitive, more elementary, more instinctive than the pleasure principle" (1990, p. 646).

the final section gestures in the direction of an awakening in the indeterminate future: "Sleep, my fancies and my wishes, /Sleep a little and wake strong" (1979, p. 151). In its own ending, the journal foreshadows endings of future readings.

MacNeice, painfully aware of the time of crisis captures the movement of this time through an abundant, almost automatic use of the preposition *and.* In the journal, *and* is repeated almost compulsively, creating a total effect of what Lejeune calls "the madness of repetition that is life itself" (2009, p. 170). *And* is MacNeice's personal obsession, a key to his arbitrary version of reality. *Ands* create constellations which display the fullness of the world, but also apprehensions defining his world. They accommodate a series of coterminous and contradictory positions. They arrest movement, establishing chasms, sites of anxiety, filled with verbal heaps and nervous suddenness, but they are also agents of good flow controlling the movement of the revival of past experiences. And unexpectedly, these excessive *ands* produce a sense of regulatory rhythm, a soothing sense of pleasure.

In Sect. X of *Autumn journal,* MacNeice uses *and* 74 times, twice more than in most other sections. It is an autobiographical entry organized between the event of a new school term as "work" for the diarist and his recollections of previous semesters as a student. "And now, in 1938 A.D,/Term is again beginning"(1979, p. 121). Between "the beginnings of other terms" and "now" there are memories of his life seen in as an "expanding ladder" until graduation, when "life began to narrow to what was done-/The dominant gerundive-" (1979, p. 120). Narrowing brings dissipation and loss of *ands* in the growing awareness of the pressure to conform to and perform in society, visibly manifest in the use of active verbs. *And* in the meditation on the early happy memories, exerts itself visibly as a connective force. It designates incidentals and separate things brought together: "dogs and cats, and plasticine and conkers," where "things" "get better and bigger and better and bigger and better" (1979, p. 120). In its unstudied accumulation, that excess is the excess of childhood for the one who remembers the "house of childhood" (1979, p. 120). It is also dynamic, vast and generates significant forward movement: "we went on/growing and growing, gluttons for the future" (1979, p. 120). The *and* of childhood and early youth enumerates not only things but also conditions: "alarm and exhilaration"; "And we had our little tiptoe minds.../And a heap of home-made dogma" (1979, p. 119). With years passing, subsequent terms bring different additions; this time their function is to force collisions of different spaces and objects: "A string of military dates for history, /And Gospels and the Acts/And logarithms and Greek and the Essays of Elia" (1979, p. 120). These heaps alert the reader to the suddenness and tangible character of everything. Because they suppress verbs, multiple *ands* mark the disappearance of a sense of time. These are the *ands* of nervousness, of worry (Gass 1986, p. 170). Dominated by nouns, the lists retard the forward movement of the mind. From then on we enter what Gass calls "the site" (1986, p. 177). "And school was what they said it was, /An apprenticeship to life, an initiation" (1979, p. 121). *Ands* in section X balance and coordinate two modes of excess in a world of bankruptcy.

The formulary of "ands", the continuity and proximity it establishes renders the process of personal gathering inconclusive. It works internally against finality.

Virginia Woolf, when "war seemed round the corner again", when she felt too "black" to "gather together" (Woolf 1984, p. 165) life and oneself, recognized intensely the necessity to oppose the dispersal. Despite everything, she wanted to "gather rosebuds while we may" (Woolf 1984, p. 165). The last sections of her *Diary* abound in "ands". These trans-individual structures are deposited within the asylum of the most personal forms of expression, the diary. Out of generational *langue*, they help form the unique parole of these diarists.

5 Conclusion: In Order to Begin

In the face of the unknown and of emptiness, to prevent forgetting but also to cope with the present moment, riven with tensions of his identity, the poet-as-diarist produces a lyrically elaborated collection, a contingent archive that he retrospectively described as an "honest" journal. This composite document contains records of experiences of private and public crises. Its archival space is held together by the rhythm of the beginnings and endings of its entries; by the natural rhythm of the changing season; and by the recorded rhythm of trains and cars, jazz songs and newspaper slogans which define the rhythm of MacNeice's powerful poetry. Most problematically, however, the internal rhythm of the whole journal is sustained by the indistinct preposition *and* with its inclusive force that allows the poet to "observe while reserving judgement" (O'Neill and Reeves 1992, p. 195). An archive gathered by means of *ands* is an archive open to the future, or as McDonald puts it "pitched into the future" (1991, p. 93).

References

Auden, W.H. 1989. *The dyer's hand and other essays.* New York: Vintage Books.

Bergonzi, Bernard. 1978. *Reading the thirties. Texts and contexts.* London: Macmillan.

Blanchot, Maurice. 1992. The essential solitude. In *Critical theory since 1965.* ed. Hazard Adams and Leroy Searle. Tallahassee: University Presses of Florida. 824–831.

Cottam, Rachel. 2001. Diaries and journal. General survey. In *Encyclopedia of life writing.* ed. Margaretta Jolly. London: Routledge. 267-269.

Encyclopedia of life writing. 2001. ed. Margaretta Jolly. London: Routledge.

Freud, Sigmund. 1990. Beyond the pleasure principle. In *Great books of the western world.* ed. Mortimer J. Adler., vol. 54. Chicago: Encyclopedia Britannica. 639–663.

Gass, William H. 1986. And: *Habitations of the word.* Simon & Schuster.

Lejeune, Phillipe. 2009. *On diary*. eds. J. D. Popkin & J. Rak. trans. Katherine Durnin. Versa Press: Manoa.

MacNeice, Louis. 1987. *Selected literary criticism of Louis MacNeice*. Oxford: Clarendon.

MacNeice, Louis. 1979. *Collected poems*. London: Faber and Faber.

MacNeice, Louis. 1996. *The strings are false: An unfinished autobiography*. London: Faber and Faber.

Mahon, Derek. 1990. *Selected poems.* London: Penguin. 76.

Marsack, Robyn. 1985. *The cave of making: The poetry of Louis MacNeice.* Oxford: Clarendon Press.

McDonald, Peter. 1991. *Louis MacNeice: The poet in his context.* Oxford: Clarendon Press.
Nussbaum, Felicity A. 1988. Toward conceptualizing diary. In *Studies in autobiography.* ed. J. Olney. Oxford: Oxford UP. 128–140.
O'Neill Michael and Gareth Reeves. 1992. *Auden, MacNeice, Spender.* London: Macmillan.
Oxford English Dictionary. 2010. Oxford: Oxford UP.
Podnieks, Elizabeth. 2000. *Daily modernism: The literary diaries of Virginia Woolf, Antonia White, Elizabeth Smart and Anaïs Nin.* McGill: Queen's UP.
Ricoeur, Paul. 2000. *Pamięć, historia, zapomnienie.* Kraków: Universitas.
Spender, Stephen. 1978. *The thirties and after.* Glasgow: Fontana.
Spender, Stephen. 1985. *Journals* 1939–1983. ed. J. Goldsmith. London: Faber and Faber.
Stallworthy, Jon. 1995. *Louis MacNeice.* London: Faber and Faber.
Stallworthy, Jon. 2007. Louis MacNeice: His own changing self. *Contemporary poetry review.* The Web. March 2009. n.pag.
Ward, John, Powell. 1991. *The English line: Poetry of the unpoetic from Wordsworth to Larkin.* Hountsmills: Macmillan.
Woolf, Virginia. 1984. *The diary of Virginia Woolf.* Vol. 5. San Diego: Harcourt Brace and Company.
Yeats, W.B. 1972. *Memoirs.* ed. D. Donoghue. London: Papermac.

Aldous Huxley and Evolving Borders of Social Space: The Changing 20th Century Society in His Essays from the 1920s

Joanna Jodłowska

Abstract The paper intends to analyse Aldous Huxley's collected essays from the years 1920–1929 as regards mentions of the way the society was, in his opinion, being reshaped in its activities, customs and concepts. In other words, the paper's aim is to analyse some of the reflections that had, in a sense, preceded the writing his famous dystopia, *Brave New World*. The aforementioned activities, customs and concepts form the borders of society—within them is enclosed its identity. A change in them, therefore, means moving the borders and instituting a new social space. It is this evolution of social space as observed and charted by Huxley that is of chief interest to this paper. However, the paper shall also link Huxley's comments on the 1920s with our society as it now is and the world Huxley had grown to warn against in *Brave New World*. The 1920s phase will be the one most strongly represented by the analysed source texts, but *Brave New World* shall figure in the analysis as a text displaying the later development of the essayistic observations in Huxley's mental landscape. The current times, on the other hand, will not be explicitly present in the paper, but since their positioning between the past of the actual 1920s and the hypothetical future of Huxley's dystopia seems natural, it is hoped that the audience will be inclined to reflect on them. Issues mentioned will include: the general atmosphere of the post-War West; the leisure culture often described by Huxley as "the Good Time"; the mechanisation of leisure and life in general with the attendant specialisation; economy; advertisement, persuasion and propaganda; aristocracy versus democracy and social engineering.

J. Jodłowska (✉)
Warsaw University, Warsaw, Poland
e-mail: joanna.jodlowska@student.uw.edu.pl

J. Fabiszak et al. (eds.), *Crossroads in Literature and Culture*, Second Language Learning and Teaching, DOI: 10.1007/978-3-642-21994-8_22,

1 Introduction

Many phenomena may be represented through the metaphor of a territory. Among others, as every physical space is, to an extent, defined by its boundaries, so a society can be understood as partly delimited by its core beliefs and practices. A shift in those can, therefore, be spoken of as a relocation of the social territory.

Aldous Huxley's position as a commentator of changing social territories is quite strong. He is the author of one of the famous Western dystopias, *Brave New World* that illustrates the destination at which, as Huxley believed, Western society could eventually arrive.[1] While seemingly his various essays from the 1920s are unlike the famous novel, critics such as Woodcock (1972, pp. 86–88) or Hillegas (1967) have demonstrated that there is an intellectual link between *Brave New World* and Huxley's earlier writing. This continuity includes the essays—for example, Frost (2006) discusses the "feelies" as a development of Huxley's perception of the cinema. Indeed, linking Huxley's non-fiction and fiction is typical of certain kinds of Huxley scholarship, since he frequently used both fiction and non-fiction to explore ideas that interested him personally—fictitious settings could, in those cases, be understood partly as devices for staging an efficient discussion (Birnbaum 2006, pp. 8–9 and Sion 2010, p. 3).

This paper intends to take the reader on a systematic journey through Huxley's 1920s social criticism in the light of the later *Brave New World*. In consequence, essays have been selected and categorized partly with reference to some of the dystopian themes at the heart of the novel, such as mechanization, specialisation or propaganda. On the other hand, the paper will also discuss some other interesting observations, particularly under the heading "Miscellaneous". Finally, a reservation is necessary: the paper consists of sections which are not always linked directly to one another. The method, while not entirely satisfactory, seems to allow for a discussion of only those essay fragments which seem most relevant, without sacrificing any interesting material.

2 The Leisure Culture

The first aspect to be discussed is what I have termed "the leisure culture," the result of a shift from recognizing and accepting leisure as one of the aspects of life, to a devotion to it. Huxley himself called this "having a Good Time" and vehemently criticised it. The criticism can be divided into three perspectives: the relation of the leisure culture to the past, its shape in the present and its hypothetical future. Huxley contrasts the contemporary perpetuity of leisure with the habits of the older generations. One instance of this occurs in his essay "America"

[1] See Huxley's comments on the potential prognostic validity of *Brave New World* in his 1945 Foreword to *Brave New World* (xvi) and in *Brave New World revisited* (2).

(from the *Jesting Pilate* collection, 1925) where he writes about the difference between the leisure of modern Los Angeles and ancient times:

> The Good Times of Rome and Babylon, of Byzantium and Alexandria were dull and dim and miserably restricted in comparison with the superlatively Good Time of modern California. The ancient world was relatively poor; and it had known catastrophe. The wealth of Joy City is unprecedentedly enormous. Its light hearted people are unaware of war or pestilence or famine or revolution, have never [...] known anything but prosperous peace, contentment, universal acceptance (Huxley 2000, pp. 552–553).

The contemporary state of leisure is described in a 1927 article entitled "Recreations":

> The jazz bands strike up, making dreary barbaric music, hour after hour. The Good Timers dance. In the blessed intervals of silence, they sit about and smoke and chatter and drink. The caterwauling begins again. His Master's Voice; obediently the Good Timers rise to their feet, begin once more to dance. And the air becomes thicker and smoker and hotter and more fetid, until at last, towards one or two in the morning, the Good Time comes to an end (Huxley 2000, vol. II, pp. 85–86).

And as for the future, Huxley expects that, given development towards less work and more free time, the "Good Time" is just going to increase. In a 1925 essay "Work and Leisure" he confronts the creators of futuristic utopias who, insofar as he understands, believe there will be a qualitative change in the society when it is given more leisure. Or to use his words, "leisured masses [...] will do all the things which our leisured classes of the present time so conspicuously fail to do." (Huxley 2000, vol. I, p. 412). Huxley does not believe in this and predicts that the leisured will simply need "more cinemas, more newspapers, more bad fiction, more radios, and more cheap automobiles" or, given a corresponding increase in education and wealth, "more Russian Ballets as well as more movies, more Timeses as well as more Daily Mails, more casinos as well as more bookies and football matches, more expensive operas as well as more gramophone records, more Hugh Waloples as well as more Nat Goulds" (Huxley 2000, vol. I, p. 413).

3 Mechanization of Leisure

This quotation from "Work and Leisure" nicely introduces the next issue: the mechanisation of leisure. The "Good Time" requires mechanized entertainment. Yet, at the time when he wrote about the potential profusion of such things, Huxley was presumably not a Luddite. He appeared to see the potential good of new technologies as well as the evil. Thus, in a 1922 essay "Music and Machinery" he writes rather disparagingly of "people [...] who disapprove, on some strangely ill-founded Wiliam-Morrisish principle, of all mechanical devices for the making or reproduction of music"(Huxley 2000, vol. I, p. 242). Huxley goes on to defend machinery, for in his opinion:

> there can be no doubt that the mechanical inventions of the last few years are doing more to create an understanding musical public than anything, short of compulsory musical education and forced concert-going could ever do (Huxley 2000, vol. I, pp. 242).

One year later he embraces the reverse side of the argument. In an essay entitled "Pleasures" he writes:

> Of all the various poisons which modern civilization [...] brews quietly up within its own bowels, few, it seems to me, are more deadly [...] than that curious and appalling thing that is technically known as pleasure (Huxley 2000, vol. I, p. 355).

He goes on to define this pleasure as the antithesis of creative entertainment past generations have had to make use of. Among other things he deals specifically with mechanized music:

> Do the democracies want music? In the old days they would have made it themselves. Now, they merely turn on the gramophone. Or, if they are a little more up-to-date they adjust their wireless telephone to the right wave-length (Huxley 2000, vol. I, p. 356).

Here the gramophone and the wireless telephone are condemned, while not so long ago they were upheld as useful. On the list of poisonous pleasures one will also find reading newspapers and "sporting vicariously", i.e. watching rather than practicing athletics.

A similar process of disenchantment can be seen in detail with respect to the movies. In 1925 Huxley writes "Where Are the Movies Moving?"—an essay praising silent cinema for its ability to be fantastic and artistic:

> On the screen, miracles are easily performed; the most incongruous ideas can be arbitrarily associated; the limitations of time and space can be largely ignored. Moreover, the very imperfections of the cinema are, in this respect, an important asset (Huxley 2000, vol. I, p. 176).

Soon, however, those "imperfections", which had given the cinema a capacity for new artistic expression, are seemingly gone. In 1929 Huxley pens a diatribe against voiced cinema, tellingly entitled "Silence is Golden".[2] It seems that, from his perspective, the society, having been first presented with a choice between destructive and creative use of entertainment technology, was steadily moving towards destructive uses.

4 Mechanization of Life

The above topic conveniently links with the next one—the mechanization of life in general. Huxley's observations and prognoses on this issue are in a sense especially interesting, because his predictions for the society partly coincide with his

[2] Laura Frost's article "Huxley's Feelies: The Cinema of Sensation in Brave New World" interestingly compares the movie description from "Silence is Golden" to scenes from *Brave New World*.

own intellectual evolution. After his visit to Japan, in *Jesting Pilate* (1925) Huxley writes something that seems at least not hostile to "progress". He comments on the increasing industrialization of the country and states that one could claim this is not the "real" Japan, after which he goes on to refute that argument:

> The future of Japan, as of every other country, depends on its 'unreal' self. Some day, [sic] in the Utopian future, when things are very different from what they are now, English and Japanese patriots, desirous of exalting their respective countries, will point, not to Cornwall or Fuji not to the country families or the descendants of the Tea Masters but to Manchester and Osaka, to the cotton spinners and the weavers of silk. [...] Progress may be defined in this connection as the gradual transformation of what we now call 'unreal' into something sufficiently noble and decent to be styled 'real'(Huxley 2000, vol. II, pp. 544–545).

It should be stressed that Huxley does not seem to bewail the substitution of Fuji by Osaka and the Tea Masters by silk weavers.

Shortly afterwards he pens a piece entitled "On Making Things Too Easy" (1926) wherein he points out one of the dangers of mechanical progress: that it "has become an end in itself" (Huxley 2000, vol. I, pp. 55). A year later he mentions that increasing specialization, which is necessary for mechanized progress, will make life nearly unlivable due to the distortion of the specialists' personalities. As a response he expects:

> a renewed interest in the Greeks or any other historical personages who may be supposed to have led a full, harmonious life as individuals, not as cogs in a social machine. As socialization and specialization increase, the Greeks will not be enough, and we may expect displays of growing admiration for primitives and savages (Huxley 2000, vol. II, p. 92).

This comment is interesting for at least two reasons. The mention of an increased fascination with savages might almost instantly bring to mind the meeting of John the Savage and the higher-class people of London in *Brave New World*. Furthermore, this prognosis foreshadows Huxley's own interests, for by 1929 he would (if only temporarily) become the apostle of what one critic called "vitalist" beliefs, focused precisely on a supposedly Hellenistic ideal of balanced life (Woodcock 1972, p. 14).

At that "vitalist" stage Huxley accuses the machine of depriving humans of creativity, which is essential for a balanced life:

> Passivity and subservience to machinery blunt the desire and diminish the power to create; pursuing the ideal of superhuman business efficiency, men mutilate the imaginative and instinctive side of their natures. The result is that they lose their sense of values, their taste and judgment become corrupted, and they have an irresistible tendency to love the lowest when they see it (Huxley 2000, vol. II, p. 332).

Despite this verdict, he believes machines cannot be banished entirely, if only because their existence has enabled an increase in population and their disappearance would condemn this surplus part of the population to death. He prescribes not destruction but consistent counteraction through de-mechanized leisure. He did not, however, believe this project to be easy, since he expected strong opposition from "film-makers, the newspaper proprietors, the broadcasters and all the rest" (Huxley 2000, vol. II, p. 332).[3]

Incidentally, Huxley's approach from that period, with its ambiguity of perceiving the machine as both destructive and necessary, is interestingly mirrored by another dystopian writer, George Orwell, who wrote that the machine should be accepted "as one accepts a drug—that is grudgingly, and suspiciously" (Orwell 1984, pp. 178 and Rees 1961, pp. 55–61). This similarity suggests that Huxley's opinion at that moment was part of a wider trend.

5 Social Engineering and the Aristocratic Ideal

A society that is mechanized requires, in Huxley's assessment, specialists. This brings the discussion to the next issue in the paper and it is again an interesting one because once more Huxley sees both good and bad possibilities, just like in the case of mechanized pleasure but in a much stronger form. On the one hand, Huxley clearly envisages a bad path the society may tread in search of its specialists. In a 1927 essay entitled "The Future of the Past" he writes:

> Our children may look forward to the establishment of a new caste system based on differences in natural ability, to a Machiavellian system of education designed to give the members of the lower castes only that which it is profitable for the members of the upper castes that they should know (Huxley 2000, vol. II. p. 93).[4]

He does predict the "contingency" that eugenics may dispose of castes and a fully democratic system may, at some point, emerge but he considers this too "remote" to be discussed.

At the same time, Huxley describes an alternative which he champions in his volume *Proper Studies* from 1927. It is an ideal not so much different in method (providing an individual with only certain knowledge) but in intention:

> the ideal educational system is one which accurately measures the capacities of each individual and fits him, by means of specially adapted training, to perform those functions which he is naturally adapted to perform. [...] but without, in the process, destroying his or her individuality (Huxley 2000, vol. II, pp. 215–216).

Thus, one can chart two potential courses of social evolution: one that the society might choose (degrading some members) and another it presumably should choose (respectfully fitting diverse individuals to social roles). The system Huxley envisions as the best in 1927 is an "aristocratic ideal". According to Huxley, it would be good for all, since to select the best, the society has to provide everything to all of its members.

---

[3] Hillegas (1967c) specifically discusses the relevance of this essay for *Brave New World*.

[4] See Bradshaw (1994, pp. 12–14), for a significantly different reading on this issue.

One is tempted to hypothesize that by the time he wrote *Brave New World* Huxley had dispensed with the aristocratic ideal, since the novel shows a total distortion of it, natural differences in ability having been superseded by controlled degradation. It may have been, however, not so much an outright dismissal as a correction, since in 1927 Huxley believed that, while the aristocratic state need not be entirely realizable, it nevertheless "may be approached in practice without involving in insoluble difficulties those who try to apply it" (Huxley 2000, vol. II, p. 226). Perhaps later he had observed that even if the difficulties are not "insoluble," they are rather grave if the society decides to choose a different understanding of "aristocracy" than he does. Indeed, in the same volume where he praises the aristocratic ideal, he also turns his attention to the dangers of eugenic reform, if it be based on the premise that whoever is currently most successful should be a model for replication. In 1929 he comes to deliver a demining verdict against eugenics because it is "working to make the world safe for Podsnappery" (Huxley 2000 vol. II, p. 386).

It is perhaps worthwhile to explore a hypothesis as to why Huxley believed in the aristocratic ideal while seeing the danger it would pose if distorted. It might have been because, as he stated:

> What interests me [...] is the particular case of the child of talent born in the lowest strata of an excessive population. [...] Brats, tout court, constitute the stuff of which our world is made. They may expire; but unless they do so on such an enormous scale as to imperil the whole fabric of society, it will make no difference to the world. Brats of talent, on the other hand, have it in them to change the world in one way or another (Huxley 2000, vol. II, pp. 521–522).

It should be noted that a certain callousness towards the ordinary "brats" most probably does not mean Huxley wished them to "expire"—he expressed distaste with the social arrangements in India that caused degradation of the coolies (despite being conscious that he is one of those whose privileged position is based on it). Here he delivers what he apparently believes to be a realistic assessment of the impact the waste of an average person has compared that of a talented one. This might be a clue as to why Huxley saw the evil potential of social engineering but nevertheless attempted to offer a supposedly better path of the same kind—because it could provide best for the talented ones.[5]

6 Advertisement and Propaganda

As has been noted, the systematic de-mechanisation of leisure that Huxley came to prescribe to the society would, he suggested, be opposed by those who profited from entertainment technologies. To those people and to their mode of

[5] See Bradshaw (1994, p. 36), for a thesis that Huxley's aristocratic sentiments were influenced by H.G. Wells and Meckier (2001, pp. 238–239) on Huxley's inadequate explanation of the screening methods in the positive aristocratic system.

communication with consumers, that is advertising, Huxley had also paid attention separately from the context of mechanization. And his stance was, again, not thoroughly condemnatory.

In a 1923 essay "The Psychology of Suggestion" Huxley notes the power of advertisement. He claims that the effectiveness of advertisements is surprising, since:

> It is only in rare, exceptional cases that the change of political opinions involves the slightest personal sacrifice. [...] [But in the case of goods] Our conviction, if it is really sincere, must find expression in the opening of pocketbooks and the signing of checks (Huxley 2000. vol. I, pp. 392–393).

Two years later, in an essay devoted strictly to advertisements, Huxley calls them, perhaps somewhat ironically, "one of the most interesting and difficult of modern literary forms" (Huxley 2000, vol. I, pp. 98). In the same year, in an essay entitled "The Country" he illustrates the power of marketing: "Till the end of the eighteenth century every sensible man, even in England, even in Sweden, feared and detested mountains. The modern enthusiasm for wild nature is a recent growth" but now even those who continue to not like it are persuaded by advertisements to participate, because this generates enormous profit for "railways, motorcar manufacturers, thermos-flask makers, sporting tailors, house agents" (Huxley 2000, vol. I, p. 447) and others.

In his later essays, the problem of being coaxed by advertisement emerges as something more sinister (either his negative sentiment developed or he extended the argument). It is seen as a systematic exploitation of feelings, which have otherwise been abandoned by the professionalized society. Once used by the Church to "harmonize" the human psyche and orient it towards a goal higher than themselves, emotions have fallen into disrepute and their fall was accompanied by the atrophy of all harmonizing rituals. What emerged in their place was the advertisement, used "by tradesmen to the end that advertisers may become rich." (Huxley 2000, vol. II, p. 273). This criticism is repeated, and perhaps intensified, in a later essay "Spinoza's Worm", where Huxley declares that compared to asceticism originating from spirituality (which is, from his current "vitalist" perspective, "bad enough") "[...] asceticism for the love of Mammon is intolerable. But it is for the love of Mammon that our modern stoics exhort us to mortify our flesh and control our passions." (Huxley 2000, vol. II, p. 330). Huxley suggests that were they allowed to create their version of utopia, the "modern stoics" would make a world in which:

> Deprived of all their distractions, shut out from all their private paradises, men would work almost as well as machines. The one legitimate desire left to them would be a desire for things—for all the countless unnecessary things, the possession of which constitutes prosperity(Huxley 2000, vol. II, p. 331).

In other words, the social frontier may move, if unchecked, towards unabated consumerism, founded on a combination of the machine as the source, the advertisement as the means and austerity in all aspects except consumption as the end. This tendency later cumulates in *Brave New World*.

Propaganda dealing with non-consumerist matters has already been mentioned but it might be worthwhile to briefly return to it. In a 1926 essay "A Few Well Chosen Words" Huxley expounds on the difference between informative and emotive words. The emotive word is not evil of itself but assumes this quality when used to produce agreement on the basis of emotions in a context in which "external, non-human truth is not indifferent" (Huxley 2000, vol. II, p. 59). The use of emotive words in such situations is far from harmless for "in the affairs of practical life [it] leads inevitably to serious practical errors" because these words "have a kind of compelling hypnotic power to make men act, without, however, telling them the real reasons for their action" (Huxley 2000, vol. II, p. 59).

An interesting usage of such an emotive, or even marketing, language in a matter that should be—strictly speaking—non-consumerist, can be found in a piece of his earlier writing. In *Jesting Pilate* Huxley recounts a religious advertisement he had seen in a Los Angeles newspaper. Various churches attempt to attract the newspaper readers with such things as:

> Dr. Leon Tucker with the Musical Messengers in a Great Bible Conference. [...] Mother's Day Flowers to all Worshippers. [...] Impersonations of Lincoln and Roosevelt. [...] N. C. Beskin, the CONVERTED JEW, back from a successful tour, will conduct a tabernacle campaign in Glendale. 'WHY I BECAME A CHRISTIAN?' Dressed in Jewish garb. Will exhibit interesting paraphernalia (Huxley 2000, vol. II, pp. 550–551).

In the context of the social territory metaphor it might be interesting to note that the advertisement may remind one of a description in H.G. Wells's dystopian novel *The Sleeper Awakes* where religious advertising briefly appears and the protagonist of the novel is scandalized by it (Wells, c2006, p. 175). One could, therefore, claim that what had been a potential negative change of the social frontier in 1899 had in 1925 become a reality and the possible further development was explored in *Brave New World*.

7 Miscellaneous

The last section discusses two unrelated issues and the first to be presented, by virtue of some continuity with the prior topic, is Huxley's opinion about the evolution of values he had witnessed during his visit to America.

In *Jesting Pilate*, in the Chicago section, he proposes a certain dynamic concerning values. A change in them may be procured either through a denial of their existence or through a "democratization". Having already explored his opinion about the aristocratic ideal, one may expect that "democratization" is understood negatively. Indeed, Huxley considers it more dangerous, because more insidious, than outright nihilism. The common people know instinctively that values exist and a flat denial shocks them. If, however, an originally high ideal is brought down to the level of average achievability, they may accept the change.

To expound on the mechanism, Huxley uses the example of an advertisement by a "mortician", i.e. an undertaker. He notes that the tradesman has changed the name of his profession and claims that the man had done this because he wished to democratically put himself on par with the best. Furthermore and more importantly, Huxley suggests that the mortician's claim to regard is made through the characteristically American understanding of Service.[6] The, originally high, Christian ideal is brought down to the level of satisfying a common social need (perhaps not even an essential one) with a suitable level of business honesty. The high word, with its high associations, is used to denote a lifestyle with drastically low requirements with the dangerous consequence that the ideal becomes too easily realizable and once realized, leaves its pursuer in a void. This creates the following alternative:"not to think but just continue to chatter and rush about as though you were doing something enormously important, or else to think, admit the world's vanity, and live cynically." (Huxley 2000, vol. II, p. 278) In other words, Huxley suggests that the process of democratizing ideals, which may initially appeal to the general public, eventually brings the disastrous effect of forced mindlessness or cynical ennui. It could, therefore, be suggested that this may be the ultimate psychological source of the leisure culture which is, in turn, the root of some other phenomena described by him; i.e. the relocation of one part of the social territory may cause further, massive, movements, leading even to a feedback loop.

The second issue in this section offers a less bleak perspective from Huxley's vantage point, although it may come as rather surprising to his readers. It is the dissolution of the multigenerational family. Huxley, quite probably because he was very concerned with individual talent, felt a huge dislike towards the families he had seen in continental Europe and, especially, in places such as India and Malaya. He goes so far as to claim that in an impoverished society, the people who form a strong familial structure, are a "living tombstone" to talent (Huxley 2000, vol. II, p. 522). Perhaps this is why, even for England which had relatively small family units, Huxley seems to celebrate the advent of a new world where:

> There will be practically no brothers or sisters to play with at home, practically no aunts or uncles, practically no cousins, practically no grandmothers even for the grandmothers will be far too busy dancing and playing bridge to pay any attention to their grandchildren. The family will be reduced to the most exiguous proportions. Father mother and child (Huxley 2000, vol. II, p. 119).

Reading this one may wonder whether some aspects of *Brave New World* were not intended as utopian. Such doubt would, however, seem unfounded, since Huxley makes two important reservations. First, he assumes that some sort of a family will always exist. Secondly, he is aware that while for him such a development would be enjoyable, those actually experiencing it might not be thrilled.

[6] Huxley believed the process to be specifically American and hopefully not transplantable to Europe—perhaps this is one of the reasons behind his conscious "Americanization" of *Brave New World* (Meckier 2002).

8 Conclusions

What do those findings suggest and for what are they useful? Firstly, this paper has attempted to provide a glance at the intellectual background against which *Brave New World* was written. That may be useful to those who would wish to explore Huxley's dystopian vision beyond the novel, but who do not have the inclination or time to read through the plethora of his essays. More importantly, the paper also attempts to be my individual contribution to the discussion of how exactly *Brave New World*-esque ideas fermented in the author's brain and finally produced the famous dystopia—somewhat different, for example, from Jerome Meckier's (2001) brief explanation in his review of the two essay volumes.

Secondly, I would suggest that the development of ideas analyzed in this paper undermines the idea that the young Huxley can be correctly described as an "amused Pyrrhonic aesthete". While Huxley had auto-identified in this way, Bradshaw (1994, p. 7) suggests that the label had resulted in a "long-standing misconception" about the writer. In my opinion, this paper corroborates evidence to support Bradshaw's diagnosis, since, as can be seen in the above essay fragments, Huxley apparently though much more seriously about various contemporary social issues and passed a more thorough judgement on them than that label would lead many to believe.[7]

Thirdly, I postulate that, while *Brave New World* is deservedly invoked in debates about the fate of humanity, its author had, in general, held a far more intellectually nuanced position than a simple reading of that novel would warrant. As can be seen, he could, for example, both praise and condemn technology, and, despite his unflattering portrayal of eugenics, he was an aristocrat at heart.

Finally, on a more philosophical and less literary level, it may be not without merit to ask whether some of the trends described in the essays have been strengthened with time. Furthermore, if they have, does that include some rather disquieting aspects Huxley had criticised back then and later expounded on in his dystopia? That too could, after all, be a question in keeping with the conference theme.

References

Birnbaum, M. 2006. *Aldous Huxley: A Quest for Values*. New Brunswick, New Jersey: Transaction Publishers.

Bradshaw, D. 1994. *The Hidden Huxley*. London and Boston: Faber and Faber.

Frost, L. 2006. Huxley's Feelies: The Cinema of Sensation in Brave New World. *Twentieth Century Literature*. 52. 4:443–473.

Hillegas, M. H. 1967. *The Future as a Nightmare: H.G. Wells and the Anti-utopias*. New York: Oxford University Press.

[7] See Meckier (2001, pp. 235–236) for a converse opinion.

Huxley, A. 1998. *Brave New World.* New York: HarperCollins Publishers.
Huxley, A. 2000. *Complete Essays of Aldous Huxley, Volume I, 1920-1925* and *Volume II, 1926-1930*. Eds. Robert S. Baker and James Sexton. Chicago: Ivan R. Dee.
Huxley, A. 2006. *Brave New World Revisited.* New York: HarperCollins Publishers.
Meckier, J. 2001. Prepping for Brave New World: Aldous Huxley's Essays of the 1920s. *Utopian Studies.* 12. 2:234-245.
Meckier, J. 2002. Aldous Huxley's Americanization of the "Brave New World" Typescript. *Twentieth Century Literature.* 48. 4: 427–460.
Orwell, G. 1984. *The Road to Wigan Pier.* Harmondsworth: Penguin Books.
Rees, R. 1961. *George Orwell: Fugitive from the Camp of Victory.* London. Secker & Warburg.
Sion, R. T. 2010. *Aldous Huxley and the Search for Meaning: A Study of the Eleven Novels.* Jefferson, North Carolina: McFarland.
Wells, H. G. *The Sleeper Awakes.* 2006. Harmondsworth: Penguin Books.
Woodcock, G. 1972. *Dawn and the Darkest Hour: A Study of Aldous Huxley.* London: Faber and Faber.

Re-vision of London in Iain Sinclair's *White Chappell, Scarlet Tracings*

Dominika Lewandowska

Abstract The aim of this paper is to explore how Iain Sinclair's *White Chappell, Scarlet Tracings* (1987) re-maps London as a literary and textual space, in which the boundaries between the past and the present, history and fiction, are removed to reveal the city's esoteric essence and allow the writer to create his own subversive mythology of contemporary urban experience. Sinclair's ideas can be considered close to Walter Benjamin's notion of "alternative historiography", proclaiming greater attention to 'other', subjective histories. The writer argues that these histories pervade the topography of the city as well as its literature and can be revealed through a technique parallel to André Breton's surrealist automatism. A writer of London, both inspired by the city and making it the ultimate subject matter of his work, Sinclair develops a concept of "geology of time", depicting contemporary London as a spiritual site out of time. The city functions as a textual structure, a palimpsest, which proves simultaneously a fluid organic entity, affecting its inhabitants. This notion can be linked with the concept of psychogeography, preoccupied with the effects of the geographical environment on the emotions of individuals, and developed by Guy Debord. Ultimately, I will argue that Sinclair's novel creates a holistic vision of an urban space which, following the cubist tradition, aims to show London simultaneously from a multitude of viewpoints. The city's topography juxtaposes and unifies the past, the present, history and fiction in a subversive and multilayered narrative—a perfect example of Bakhtinian carnivalesque.

D. Lewandowska (✉)
Warsaw University, Warsaw, Poland
e-mail: dominika.lewandowska@o2.pl

J. Fabiszak et al. (eds.), *Crossroads in Literature and Culture*, Second Language Learning and Teaching, DOI: 10.1007/978-3-642-21994-8_23,

1 Introduction: Iain Sinclair and the Literary Re-vision of Urban Space

The complexity of Iain Sinclair's writing eludes categorisation. A natural consequence of his frenetic and rather obsessive preoccupation with textuality and self-referentiality is that many may feel the temptation to attach the label of postmodernism to his work. However, it soon turns out that, as so often happens in Sinclairian narratives, "the outwrappings bear no direct relation to the contents" (Sinclair 2004, p. 16). While Sinclair's urban space is built on the assumption that "Nothing is written, everything re-written" (Sinclair 2004, p. 54), this does not imply a literary (re-)production that is ultimately futile. On the contrary—it should be viewed as an attempt at constructing, or rather revealing and opening up a textual space where patterns can be traced and recognised as a part of the whole which itself will never be fully grasped. In *White Chappell, Scarlet Tracings* (1987), his first London novel, Sinclair is preoccupied not with establishing fixed connections but pointing to momentary (and yet eternally recurrent) convergences, which reflect and provide insight into some profound underlying meaning. Thus, it can be argued that his text constitutes an attempt at rediscovering and re-imagining London through literature, which ultimately conceives the capital's cityscape anew. This paper will explore this literary re-vision of the city by tracing some of the influences on Sinclair's narrative, analysing it as a textual space of dissolved boundaries, and considering the new realm of London as setting a template for the artistic and cultural concerns which mark the writer's subsequent works of both fiction and non-fiction.

2 Literary Genealogy of Sinclair's London: Modernist Myths, Blakean Mysticism

While *White Chappell, Scarlet Tracings* should not be considered in postmodern terms, the narrative, as well as Sinclair's literary practice at large, do seem to be largely informed by a modernist sensibility. This connection may be traced in the intellectual complexity of his texts and the wealth of literary and cultural references, as well as an interest in different fields of knowledge. The writer, perhaps somewhat self-mockingly, points to his affinities with modernism. In *Dining on stones* (Sinclair 2005, p. 370) Norton, one of Sinclair's notorious anti-heroes, advises: "When a natural climax arrives, a crisis in the narrative, subvert it: pick up a book. Go with the old modernist strategy: quotation. Eliot, Pound. Yeatsian dictation." Most importantly, however, the modernist influence can also be traced in Sinclair's mythologisation of the urban space as a means of grasping another, alternative reality of contemporary London. Following T. S. Eliot's argument, presented in his 1923 essay *Ulysses, Order and Myth* (2005, pp. 165–166), that traditional narratives can no longer encompass the reality of the modern world, and that this world must be instead translated into myths, Sinclair creates his own mythology of contemporary urban experience. As he explains in *Suicide Bridge* (Sinclair 1995, p. 148):

> The essential myths of beginning are followed by myth as subversion. [...] Myth emerges as a weapon, a tool of resistance. It emerges in the hands of men wanting to maintain a contact with the previous, with the era of power & high function.

Consequently, Sinclair's modernist quest becomes an attempt at subverting the official, canonical space of London and myth becomes the only form capable of grasping the energies of the writer's 'other' version of the city.

This preoccupation with the mythic, as well as the mystic, links Sinclair with another great magus of London, William Blake, whose name is repeatedly invoked in his texts. One could argue that it is in fact this connection that allows for a better understanding of the nature of Sinclair's literary practice. As Alex Murray explains, Sinclair views Blake's writing "as providing a model for a form of engagement with the mythology of place that can underpin the 'geometry of opposition' that the secret history of the city can provide" (Murray 2007, p. 79). The two artists share several elements both creatively and ideologically, but the most significant and crucial of those appears to be a notion of 'vision'. Such visionary writing has become a rather common trend in contemporary literature of England's capital. As Michael Moorcock argues in his introduction to Sinclair's *Lud Heat and Suicide Bridge*, "At its best London fiction has, in the past twenty years, become characteristically a visionary medium" (Sinclair 1995, p. 3). However, despite great prolificacy of visionary texts, it is Sinclair that Moorcock considers a truly original and significant voice of this new phenomenon.

3 A Narrative Space of Convergences

Sinclair's London, envisioned in the realm of myth, is a space where all planes—of time, space, history—converge. The novel not only crosses, but actually removes the divisions between the present and the past, history and fiction. The following subsections will explore how these concepts permeate one another to account for Sinclair's idea of the city.

3.1 The Past and the Present

One of the fundaments of the urban re-vision realised and explored in *White Chappell, Scarlet Tracings* is the dissolution of boundaries between the past and the present. The merging of temporal planes is indicated already by the novel's epigraph, a slightly altered quote from Charles Dickens's *Our Mutual Friend*: "then I tell you what change I think you had better begin with, grandmother [sic].

You had better change Is into Was and Was into Is, and keep them so" (Dickens 2002, p. 410). This decision is dictated by the belief that strict divisions into 'now' and 'then' are in fact artificial, as the two are constantly coexisting, interacting and influencing each other, combining into one temporal sphere. Sinclair introduces a mythical cyclical notion of time, in which certain patterns repeat themselves endlessly. In the novel, the reader is presented with three intertwined narratives—two taking place at the end of the 20th century, the third happening a century earlier. These three plot threads are often juxtaposed to a point where it is difficult to tell them apart. They prove irrevocably linked: the modern protagonists investigate the subject matter of the other, Victorian, plot—the mystery of Jack the Ripper murders, thus reviving it; at the same time, the events and themes from the past recur in the present thus haunting the modern narrative—and so the Victorian mystery shapes and defines today's London. As 'Sinclair' (the novel's narrator) puts it, "You allow yourself to become saturated with this solution of the past, involuntary, unwilled, until the place where you are has become another place; and then you can live it, and then it is" (Sinclair 2004, p. 22). Moreover, at one point in the novel, Joblard, one of the contemporary protagonists, wonders: "As the century dies will another pattern of sacrifice be demanded? Do we slowly begin to understand only because we are about to become performers in the same blind ritual?" (Sinclair 2004, p. 48). It may be argued that this removal of temporal boundaries poses a challenge to any straightforward division into causes and effects, an idea which will be further explored in terms of the convergence of reality and fiction.

3.2 Fact and Fiction

Sinclair's notion of simultaneity and cyclicality relies upon the writer's belief in 'the heat', a hidden energy that drives reality. Thus, the novel (and in fact all of Sinclair's writing) is driven by a search for that one flickering moment when everything will fall into place. This explains why Sinclair obsessively revisits the same sites, the same stories, trailing that heat—the essence of the 'other' London. In Sinclair's urban mythology, this eternal force manifests itself in literature, endowing it with a revelatory as well as prophetic quality. Consequently, literary texts become the most valid source of history (or histories) of the city, a means of resuscitating the past and then casting it into an entirely new form (Gasiorek 2009, p. 203), exorcising it through the act of writing and reading. These ideas come close to Walter Benjamin's notion of "alternative historiography", since it could be argued that Sinclair's project echoes Benjamin in his conviction that dominant historical narratives constitute a key element in the destructive course of progress, and proclaims with him the need to "explode oppositional histories out of the continuum that has attempted to smother them" (Murray 2007, p. 73). *White Chappell, Scarlet Tracings* pursues the uncovering of precisely these 'other' histories. The text which serves in the novel as the main "tunnel into time"—Arthur

Conan Doyle's *A Study in Scarlet*—while in a way iconic, hardly belongs to the realm of 'high' literature. Here, however, its unique copy becomes "a piece of history, a true splinter of the 1880s", an *actual piece* of that reality, and is treated as one of those visionary narratives that "got out into the stream of time, the ether; they escaped into the labyrinth. They achieved an independent existence. The writers were mediums; they articulated, they gave a shape to some pattern of energy that was already present" (Sinclair 2004, p. 117). Sinclair introduces here the figure of the writer as a visionary, whose literary production is necessitated by the 'heat' and becomes a form of possession. In the words of the novel's narrator, "the tale must be told, no rest from it" (Sinclair 2004, p. 94) and it is the artist's predestination to process and convey it. This conception of artistic production may at first seem to follow the Romantic tradition, but in fact the semi-automatic writing technique involved in the creation of such metaphysically inspired texts seems more related to the already mentioned Yeatsian dictation as well as the surrealist concept of automatism. As theorised in his *Manifestoes of Surrealism* (Breton 1972), by opening oneself up to André Breton's pure psychic automatism, one gains access to some hidden truth, the 'other' reality. In *White Chappell, Scarlet Tracings*, this 'other' reality inspires, guides and controls the writer, who in turn constructs his own literary space, a notion hinted upon by another character, James Hinton, when he notes that "Something *thinks* us, that's evident. Something dictates my dictation" (Sinclair 2004, p. 82).

However, Sinclair's re-vision of London sees the significance of literary texts go beyond their revelations of the past, as the writer proclaims a further, more complex interdependence between history and fiction. It soon becomes apparent that in this urban mythology not only do works of literature record and communicate the city's history—they also cause it to happen. William Gull had to commit his crimes due to the fact that they were depicted in literature: in his *Illuminations*, Rimbaud described "the elements of the Whitechapel millennial sacrifice. And by describing, *caused* them" (Sinclair 2004, p. 120). Moreover, 'Sinclair', the narrator, claims that it is enough to "peel down *Study in Scarlet* or *Jekyll and Hyde* or *Mystery of a Hansom Cab* and out come the prophetic versions. Beneath the narrative drive is a plan of energy that can, with the right key, be consulted"(Sinclair 2004, p. 49). As a consequence, what the writer constructs is a mythology in which history and fiction permeate each other and form a unity of complex correlations. This is achieved through a prose technique which, according to Robert Bond, displays "an associative method relating historical and textual evidence rather than a linear method recognising the distinction between causes and effects" (Bond 2005, p. 105).

The concept of literature's power to shape and determine reality calls for a writer figure who is not only an inspired visionary, but also a shaman, capable of casting a narrative spell on the life of London. This notion has been extensively explored by Sinclair in his essay *The Shamanism of Intent*. It postulates the existence of artists who have the power to enchant the cityscape and its culture, healing it in the process. As the writer explains,

> The will to continue, improvise upon chaos, could be defined as "intent": a "sickness-vocation", as Eliade has it, an elective trauma. The health of the city, and perhaps of the culture itself, seemed to depend upon the flights of redemption these disinherited shamans (…) could summon and sustain (Sinclair 2003, p. 240).

Thus, the creative activities of these 'chosen' artists, cursed and blessed with the unique vision of the city, become a form of a mystic ritual meant to undermine and counter the negative (past, present, historical, fictional) energies of London.

3.3 Time, History, Space

Another essential component of Sinclair's urban vision, inseparable from the notions of time, fiction and history, and one that the writer is primarily preoccupied with, is that of space. Sinclair is indisputably a writer of London. Both inspired by the city and making it the ultimate subject matter of his work, "[he] participates in the simultaneous act of reading and writing London" (Murray 2007, p. 68), a practice which naturally entails writing *himself* into his narratives, hence the introduction of 'Sinclair', the narrator. Sinclair walks, sees and maps the urban space as a carrier of its history, a notion reflected in his concept of "geology of time". According to this notion, the very bricks with which the city has been built are soaked with its past, and touching them allows history to be momentarily revived (Sinclair 2004, pp. 42, 102). As a result, the cityscape functions as a textual structure, a palimpsest, where we find "dry pages of the trees" (Sinclair 2004, p. 73), where "fields were blank pages" (Sinclair 2004, p. 110), "The walls are calcined books" (Sinclair 2004, p. 181), and interiors have "a narrative quality" (Sinclair 2004, p. 54). At the same time, however, it constitutes a living, organic entity, a space of fiction and history enveloping those who venture into its territory. As *White Chappell, Scarlet Tracings'* 'Sinclair' explains, in London,

> it is possible to walk back into the previous, as an event, still true to this moment. […] The past is a fiction that absorbs us. It needs no passport, turn the corner and it is with you. […] Detached from this shadow you are nothing, there is nothing. You have no other existence (Sinclair 2004, p. 53).

As this passage indicates, certain locations are sites saturated with the past but also with different fictions, spaces reworked through literature. Sinclair discusses this idea in *Lights Out for the Territory* (2003, p. 142), where he proclaims that everyone is welcome to divide London according to their personal anthology. The statement is followed by a long passage listing a possible reading of the capital: it begins with J. G. Ballard's Shepperton, Michael Moorcock's Notting Hill and Angela Carter's south of the Thames, and ends with "Stewart Home commanding the desert around the northern entrance of the Blackwall Tunnel; Gerald Kersh drinking in Fleet Street; Arthur Machen composing *The London Adventure or the Art of Wandering*." Thus, Sinclair endows the palimpsest of the city with a new resonance as well as giving new meaning to the concept of *actually* rewriting the city.

The notion of an urban space as pervaded with certain energies can be (and has often been) related to psychogeography, preoccupied with "the study of the precise laws and specific effects of the geographical environment consciously organised or not, on the emotions and behavior of individuals" (Debord 1955, n.pag.). The nowadays increasingly fashionable concept has largely lost its significance, but in its original form, introduced by the Letterist International in Paris of the 1950s, and further developed by Situationists, it has become central to the prose practice of Sinclair. The writer, as well as his protagonists in the novel, discover London's narrative history and its sites of energy by walking it: "They are on the street and the heat of the old story begins to work with them" (Sinclair 2004, p. 42). Of all ideas central to psychogeography, the one which appears to be of particular significance to Sinclair's praxis is that of the dérive, or the 'drift'. As Guy Debord explains in his essay *Theory of the Dérive* (1958),

> In a dérive one or more persons during a certain period drop their relations, their work and leisure activities, and all their other usual motives for movement and action, and let themselves be drawn by the attractions of the terrain and the encounters they find there. Chance is a less important factor in this activity than one might think: from a dérive point of view cities have psychogeographical contours, with constant currents, fixed points and vortexes that strongly discourage entry into or exit from certain zones (n.pag.).

The premise which holds that walking serves as a source of insight into the alternative, buried reality of London, yet again links the Sinclairian narrative with Walter Benjamin—this time due to his preoccupation with the figure of the flâneur. An observer of the modern urban space, he was first conceptualised by Charles Baudelaire as "a person who walks the city in order to experience it" and in the 20th century revived by Benjamin as an observer of the old Paris disappearing with the rise of consumer society. In *White Chappell, Scarlet Tracings*, both the writer and his characters become the contemporary reincarnations of the flâneur seeking out narrative scraps of the 'other' realm. Thus, walking the city functions as a narrative practice as Sinclair "reveals himself to be a type of textual ragpicker resurrecting texts (including his own) that official cultures have discounted, or omitted" (Seale 2007, p. 106). What the reader is presented with in the novel is a conjunction of the concepts of inspired psychogeography and observant flânerie, which may be argued to produce Sinclair's own version of surrealist automatism—automatic walking—where the artist and his characters yield to the energy of the city, allow it to direct them, but at the same time, through experiencing it, reinvent London in their multilayered fictions.

4 Crossing the Boundaries of the Text

As a narrative space of convergent ideas, Sinclair's novel reflects the textual complexity of the city it conceives, and itself resonates with a myriad of voices and images. The multilayered-ness of *White Chappell, Scarlet Tracings* will be explored in the following subsections.

4.1 Textual Carnival

The novel's holistic, palimpsestic re-vision of London's cityscape, juxtaposing and unifying past and present, history and fiction in a subversive and complex text, can be regarded as a perfect example of Mikhail Bakhtin's carnivalesque, explored in his essays *Problems of Dostoevsky's Poetics* (1984a) and *Rabelais and his World* (1984b). Sinclair's narrative sees the well-known mainstream literary space of London become estranged and re-envisioned in the form of a textual labyrinth, through which the reader is led by a group of unlikely guides—a pack of seedy, at times grotesque and often darkly hilarious bookdealers, the kind of protagonists who are usually kept off the literary map. Through, and in, the novel's subversive mythology, parallel to the Bakhtinian interpretation of the carnival, the established image of the city is overturned, revealing the 'other' territory dominated by hitherto suppressed voices, and the official truths become contested by means of interweaving "the "High Occulting" of cosmologists, modernist poets, alchemists [...] with the low and pecuniary world of pulp fiction" (Sutton 2007, p. 58). Consequently, Sinclair creates in the novel the site of resistance to authority, thus realising through his praxis Bakhtin's conception of literature's cultural potential. Moreover, the notion of the intertwined narratives, which constantly interrelate with one another, as well as with other texts, may be read in terms of Bakhtinian dialogism. Defined in *The Dialogic imagination* (Bakhtin 1981), the dialogic work, as opposed to the monologic one, is identified by its continual interaction with other texts and authors. Since this interaction takes the form of a dialogue, the influence goes both ways and all parties become affected in the process. We may therefore argue that not only is Sinclair's novel informed by other narratives, but it also informs, challenges and alters those texts.

4.2 Visual Collage

Since Sinclair's textual kaleidoscope of narratives, histories and literary traditions, explored through walking, writing and reading merges different perspectives to create a new eclectic vision of London, it may be associated with Cubism and its artistic technique of showing the subject from a multitude of viewpoints to represent it in a broader context. It should be noted that Sinclair retains a very strong interest in the visual arts, particularly painting and photography.[1] This seems perfectly logical: after all, his literary practice is itself largely visual, since his writing stems from the actual observation of the city and thus becomes a panorama of his preferred London sites through space and time. Furthermore, this particular correspondence is acknowledged by 'Sinclair', the narrator, as he explains: "Our

[1] In *The Verbals* (Jackson 2003, p. 83), Sinclair proclaims photography and painting the two forms of art which have greatly influenced his writing.

narrative starts everywhere. We want to assemble all the incomplete movements, like cubists" (Sinclair 2004, p. 51). Sinclair's interest in the movement is particularly understandable in view of the fact that, in the words of Daniel-Henry Kahnweiller:

> This [...] artistic language has given painting an unprecedented freedom. It is no longer bound to be the more or less verisimilar optic image which describes the object from a single viewpoint. [...] Coloured planes, through their direction and relative position, can bring together the formal scheme without uniting in closed forms. [...] Instead of an analytical description, the painter can, if he prefers, also create in this way a synthesis of the object, or in the words of Kant, 'put together the various conceptions and comprehend their variety in one perception' (qtd. in Bond 2005, p. 107).

The Cubists' insistence that despite the holistic nature of their formal scheme the outcome of their artistic production remains ultimately open and liberated, corresponds with the project of Sinclair, who aims at imagining through his text a synthesis of the city which remains fluid, dynamic, dialogic.

5 Conclusion: The New Territory

As we have seen in the previous sections of this paper, *White Chappell, Scarlet Tracings* constitutes a narrative endeavour to revise and re-envision the literary space of London. Although extensively drawing on the traditions of modernism and the avant-garde, Sinclair's text is never derivative. Instead, it takes certain elements of these legacies and transfigures them into a new and thoroughly original cityscape, reinventing what has been, so that it becomes what is (Sinclair 2004, p. 114). Sinclair's influences are very far-reaching, sometimes obscure, and eclectic often to the point where they produce a semblance of narrative chaos, but in fact they prove highly consistent. He is rewriting the same novel: his novel. Thus, the counter-site to the official territory of London is revealed to be a personal space, a personal vision. And just as all these varied elements fall into place within this narrative, so does the novel fit in with other texts by Sinclair, combining into his mythology of the city. It is not surprising that *White Chappell, Scarlet Tracings* is considered both the last book of one trilogy and the first one of another: some critics and readers regard it as the last part of the trilogy consisting also of *Lud Heat* and *Suicide Bridge*, while others believe it to be the first novel in the trilogy which encompasses also *Downriver* and *Radon Daughters*. However, in view of Sinclair's visionary project, it would probably be most accurate to consider it both.

References

Bakhtin, M. 1981. *The dialogic imagination*. Austin : University of Texas Press.

Bakhtin, M. 1984a. *Problems of Dostoevsky's poetics*. Minneapolis : University of Minnesota Press.

Bakhtin, M. 1984b. *Rabelais and his world*. Bloomington, Ind.: Indiana University Press.

Bond, R. 2005. *Iain Sinclair*. Cambridge: Salt Publishing.

Breton, A. 1972. *Manifestoes of surrealim*. Ann Arbor: University of Michigan Press.

Debord, G. 1955. Introduction to a critique of urban geography. http://library.nothingness.org/articles/SI/en/display_printable/2. Accessed 10 April 2010.

Debord, G. 1958. Theory of the Dérive. http://library.nothingness.org/articles/SI/en/display_printable/314. Accessed 10 April 2010.

Dickens, C. 2002. *Our Mutual Friend*. Ware: Wordsworth Editions Ltd.

Eliot, T. S. 2005. *Ulysses*, order and myth. In *Modernism: an anthology*, ed. L. Rainey, 165-167. Oxford: Blackwell Publishing.

Gasiorek, A. Postmodernisms of English fiction. In *The Cambridge companion to the twentieth-century English novel*, ed. R. L. Caserio, 192-209. Cambridge: Cambridge University Press, 2009.

Jackson, K. 2003. *The verbals: Iain Sinclair in conversation with Kevin Jackson*. Kent: Worple Press.

Murray, A. 2007. Literary genealogy and the aesthetics of critique in the work of Iain Sinclair. In *City visions. The work of Iain Sinclair*, eds. R. Bond and J. Bavidge, 68-81. Newcastle: Cambridge Scholars Publishing.

Seale, K. 2007. Iain Sinclair's Textual *Obscenery*. In *City visions. The work of Iain Sinclair*, eds. R. Bond and J. Bavidge, 99-107. Newcastle: Cambridge Scholars Publishing.

Sinclair, I. 1995. *Lud Heat and Suicide Bridge*. London: Vintage.

Sinclair, I. 2003. *Lights Out for the Territory*. London: Penguin Books Ltd.

Sinclair, I. 2004. *White Chappell, Scarlet Tracings*. London: Penguin Books Ltd.

Sinclair, I. 2005. *Dining on Stones*. London: Penguin Books Ltd.

Sutton, P. 2007. High Occulting and Pulped Detectives: Sinclair Sells Out? In *City visions. The work of Iain Sinclair*, eds. R. Bond and J. Bavidge, 56-65. Newcastle: Cambridge Scholars Publishing.

Philip K. Dick: One Man's Illusion Might Invade the Reality of Others

Artur Skweres

Abstract The rules governing reality were a major concern of the prolific American science-fiction writer, Philip K. Dick. In his works, the cohesion of the constructed world was not a constant. It could suddenly or imperceptibly dissolve, leaving the protagonists of the novel wondering, whether what they were experiencing was illusory or real. Dick speculated that every person might perceive the world differently, and as a result live in a singular, separate world of his or her own. What is more, Dick prided himself on being able to construct fictitious worlds which would stand the test of time, yet he admitted that he found enjoyment in their falling apart. Consequently, his characters would have to face an unusual threat—that of the invasion of the mind. The worlds Dick destroyed were the small, private universes of his characters, invaded by an external, often malicious force or agent that would ensure their decomposition. Dick's heroes saw their intimate, inner worlds constantly invaded, and the boundaries of what they considered reality breached. The purpose of this paper is to explore the relationships between characters in Philip K. Dick's fiction, whose interactions lead to a change in their sense of reality or identity.

1 Introduction

Philip Kindred Dick defined himself not as a novelist, but as a "fictionalizing philosopher" and has left an impressive heritage of 44 published novels and over 120 short stories (Sutin 2005, p. xvii). Although they shared many common

A. Skweres (✉)
Adam Mickiewicz University, Kalisz, Poland
e-mail: askweres@amu.edu.pl

J. Fabiszak et al. (eds.), *Crossroads in Literature and Culture,* Second Language Learning and Teaching, DOI: 10.1007/978-3-642-21994-8_24,

themes, they were seldom repetitive, retaining the sense of originality and novelty, a feature he considered crucial to the field of science-fiction (Dick 1995a, p. 44). Nonetheless, certain patterns can be observed. According to Brunner, Dick's main subject was to trace the, often contradictory, relationships between our perceptions and reality (2008, p. 13). The purpose of this paper is to explore the relationships between characters in Philip K. Dick's fiction, whose interactions lead to a change in their sense of reality or identity. It will be demonstrated that in Dick's stories reality was subject to change under the influence of an external or internal agent. The result was a sense of instability, danger, and seemingly irrational fears of Dick' characters.

2 The Definition of Reality According to Philip K. Dick

Philip K. Dick feared loneliness. His horror of being alone might have stemmed from the solitary character of the writing profession, which cost him the relationships with his five wives or, less directly, it might have been caused by the early loss of a twin sister and a divorce of his parents. In his novels isolation often leads to suffering and death. On the other hand, the willing seclusion of his characters is caused by another fear—that of intrusion. In Dick's stories the personal space has shrunk from the inviolability of one's home or possessions to the last line of defense—the human psyche. The interaction between the characters and their surroundings is dynamic and intimate to the highest degree, since their thoughts can be intruded upon, perceptions altered, memories modified.

As it has been stated, one of the constant themes of PKD's fiction is the uncertainty of the reality. According to Stanisław Lem:

> It is Ph. Dick who deals with creation of worlds that frustrates the distinctions between the facts and artifacts of existence. His worlds create a rich spectrum of nightmares, whereby common to them is a progressive adhesion of reality and hallucination, fatal because the diagnoses of the state of affairs (real or imaginary) become seemingly dependable—only to crumble in downright cruelty.
>
> The element of decomposition dominates Dick's worlds, being nearly monstrously rampant, is the blindly pillaging entropy, frightening even time, whereby in the mainstream of Dick's writing each book is another incarnation of the unstoppable agony.[1]

[1] The English translation is mine. The Polish version in is as follows: Właśnie Ph. Dick zajmuje się światotwórstwem udaremniającym rozróżnianie pomiędzy faktami i artefaktami egzystencji. Światy jego tworzą bogate spektrum koszmarów, przy czym wspólny im jest postępujący zrost jawy z majaczeniem, fatalny przez to, że diagnozy stanu rzeczy (jawa czy zwid) zdobywają pewność pozorną—po to, by tym okrutniej runąć.

Dominantą światów Dicka jest żywioł rozkładu, niejako rozpędzony monstrualnie, ślepo buszująca entropia, przerażająca nawet czas, przy czym w głównym ciągu dzieł Dicka każda książka stanowi inne wcielenie lawinowej agonii] (Lem 2009a, pp. 83–84)

The distinction between reality and the imaginary is far from clear in Philip K. Dick's fiction. To Philip K. Dick, the change of perceptions constituted the change of reality. Having defined reality as "that which, when you stop believing in it, doesn't go away" (Dick 1995c, p. 261), Dick subordinated it to the individual perceptions and beliefs. Hallucinations can change the way reality is perceived by the affected person, yet they do not change the world as it is perceived by others. What changes, even if only temporarily and illusorily, is this one person's *inner world*. A distinction, therefore, is needed between the inner world and the outer world which is shared by everyone. Although this idea was expressed first by Heraclitus, one of the Pre-Socratics who fascinated him, Philip Dick made this distinction his own. As he writes in a private letter:

> For each person there are two worlds, the *idios kosmos*, which is a unique private world, and the *koinos kosmos*, which literally means *shared* world (just as *idios* means private). No person can tell which parts of his total worldview is *idios kosmos* and which is *koinos kosmos*, except by the achievement of a strong empathetic rapport with other people... The *koinos kosmos* has, in a certain sense, the support of three billion human beings; and *idios kosmos* the support of only one (Dick 1975, pp. 31–32).

Every person might perceive the world differently and as a result live in a singular, separate world of his or her own. Yet to Dick the private world was by no means unassailable. Consequently, his heroes are highly individualistic, very protective of their privacy. What is surprising in Dick's fiction is that his heroes' protectiveness and proclivity for self-isolation is largely justified, since upon it hinges the safety of what they consider reality. If they fail to protect themselves from a covert, outside influence on their *idios kosmos*, the results may be disastrous.

3 External Influences Imperceptibly Changing Reality

The reality in Philip K. Dick's fiction could change imperceptibly, influenced by an agent—from an item, as the perfect work of art in *The man in the high castle*, to a drug in *Flow my tears, the policeman said*. In *The man in the high castle*, a novel dealing with an alternate reality in which America lost World War II, the characters are fascinated by historical, original items. As it turns out, most of them are fake. The reality itself is fake and can only be glimpsed at in two instances, both involving art. First the hero of the novel, Mr. Tagomi, meditates looking at a perfect piece of jewelry and briefly moves into a parallel reality where America is no longer dominated by the Japanese. Another instance is experienced by the eponymous, secluded author within the novel, Hawthorne Abendsen, who manages to describe the alternate world where history had taken a different course and the allies have won the war with the German-Japanese Axis. He does not arrive at this idea alone but with the help of an oracle, I-Ching, which he uses to plan the plot of his novel (Dick 2001).

In *Flow my tears, the policeman said*, the agent introducing a change in the perception of reality is both a substance and a person. The main character of the novel, Jason Taverner, a famous singer adored by millions, wakes up one day in a seedy hotel-room, without any of his possessions or documents. In a police-state in which he is living it constitutes a death sentence. What is more, to his surprise, within one night he lost his status of celebrity, since no one remembers him. The sole exception is the sister of the General of Police, Alys Buckman. As it transpires, the change which erased all records of Taverner's existence was caused by her. Alys experimented with a drug, KR-3, which influenced the perception of time and reality. Taken in overdose, it affected not only Alys, but also those around her, including the person whom she most admired, the singer Jason Taverner:

> Taverner wasn't the one who took the KR-3. It was Alys. Taverner, like the rest of us, became a datum in your sister's percept system and got dragged across when she passed into an alternate construct of coordinates. (…) We occupied two space corridors at the same time, one real, one irreal. One is an actuality; one is a latent possibility among many, spatialized temporarily by the KR-3. But just temporarily (Dick 1993, p. 211).

In *Flow my tears, the policeman said* the change which occurred within one person's inner world, yet in the course of the novel, it affected the perception of reality of everyone around her, whereby the usual is the opposite—the group affects the individual. Such handling of the changes in reality may be revealing of Dick's latent desire to influence other people in a profound yet inconspicuous manner.

Philip K. Dick's short story, the "The electric ant" presents a surprising example of such a reverse interaction, where the individual effectuates changes in his *idios kosmos* which affect the *koinos kosmos*. After an accident, the hero of the story is discovered by his doctor to be an automaton, a so-called *electric ant.* His status changes immediately; as a non-human he is no longer addressed as Mister Poole, but as Poole. He realizes that he is not a full member of society because all his thoughts have been preprogrammed and that he had never been free in his actions. Assailed by suicidal thoughts, he resolves to look for his programming mechanism to finally gain full control over his life and perceptions. Once he has modified his reality controlling mechanism, he and his companion observe unexpected events, such as a flock of birds appearing in the middle of the room. Finally, he destroys the mechanism and dies. Unbeknown to him, the changes affected not only him but everyone else as well, and with his death, the world ends. Once again, a single, seemingly insignificant individual can be seen to inadvertently modify the world and even put an end to its existence.

A more deliberate attempt to modify the reality of other people, with the malevolent purpose of enslaving them in one's own, controllable universe, can be observed in *The three stigmata of Palmer Eldritch.* Its eponymous hero is a merchant selling a drug called Chew-Z, which he brought from an alien solar system. The drug promises to relieve the off-world colonists of their daily concerns by enabling them a temporary visit to a dream world, which they could visit together. Yet they soon discover that the illusion may trap them forever under an absolute control of Eldritch. Even when he is killed, his presence is still looming ominously, and anyone who ever came under his control is marked by the appearance of his stigmata on their bodies: artificial eyes, teeth, and arms (Dick 2008c, p. 116). Well after the drug has ceased to exist, its illusion still affects the reality of its victims.

The problem of the inability to distinguish between illusion and reality is also pronounced in *Ubik*. The novel introduces the institution of a *prudence organization* which prevents "unauthorized intrusions" of telepaths and mind-readers into the mental processes of their clients. They warn of the possibility of invasion of the mind by another person, which happens in the course of the novel. A group of employees of one of the prudence organizations dies in an accident; their consciousness is artificially kept alive by means of a *cold-pack*. The deceased are unaware of what transpired and oblivious to the fact that they reside in an illusion of a real world. Yet soon this world starts to crumble, giving way to all-consuming decay. The surrounding objects start to revert in time to their earlier forms; finally, the group members start to die, due to an accelerated aging process. The group realizes that there is a malevolent force which is trying to kill them. At the same time they perceive a counter-force, exerted by their surviving employer, Glen Runciter, who is attempting to stop the decomposition: "the pair of opposing forces were at work. Decay versus Runciter, Al said to himself. Throughout the world. Perhaps throughout the universe. Maybe the sun will go out, Al conjectured, and Glen Runciter will place a substitute sun in its place. If he can" (Dick 2008a, p. 574). Runciter is perceived as the savior with semi-divine powers, able to stop the changes performed in the reality of his late subordinates. The other intrusion is that of Jory, a particularly powerful individual kept in cold-pack who is revealed to have generated the illusory realities in order to consume their life force. Jory acts as death and the devil, the tempter and the prowling lion, first isolating his victims and then devouring their vitality to prolong his own existence.

As it could be observed, the problem of intrusion in Philip K. Dick's fiction can be seen as both positive and negative, as it can bring either loss or gain. The problem will appear more complex in the following examples where the invasion comes not by an external agent, but from within. As a result, it can affect the heroes' perception of reality just as it can effectuate a change of their identity.

4 Changes Effected From Within

Philip K. Dick claimed that some of the anxieties present in his fiction stemmed from his personal life, caused by events such as the traumatic loss of his twin. The incorporation of characters and events from his real life into his works under different guises is well documented (Sutin 2005, Jedrzejak 2007). Soon before his death, Dick admitted that his writings revealed so much of his personal life that "[f]uture biographers will find their job done for them before they start. My life's an open book and I myself wrote the book" (Galen 1986, as quoted in Sutin 2005,

p. 234). One of the recurring themes which may be seen as reminiscent of Dick's private life is the death of a young woman or a twin.

Philip K. Dick and his sister Jane were born in 1928, whereafter his twin sister died. Dick always referred to the loss as having a major influence on his life. In his journal, *Exegesis*, Dick writes: "My sister is everything to me. I am damned always to be separated from her/& with her, in an oscillation. Very fast. Both: I have her in me, and often outside me, but I have lost her; 2 realities at once yin/yang" (as quoted in Kendal 2002). As a child, Dick used to talk about an imaginary friend who had all the hallmarks of his late twin—dark hair and eyes, slight built. He imagined her to have a lively character and to be a help in times of trouble (Carrère 2004, p. 9). Such a woman also frequently appeared in his novels, yet her appearances often had the character of an unwelcome invasion; they bode badly for the hero of the novel, producing unexpected and unfortunate results, for instance the irresponsible and provocative Alys Buckman who slept with her brother in *Flow my tears, the policeman said,* or the femme fatale Pat Conley in *Ubik.* Although one might conclude from these examples that the appearance of such a literary incarnation of Philip K. Dick's twin was generally perceived a bad omen, the yearning for her presence is easily discernible. One such instance can be found in Dick's *Valis*. The theological system of the novel describes a deity who cannot forget his sister:

> This woman, who died long ago, was one of the primordial twins. She was one half of the divine syzygy. The purpose of the narrative is the recollection of her and of her death. The Mind does not wish to forget her. (…) All the information processed by the Brain—experienced by us as the arranging and rearranging of physical objects—is an attempt at this preservation of her; stones and rocks and sticks and amoebae are traces of her. The record of her existence and passing is ordered onto the meanest level of reality by the suffering Mind which is now alone (Dick 2009a, pp. 201–202).

What can be observed here is possibly an elaboration of the Gnostic idea of two deities, one of which is imperfect and causes evil, and the modification of Heraclitus's theory of the search for the lost self.[2]

Another, example of a dormant, hidden personality can be found in *Dr. Bloodmoney*. The novel portrays the results of a nuclear apocalypse, one of which are numerous mutations which people and animals undergo. One of them affects Edie Keller, who appears to have an imaginary brother. In reality her brother indeed exists, living within her body. Bill, a sentient fetus in fetu, frequently converses with Edie telepathically. Although he is devoid of sensual perceptions, he is also able to communicate with the dead, which accentuates his position of a semi-alive, semi-dead being. Apart from relating his sister what happens in the "tomb world," Bill is eager to hear what the world perceived by Edie looks like. Lacking developed eyes and ears he resembles the slave from

---

[2] According to Yamakawa, Heraclitus's "real self was once lost and had to be found again. Anyone who loses his sight of himself does so by falling into his own *idios kosmos*"(Yamakawa 2002, p. 18), which mirrors the extrapolations by Philip K. Dick.

Plato's allegory of the cave: apart from the indistinct shadows and sounds of the outside reality that reach him, he cannot fully comprehend how the world looks like by the descriptions alone. Dick portrays him thus: "Someday the girl would die and they would open her body, perform an autopsy; they would find a little wrinkled male figure, perhaps with a snowy white beard and blind eyes... her brother, still no larger than a baby rabbit." (Dick 1965, p. 116) The story greatly resembles Dick's descriptions of a latent presence of his dead sister, yet with a more physical approach. It is also unique, because it is the sister who survives and the brother, half blind and deaf, accompanies her, oblivious of the reality surrounding him.

A particular situation of a stranger residing within one's body and gradually taking it over appears in *A scanner darkly*. In the novel the danger from within is foreshadowed in a number of ways. The narcotics department in the novel can barely contain the spread of a new form of drug, Substance D. Due to rampant corruption and treason, utmost secrecy is preserved at all levels within the police force. One such measure are *scramble suits,* worn by everyone, which enable the preservation of absolute anonymity. The suit modifies the voice patterns and appearance of its users, acting as an ever-changing mask, completely obscuring the identities of police officers, also from each other. As a result of such distrust prevalent within the body of the police force, it becomes possible for individual police officers to suspect one another of being criminals.

The hero of the novel, Robert Arctor, is a narcotics agent working to uncover the trade of Substance D. In order to protect his cover, he starts to take the drug himself. Feeling that the substance is damaging his brain beyond repair, yet unable either to withdraw from the habit or abandon the mission, he feels that he needs to resolve the case as soon as possible. When the authorities, unaware of his identity, ask him to spy on his drug-addict persona, he gradually starts to share their suspicions, finally completely failing to see any connection between himself as an agent and his drug-riddled alter ego. Consequently, in his body appear two personalities: one of an anxiety-ridden junkie fearing authorities and one of a righteous policeman who wants to bring Arctor to justice.

In the novel, the enemy, disclosed by the authorities, resides within the hero. Unable to conform to the demands placed before him, he undergoes a conflict of identity, a conflict which he loses. The clash is not only that of principles—it is the battle of character between the drug-taking, weaker ego and the superego, which wants to conform to law and the expectancies of the society. *A scanner darkly* contains a famous quotation from Goethe's *Faust*, which sheds more light on the split of personality of the main character, suggesting that the conflict should also be seen in the dimension of ambition:

> Two souls, alas, are housed within my breast,
> And each will wrestle for the mastery there.
> The one has passion's craving crude for love,
> And hugs a world where sweet the senses rage;
> The other longs for pastures fair above,
> Leaving the murk for lofty heritage.

(Goethe 2005, p. 67)[3]

In this fragment Faust wants to ascend above the ordinary, towards greatness, but he also utters a cry for the separation of the divine from the mundane, the artist from an ordinary man, the saintly soul from the sensual body; such expression of the duality of human nature, which was characteristic of German Classical literature (Gillies 1957), finds a different meaning in Dick's *A scanner darkly*. Here it is taken literally, as one person houses two personalities.

The other source of dichotomy in Arctor is the relation between the drug subculture—the *dopers*, and the rest of society, the so called *straights*. The difference between the two can be readily perceived by the appearance of the *straights*: "in their fat suits, their fat ties, their fat shoes," they lack empathy and overstress their affluence, since they "live in their fortified huge apartment complexes guarded by their guards, ready to open fire on any and every doper who scales the wall with an empty pillow-case to rip off their piano" (Dick 2008b, p. 666). The *straights* lack sympathy not only for the drug-addicts, but also for helpless animals. As Arctor realizes:

> Addicts, in his experience anyhow, rarely hurt animals. He had witnessed junkies feeding and caring for injured animals over long periods of time, where straights probably would have had the animals "put to sleep," a straight-type term if there ever was one—and also an old Syndicate term as well, for murder (Dick 2008b, p. 715).

The reason for the *doper*'s care for the fate of animals may stem from their own inability to cope. Unemployed, they have to depend on the charity of others to survive and get another portion of the drug. Their submission to the deadly addiction kills their intellect and reduces their existence to the pursuit of the most basic urges, binding them to the cravings and passions until death, reducing their ambitions and relationships to the purposes directed by the survival instinct. Therefore, Arctor's split into two personalities: one of a straight and one of a doper reveals the preexistent tension between the his sense of duty and social ambitions and the conflicting, earnest helplessness in the face of unconquerable addiction.

Another instance of a personality split which is triggered by an outside influence is present in the short story "Imposter." Its hero, Spence Olham, is one of the scientists responsible for the Earth's defense systems in an uneven war with alien intruders. The humankind's situation is dire, yet Olham remains loyal and is sure that human race will prevail (Dick 2002a, p. 45). As he meets with a friend on his way to work he is suddenly assaulted and declared a spy. Convinced of his innocence, Olham manages to escape his pursuers and strives to prove the falsity

---

[3] Translation from German into English is mine. The original German version s as follows:

> Zwei Seelen wohnen, ach! in meiner Brust,
> Die eine will sich von der andern trennen:
> Die eine hält, in derber Leibeslust,
> Sich an die Welt mit klammernden Organen;
> Die andere hebt gewaltsam sich vom Dust
> Zu den Gefilden hoher Ahnen. (Goethe, as cited in Dick 2008b, p. 775)

of the allegations. In the final twist of plot, he is astonished to realize that he is indeed an android oblivious of the fact that he has killed the real Spence Olham and has taken over his identity. Once he uncovers the truth a bomb which he is carrying within his body explodes.

Spence Olham is proven to be dangerous to himself and others despite his good intentions. When he discovers the role planned for him, that of a Trojan horse, he is unable to redeem his error, as the very realization of his false identity brings doom to himself and others. His greatest fear is that of being seen as different and, consequently, as someone dangerous. This resembles Philip K. Dick's own anxieties, as he remarked in an interview:

> I used to believe the universe was basically hostile. And that I was misplaced in it, I was different from it... fashioned in some other universe and placed here, you see. (...) I had a lot of fears that the universe would discover just how different I was from it. My only suspicion about it was that it would find out the truth about me, and this reaction would be perfectly normal: it would get me. (Williams 1986, p. 155)

Similarly, Olham becomes a wanted man because he is thought to be hiding his true identity. From an ordinary person accepted by others, he is suddenly shunned by the closest friends and chased by the authorities. As long as he was ignorant of the knowledge of his real identity, the bomb did not explode. Only when his otherness is revealed to be true, the ersatz identity which was established for him by unknown perpetrators crumbles, with the physical death immediately following.

5 Conclusion

To conclude, the rules governing reality were a major concern of Philip K. Dick. Dick speculated that every person might perceive the world differently, and as a result live in a singular, separate world of his or her own. His characters would have to face an unusual threat—that of the invasion of the mind. The worlds Dick destroyed were the small, private universes of his characters, invaded by either internal or external forces that would ensure their doom. Despite their efforts to separate themselves from others, Dick's heroes saw their intimate, inner worlds constantly invaded. As a result, they suffered from an identity crisis, and the once clear boundaries of what they considered reality became obscure. Consequently, the works of Philip K. Dick present an intricate mechanism of setting and overstepping boundaries between one's world, one's reality, and the reality of others.

References

Brunner, John. 2008. Rzeczywistość Philipa K. Dicka [The Reality of Philip K. Dick]. In Philip K. Dick *Opowiadania najlepsze* [The best stores of Philip K. Dick], ed. Małgorzata Chwałek, 7-13. Poznań: Dom Wydawniczy Rebis.

Carrère, Emmanuel. 2004. *I am alive and you are dead: a journey into the mind of Philip K. Dick.* New York: Picador.

Dick, Philip Kindred. 1965. *Dr. Bloodmoney: or how we got along after the bomb,* New York: Ace Books.

Dick, Philip Kindred. 1975. "Letter of Comment." (31-32) In Bruce Gillespie, (ed.) 1975. *Philip K. Dick: Electric shepherd.* Norstrilia Press: Melbourne.

Dick, Philip Kindred. 1993. *Flow my tears, the policeman said.* New York: Random House USA Inc.

Dick, Philip Kindred 1995a. "Philip K. Dick on philosophy: A brief interview," Conducted by Frank C. Bertrand. In: *The Shifting Realities of Philip K. Dick. Selected Literary and Philosophical Writings,* ed. Lawrence Sutin, 44-53, London: Vintage Books.

Dick, Philip Kindred. 1995b. "The short happy life or a science fiction writer." In: *The shifting realities of Philip K. Dick. Selected literary and philosophical writings,* ed. Lawrence Sutin, 44-53, London: Vintage Books.

Dick, Philip Kindred. 1995c. "How to build a universe that doesn't fall apart two days later" In: *The shifting realities of Philip K. Dick. Selected literary and philosophical writings,* ed. Lawrence Sutin, 259-280, London: Vintage Books.

Dick, Philip Kindred. 2001. *The man in the high castle*. London: Penguin Books.

Dick, Philip Kindred. 2002. "The electric ant." In: Sutin, Lawrence and Lethem, Jonathan and Galen, Russell (eds.) *Selected stories of Philip K. Dick*, 382-400, New York: Pantheon Books.

Dick, Philip Kindred. 2002. *Minority Report*. London: Orion Publishing Group.

Dick, Philip Kindred. 2002a. Imposter. In *Minority report*, 45-60. London: Orion Publishing Group.

Dick, Philip Kindred. 2008. *Five great novels*. London: Gollancz.

Dick, Philip Kindred. 2008a. Ubik. In *Five great novels*, 495-648. London: Gollancz.

Dick, Philip Kindred. 2008b. A scanner darkly. In *Five great novels*, 649-841. London: Gollancz.

Dick, Philip Kindred. 2008c. The three stigmata of Palmer Eldritch. In *Five great novels*, 1-158. London: Gollancz.

Dick, Philip Kindred. 2009. *Philip K. Dick Valis and later novels,* New York: The Library of America.

Dick, Philip Kindred. 2009a. *Valis*. In: *Philip K. Dick Valis and later Novels,* Philip Kindred Dick,171-398, New York: The Library of America.

Galen, Russell. 1986, June. Interview. In: Lawrence Sutin (ed.) *The Shifting Realities of Philip K. Dick. Selected Literary and Philosophical Writings*, 234. Vintage Books: New York.

Gillespie, Bruce (ed.) 1975. *Philip K. Dick: Electric shepherd.* Norstrilia Press: Melbourne.

Gillies, Alexander. 1957. *Goethe's Faust. An Interpretation* http://wolfenmann.com/gillies/gillies4.html. Accessed 11 June 2010.

Goethe, Johann Wolfgang von. 2005. *Faust, part one,* translated by Philip Wayne, London: Penguin Classics.

Jędrzejak, Tomasz Marcin. 2007. *Dick Philip kindred. Życiotwórczość.* Warszawa: Szkoła Wyższa Przymierza Rodzin.

Kendal, Rachel P. 2002. "Philip K Dick and the science of fiction". http://www.authorsden.com/visit/viewArticle.asp?id=7527. Accessed 11 June 2010.

Lem, Stanisław. 2009. *Mój pogląd na literaturę* [*My views on literature*]. Warszawa: Agora SA.

Lem, Stanisław. 2009a. O niekonsekwencji w literaturze [About inconsequence in literature]. In: *Mój pogląd na literaturę* [*My views on literature*], Stanisław Lem, 77-96, Warszawa: Agora SA.

Robinson, T. M. and Westra, Laura (eds.). 2002. *Thinking about the environment: our debt to the classical and medieval past*, Oxford: Lexington Books.
Sutin, Lawrence (ed.). 2005. *The shifting realities of Philip K. Dick. Selected literary and philosophical writings*. Vintage Books: New York.
Sutin, Lawrence and Lethem, Jonathan and Galen, Russell (eds.). 2002. *Selected stories of Philip K. Dick*, New York: Pantheon Books.
Sutin, Lawrence. 2005. Introduction In: Lawrence Sutin (ed.) *The Shifting Realities of Philip K. Dick. Selected Literary and Philosophical Writings*, ix-xxix. Vintage Books: New York.
Williams, Paul. 1986. *Only apparently real/the World of Philip K. Dick*. New York: Arbor House.
Yamakawa, Hideya. 2002. Order, harmony, and justice: the Greek Conception of the Environment, In: *Thinking about the environment: our debt to the classical and medieval past*, ed. T. M. Robinson, Laura Westra, 15-24, Oxford: Lexington Books.

The Idea of Ambition as a Social Process Based on Role Transition Theory: Dyadic Power Relations in *Coriolanus* and *Macbeth*

Natalia Brzozowska

Abstract Ambition is a good servant but a bad master. William Shakespeare often analysed the complications stemming from "sick ambition", one of the greatest fears of the Elizabethan age. Interpretations of the Bard's ideas are rooted in many sciences, but the sociology of literature still represents a new frontier to many literary scholars. It is my belief that Shakespearean ambition should be considered as a social phenomenon; as a process by which new territories are sought with a view to gaining access to new resources. This may involve either the expansion of certain roles or the complete transition of roles. Various role processes occur if social relations are further complicated by the involvement in dyadic interactions: strong role-based collaborative systems. This paper addresses the problem of two dyads where the driving force is ambition: Coriolanus and Volumnia in *Coriolanus*, and Macbeth and Lady Macbeth in *Macbeth*. The role of Coriolanus as a successful son is excessively fetishised by Volumnia, and, what is more, her inadequate definitions of roles force her son to transgress many of the boundaries set by his official roles. Macbeth's and Lady Macbeth's actions are an example of an unsuccessful role transition: they become excessively infatuated with the role of the king and follow this obsession into misrepresentation, and then the desire to maintain the front at all costs. This study has been influenced by Erving Goffman, as well as more recent developments in role theory. This viewpoint also considers power relations through the works of classics: new takes on the philosophies of Max Weber, Georg Simmel and Theodor Adorno.

N. Brzozowska (✉)
Adam Mickiewicz University, Poznań, Poland
e-mail: nbrzoz@amu.edu.pl

J. Fabiszak et al. (eds.), *Crossroads in Literature and Culture*, Second Language Learning and Teaching, DOI: 10.1007/978-3-642-21994-8_25,

1 Introduction

Both the ambiguity and depth of Shakespeare's dramatic works, as well their importance for the literary cultures of the Western world, explain the continuous appeal of the Bard's creations. This in turn allows for many critical possibilities from various standpoints: from psychoanalysis to semiology or history. Sociology, however, has been somewhat neglected, perhaps unjustly, as it is often said that the social dimension is often foremost in Shakespeare's works. It is worth mentioning that the sociological theories applied here are derived, for the most part, from classical sociology, based on the social reality of the industrial era. One must retain a certain degree of caution when applying these approaches in a more creative manner. The stress on economic factors and work ethics of the past, as well as an invalid (to some) approach to 'mass' culture, pose some difficulty for the modern critic. However, the theory of social roles does not necessarily connote industrial roles. Furthermore, classical theories are still re-analyzed and rewritten in modern-day sociology, which suggests that they are far more than only historical constructs of a previous era. Roles may be used in a flexible way, for instance—the approaches of the Sixties (Goffman) and the innovative, most recent studies of roles in the film industry (e.g. by Kristina Vaarst-Andersen), not to mention the socio-psychological role theories of the modern classroom. This article shall seek to show that ambition and social roles have particular ties in Shakespearean drama, and that the former may be analysed using the theories of sociology. The theory of social roles, role transition and Simmel's theory of dyad in particular will be employed here as analytical tools.

2 A Sociological Approach to Ambition

2.1 The Key Concepts of Socialisation and Social Role

What differentiates sociology from most other disciplines is that its primary aim is to recognise collective aspects of behaviours, and that almost nothing may be analysed outside the social context. Those following the so-called realistic conception of sociology base themselves on the essential ontological statement that man is an entity that strives towards collectiveness, and certain human skills cannot develop outside a communal environment (Turowski 2001, pp. 13–14). Therefore, one of the main aims of sociology is to study and explain the patterns of associations. The various goals of the discipline revolve around the interactions and the organization of society, as well as the limitations and sanctions that society brings upon itself (Turner 1994, p. 8).

Ambition may be considered a socio-psychological concept associated with evolution and social mobility. As Harris (1966, p. 351) noted, it is an anticipatory and goal-orientated phenomenon that is influenced by the culture and social structure of a particular society. In sociological terms, ambition is a process. Pursuing a particular station in society also means pursuing the access to new resources, new opportunities for individual development: wealth, respect, popularity (the keywords used in sociology to denote goods, as well as the basis of social inequalities). Ambition involves a certain degree of societal imagination. This imagination is firmly founded in the cultural background of the individual and his or her socialisation. It could also be the common desire of a particular group or individual that becomes a point of reference.

Socialisation is one of the key concepts of sociology. The view widely shared by most sociologists is that without proper socialisation the human being will not be able to function in society. Primary socialisation takes place during childhood, with the family as the educator; secondary socialisation sees the adult prepare for new tasks and roles. In many ways this preparation, this internalisation of values which construct roles, will determine the future social actions of the individual. The natural temperament of man is to strive towards the optimal position for himself, and, in turn, the best set of roles associated with it (Łoś 1985, p. 123). To many classical sociologists and their followers 'social structure' denotes the combination of such roles and statuses within society, as well as "the organized pattern of the interrelated rights and obligations" (Theodorson and Theodorson 1969, p. 395). 'Rights and obligations' are terms crucial to an understanding of the idea of social role, although more recently this view has been challenged by sociologists such as Baker et al. (1991) and Callero (1994), who maintain a more individualistic view that roles are also "cultural objects that serve as resources in interaction" (role-using) (DeLamater 2006, p. 89). The greatest contributors to role theory: functionalist and symbolic-interactionist theories, though differing in their approach to roles, find a common denominator in "explain[ing] roles by presuming that persons are members of social positions and hold expectations for their own behaviors and those of other persons" (Biddle 1986, p. 67). In order effectively to fulfil a role, one must attain vital knowledge of what the role stands for (what course to take, as well as the requirements and margins of the role, i.e. the degree of flexibility one may allow for), the role's place in society and its impact.

2.2 Ambition as a Result of Dyadic Social Actions

Ambition becomes problematic when considering role boundaries, as it is associated with a certain degree of mobility. There is always the possibility of facing the unknown: a lack of information as to the acceptability of transgressing norms, undefined new roles or problems in maintaining the appropriate performance. Ambition, as has been noted, is not only a psychological desire but also a notion firmly rooted in the social structure. The most immediate background is the individual's social group. A consensus has yet to be reached as to the minimum number of people that constitute a social group. It is still argued that two people do not form a group because of the lack of certain processes including mediation and clique-forming. However, the classic theorist Georg Simmel, whilst bearing in mind group processes, proposed a distinction based on slightly altered terms and introduced the theory of dyads and triads, mentioning the differences between these relationships. It occurred to him that the key may also lie in the particular number of people involved. It is not really the question of the number itself, but the reciprocal structure and specific processes that take place that are unlikely to be repeated in the same way in structures consisting of different numbers of implicated members (Szmatka 2007, p. 63). The association of two will undoubtedly vary from all others: a withdrawal of one element destroys the structure; however, in a dyad it is still possible to maintain individuality; responsibility for actions cannot be shifted onto other people in the group; the whole process relies on immediate reciprocity. The emotional intensity of dyads is stressed often.

Dyadic relations are the focal point of this study, as the deep intimacy of this relation has an influence on ambition. It is also worth remembering that dyads may be based on the idea of power. In accordance with Max Weber, the idea of power employed here is the ability to realise one's own will in communal action, even against the will of others (Weber 2006, p. 470). It is the possibility of exhorting influence (Scott 2006, p. 12). Much relies on the subjective insight of the subordinate side of the relationship and the method of influence employed by the dominant side. The depth of the relationship between Macbeth and his wife, and Coriolanus and his mother, based on the idea of the dyad and perhaps even the small social group, may explain the way ambition is fuelled by their relationship and the power relations within it. As with dyads, the point of reference is often the other person in the relationship. Both of those dyads employ a variety of power-related influences. Lady Macbeth's husband becomes her whole world, and Macbeth, for a long time, is greatly influenced, even subconsciously, by fear of his wife. Coriolanus is dominated by Volumnia up to the point where he cannot escape his primary role of the soldier-son. Both dyads find proper role transition problematic.

3 Dyadic Power Relations in Selected Tragedies of William Shakespeare

3.1 Macbeth and Lady Macbeth

In family dyads, such as the Lady Macbeth–Macbeth dyad, the relationship is close and deep. Value systems, if they are dyadic constructs (and so likely to occur in certain family relationships) are hard to verify and reject without breaking up the original association. The couple, at one point, have no-one to turn to but each other (they treat their crime as a bonding element) and this results in a growing pressure to keep the dyad alive. Eventually, the structure cannot be saved and falls apart. The problem of the dyad is that its isolation and exclusivity are only apparent (Simmel 1975, p. 159). Two people cannot function outside the world and its norms and expectations. This is, perhaps, one of the causes of Macbeth's punishment and Lady Macbeth's death.

Here, colossal responsibility lies in the complimentary roles of man and woman, husband and wife. Zbierski (1988, p. 458) mentions that Lady Macbeth and Macbeth have little internal insight regarding themselves, but they know each other all too well. Moreover, they also know what is socially expected of the other half in the relationship. Macbeth is sensitive to his wife's role, and Lady Macbeth knows fully the king role and the man role. However, they have difficulty playing their own roles. This mutual ability to sink into each other's mind paradoxically makes them perfectly suited partners, but it is also what gives rise to their crimes (Zbierski 1988, p. 458). They are fuelled by their expectations of each other. The sociometric relationship falls apart when Macbeth decides to alienate himself completely and Lady Macbeth begins to show signs of madness, and, therefore, is barely acknowledged by her social group (or dyad) due to her inability to function in it. This is the beginning of the end of Macbeth's internal moral disputes, and the rise of his narcissism: "Macbeth alone will judge men and impose sentences upon them. In this way he places himself at the center of creation; the community falls away. Rejecting all laws and customs outside his own being, Macbeth is no longer cognizant of conscience" (Rosen 1960, p. 91). However, before all this happens, they constitute a dyad whose prime connection is their decision to go ahead with the crime of murdering King Duncan.

The couple's problem in role transition is the lack of guidance and knowledge as to the new roles they must adopt. These are not individualised roles, in accordance with Callero (1994), but roles strictly set in the everyday world. This vagueness is not without purpose; the gaps in Macbeth's knowledge will be filled by Macbeth himself; they are aware that this may lead him to crime. They address him with titles he is puzzled with. In the process they become role-senders and mark him with a role that he has perhaps not been expecting. Macbeth's evil may, in Goffman's terms, be attributed to a lack of clarity as to the role he is about to perform. He is puzzled by the Witches' prophecies, but attracted at the same time. The Witches' manipulative strength attracted him because they also passed on information that was true at the time (he was, with Sinel's death, actually the thane of Glamis, and would soon discover that he had been made thane of Cawdor). The Witches' presentation of the mix of new ascribed and achieved roles make their story exciting, yet believable enough not to be dismissed. Macbeth exclaimed "Would they stayed!" (Shakespeare 1983, *Macbeth* 1.3.82) because, though the Witches had aroused his curiosity, he was unsure of the way to interpret the new role he had been given, and seeks their help and guidance. Symbolic interactionists such as Goffman call role ambiguity cases when there is little clarity as to the expectations and norms surrounding a role, especially a new role. Role margins are not easily definable, so there is a danger of deviation when the new member transgresses certain boundaries. This usually takes place within the realm of achieved roles, for which an individual may be less equipped (insufficient socialisation). The problem of Macbeth and the Witches is the reverse: while he is more or less prepared mentally for his achieved roles, the best (i.e. most desirable) and the most surprising one they mark him with stands more as an ascribed role. The role of the king is unexpectedly bestowed upon him. He will be forced to act out the role he knows he has not been adequately prepared for. With a great degree of ambiguity surrounding the role (as mentioned before, knowledge of 'role' is comprised of several elements, including understanding which course to take), this path may lead to easier rationalisation of faulty or deviant behaviours.

It is then natural that Macbeth and Lady Macbeth make mistakes during role transitions. This happens often if the individuals do not achieve the role naturally and smoothly. However, they try to maintain the performance. The first element of the performance is their front. It is "that part of the individual performance which regularly functions in a general and fixed fashion to define the situation for those who observe the performance (…) the expressive equipment of a standard kind intentionally or unwittingly employed by the individual during his performance" (Goffman 1990, p. 32). Macbeth especially becomes involved in obsessive front maintenance, erasing all that might not fit his new world, eradicating any questioning thought. The couple quickly realises that they are playing the incorrect roles, and cannot play them well. Callero (1994) would have called this the failure of an identity claim: "roles are not uniformly available as identity claims" (DeLamater 2006, p. 89), though roles may be claimed anyway for the resources they bring. This role-as-resource approach of Callero is linked to ambition as a process leading to access to a new frontier. Claiming a role may seem innocent in itself, but here the couple's manipulation is cynical. Misrepresentation, the acting out of a part which one is certainly not entitled to perform, is a notion linked to Goffman's performance theory of roles. Pretending to be someone one is not may be false, but not necessarily 'evil', however, the social reactions to such misrepresentation may range from acceptance to revulsion depending on the 'sacredness' of the part, the skill of the 'impersonator' and the possible gains of the actor (Goffman 1990, pp. 66–67). The sacredness of the king role is incomparable with anything else, especially during Elizabethan and Jacobean times. Macbeth, as an impostor, a tyrant and a usurper (all images placed at the very bottom of the Elizabethan value system), is the embodiment of 'sick ambition'.

3.2 *Coriolanus and Volumnia*

The dyadic relationship of son and mother, on the other hand, is one of the strongest in nature and one of the most oft-discussed in literature. As with Macbeth and his wife, this is also a power relation, although more uneven than marriage: despite the fact that Volumnia is, role one, a woman (therefore by Jacobean standards not the power in the relationship), she is also, role two, a widowed mother and, role three, the daughter of a prominent Roman, whom she invokes in her conversation with the tribunes. Her noble background and widowhood in addition to her personality make her a difficult woman to dismiss. The socialisation of offspring by such 'special' people in such circumstances can produce a very specific mixture of qualities : "I had rather had eleven die nobly for their country, than one voluptuously surfeit out of action" (Shakespeare 1993, *Coriolanus* 1.2.24–25).

Coriolanus' socialisation equips him well for one role alone, the soldier. He is taught to value one trait above all, boldness. Of course, Coriolanus has many roles. Society demands this from him: he is a father, a husband, a politician, a hero, a military leader, etc. However, his problem is that one role takes precedence over the others: the role of the son. This hierarchy of the role-set is very clearly indicated in the beginning: "He did it to please his mother" (*Coriolanus* 1.1.38). The case of Coriolanus is slightly different from that of Macbeth, as he does not seem to be so sick in his ambition for power itself. His main goal seems to be maintaining the status quo for himself as a valiant soldier. He may be ego-sensitive, but not Machiavellian in his desires. As has been mentioned, however, he is involved in a dyadic struggle with his mother. It is she who seems to be more ambitious. She desires him constantly to stretch the margins of his role, disregarding the fact that one must be well prepared to handle such processes:

> To see inherited my very wishes,
> And the buildings of fancy: only
> There's one thing wanting, which I doubt not but
> Our Rome will cast upon thee.
> (*Coriolanus* 2.3. 213–216)

Coriolanus' inability to cope stems from this insufficient socialisation and that his main role of soldier-son conflicts with other roles. What is worse, he feels that he is perfectly capable of achieving anything solely because of his virtues. He was prepared to disregard social limitations on the grounds that as long as he is valiant and heroic in his deeds on the battleground, he deserves everything that may lie off the battlefield. His misinterpretation of roles and reality is the cause of his demise. The problem of ambition as related to role transition is manifold. Firstly, there is the impossibility for Coriolanus to move smoothly from one role to another. He seems to struggle when the need arises to move from the soldier role to the role of the politician and back again. Even though he has the ambition to handle all these roles, he fails. His construction is similar to what Theodor Adorno called the "authoritarian personality". This role-related stiffness, the domination of one overblown role and the inability to move flexibly from one to another is part of this construct. His traits include love for authority and the disregard for anything and anyone 'weak', combined with the consciousness of standing above the law that governs ordinary people, even laws that most people accept as natural through primary socialisation (an example is his rejection of any dialogue with the common folk) (Szacka 2003, p. 140). Even though Coriolanus is aware of these laws, he remains 'tough', cynical, superstitious, and, unfortunately, very destructive (Adorno 1950, p. 228). When people reject him, he does not reflect on his traits, but states "There is a world elsewhere" (*Coriolanus* 3.3.134). This array of characteristics may suggest that Adorno's authoritarians are not far from the dramatic idea of *hubris*. Furthermore, there is the desire of Volumnia for Coriolanus to continue to rise through the social system. While she wants him to advance, she never prepares him for the transition of roles, keeping him locked instead in one dominant role, the son role. Her secondary socialisation of Coriolanus comes too late. She undergoes a process role theorists such as Erving Goffman, called the fetishisation of a role, the excessive infatuation with a role, which symbolises the degree of comfort one associates with one role. The role becomes a symbol of values, goods and aspirations and, when it is attained, it is relished and played out with super-conformist flair. The role may even take over someone's life and stretch widely outside socially permitted role margins. The role, if very significant and cherished by the actor, is likely to become a ritualised, super-conformist fetish to be sustained at all costs. Fetishisation often occurs with roles attributed to the more optimal distribution of goods (wealth, power, respect), especially in formalised hierarchies. Complications stem from the Coriolanus-Volumnia dyad. Her excessive infatuation with the role of the successful son is transferred to Coriolanus himself, who equates his value to the degree of achievement of his mother's ideals. Maintaining 'dignity' and 'valour' also becomes his fetish. He has only one point of reference: his mother. The father is dead, and there seems to be no other family member or friend approaching Volumnia in terms of importance. Coriolanus believes that as long as he plays that one achieved role his mother most approves of, everything should naturally fall into place. However, this cannot happen in the world he lives in. He fails to recognise that the constant stretching of role boundaries (the military hero) is not a process that society will accept. The most painful outcome of role-related problems is role failure, when the individual becomes extremely unsuccessful in terms of meeting societal expectations. This occurs when certain elements of role—those without which it is not possible to act out a satisfactory performance—are not developed or when the person has not been adequately prepared for the role (Horton and Hunt 1980, p. 120). It is worth mentioning that Volumnia distances herself from the negative outcomes of her son's socialisation, but this decision is long overdue. The final blow and the revelation of the relative weakness of the image of Coriolanus seems to come not when Aufidius calls Marcius a traitor, but when he is called a "boy of tears" (*Coriolanus* 5.5.100). Coriolanus' problem is that he cannot escape the son role. He remains, in accordance with his mother's wishes, primarily the 'warrior son' of Volumnia.

4 Conclusion

Ambition as role transition, in both cases, desired but unsuccessful, is the driving force behind the two dyads, both of which are based on standard, natural associations: husband and wife, and son and mother. If value systems are derived mostly from these associations, the intensity of such collaborative systems may be difficult to overcome. It may be beneficial to use sociological theories when analysing literature, including classical work, in an interdisciplinary manner. Simmel's dyad theory helped explain the strength of two toxic relationships, and role theory named particular processes that may be associated with ambition as a social construct.

Adorno's concept of authoritarian personality may not necessarily be limited to his observations of World War II fascists, but can also apply to other figures, including fictional characters such as Coriolanus. What is more, the oft-repeated cliché may be true: Shakespeare was a consummate social observer. The study's final conclusion is that social relations perhaps determine much of what people call ethical reasoning, as ambition was definitely a sensitive ethical matter to the Elizabethans. As has been shown, this is a notion that Shakespeare himself noted and integrated into his plays, including the two tragedies *Macbeth* and *Coriolanus*.

References

Adorno, T. and E. Frenkel-Brunswik, D. Levinson and N. Sanford. 1950. *The authoritarian personality. Studies in prejudice series, Volume 1*. New York: Harper & Row.

Baker, W., E. Faulkner and R. R. Faulkner. 1991. Role as resource in the Hollywood film industry. *The American journal of sociology*, Vol. 97, No. 2, 279-309.

Biddle, B. J. 1986. Recent developments in role theory. *Annual review of sociology* 12:67-92.

Callero, P.L. 1994. From role playing to role-using: Understanding role as a resource. *Social psychology quarterly* 57(3), 228-243.

DeLamater, J. D. ed. 2006. *Handbook of social psychology*. New York: Springer Science and Business Media.

Goffman, E. 1990 [1959]. *The presentation of self in everyday life*. London: Penguin Books.

Harris, E. E. 1966. Some comparisons among negro-white students: social ambition and estimated social mobility. *Journal of Negro education* Vol. 35, No. 4, 351-368.

Horton, P. B., C. L. Hunt. 1980. *Sociology*. New York: McGraw-Hill.

Łoś, M. 1985. Role społeczne w nowej roli [The new roles of social roles]. 1985. In: *O społeczeństwie i teorii społecznej. Księga poświęcona pamięci Stanisława Ossowskiego* [On society and social theory. A book in the memory of Stanisław Ossowski], eds. E. Mokrzycki, M. Ofierska and J. Szacki, 123-144. Warszawa: Państwowe Wydawnictwo Naukowe.

Rosen, W. 1960. *Shakespeare and the craft of tragedy*. Cambridge: Harvard University Press.

Scott, J. 2006. *Władza*. Warszawa: Wydawnictwo Sic!

Shakespeare, W. 1983. *Macbeth*. (Edited by Bernard Lott. New Swan Shakespeare edition). London: Longman.

Shakespeare, W. 1993. *Coriolanus*. (The Illustrated Stratford Shakespeare). London: Chancellor Press.

Simmel, G. 1975. *Socjologia*. (translated by Małgorzata Łukasiewicz.) Warszawa: PWN.

Szacka, B. 2003. *Wprowadzenie do socjologii* [Introduction to sociology]. Warszawa: Oficyna Naukowa.

Szmatka, J. 2007. *Małe struktury społeczne* [Small social structures]. Warszawa: Wydawnictwo Naukowe PWN.

Theodorson, G. A. and A. G. Theodorson.1969. *A modern dictionary of sociology*. New York: Thomas Y. Crowell Company.

Turner, J. H. 1994. *Socjologia: koncepcje i ich zastosowanie*. (Translated by Ewa Różalska.) Poznań: Zysk i Spółka.

Turowski, J. 2001. *Socjologia. Małe struktury społeczne* [Sociology. Small social structures]. Lublin: Towarzystwo Naukowe Katolickiego Uniwersytetu Lubelskiego.

Weber, M. 2006. Typy panowania. In *Socjologia. Lektury* [Sociology. Collected excerpts], eds. P. Sztompka and M. Kucia, 470-492. Warszawa: Wydawnictwo Znak.

Zbierski, H. 1988. *William Shakespeare*. Warszawa: Wiedza Powszechna.

Part VI
Across Literary Epochs, Bringing Together Writers

Genius and Madness Mirrored: Rossetti's and Yeats' Reception of William Blake

Anna Budziak

Abstract It took over a hundred for the oeuvre of William Blake to be fully appreciated both for its thought and form. Nineteenth-century historians of literature, interpreters and fellow poets, struggled to comprehend his work, which attempts were rarely rewarded with success. The nineteenth-century critics saw in Blake a "genius tinctured with madness," a "strange" poet, or a poet lacking the means of expression, literally "illiterate." Those differences in opinion resulted from an apparent incongruity in Blake's work which combines a striking simplicity of style with philosophical profundity and imaginative scope. His critics saw his imperfections and were struck by his genius. But his genius was not easy to explain; among those who tried, the most prominent were D. G. Rossetti, and W. B. Yeats. This paper aims at sketching the vicissitudes of Rossetti's and Yeats's attempts to make Blake's work accessible to a wider readership. And since the "mark" of genius, as stated by Arthur Koestler, is in "opening of the new frontiers," rather than in "perfection," some intellectual daring, rather than flawlessness, was needed of his interpreters. Such were the intellectual adventures of Rossetti and Yeats: Rossetti, by imitating Blake's style, put his own reputation at stake, and Yeats, by seeing his own esoteric system mirrored in Blake's, not infrequently erred. Finally, then, this paper specifies where they succeeded and where they failed by asserting or denying their own spark of genius in the realm of interpreting.

Next to possessing genius one's self is the power of appreciating it in others
—Mark Twain.

A. Budziak (✉)
Wrocław University, Wrocław, Poland
e-mail: budziaka@gmail.com

J. Fabiszak et al. (eds.), *Crossroads in Literature and Culture,* Second Language Learning and Teaching, DOI: 10.1007/978-3-642-21994-8_26,

1 Introduction: Victorian Critics

The Blakean mixture of imaginative poetic genius and child-like simplicity in style has proven uneasy for William Blake's interpreters and imitators. To the unaccustomed eye, his genius might seem to comprise what is poetic, insane and infantine, thus apparently substantiating the belief that the poet, the child, and the madman have something they share. The notion of such a triumvirate, though both facile and false,[1] seems to have confused the nineteenth-century reception of Blake. First, his Victorian critics labeled him as "strange," even bordering on the insane. But in the 1890s he was regarded as an unquestionable genius. It is interesting to see how this subtle border between Blake's alleged insanity and genius would pose challenges to Dante Gabriel Rossetti and William Butler Yeats, how these two poets, trying to steer the reception of Blake over the frontier of madness and into the haven of genius, themselves went a little astray.

Undoubtedly Blake was difficult to assess and to accommodate within the English literary tradition. He wrote in relative obscurity[2] and was buried in an unmarked grave, but then he would rise to the status of hero. The first significant work that focused exclusive attention on Blake appeared in 1863: it was Blake's biography written by Alexander Gilchrist (though the last chapter was written by Dante Gabriel Rossetti after Gilchrist's death), entitled *Life of William Blake: Pictor Ignotus*. Rossetti not only cooperated in writing this biography; he also provided its author with the manuscript of Blake's poetry, nowadays know as the Rossetti Manuscript, found in the British Museum in 1847 (Vries 1919, p. 39). In his biography of Blake, as indicated by Deborah Dorfman, Gilchrist presented the "prophet-hero." Such hero-fashioning stemmed from Gilchrist's fascination with Thomas Carlyle and with his ethics of hero-worship (quoted in Hilton 1988, p. 136), which highlighted individual heroism as pitched against the mechanized and interest-driven world of a growing capitalist economy. Whatever the evaluation of Blake's work might be, Gilchrist would "only speak of him… with

[1] For an illuminating critique of the tendency to confuse the languages produced by "the grotesque trinity of child, poet, and madman" see Gilles Deleuze's "The Schizophrenic and Language." Deleuze warns against simplistically collapsing these three apparently similar discourses into one and explains that surface similarities (for example, in the creation of portmanteau words) actually result from completely different thought processes. In his structuralist analysis, Deleuze points out the crucial differences between, on the one hand, the grammatical (or surface) qualities of the language created by Lewis Carroll in his *Alice in Wonderland* and, on the other hand, the anti-grammatical (or in-depth) language created by Antonin Artaud, writing in the asylum, and the vocabulary of a young schizophrenic patient, for whom "sounds […] [are] confused with the sonorous qualities of things, with the noisiness of bodies, with their actions and passions […]" (1979, p. 284).

[2] Robert Ross would claim that the opposite is true, that "[t]he real wonder is that anyone so inarticulate should have found the many contemporary admirers as he did" (1909, p. 85).

praise" (quoted in Hilton 1988, p. 136).[3] But as the author of prophetic books, Gilchrist's Blake dangerously approaches the brink of madness. The allusions to Blake's alleged insanity were also present in earlier biographical works. In particular, the references to Blake's alleged insanity could be found in the 40-page biographical sketch by Allan Cunningham, in his *Lives of the Most Eminent British Painters, Sculptors, and Architects* (1830), where Blake is shown as a split personality comprising two personae: one manifest in the daylight, and the other being its nocturnal counterpart. As put by G. E. Bentley (the editor of Cunningham's sketch), Cunningham lays "emphasis… upon a schizophrenic interpretation of Blake's personality, showing him sane and engraving other men's work by day, and mad and composing his own [utterly wild] poetry by night" (quoted in Hilton 1988, p. 135).

2 An Echo in Rossetti: A "Tinge Of Insanity"

The allegations of madness caused Rossetti to approach Blake's works with caution and reserve. Rossetti would never let himself go. As noted by Christina Emerentia Bassalik-de Vries, Rossetti always kept his reading public in mind,[4] to the extent that he would seek psychological identification with his potential reader rather than explore a challenging sensibility and a new mode of expression. He would meet his readers' needs halfway rather than risk crossing a new frontier in the hope that they would follow. But such prudence was, perhaps, justified in a poet who himself had suffered severe criticism and ridicule. The first blow came from the outraged defenders of the principles guarded by the Royal Academy of Arts: Rossetti suffered attack when the aims of the Pre-Raphaelite Brotherhood were revealed and their rebellion against the aesthetic norms, represented by the muted paintings of Joshua Reynolds, became apparent. The next, and rather low, blow was dealt to Rossetti in 1871 when the enraged spasmodic poet, Robert Buchanan—infamously commemorated in the vitriolic verses of Pound as "Foetid Buchanan"[5]—dubbed Rossetti's poems the "fleshy school" of poetry (Buchanan 1871).[6] Twice bitten, Rossetti would approach Blake's poetry—which he thought to be marked by a

---

3 "Gilchrist's reviewers… almost unanimously accepted Blake as Hero. They relinquished mad Blake and took to their hearts the gentle and frugal engraver who chanted hymns on his deathbed and died with his debts paid. …" (Dorfman qtd. in Hilton 1988, p. 136).

4 In his correspondence, Rossetti states: "Above all ideal personalities… with which the poet must learn to identify himself, there is one supremely real… namely that of his reader" (qtd. in Vries 1910, p. 32).

5 See Ezra Pound's lyric "Yeux Glaques" in "Hugh Selwyn Mauberley": "When John Ruskin produced/'King's Treasuries'; Swinburne/and Rossetti still abused./Foetid Buchanan lifted up his voice…" (1957, p. 65).

6 See Robert Buchanan's "The Fleshy School of Poetry: Mr. D. G. Rossetti."

"slight tinge of insanity" (quoted in Vries 1910, p. 32)—with great caution.[7] In 1874, the Rossetti brothers, William Michael and Dante Gabriel, published *The Poetical Works of William Blake: Lyrical and Miscellaneous*, which by 1890 had been reprinted six times. However, Blake's *Marriage of Heaven and Hell* was not included in this volume since, even if the Rosettis appreciated the poem, they resolved not to shock their conservative public (e.g., Hilton 1988, p. 137).

Nevertheless, although Dante Gabriel Rossetti distanced himself from the philosophical content of Blake's work, Blake's thought exerted some influence on him. For instance, as explained by Vries, the impact of Blake manifested itself in the way Rossetti represented human selfhood. For Rossetti, as for Blake, the self was manifold; we might say it was anti-essentialist. It consisted of passing moods (*Stimmungen*) which Rossetti expressed through allegorical personifications, and which can be seen as reflecting Blake's idea of *emanations*. Those moods, *Stimmungen* or emanations, are visible in the imagery of Rossetti's poetry (for instance, in "The Lost Days") and in his pictures (in *How they Met Themselves* and *The Gates of Memory*). A community of thought between Rossetti and Blake is also manifest in their mystical leaning; however, the nature of Rossetti's mysticism evidently differed from Blake's. For Blake, mysticism was an expression of his deep religious belief; for Rossetti its attraction lay in irreducible mystery, as he was drawn to the mysterious, the veiled, the macabre (1910, pp. 20–25). Through his mysticism, Blake strived to reach outside himself to the Supreme Being; Rossetti, on the contrary, while celebrating mysteriousness, was rather introspective and aimed at fathoming the hidden secrets of his self.

While replacing Blake's theology with psychology, Rossetti, nevertheless, unreservedly imitated Blake's style.[8] This imitation consisted either in echoing the rhythms of Blake's poetry or in direct borrowings. Vries notes that Rossetti's best poem, "The Blessed Damozel" (1850), is full of such stylistic mirroring: it repeats both the inscription from Blake's Illustrations for the Book of Job and the refrain from Blake's lyric (from *Song of Innocence*) entitled "Laughing Song."[9] The inscription from the Book of Job (38:7; Plate 14) is reflected in the lines which read: "… and now she spoke as when/*The stars sang in their spheres*" (Vries 1910, pp. 36–39). The stylistic echo, which shows, for instance, in the line consisting entirely of a list of female names, is rather less fortunate; and it attracted the critical attention both of Vries and of the uncompromising critic of Decadence, Max Nordau. Blake's "Laughing Song"—short, simple, and unpretentious—includes a musical dactylic tetrameter: "When Mary, and Susan, and Emily." Rossetti imitated it with a difference which can hardly be considered an improvement on Blake: he extended this line to "Cecily, Gertrude, Magdalene,

[7] For instance, in Rossetti's opinion, Blake's psychological imbalance would account for his colouring of the tiger red and green. It was only in the 1890s that Yeats interpreted these colours as symbolical of passion and nature.

[8] Vries notes that, unlike Blake who did not limit his allegorical figure to one sex, Rossetti employed only female figures for his personifications (1910, p. 24).

[9] See Blake 1966, p. 125.

Margaret and Rosalys" (Rossetti n.d., p. 9). In his devastating critique, delivered in 1895, Nordau (quite rightly) points out that this string of names—"arous[ing] gliding shadowy ideas of beautiful young maidens" (1993, p. 90)—has no justification in terms of the poem's coherence; rather it is characteristic of echolalia, as is the insistent use of repetition in another of Rossetti's poems, "Troy Town," described by Nordau as symptomatic of "an obsession, which the patient cannot suppress" (1993, p. 92). "Troy Town," as well as the poem entitled "Chimes," is also critically reviewed by Vries. She quotes the unfortunate tongue-twister from "Chimes"[10] ("Lost love—labour and lullaby/and lowly let love lie") next to the refrain from "Troy Town." Indeed, the heavily alliterated line "Oh Troy town, Troy's down! Tall Troy's on fire" is repeated fourteen times, which even according to Vries, hardly as severe a critic as Nordau, seemed as "too much of a good thing" (1910, p. 35).

Rossetti strived to reproduce Blake's euphonies, so he filled his poetry with forms of expression and rhythms typical of nursery rhymes: monosyllabic words, frequent alliteration, repetitions and refrains.[11] However, the effects of simplicity in Blake and Rossetti are different: Blake's style is unadorned vernacular, the infantine, whereas Rossetti's approaches the infantile. Blakean simplicity finds justification in his genius, in what is genuine and borders on child-like earnestness. This combination of the genuine and ingenious finds a good explanation in Robert Browning. In his famous lines from *Prince Hohenstiel-Schwangau*, Browning asserts: "Genius has somewhat of the infantine: but of the childish, not a touch nor taint" (2009, p. 127).

With his continuous awareness of the reader's scrutinizing eye, Rossetti lacked this Blakean child-like earnestness. In the words of Vries, "Rossetti's poems gain in clearness and construction, but lose in freshness and spontaneity when compared to Blake's" (1910, p. 33). Paradoxically, through this unfortunate imitation of style—but divorced from Blake's independent and revolutionary spirit—Rossetti laid his poetry open to Nordau's ridicule. Nordau especially criticized Rossetti's inclination to force melody at the cost of sense. In his *Degeneration*, Nordau condemned Decadence from the point of view of a physician: he saw in it a

[10] For a commentary on the repetitions in "Chimes" and "Troy Town," see Vries 1910, pp. 34–35.

[11] For a characterization of Blake's "simple" style, see Vries, who describes "Songs of Experience" as "always the same: the words for a greater part of Teutonic origin, are very simple and often monosyllabic. In the meter also the same tendencies can be observed: generally short-line stanzas rhyming in couplets are used. Besides end-rhyme, interlinear rhyme occurs where the lines are prolonged… Another peculiarity of style to these poems consists in the repetition of the same words… [s]ometimes a whole line is repeated… Occasionally even a whole stanza is repeated" (1910, pp. 28–30).

pathological stage of culture and society, a threatening disease,[12] and not "a new and beautiful and interesting disease" as Arthur Symons would want it (1923, p. 97). Of "The Blessed Damozel" Nordau says that the meaning there is non-specific, words are valued for their associative power and vocal qualities (1993, pp. 87–91). The deployment of needless repetition and alliteration is even more striking in "Stratton Water" and in "My Sister's Sleep," where words are strung together, not because they make any intelligible sentences, but merely because they rhyme: "so wet she comes to wed," in "Stratton Water," and "the hollow halo it was in/Was like an icy crystal cup," in "My Sister's Sleep" (1993, p. 93).[13]

3 A Reflecting Mirror to Yeats: "No One will Call Him Mad Again"[14]

In his unfortunate appropriations of Blake, Rossetti was not an isolated figure; other nineteenth-century misrepresentations of Blake can be found in Swinburne, Symons and Yeats. Their interpretative mistakes were mocked by a devoted friend of Oscar Wilde, Robert Ross. Ross devoted two essays to Blake: in the first one (published in 1906) he emphasizes Blake's merit as a painter,[15] in the other

[12] Nordau's criticism of Decadence appeared some 30 years after the publishing of "The Blessed Damozel." It was included in his famous refutation of Decadent art and mentality, *Degeneration*, which nowadays, is treated as a historical document rather than an opinion-forming treatise. In his criticism, Nordau lacks any restraint: for instance, in Pre-Raphaelites, he denigrates the attitude which he calls "mysticism"—the propensity in a "mystic" to create and "mix[…] his ambiguous, cloudy, half-formed, liminal representations with… his immediate perceptions" (1993, p. 69)—and which he regards as the reverse of the healthy. He states that Decadent Catholicism (as in Huysmans, Pater, Wilde, Thomson, and Johnson) involves "the ebullitions of piety… accompanied by a seriousness which often amounts to lasciviousness" (1993, p. 73). He links the English tendency to mysticism, which he views as an infallible sign of degeneration, with physical decline and exhaustion caused by living in big cities and by industrialization.

[13] Cf. also Morris's "Earthly Paradise": "O Margaret sitting glorious there/In glory of gold and glory of hair/and glory of glorious face most fair" (qtd. in Nordau 1993, p. 92).

[14] The phrase comes from Yeats' letter to John O'Leary, 7 May 1889 (qtd. in Mann 2009).

[15] Ross sees Blake's genius in Blake's art of illustration. Blake's individualism shines through his own original technique (water colour on a plaster ground) and through his aesthetic revolt against "classicality" or "pseudo-classicism" in English painting. He views Blake as a solitary genius who cannot easily be accommodated within the tradition of the English arts: Blake "constitutes no link in English painting" (1906, p. 151). If there is any tradition to which he belongs, it is medievalism. So, he comes closest to "the gothic revival"; he is seen as belonging to the line of writers which includes Walpole, Beckford, Chatterton, and Percy (1906, p. 156; 1909, p. 85).

Blake's leanings towards Gothic art were also stressed by Vries, who claims that he acquired the taste for Gothic when serving as an apprentice to James Basire. Beginning in 1773, he spent five years making drawings of the monuments in Westminster Abbey. The influence of Gothic on his art is reflected by the "brilliancy" of colours he used, especially in the Illustrations accompanying the Book of Job (1910, p. 45).

(published in 1909) he engages critically with the Victorian reception of Blake as a poet. Ross is very skeptical of the late Romantic poets who were seeking their pre-Romantic antecedents in the work of Blake. He disapproves of virtually all existing critiques:

> With the exception of Gilchrist,... nearly all the writers on Blake have thitherto treated him as a mirror in which they have sought their own anticipated reflection, or at all events a mirror before which they could indulge in diverting literary contortions... Mr. Swinburne was the first to try the 'All Hallow E'en' experiment... He saw, or thought he saw in the glass a mid-Victorian of the eighteenth century, an eloquent atheist of the sixties, ... he saw Victor Hugo and Shelley and a great many other things. But he obviously never saw William Blake... Mr. Arthur Symons... stepping deftly over the debris of Messrs. Ellis and Yeats, sees in Blake only a predecessor of a rather languid English school of symbolist..." (1909, p. 85)

A special place in this hall of mirrors is occupied by Edwin Ellis and W. B. Yeats, who "looked in the mirror and saw a continuation of an Irish poet and an American medium" (1909, p. 85).

Ross's comment leads us onto the third phase in the reception of Blake: to the recognition of Blake's prophetic works. Indeed, into the 1890s, the idea of Blake's strangeness was strong. The adjective "strange" became "[t]he key word in [Margaret] Oliphant's account" of Blake (in 1882), and it was persistently used to describe Blake by other historians of literature, among them, Henry A. Beer (in 1899) and George Saintsbury (in 1896) (Hilton 1988, pp. 137–138).[16] But in 1893, when Ellis and the young Yeats's three volumes of *The Works of William Blake, Poetic, Symbolic, and Critical*[17] were published, a kind of a watershed was reached. Blake would no longer be regarded as "tinged" with madness; his prophetic works and longer poems, such as *Jerusalem* and *Milton*, finally found their publishers. Yeats never doubted the soundness of Blake and the integrity of his works. On the contrary, he devoted "hundreds of pages" to commentary and paraphrase with the purpose of proving "'the solidity of [Blake's] myth, and its wonderful coherence'" (quoted in Hilton 1988, p. 139). But through his commentary and paraphrase, Yeats had made Blake into his own image, so that Blake,

[16] Hilton refers to William Minto's *The Literature of the Georgian Era* (1895), George Saintsbury's *History of the Nineteenth Century Literature* (1780–1885) (1896), and Henry A. Beer's *A History of English Romanticism in the Eighteenth Century* (1899).

[17] For the impact of this phase on later criticism of Blake, see Hilton 1988, pp. 139–141. The edition of Blake's works by Ellis and Yeats was seminal in that it attracted attention to his philosophy; this interest reached its peak between the years 1924 and 1926. In 1924, S. Foster Damon published *William Blake: His Philosophy and his Symbols*, "the first, still widely quoted critical study"; and in 1926, an all-embracing "Index of Blake's Symbols" appeared, accompanying D. J. Sloss and J. P. R. Wallis's Oxford edition of *The Prophetic Writings of William Blake*. Blake's reputation was finally established as indisputable by Northrop Frye's *Fearful symmetry: A Study of William Blake* where "the second half of the eighteens century in rechristened 'the age of Blake'" (Hilton 1988, p. 140).

to use Ross's metaphor, would hardly recognize himself as "mirrored" in Yeats. Certainly, as a Christian he would find it hard to see his views reflected in Yeats's occult science. So, what was Yeats's Blake like? In short, he was fashioned from Yeats's patriotic feelings for Ireland and from his fascination with esotericism. In the letter sent to John O'Leary, Yeats expressed the belief that Blake "was of Irish extraction," and that "his grandfather was an O'Neal who changed his name for political reasons" (quoted in Antonielli 2008, p. 20). In *The Works of William Blake*, Yeats states: "but if the old O'Neil's [sic] origin was hidden, the wild O'Neil blood showed itself strongly in the next generation" (quoted in Antonielli 2008, p. 20). Thus, Blake's alleged Irishness showed itself in his imaginative powers.

In his interpretation, Yeats put before Blake a mirror which reflected a Kabbalist and Occultist[18] rather than an ardent Christian believer. His misrepresentations of Blake leaning in the direction of Kabala and Occult are discussed by Adrianna Antonielli, who claims that by de-emphasizing the role of Swedenborg and Boehme in the creation of Blake's philosophical symbolism and by stressing Blake's links with "'the Eastern mystics and the medieval Kabbalists,' [Yeats] is perhaps too audacious" (20). Other critics were unanimous as to Yeats's misinterpretation of Blake. The negative remarks about Yeats's misappropriation of Blake include the comment made by the editor of Ellis and Yeats's work, Geoffrey Keynes, who classified it as "... a work of enthusiasm rather than of accurate scholarship." This view was expressed in 1938 and confirmed by Harvard Adams 30 years later. In 1947 Northrop Frye simply claimed "that Ellis and Yeats 'approached Blake... from the wrong side of Blavatsky" (quoted in Antonielli 2008, p. 11).

However, in intellectual terms Yeats's debt to Blake is immense. He absorbed Blake's theories of two opposing forces governing the universe—the forces of expansion and contraction—and made them into the basis of his historiosophy and his theory of the self, relying on the symbols of expanding and contracting gyres and on the concept of the growing and waning self and anti-self (e.g., Antonielli 2008, p. 14). Also, his idea of the co-existence of contradictions—or rather of gradual transmutation of a thing into its opposite—as Yeats admits in *A Vision*, came to him not from Hegel, but from Blake (quoted in Mann 2004). In one of his *Last Poems*, "An Acre of Grass," Yeats resolves "Myself must I remake/Till I am Timon or Lear/Or that William Blake/Who beat upon the wall/Till Truth obeyed his call" (1982, p. 347). However, if Yeats modeled his theories on Blake's, he also did the reverse: in his daring paraphrase of Blake's symbolism, he "remade" Blake into his own image.

Seeking in Blake a justification for his own esoteric system, Yeats became responsible for many semantic shifts in Blake's symbolism, such as that from

---

[18] Yeats systematized, explained and paraphrased Blake's work from a point of view which had been shaped by the Theosophical Society, which he joined in 1887 under the influence of the Russian medium Helena Petrovna Blavatsky.

Blake's negative symbol of the vortex to Yeast's positive symbol of a gyre.[19] As argued by Antonielli, Yeats adapted Blake's symbolism to his own numerology[20] and astrology by connecting Blake's twenty-seven Churches with the Zodiac signs (Antonielli 2008, pp. 19–20, p. 25; see also Mann 2004). Rather imaginatively, Yeats delivered a symbolical interpretation of the sounds in Blake's works, claiming, for example, that "the word Orc,... 'conveys its passionate and violent meaning'" (quoted in Antonielli 2008, p. 25). Further, Yeats extended Blake's "fourfold symmetries" (Mann 2004) so that they would constitute a foundation not only for the universe but also for the human self. The number four is seen as reflected in four Zoas and in four "atmospheres" (Antonielli 2008, pp. 17–19), as well as, to use the words of Yeats, in "the wheels of Ezekiel,... the four beasts of Apocalypse, and... Raphael, Michael, Gabriel and Uriel, the Kabalistic regents of the cardinal points... presid[ing] over psychic and bodily affairs" (quoted in Mann 2004). Significantly, the Supreme Being in Blake is also explained by the number four, not by three: it is understood by Yeats in terms of Boehme's tetrad rather than as the Holy Trinity. According to Yeats, Blake "[l]ike Boehmen and the occultists generally, postulates besides the Trinity, a fourth principle, a universal matrix or heaven or abode, from which, and in which all have life" (quoted in Antonielli 2008, p. 17). In Yeats's metaphorical language, this "fourth principle" is also "the looking glass" or "mirror" (Mann 2004), "the mirror of Sophia, or wisdom" which reflects the Father, the Son, and the Holy Spirit in their unity (Antonielli 17). As put by Yeats, "God looking into this mirror, ceases to be mere will... and enters on that eternal meditation about Himself which is called the Holy Spirit," which is seminal and manifests Itself in "the numberless thought-forms of the great mirror, the immortal or typical shape of all things, the 'ideas' of Plato" (quoted in Mann 2004). All in all, it seems that Yeats, looking in the mirror of Blake's poetry, saw the integrity of his own esoteric system. In Blake's symbolism, he saw the "matrix" and "abode" for his own imagery, and in Blake's genius he saw the reflection of his.

4 Conclusion: Ironies

Rossetti and Yeats deserved their greatest credit for recognizing Blake's genius in the century which dubbed him as mad. But their attempts to engage with that genius, through imitation and explanatory paraphrase, were not wholly successful.

---

[19] Mann (2009) exemplifies those shifts with the change of meaning from Blake's "vortex" to Yeats's "gyres": "Blake's vortexes are negative parodies of Cartesian mechanistic thinking rather than Yeats's gyres which are close 'to those contraries which give life' such as male and female, active and passive, God and the mirror...".

[20] On Yeats's interpretation of the symbolism of Blake's twenty-seven Churches, see Antonielli 2008, pp. 22–25. Mann notes that the number twenty-seven, which had negative associations in Blake, corresponds to Yeats's twenty-eight, which is the number of phases in the development of personality.

By the end of the 19th century, Blake's genius remained inimitable, though copied by Rossetti, and unexplained, although extensively interpreted by Yeats; it stayed beyond this frontier which is set by the mediocre—or, perhaps one should say, just the middling us—to e*strange* the poetic genius as mad. Yeats explained what he understood as marked by genius in his commentary to the works of Blake. For him "the genius or the central part of man," thus the genius in the sense of *daimon*, is identified with the "poetic genius, or 'universal mood'" (Antonielli 2008, p. 16). Yeats's idea of genius bears a clearly Emersonian mark in his belief that a spark of genius lights the fire of brotherhood. But at the same time, his idea is not free from more elitist leanings in that he links genius with the flight of poetic Imagination. In his commentary to the works of Blake, Yeats claims that "[e]ach man opens his own mind inwards into the field of Vision and there, in this infinite realm, meets his brother-man." In this "infinite realm," as Yeats believed, he met Blake through "brotherliness of Imagination" (Yeats quoted in Antonielli 2008, p. 16); but, in fact, he imaginatively paraphrased this eighteenth-century poet to adapt his ideas to his own esotericism. Indeed, in Blake he might have found a powerful resonance for his own genius, but, as his critics seem to suggest, he might have also lost the specificity of Blake's.

Rossetti's plight, after his attempt at steering Blake's work into the mainstream of literature, was even more precarious. Wary of putting himself at risk, Rossetti did not engage himself with Blake's theories, but he imitated Blake's style for its naturalness and simplicity. Effectively, then, Rossetti divorced Blake's expression from Blake's prophetic, moral and imaginative genius; he severed Blake's "fiery thought" from the "simple words"[21] through which it burnt. Imitating Blake, he deadened his own style by monotonous repetitions, and thus, inadvertently, he gave force to Emerson's warning that "imitation is suicide" (1841, p. 53). Paradoxically, by playing it safe—having stopped short of publishing those of Blake's works which might be regarded by his contemporaries as "strange" or insane—Rossetti managed to provoke Nordau's scornful criticism for having written "nebulous" and "obsessive" verse. Eventually, the intellectual and stylistic balancing on the border of Blake's genius and alleged madness proved uneasy for Rossetti and Yeats: Yeats slipped from interpretation to arrogation, and Rossetti missed the frontier between deadening imitation and inspiring influence.

References

Antonielli, A. 2008. William Butler Yeats's 'The Symbolic System of William Blake.' *Estudios Irlandeses* 3:10-28.

Blake, W. 1966. *Complete Writings*, ed. Geoffrey Keynes. Oxford: Oxford University Press.

Browning, R. 2009. *Prince Hohenstiehl-Schwangau.* N.p.: BiblioLife.

[21] Cf. Emerson's "The Poet": "To clothe fiery thought/In simple words succeeds,/For still the craft of genius is/To mask a king in weeds" (1904, p. 292).

Buchanan, R. 1871. The Fleshy School of Poetry: Mr. D. G. Rossetti. *The Contemporary Review* 18. The Victorian Web. Literature, History, and Culture in the Age of Victoria. http://www.victorianweb.org/authors/buchanan/fleshy.html/. Accesssed 15 March 2010.
Deleuze, G. 1979. The Schizophrenic and Language: Surface and Depth in Lewis Carroll and Antonin Artaud. In *Textual Strategies: Perspectives in Poststructuralist Criticism*, ed. and introd. J. V. Harari, 277-295. Ithaca, NY: Cornell Univeristy Press.
Emerson, W. R. 1904. Poet. In *Poems: The Complete Works of Ralph Waldo Emerson*. Vol. 9. Introd. and notes. E. W. Emerson, 292. Boston and New York: Houghton, Miffin; Cambridge, US: The Riverside Press.
Emerson, W. R. 1954. Self-Reliance. In *Essays: First Series*. 1841. *Basic Selections from Emerson: Essays, Poems and Apothegms*, ed. E. C. Lindeman, 53-73. New York: New American Library.
Hilton, N. 1988. Blake and the Apocalypse of the Canon. *Modern Language Studies* 18.1: 134-149.
http://www.estudiosirlandeses.org/Issue3/Issue3Pdfs/pdfAriannaAntonielli.pdf/. Accessed 15 Oct. 2009.
Mann, N. 2004. Yeats's *Vision* and William Blake. http://www.yeatsvision.com/Blake.html/. Accessed 27. Jan. 2009.
Nordau, M. 1993. *Degeneration*, introd. G. L. Mosse. Lincoln: University of Nebraska Press.
Pound, E. 1957. Hugh Selwyn Mauberley. In *Selected Poems of Ezra Pound*. New York: New Directions.
Ross, R. 1909. A Recent Criticism of Blake. *The Burlington Magazine for Connoisseurs* 16.80: 84-87.
Ross, R. 1906. The Place of William Blake in English Art. *The Burlington Magazine for Connoisseurs* 9. 39:150-167.
Symons, A. 1923. Decadent Movement in Literature. In *Dramatis Personae*, 97-117. Freeport, NY: Books for Libraries Press.
Rossetti, D. G. n.d. The Blessed Damozel. In *The Blessed Damozel and Other Poems*, 1-13. New York and London: Putnam and Sons.
Vries, J. C. E. (Bassalik-de). 1910. William Blake in his Relation to Dante Gabriel Rossetti. Doct. Diss. University of Zürich. http://openlibrary.org/b/OL14048369M/William_Blake_in_his_relation_to_Dante_Gabriel_Rossetti/. Accessed 20 Oct. 2008.
Yeats, W. B. 1982. *The Collected Poems of W. B. Yeats*. London: Macmillan.

Between Geoffrey Chaucer's *Canterbury Tales* and Peter Ackroyd's *Clerkenwell Tales*: A Dialogue of the Contemporary Novel and Medieval Literary Conventions

Joanna Bukowska

Abstract Ackroyd's novel might be viewed as a narrative territory, emerging out of the dialogue between the contemporary and medieval cultures, their literary texts and alternative concepts of history and fiction. The paper analyses the multiple ways in which the novelist rewrites *The Canterbury Tales*. If Chaucer's poem presents a microcosm of late medieval English society, Ackroyd's novel is a collection of vignettes, making up the image of London in 1399, when the integrity of the city is shown as threatened by political intrigues and heresy. Ackroyd also remodels Chaucer's pilgrims into such characters which suit the demands of the contemporary narrative and help him to develop his fictitious plot. Despite the apparent structural similarity of both texts, the concept of pilgrimage as a linear and movable frame of Chaucer's poem is replaced in *The Clerkenwell Tales* by that of the city in turmoil, whose image, focalised by individual characters, and emerging out of a mosaic of various perspectives, appears as both solid and elusive. The literary debt is demonstrated, therefore, to be less straightforward than Ackroyd suggests. He replays certain tones of Chaucer's poem but what he borrows is usually given a new form and endowed with new implications. This simultaneous assertion of the dependence on the medieval material and its transformation seems to enact at the level of the narrative discourse a warning against unquestioning acceptance of any overtly pressed meaning, which, as the implications of the story indicate, might be potentially dangerous at any time.

J. Bukowska (✉)
Adam Mickiewicz University, Kalisz, Poland
e-mail: bjoanna@amu.edu.pl

J. Fabiszak et al. (eds.), *Crossroads in Literature and Culture*, Second Language Learning and Teaching, DOI: 10.1007/978-3-642-21994-8_27,

1 Introduction

The creation of medievalist fiction has always involved the transcendence of historical and cultural distance. Whether the Middle Ages have been treated as a rich reservoir of high ideals and sentiments or whether their distinct cultural phenomena and literary motifs have appealed to subsequent generations, the patterns of medieval life and literary conventions have usually been rewritten in accordance with the medievalist writer's own mindset. In his own attempt to revive the spirit of fourteenth century London Peter Ackroyd combines creative energy with his concern for mimetic detail and the postmodernist preoccupation with literary games in which *The Clerkenwell Tales* engage with Chaucer's *Canterbury Tales*.

2 Ackroyd's Literary Debt

The literary debt which is alluded to in the title of Ackroyd's novel is manifested by the paratexts, which overtly present Chaucer's work as the primary intertext of *The Clerkenwell Tales*. The close similarity between the lists of contents of Ackroyd's novel and of *The Canterbury Tales* seems to suggest that this close dependence might also be revealed by the structure and the characters of the contemporary novel. When this assumption is verified in the reading process, it becomes apparent that Ackroyd rewrites *The Canterbury Tales* in a number of different ways. The contemporary writer not only changes the sequence of tales but also asserts his own voice by replacing the tales presented in the original by Chaucer-the pilgrim, that is "The Tale of Sir Thopas" and "The Tale of Malibee", with "The Author's Tale," consisting of end notes, which blur the distinction between history and fiction. The similarity of structure is, therefore, superficial. Although Ackroyd retains the division into tales, they are not thematically independent but constitute splinters of a single, though fragmented, plot. The absence of "The General Prologue" seems to be compensated for by the profuse use of direct description. The Chaucerian characters with whom Ackroyd populates his novel are transformed into the citizens of London and do not function as narrators but rather foculisers of the narrative process. Like Chaucer's pilgrims they are primarily defined by their professions and social rank, though they are also individualized by being granted a name and a surname, as befits the characters in the novel.

Although the characters of *The Clerkenwell Tales* only distantly echo the silhouettes of the original pilgrims, Ackroyd purposefully directs the reader's attention to *The Canterbury Tales*. To validate his appropriation of Chaucerian characters he quotes William Blake's opinion concerning the timeless relevance of Chaucer's analysis of human nature. Emulating the literary strategies of medieval poets, who invoked the views of the great *auctores*[1] of the past, Ackroyd place Chaucer in a position of authority, meaning his work needs to be continued and transmitted. In the Middle Ages, the words of others could be invoked "for purposes of confirmation" (Burrow 1990, p. 30) but there were also other factors, which conditioned "certain intertextuality or interdependence of texts (Burrow 1990, p. 34)". Even the texts which were not directly related to other texts exhibited an illusion of such dependence (Burrow 1990, p. 34). Ackroyd seems to follow this medieval practice when he claims to borrow from Chaucer more than he actually does. Ackroyd's novel reconstructs, thus, not only the medieval cultural phenomena and historical details but to a certain extent selected literary strategies also.

Ackroyd's suggestion that the universality of Chaucer's poetic view on human nature justifies his incorporation of Chaucerian characters can be perceived as a manifestation of the writer's attempt at such a representation of the medieval life, which would provide commentary on the contemporary world. Since Ackroyd situates the plot of his novel in the last year of the fourteenth century his indication of the universal implications of his narrative can be perceived as modeled on Barbara Tuchman's (1978) suggestion that the fourteenth century, due to its calamitous character, can be perceived as a distant mirror of the twentieth century. Tuchman (1978, pp. xiii, xiv) compares the instability of both centuries and points out that due to the effects of "plague, war, taxes, brigandage, bad government, insurrection, schism in the Church", fourteenth century society had already witnessed "a distraught age whose rules were breaking down under the pressure of adverse and violent events". The social and political unrest of the period is reflected by the plot of Ackroyd's novel. The mad prophetic visions of Sister Clarice from Clerkenwell nunnery enhance the atmosphere of doom and prepare the ground for the political intrigues of a secret society, Dominus, whose manipulation and abuse of religious fervor lead to the deposition of Richard II. Although the political questions are downplayed in *The Canterbury Tales*, Chaucer also reflects the unstable character of the end of the century in his portrayal of late fourteenth century society. As suggested by Cooper (1989, p. 5): "The ethical questions it concerns are both the timeless ones of fallible human beings living in an unstable world and those raised by the frightening breakdown of hierarchy and stability in late fourteenth century". It is this concoction of the universal and specific aspects of Chaucer's representation of the late fourteenth century that Ackroyd invokes by the integration of Canterbury pilgrims into his novel.

The list of dramatis personae with which Chaucer's characters are ushered into the narrative world of *The Clerkenwell Tales* makes it evident that Ackroyd is playing both with the impression of his dependence upon *The Canterbury Tales* and with the generic conventions of the novel. The incorporation into his novel of a paratext typically found in drama, might be considered as an allusion to the

---

[1] As argued by Burrow (1990, p. 32) in his analysis of Middle English literature: "Authority belongs to the *auctor*...To be an *auctor* is to *augment* the knowledge and wisdom of humanity".

dramatic element, which in *The Canterbury Tales*, in particular in the prologues and epilogues to the tales, results from the interplay between the pilgrims, from the clash between their divergent social outlooks and also from what Cooper (1989, p. 16) describes as the refusal of many of them "to stay humbly in their social place". In contrast, the dramatic quality in Ackroyd's novel can be discerned only in the way the element of the dramatic personae casts the characters of *The Clerkenwell Tales* in the role of Chaucer's pilgrims, although this extension is only partly realized at the level of narrative. Unlike Chaucer's characters, Ackroyd's Londoners engage in various pursuits all over the city as they become entangled in or affected by the surreptitious dealings of the predestined and Dominus. Ackroyd replays, therefore, certain tones of Chaucer's poem but gives to them a new form.

The juxtaposition of the dramatis personae and "The Author's Tale" reveals the multiple layers of Ackroyd's literary game. Although most of the characters in *The Clerkenwell Tales* are identified as the figures from Chaucer's poem, in his final tale, Ackroyd (2004, p. 212) claims to have derived their names from a letter, allegedly found in 1927 "within a bundle of ecclesiastical documents in the library of Louvain Cathedral". This letter, attributed to one of the characters, William Exmewe, is said to have contained the list of the members of Dominus and the predestined. The historical value of this data is however, difficult to verify in contrast to many other overtly historical events incorporated into the narrative world of the novel such as, for example, Henry Bolingbroke's campaign against Richard II. It might be argued, therefore, that the majority of characters of *The Clerkenwell Tales* belong to the world of literary fiction which imitates both history and another literary world.

3 The Transformation of Medieval Characters

The degree of correspondence between the characters of *The Clerkenwell Tales* and their Chaucerian models varies for each character. There are characters resembling Chaucer's pilgrims and characters (both fictitious, for example Sister Clarice or Hamo Fulberd, and historical, for example Richard II, Henry Bolingbroke) that do not have their counterparts in *The Canterbury Tales*. Nevertheless, even in the case of characters who retain the impression of similarity, the details of their representation are deliberately altered. The sense of resemblance stems from the fact that Ackroyd often hints at a familiar feature of a chosen Canterbury pilgrim, while the disparity results from the transformation of this trait and the exposition of its new implications. As in the case of the prioress, what appears to be similar, turns out to be different:

> Her monkey sensing her melancholy, clambered upon her shoulders and began with the gold ring suspended on a silken thread between her breasts. She sang to it a new French song…and then played handy-dandy with a hazelnut. She had entered the House of Mary when still a young girl, and had somehow retained the dazed demureness of her childhood (Ackroyd 2004, p. 2).

It is the absence in her portrayal of such ecclesiastical virtues as piety, detachment from worldliness or renunciation of self, that creates a correspondence between Ackroyd's prioress and her Chaucerian prototype. In the likeness of Madam Eglantyne, Dame Agnes de Mordeaunt manifests her predilection for jewelry and rich clothes. Nevertheless, where Chaucer exposes his Prioress's lack of restraint by ironically praising her "conscience and tender herte" (Chaucer 1996, p. 9, v. 150) which is to be seen in her treatment "of smalle houndes ...that she fedde with roasted flesh, or milk and wastel breed" (Chaucer 1996, p. 9, v. 146–147), Ackroyd exposes Agnes's childish behaviour and the excessive endearment with which she treats her monkey. The provincial French of Madam Eglantyne spoken "after the scole of Stratford-atte-Bowe" (Chaucer 1996, p. 8, v. 125) contrasts also with Agnes's familiarity with new French songs. These discrepancies concern both the descriptive details and the manner in which they are presented. Even when Ackroyd delineates the prioress's silhouette with the same mild irony which characterizes her presentation in Chaucer's "General Prologue", this ironic effect is generated, in *The Clerkenwell Tales*, in an entirely different way. If, in the Middle English poem, it results from the direct juxtaposition of the nun's affected manners and her ecclesiastical office, in the contemporary narrative, it is produced by the explanatory statement revealing her social background and the manner in which ecclesiastical offices were transferred within noble families: "Her aunt had been prioress before her, and she assumed familial responsibility for the acres as well as souls under her care" (Ackroyd 2004, p. 1). Although in the broadest contours the character seems to derive directly from Chaucer's poem, it is different in terms of details.

Ackroyd not only remodels the inherited material but also introduces new individualising features expanding, thus, the spectrum of character traits, and reducing the representative dimension which in Chaucer's poem coexists with the presence of an individualising detail. The prioress in *The Clerkenwell Tales* is inquisitive enough to suspect a manipulative policy behind Sister Clarice's mad babbling: "Agnes looked at her with distaste. She suspected that there was much subtler craft in the young nun's demeanor, but she could not prove it" (Ackroyd 2004, p. 12). Her predilection for green, the colour of the underworld spirits, clearly evokes medieval symbolism but together with her tender stomach and quick-temper it simultaneously forms an amalgam of differentiating qualities that align her character with contemporary rather than medieval notions of subjectivity. As explained by Vitz (1992, pp. 15–17), in the medieval literature characters were distinguished by their display of features pertaining to the universal ideal in a particular quantity. In modern narratives, in turn, characters are individualized by the incorporation of distinctive, unusual character traits, since modern concepts of selfhood define individuality not in terms of quantity but quality.[2]

[2] Vitz (1992, p. 15) illustrates this distinction by placing the medieval characters along a vertical, "hierarchical or evaluative axis", establishing "the range between superiority and inferiority" in contrast to contemporary characters built along "a horizontal axis" or "the axis of differentiation", representing "normal as opposed to unusual, unlike" (Vitz 1992, p. 16).

The extension of the range of features in the case of some characters borrowed by Ackroyd from Chaucer produces, however, an effect which entirely alters the implications of the original text. Ackroyd's clerk, Emnot Hallyng, though he is also like Chaucer's pilgrim "the clerk of Oxforde", "even leaner than his horse", with "his pale face and spare frame" (Ackroyd 2004, p. 41), falls short of the ideal erected by Chaucer. Emnot's theological studies are outweighed by his scholarly pursuits of numbers, patterns and alchemy, the knowledge of which enable him to produce the gunpowder, used by the predestined in their assaults on London churches. Still another method used by the contemporary novelist to rewrite Chaucerian characters involves the transformation of narrative strategies present in *The Canterbury Tales*. For example, although Ackroyd retains the polarization between the figures of the Friar depicted as the embodiment of ecclesiastical corruption and the Parson as an ideal clergyman, he replaces the conventionally satirical portrayal of Chaucer's lascivious and abusive Friar with the figure of William Exmewe, an ultimately debased Machiavellian hero who does not shy away from intrigues, blackmail and murder to achieve his religious and political aims. Despite preserving the appearance of a regular friar, Exmewe leads a heretical sect of the predestined or foreknown, who reject the beliefs of the established church, replacing them with their own conviction about being "prompted wholly by the spirit of God", and bound to "satisfy the promptings of their nature" without any remorse (Ackroyd 2004, p. 37). Since they "eagerly sought for the day of doom" (Ackroyd 2004, p. 38), Exmewe "had persuaded them …that five London churches or sacred places must be visited by fire or death. Only in this manner could the day of doom be delivered" (Ackroyd 2004, p. 39). Ackroyd rewrites therefore the figures of Chaucer's pilgrims so that their remodeled character constructs could help him to develop his fictitious plot and could suit the demands of the contemporary narrative.

4 The Image of the City in Turmoil as an Alternative to the Narrative Role of Chaucer's Pilgrimage

Unlike in the Middle English poem the common ground between the characters in *The Clerkenwell Tales* is not established by the shared experience of the pilgrimage to Canterbury but by their active participation in or just being a witness to the dramatic events leading to the deposition of Richard II and the succession of Henry Bolingbroke and by their status of being Londoners. The setting plays, therefore, in *The Clerkenwell Tales* an essential role. As observed by Higdon (2005, p. 218), Ackroyd's revival in his novels of London from different historical periods supplies "the thematic energy of his urban art". In *The Clerkenwell Tales* the novelist creates the image of London in 1399, when the peace and order of the city is shown to be threatened by political intrigues, heresy and Sister Clarice's prophecy of doom. At first, the incorporation of the vision of late medieval London into the narrative model provided by *The Canterbury Tales* might suggest a clear departure from Chaucerian tradition, since Chaucer, as Ellis (2005, p. 75) observes, does not describe London directly or at length.[3] The illusory nature of this impression becomes, nevertheless, evident in the light of evidence gathered by Ellis (2005, pp. 75, 77) who demonstrates that London "inhabits and influences" Chaucer's work by being translated into "the variety, energy and openness of Chaucer's literary forms and diversity of voices" (Higdon 2005, p. 77). The dissenting voices are also present in *The Canterbury Tales*, where they may be seen as overtly representing the microcosm of late medieval English society, and as indirectly encoding "the contesting voices of urban experience" (Ellis 2005, p. 79). Although Ackroyd's representation of London in a specific historical moment is of his own making, and the time of his novel's action actually postdates the composition of *The Canterbury Tales*, the extensive space given to the image of a conflict-ridden city in the contemporary novel might be also perceived as an effect of Ackroyd's particularisation of certain tendencies present within Chaucer's poem.

The Clerkenwell Tales present the topography of medieval London through the lens of individual characters who focalise their fragments of the London narrative. For the miller accustomed to the sound of water rushing under his mill London is "a city of springs and streams" (Ackroyd 2004, p. 121). The representation of London is the sum of various attitudes and its topography is structured by an individual gaze. Dame Agnes de Mordeaunt, the prioress, "sitting in the window of her chamber" (Ackroyd 2004, p. 2) is shown to contemplate the urban landscape:

> In the open fields beyond the walled garden, stretching down to the river, she could see the molt-house with its dovecote, the familiar cart-house, and the turf-house beside the stables. On the western bank of the Fleet river stood the mill-house and, on the other side, a cottage of whitewashed walls and thatched roof which belonged to the bailiff of the convent.(...)Rumors of the strange events within the convent had already reached as far as the cookshops of East Cheap and the fish-stalls of Friday Street; although Agnes had not been informed of these somewhat garbled reports, she was aware of the strange disquiet in the vicinity and felt uneasy (Ackroyd 2004, p. 2–3).

The fragment of London that she can see through her window becomes integrated in her mind with the matters of her convent's future, with which she is most anxiously concerned.

Individual tales personalize not only the view of the city landscape, its topography and landmarks, but also the images of London life, its noises and

[3] As Ellis (2005, p. 75) indicates, the pilgrimage to Canterbury begins in Southwark outside the city walls and the status of several pilgrims as residents of London is not entirely clear. The evidence gathered by (Ellis 2005, p. 75) shows that, apart from Chaucer and the Cook, directly identified as Londoners, the other pilgrims can only be treated as possible London citizens (Merchant, Sergeant of the law and Guildsmen), or possibly dwell outside the city walls (the Host, the Prioress, the Pardoner, the Manciple) or, as the remaining group, may not even be related to the experience of the city.

smells, which are filtered through the senses of particular characters. The voices of London can be heard, for example, in the description of the carpenter's flight from the spot where Hamo Fulberd, the illuminator, struck a tooth drawer in the carpenter's defence:

> Richard Marrow left the tooth drawer at the mercy of the people, and managed to make his way down St Martin into Old Change. There was much building work here, in the precincts of St Paul's and the street was filled with cries of Yous! and Yis! and Hoo! The builders' carts were pulled by horses or by mastiff-hounds, and the labourers played football or sang over their cups in their brief if frequent intervals of rest. It was the way of London (Ackroyd 2004, p. 22).

Both the voices and the urban odors become apparent to the friar, William Exmewe on his entry into Smithfield:

> Now that the rain had cleared, Exmewe opened the wicket gate of the priory and walked out into Smithfield. It was not market day but the open space was churned up by horses and carts and wagons of every description; pigs were rooting among the rubbish, and black kites wandered among discarded bones as if they were mourning for London itself. The name of God was all around them—"God save you", "God's speed", "God give you grace"—muttered casually and under the breath, or cried aloud in greeting, like some susurrus of benevolence from the divine world. The smell of slaughtered animals, coming from the shambles, mingled with human scents as they passed the Broken Seld, the Bell on the Hoop, the Seresinshed, and the Cardinal's Hat (Ackroyd 2004, p. 18).

Ackroyd's frequent references to these unsavory details of medieval urban life provide a sharp contrast with the eulogizing tone and rhetorical character of medieval descriptions of the city written, as (Ganim 1998, p. 81) indicates, "within the tradition of *encomium civitas*". What *The Clerkenwell Tales* possibly share with the famous twelfth century *Descriptio of London* by William Fitz Steven is the portrayal of "the city as a source of dynamism and change" (Ganim 1998, p. 79–80). Ackroyd's novel achieves this effect by defining the space of fourteenth century London in terms of the characters' itineraries.

The limited space of the walled city is in his novel almost mapped by the footsteps of Londoners. As with the route of the second nun, who accompanies Sister Clarice to a secret meeting of Dominus, Ackroyd lists each street the characters pass and each turning they take on the way to their destination:

> They were traveling south, across Smithfield, along Little Britain and down St Martin's; as a girl Bridget [the second nun] had walked through these streets with her nurse and companion, Beldame Patience, and their perpetual activity never failed to reassure her. She knew every shop and shack, every shop and tenement, but she was always surprised by the city's endless life. Then she had been obliged to enter the convent (Ackroyd 2004, p. 205).

The steps the second nun takes also constitute a reenactment of her past experience when she was also a part of the dynamic life of the city. What, in turn, captures the attention of the squire, Gybon Maghfeld, when he is summoned to the city hall by Henry Bolingbroke to discuss "the policy for this unsettled time" (Ackroyd 2004, p. 162), is the anxiety involved in Londoners' mobility:

> There were citizens moving about from street to street, or from lane to lane, with intense looks of fear and amazement. He observed their faces as he passed them, but he recognised none of them. He was then struck by the curious possibility. What if the figures were created out of panic fear, out of the anger and excitement of the city itself? They might emerge at times of fire or of death, a visible group of walkers in the night. They might appear on the same London streets through all of the city's history (Ackroyd 2004, p. 163).

Although the squire's mind endows what he sees with an almost phantasmagorical dimension in which London as a living organism structures and shapes the dynamics of its citizens' movement, it is also possible to trace in this description the same sense of the city as "defined by the movement of people through it" that Ganim (1998) singled out as a characteristic trait of such early portrayals of London, as those of William Fitz Steven or Langland. Ganim (1998, p. 85) discerns in these descriptions "the sense of the pulse of street life, of crowds and their random patterns, of a city enlivened by its small rather than its grandest designs". He attributes this sense of energetic circulation to "London's casual relation to its suburbs and its indifference to the strong relation between city centre and city walls so characteristic of continental city plans" (Ganim 1998, p. 85). By presenting his characters as hectically circulating all over London Ackroyd not only manages to recreate the tension of a specific historical moment but might be seen also as modelling his description on the medieval literary conventions.

If London, shown as a city in turmoil, functions as a frame for the individual Clerkenwell Tales, it also outgrows this role to emerge as an independent narrative entity. As suggested in the already quoted account of the squire, London might be also perceived as a living organism. The city is described as hard and merry, sick and sinful. During her cross-examination, Sister Calrice speaks about: "The old, foul and sick sins of London…"(Ackroyd 2004, p. 160) and is warned by John Duckling, the nun's priest, that she "will put London in a fury" (Ackroyd 2004, p. 65). While planning the five attacks on the churches, Exmewe refers to them as "the five wounds of London" (Ackroyd 2004, p. 39) and at this time of test and trial the city is shown as "bracing itself for a fever" (Ackroyd 2004, p. 163). Endowed with organic features, almost personified into yet another character in the novel, the city wakes to the sound of its bells, sighs, braces itself for a fever and is almost brought to the edge of doom by the wounds inflicted by the predestined.[4] In *The Clerkenwell Tales*, London represents both the spatial frame of the fragmentary narrative and the collective experience of the characters in this novel. It is a territory whose fate is determined by its inhabitants, but which also wields its power over those who inhabit it.

The identity of London is constructed in *The Clerkenwell Tales* in deliberately elusive terms. While using his dialectics to persuade Hamo, the young illuminator, to set the church of St Sepulchre on fire, Exmewe depicts London as "no more than a veil, a pageant cloth, which must be torn asunder to see the face of God

[4] It is the fleshy, anthropoid image of London, which Ackroyd also reflected upon, when he chose to write its *Biography* (Ackroyd 2003), rather than history. Ackroyd is also reported to have said that in all his novels London "becomes a character—a living being" (Higdon 2005, p. 218).

shining. But at times like this the city seemed real enough". London represents both an abstract ideological concept and a real city. The ambiguity of its description results from the proliferation of forms and features ascribed to medieval London, both "a city of janglers and a city of God"[5] (Ackroyd 2004, p. 23). Its actuality is rooted in recognisable landmarks and topographic details, its abstract images depend upon ideological projections made by various factions, its organic aspect results from the identification of the city with its citizens, with their individual features and with their collective experience.

It is within such a context that the individual paths of the characters cross to overcome fragmentariness of the narrative world and to establish a web of mutual interdependencies. As the image of London in turmoil replaces the linear frame of pilgrimage, underpinning the structure of Chaucer's *Canterbury Tales,*[6] it appears possible that the narrative conundrum resulting in *The Clerkenwell Tales* from the manipulative ways in which political factions use heretical groups to achieve their aims can be metaphorically represented by the motif of a maze. Both the physician, Thomas Gunter and the clerk, Emnot Hallyng use the metaphor of a maze (Ackroyd 2004, pp. 133, 153) to describe their state of complete confusion resulting from their discovery of the fragmentary evidence concerning the nocturnal meetings of "the guardians' of the city" and their support for the violent activity of the heretical groups. The movement through a maze might be experienced as an erratic and confusing activity. In this sense it reflects both the experience of Ackroyd's characters and his readers, both of whom struggle to discover the meaning of the mysterious connections between individual events, as if trying to find their way out of the maze. Both Sister Clarice's madness and her persecution turn out to have been staged. She manages to stir up the city to initiate the transfer of power and on the evening of Henry Bolingbroke's coronation she delivers a self-congratulatory speech:

> It has come to pass as we have wished. Exmewe has been expelled, and will not speak. He plotted with heretics and he has gone out with the winds. The predestined men have been scattered, and nothing further will be known concerning them. Yet they have left a comfortable inheritance. This new king is not holy…We will rule behind the king…Dominus rises (Ackroyd 2004, p. 206).

[5] As Ackroyd explains in *London: The biography* Ackroyd (2003, p. 40), the double nature of London as "a city of God as well as a city of men" can be seen as embodied in the character of Clerkenwell, both a location of the Clerk's Well, one of the ancient holy wells of healing, and a staging ground of both "miracle plays as well as more secular bouts of wrestling and jousting". In Ackroyd's opinion Ackroyd (2003, p. 40) this concoction of *sacrum* and *profanum* makes the area representative of whole London, and the novelist utilises this parallelism in the title of *The Clerkenwell Tales.*

[6] The pilgrimage motif is, nevertheless, repeatedly evoked in *The Clerkenwell Tales*: in reference to April as a month "when folk longed to go on pilgrimages" Ackroyd (2004, p. 35), which is almost an exact quotation from Chaucer's "Prologue"; in Garret Barton's (the franklin's) reflection on St Paul's Cathedral (p. 56); in Jolland's (the monk's) lesson on providence and destiny, in which the general motif of the pilgrimage of life is replaced with "a pilgrimage to Canterbury"(p. 99), in his quotation of from Chaucer's "Ballad of Good Counsel" (p. 102), and also in reference to punishment imposed on the sergeant of law (p. 197).

Sister Clarice's London is, therefore, the world of deceitful appearances. In such a world it is more difficult to make a right choice and many of her, or Exmewe's, followers are led in a false direction.[7]

If it is correct to assume that the motif of a maze may be seen as a metaphor for the web of interrelations among the characters in *The Clerkenwell Tales* as well as for the structure of this novel, then it is also interesting to consider why Ackroyd chooses this particular concept to replace the idea of Chaucer's pilgrimage. In medieval literature and culture both motifs functioned as metaphors of human existence. The mutual implications of these notions are indicated by Sikorska (2005, pp. 137, 154) who shows that while the concept of pilgrimage encapsulated the journey of man from birth to death, and "a quest for God", the labyrinthine quality of this journey resulted from the endless necessity of having to choose between good and evil. Yet these two representations of human existence, the pilgrimage of life and the labyrinth of life, could also retain their distinct characteristics and serve to underline the difference between Christianity and heresy:

> On the one hand, Christianity conceptualised the human journey to Christ as a straight path, where a just and honest man chooses rightly, on the other hand heresy typifies straying from the right path, and various temptations symbolise the wrong choices (Sikorska 2005, p. 146)

As Sikorska demonstrates, heretical doctrines, represented in the Middle English texts, beguile their victims with the "seeming rightfulness of the argument" (Sikorska 2005, p. 146) and "subscribe to the labyrinthine aesthetics" (Sikorska 2005, p. 145). Although Ackroyd predominantly uses the motifs of mazes and labyrinths as metaphors for confusion, he may also be familiar with their medieval implications, since it is possible to find in his novel an example of a dialogue in which a labyrinth stands metaphorically for heretical activity. Thomas Gunter, the physician, questioning the sergeant-at-law about his secret connections with heretics, responds to Vavasour's pretence of being "in a maze", that is confused, in the following way: "…I believe you to be in a labyrinth. But it is one of your own making" (Ackroyd 2004, p. 170). In *The Clerkenwell Tales* politics secretly intertwine, therefore, with heresy. Since the characters are subject to double manipulation, that is, they are ensnared with the heretical views of Exmewe, who, in turn, is eventually revealed, to be a pawn of politically oriented Dominus, the choice of the image of a maze as a representation of confusion, arising at the level of both the story and the narrative discourse, seems particularly apt.[8]

---

[7] Medieval London could not be literally envisaged as a maze. In *London: Biography*, Ackroyd (2003, p. 2) uses, however, this image to represent the topography and history of contemporary London: "London is labyrinth, half of stone and half of flesh."

[8] Ackroyd's preoccupation with heretical activity of the predestined makes it impossible for pilgrimage to form a valid background of their intrigues, since heretics, and in particular Lollards, with whose beliefs Ackroyd compares and contrasts the views of the predestined in "Author's Tale", were, as Finucane (1977, p. 200) points out, severe critics of both image-worship and pilgrimages.

The concept of a maze is sufficiently different from that of pilgrimage to express the narrative purposes of Ackroyd's novel, yet the two images remain closely related, due to the correspondences established between them within the medieval culture. It might be said, therefore, that Ackroyd simultaneously hints at Chaucer's work and rewrites his strategies by giving them a different shape and an entirely new meaning. If *The Clerkenwell Tales* are related to *The Canterbury Tales*, then it seems also important to acknowledge a double nature of this relation. Ackroyd's overt suggestions of his direct dependence on Chaucer's poem are in direct contrast with his purposeful alteration and modification of the borrowed material. This simultaneous preoccupation with opposing or contradictory purposes is, as Bran (2009, p. 16) argues, repeatedly identified in postmodern theory (Eco, Jencks, Hutcheon) with an ironic mode, used to expose "the constructed, discursive and mediated nature of reality." Shaped by a similar narrative strategy, the relation between Ackroyd's novel and Chaucer's poem, reinforces, the warning inherent in the plot of *The Clerkenwell Tales*, in which it is the acceptance of reality at face value which makes the novel's characters susceptible to dangerous ideologies and manipulation.

5 Conclusion

Ackroyd's reconstruction of fourteenth century London represents an example of contemporary fiction, which both incorporates and imitates medieval history and medieval literary models. The multi-levelled literary game already begins with the title of the novel and the following paratexts, purposefully exposing the dependence of *The Clerkenwell Tales* on *The Canterbury Tales*. The incorporation of Chaucerian characters into Ackroyd's novel is presented as imparting a universal quality to its representation of dangers inherent in secret manipulation and the utilisation of religious ideals for political purposes. The literary debt turns out, however, to be less straightforward than is suggested by the author. Ackroyd replays certain tones of Chaucer's poem but what he borrows is usually given a new form and is endowed with new implications. The characters resembling Chaucerian prototypes in the broadest contours differ with respect to the details of their representation, wile the characters retaining the distinctive features of their medieval counterparts have their overall silhouette deliberately altered by the extension of the spectrum of differentiating traits. In short, although Ackroyd creates the impression of directly borrowing from the medieval poet, he remodels Chaucer's pilgrims into such character constructs which suit the demands of the contemporary narrative and help him to develop the fictitious plot of his novel. The structural similarity of both texts is also only ostensible. Even though the division of the contemporary novel into tales resembles the Middle English poem, *The Clerkenwell Tales* replace Chaucer's set of independent stories embedded in the frame of pilgrimage with a collection of vignettes presenting a splintered, fragmentary image of medieval London affected by the events of 1399. Mapped by the itineraries of individual characters and rendered through a mosaic of perspectives the representation of London as a city in turmoil escapes singular definition. The contemporary perspective intersects, therefore, with the medieval vision. While establishing, in his portrayal of the late medieval world, a link with the medieval literary tradition by hinting at certain distinctive elements of Chaucer's poem and a range of selected literary conventions, Ackroyd carefully rewrites and alters the medieval material. This simultaneous assertion of similarity and difference creates an ironic distance, which introduces a sense of alertness towards any manipulation. The literary game in which *The Clerkenwell Tales* engage with the related sources, in particular with *The Canterbury Tales*, indicates the need to treat with suspicion what is professed at the level of both the narrative discourse and the presented story, and, thus, to remain immune to any seductive ideologies, which in the novel wrought havoc with the fourteenth century world, but which may pose a threat also to the contemporary world.

References

Ackroyd, Peter. 2003. *London: The biography*. New York: Anchor Books. A division of Random House, Inc.

Ackroyd, Peter. 2004. *The Clerkenwell Tales*. London: Vintage.

Bran, Nicol. 2009. *The Cambridge introduction to postmodern fiction*. Cambridge: Cambridge University Press.

Burrow, J.A.1990. *Medieval writers and their work. Middle English literature and its background* 1100—1500. Oxford–New York: Oxford University Press.

Chaucer, Geoffrey. 1996. *The Canterbury Tales*. London:Penguin Books.

Cooper, Helen. 1989. *Oxford guides to Chaucer: The Canterbury Tales*. Oxford–New York: Oxford University Press.

Ellis, Steve. 2005. *An Oxford guide: Chaucer*. Oxford: Oxford University Press.

Finnucane, Roland C. 1977. *Miracles and pilgrimages*. Popular beliefs in medieval England. New York: St. Martin's Press.

Ganim, John M. 1998. The experience of modernity in late medieval literature: Urbanism, experience and rhetoric in some early descriptions of London. In *The performance of Middle English culture: Essays on Chaucer and drama in honor of Martin Stevens*, ed. by James J. Paxon et al., 77-97. Cambridge: D.S. Brewer.

Higdon, David Leon. 2005. Colonising the past: The novels of Peter Ackroyd. In *The contemporary British novel* ed. by James Acheson–Sarah C. E. Ross. Edinburgh: Edinburgh Uniersity Press.

Sikorska, Liliana. 2005. The labyrinth of life: St. Augustine's quests and Mergery Kempe's pilgrimages. *In Naked wordes in English*, ed. by Marcin Krygier– Liliana Sikorska, 137-157. Berlin–New York–Oxford: Peter Lang.

Tuchman Barbara W. 1978. *A distant mirror. The calamitous Fourteenth century*. New York: Random House Publishing Group

Vitz, Evelyn Birge.1992. *Medieval narrative and modern narratology: Subjects and objects of desire*. New York: New York University Press.

Thomas MacGreevy and Samuel Beckett. Affinity and Controversy

Wacław Grzybowski

Abstract Acquainted with Beckett in (1928) Parisian École Normale, Thomas MacGreevy soon became his confidant and a literary mentor introducing him to James Joyce and Richard Aldington. Their artistic interests took form, among others, of the common declaration "Manifesto. Poetry is Vertical" (1932), signed by them and several other poets associated with Jolas's *transition*. However, the intellectual attraction between the two was, at the same time, disturbed, or, so to say, spiced, by the tension between Beckett's agnosticism and MacGreevy's Catholicism. Sean Kennedy's illuminating article ("Beckett Reviewing MacGreevy: A Reconsideration," in; The Irish University Review, September 2005, pp. 273–288) notices a number of ambiguities in Beckett's praise of MacGreevy in his "Humanistic Quietism" (1934), devoted solely to his friend's first and only volume of poems, as well as in "Recent Irish Poetry" (1934). In "Humanistic Quietism" Beckett describes MacGreevy's poetry as "the adult mode of prayer syntonic" to his emotions ("Humanistic Quietism" p. 11), which suggests, according to Kennedy, that MacGreevy's faith, expressed in his contemplative metaphors, is always subjective and blind. "He sees what he wants to see. ... As such, it is always likely to be consonant with one's own needs. MacGreevy's own preference is decisive." (Kennedy 278) The description of MacGreevy as "the Titchener of the modern lyric" in "Recent Irish Poetry" is to confirm such evaluation, since the nineteenth century psychology of Titchener represents the immanent biological concept of human psyche. However, the correspondence between Beckett and MacGreevy, quoted by Kennedy, opens the space for more complex interpretation of this controversy. Without excluding the tension between the two, my article offers another explanation of these critical allusions.

W. Grzybowski (✉)
University of Opole, Opole, Poland
e-mail: wgrzybowski@wp.pl

J. Fabiszak et al. (eds.), *Crossroads in Literature and Culture*, Second Language Learning and Teaching, DOI: 10.1007/978-3-642-21994-8_28,

By combing the analysis of their letters with essential fragments of their poems it shows the difference of their approach to human consciousness.

1 Introduction

The comparison of Samuel Beckett with Thomas MacGreevy, the former's closest friend from Parisian École Normale Supérieure, is significant not only for literary biographies of the two, but also because of the meaning of the concept of consciousness for literature. For what brings Beckett and MacGreevy together, in intellectual sense, is, among others, the theme of consciousness underlying their writings. The direct expression of this topic can be found in the Verticalist declaration "Manifesto. Poetry is Vertical" (published in 1932), signed by both of them, as well as in Beckett's review of MacGreevy's *Poems*, "Humanistic Quietism" (published in 1934). At the same time, the idea of consciousness is also a catalyst which provokes the increasing differentiation of their literary attitudes.

As Antoni Libera (2004, p. 27) claims, Joyce's influence made Beckett accept Giambattista Vico's idea of the human poetic auto-creativity as foundation of human existence. Beckett's discovery of Joyce and Vico was preceded, however, by his study of Descartes, Berkeley, Schopenhauer and Kant, representing the "Copernican turn" in philosophy that shifted the stress from the object of cognition to the very awareness of the cognitive subject (30). Thus Vico's notion of the poetic auto-creativity was translated by Beckett into the modern concept of consciousness. This was inscribed by him into the avant-garde project of art intended to "emerge into existence and pass away into death," as a result of being "imbued with the ephemeral and contradictory logic of reintegration of 'art into the praxis of life'" (Bouchard 1997, p. 138). MacGreevy's understanding of consciousness was more linked with the classical notion of three transcendentals and with Catholic spirituality, in particular with St. Francis de Sales, to whom he ascribed the initiation of the neoclassical period in French literature (MacGreevy 1943, p. 2). As he claimed in his famous essay on the art of Jack B. Yeats, the knowing of the transcendental reality of truth, good and beauty constituted the core of human spiritual awareness (MacGreevy 1945, p. 7). If Beckett's poetry and prose suggested an immanent character of consciousness creating its own world by transcending from passion to pain, from one indefinite state to another, MacGreevy's poems and essays spoke of a different immanence and transcendence, i.e., about the immanence of artistic consciousness seeking the reflection of its self-image, open, at the same time, to order of Being and thus participating in the transcendence of truth, good and beauty. If MacGreevy was praised by Beckett as a metaphysical poet, it was not only because of his theological allusions, rare though they were, but most of all because his poems conveyed a powerful expression of the intuition of the non-physical ground of Being.

2 Affinity

Thomas MacGreevy was born in 1893, in Tarbert, Co. Kerry. After several years of work for English administration in Dublin and London, he joined Royal Field Artillery to face the trauma of the Belgian death fields of World War I. Afterwards he was offered two semester course of political science at Trinity College. All in all he chose the vocation of a poet, an art critic, an essayist, and finally, in 1950, was appointed a director of National Gallery in Dublin.

James Knowlson (1996, p. 96–101) mentions him as Beckett's closest friend and confidant at École Normale. MacGreevy was probably the one who brought the news about the literary competition, founded by Richard Aldington, the author of famous *Death of a Hero*, that brought Beckett's first successful poem, the very "Whoroscope." (*Schreibman* 1994a, p. 120)[1]. However, that was not the only MacGreevy's contribution to the derailing of Beckett off the main track of academic career into rugged embraces of literature and art. MacGreevy's biographer, Susan Schreibman (1994b, p. 1) points to intensity of correspondence between the two in which Beckett shares with his older friend numerous "observations on art, music and literature in several languages." This implies a significant impact on "the normally reserved Beckett" of "this outgoing Kerryman, whose friends and acquaintances included Richard Aldington, Harry Clarke, Salvador Dali, James Joyce, Lennox Robinson, and the Yeats brothers: Jack and W.B." (2). An academic sceptic, young Beckett seemed to be at a loss with MacGreevy's unexpected erudition, exquisite taste, extrovert attitude and humanistic religiosity. However, his closeness to Joyce and Aldington seemed to attract Beckett. "Within weeks, MacGreevy's life in the Quarter became Beckett's. Through some fault of MacGreevy's, Beckett got diverted from his academic work. Rudmose Brown (Beckett's academic supervisor) was foiled a second time. Beckett never finished his thesis." (2)[2] And, as Schreibman writes, "the rest is history" (2).

[1] *Schreibman writes there*: "It seems most likely that Aldington, who probably saw MacGreevy on the closing day of the competition, told him that of all the entrants he found none worthy of the prize. MacGreevy it would appear mentioned this to Beckett upon returning to the École (École Normale) (where they were both living), and Beckett, prompted most likely by his need for cash, sat down and wrote. Many years later Beckett himself recalled the genesis of the poem in a letter to Nancy Cunard: *Whoroscope* was indeed entered for your competition and the prize of I think 1,000 francs [10 pounds]. I knew nothing about it until the afternoon of the last day of entry, wrote first half before dinner, had a guzzle of salad and chambertin at the Cochon de Lait, went back to the École and finished it about three in the morning".

[2] The first time was, as Schreibman explains there, when: "MacGreevy should have given up his post in October 1927 to make way for the new candidate from Trinity, but officials at the Ecole were so pleased with his work that they, in a most unusual step, allowed MacGreevy to retain his appointment for another year, and recommended to Trinity's Board of Trustees that they might send their new candidate to a provincial University instead. An enraged Professor Brown (who was never very fond of MacGreevy from MacGreevy's own days at Trinity College) would not hear of sending his star pupil to a provincial institution, and instead found him an appointment at Belfast's Campbell College. And although Beckett must have been grateful to Brown for securing

As a part of the early Joycean circle in Paris, both contributed to a collection of essays *Our Exagmination Round His Factification for Incamination of "Work in Progress"* defending the initial version of *Finnegan's Wake*. However, their texts showed essential differences between them. Beckett's "Dante... Bruno. Vico ... Joyce" (1929, p. 7) focused on Joyce's inspiration with Vico's *New Science* and the thesis about the immanent poetic beginning of human civilization suggesting the power of human consciousness to create the deities of ancient religions. While MacGreevy's "*The Catholic Element in Work In Progress*" (1929b, p. 121) suggested Aristotelian and Thomistic sources of Joyce's aesthetics, and emphasized his classicism, exceptional in English literature, therefore provoking resistance of critics both in England and in Anglicised Ireland. His review *pointed to the classical-scholastic* "splendour of order" *of Joyce's text, i.e. the inter-relation between wholeness, harmony, and suggestive clarity of its meanings.* Although MacGreevy was aware of Joyce's apostasy, he noticed the residue of classical and Catholic ethics in Joyce's depiction of his Daedalus as "the most bitterly moral figure in modern English literature" (1929a, p. 475).

Obviously, this was not a controversy. In fact, the beginning of the 1930s brought a growing literary closeness of the two friends expressed, among others, in their contribution to the Verticalist declaration "Manifesto. Poetry is Vertical" (Arp et al. 1971, p. 73–74). Together with a group of writers linked with Eugene Jolas's transition, where the text was published in 1932, they announced repudiation of "the hypnosis of positivism" and "the delineation of a vitalistic world," along with the classicist "sterilization of imagination" (73). On the one hand, the Verticalist idea of "the hegemony of the inner life over outer", and of "the creative personality" rooted in "the transcendental 'I'", suggested the affinity of Beckett's and MacGreevy's focus on the artistic consciousness. On the other, MacGreevy's attraction to some forms of classicism, e.g., of Pierre Corneille's dramas and St. Francis de Sales's *Treatise on the Love of God*, together with his pursuit of the three classical transcendentals and the modernist epiphany, suggested his distance to the Verticalist rejection of classicism. In other essays, he could criticize rigidity of other forms of neoclassicism (MacGreevy, *T. S. Eliot: A Study*, 1931, p. 22), but in general, contrary to what "Manifesto" stated, MacGreevy's poems showed that the knowing of truth, good and beauty, in "the creative act," constituted the essence of human consciousness. The avant-garde thesis about "the final disintegration of the 'I' in the creative act" seemed to have only figurative meaning for MacGreevy, whose poetic texts and essays—such as *Jack B. Yeats: An Appreciation and an Interpretation* (MacGreevy 1945) and "*Art Criticism and Science*" (MacGreevy 1960, p. 161)—pointed to inner liberty of the

(Footnote 2 continued)
him full-time employment, he was not comfortable teaching and was profoundly unhappy there. When term ended he could not get out of Ireland fast enough. During the following spring, the matter was brought up again by TCD, and this time Beckett's appointment was approved. Beckett most probably arrived in Paris in August 1928 with the intention of simply moving into the rooms reserved for the English Reader and settling into a quiet life of scholarship".

poetic self, selflessly generous, but integrated and consolidated in the acts of artistic creation and poetic epiphany.

3 Anti-epiphany

Beckett's identification with the avant-garde art was much deeper. Starting with his poems, his style of writing frustrated all expectations for epiphany. A brief paradoxical meditation "The Vulture" illustrates this peculiar state of artistic and existential consciousness:

dragging his hunger through the sky of my skull shell of sky and earth
stooping to the prone who must soon take up their life and walk
mocked by a tissue that may not serve till hunger earth and sky be offal (Beckett 1984, p. 9)

Here the guiding leitmotif is the pain of hunger, physical, cognitive, existential and aesthetic, ascribed to an abstract bird of prey and at the same time obscurely emerging within the speaker's awareness as the omnipresent tension, as the "recollection of being as contradictory and transitory *Ereignis*" (Bouchard 1997, p. 139), where *Ereignis* means "an event" of becoming, revealing and being appropriated (Polt 2006, p. 73). In a way it anticipates his later existentialist enunciations, such as the one from "First Love" (Beckett 1974, p. 9–37) of the "omnidolent, impious dream" to understand pain, "to be nothing but pain" (20), with pain as the closest equivalent of the contradictions contained within *Ereignis*.

4 Epiphany

Despite Beckett's attraction to anti-epiphanic art, his succinct review (Beckett 1971, p. 11–13) of MacGreevy's *Poems*, "Humanistic Quietism" (from the 1934 *Dublin Magazine*), expressed appreciation, at least partial, of his friend's poetic epiphanies. Beckett emphasised there a particular "radiance," *claritas* of his friend's texts, "without counterpart in the work of contemporary poets writing in English" (11). As he wrote, MacGreevy's poetry derived "from the nucleus of endopsychic clarity, uttering itself in the prayer that is a spasm of awareness," (12) the prayer being "no more (no less) than an act of recognition"(11). A few perfect examples of MacGreevy's poetic-contemplative epiphanies were to highlight the poet's "self-absorption into light," such as "Gloria de Carlos V," a meditation upon Titian's vision of the Holy Trinity in the famous painting *La Gloria*:

But a moment now, I suppose,
For a moment I may suppose,
Gleaming blue
Silver,

Gold,
Rose,
And the light of the world. (MacGreevy 1991), p. 36[3];

and a fragment of the longest and most experimental poem "Cron Trath na nDéithe," its title being Gaelic paraphrase of "twilight of gods":

The brightness of brightness
Towering in the sky
Over Dublin…, (1991, p. 19)

and closing lines of "Nocturne of the Self-Evident Presence," a fruit of the poet's journey to Switzerland, which, according to Susan Schreibman, was the most powerful and intriguing example of the epiphanic "radiance" (Schreibman 1991, p. xviii). As such it deserves full presentation:

Fortunate,
Being inarticulate,
the alps
Rise
In ice
To heights
Of large stars
And little;
To courts
Beneath other courts
With walls of white starlight.
They have stars for pavements,
The valley is an area,
And I a servant,
A servant of servants,
Of metaphysical bereavements
Staring up
Out of gloom.
I see no immaculate feet on those pavements,
No winged forms,
Foreshortened,
As by Reubens or Domenichino,
Plashing the silvery air,
Hear no cars,
Elijah's or Apollo's
Dashing about
Up there.
I see alps, ice, stars and white starlight
In a dry, high silence (MacGreevy 1991, p. 42–43)

What attracted Beckett's attention must have been not only the poem's paradoxical conclusion, but also its perfect rhetorical structure of condensed,

---

[3] All the quotations of MacGreevy's published poems in this book are taken from *Collected Poems of Thomas MacGreevy: An Annotated Edition* (New York: Anna Livia Press & The Catholic University of America Press, 1991), ed. Susan Schreibman.

metaphoric meanings, rhymes, assonance, and poetic retardation. However, the poem's metaphysically ascetic and realistic return to the things in themselves offered powerful counterpoints to Beckett's vision of the flow of consciousness, hunger and pain in "Vulture." Where Beckett's metaphor reduced, in an intriguing manner, "the sky", to the speaker's "skull shell of sky and earth", and identified reality with immanent appropriation of it in human consciousness, the voice in MacGreevy's poem unfolded the perspective of cognitive participation in beauty of the cosmos. Existential humility of Beckett's speaker made him stoop over others trapped in the same ironic contradiction of desire of life, hunger and physicality, while MacGreevy's speaker derived his humility not only from his own gloom, which was not far from Beckett's pessimism, but also from the sense of mystery underlying the cosmic harmony reflected in the nocturnal landscape of mountains. Finally Beckett's metaphor of "vulture's hunger" pointed to the essential current of existence: *Sein zum Tod*, life towards death; while MacGreevy's vision revealed a different core of reality: "the dry, high silence" of the royal presence of the Creator. Physically unprovable, but logically and intuitively knowable, the source and ground of Being was announced, by the poem's title and conclusion, as real and perceptible, despite the veil of silence, in the tension between the generous beauty of the universe and economy of God's self-revelation. What could be important for Beckett was the fact that the renunciation of the comforting visualizations of religious art gave the concluding lines a particular power of expression and delineated analogy to his own existential realism.

5 Conclusion

Characteristically Beckett's "Humanistic Quietism" emphasized the value of humility in MacGreey's poetry. He stated: "To the mind that has raised itself to the grace of humility 'founded'—to quote from Mr. MacGreevy's *T. S. Eliot*—'not in misanthropy but in hope,' prayer is no more no less than an act of recognition" (1971, p. 11). In his essay on Eliot, to which Beckett alluded, MacGreevy described it as "that penitential Catholic virtue, founded not on misanthropy but on hope, that is so utterly alien to the puritanical mind" (1931, p. 16). Humility and hope are put together by the Irish poet as essential features of the consciousness representing a synthesis of contemplative spirituality and cultured humanism, as opposed to the puritanical mind. As Sean Kennedy claims: "Given Beckett's impatience of religious orthodoxy, it is doubtful that his recovery of this distinction in his review is wholly sympathetic" (2005, p. 277). In fact, side by side with words of praise, Beckett's essay contains also ambiguous comments, as if signaling a slight irony and different approach to MacGreevy's.

In a broader review of the poetic Irish modernism in 1934, "Recent Irish Poetry," Beckett speaks of MacGreevy's position as "independent" and "intermediate" (1983, p. 74). He describes him as "an existentialist in verse, the Titchener of the modern lyric" focused on "the act and not the object of perception" (74). For Kennedy, this suggests that Beckett excludes the author of *Poems* from the more essential movement of "new thing" in Irish poetry and confines the "nucleus of a living poetic" to their other friends, Denis Devlin and Brian Coffey (2005, p. 278). What is more, the allusion to Titchener, a structuralist psychologist, is assumed to be also of negative character, alluding to the suspected solipsism of MacGreevy's faith. For Titchener viewed consciousness as connected "not with the environment, but only with processes occurring within the organism, especially in the nervous system" (Kennedy 2005, p. 274).

Curiously enough, a year after publication of "Recent Irish Poetry," answering to MacGreevy's advice to read Imitation of Christ and the Bible, Beckett answers by describing his own existentialist attitude in terms allusive of Titchener's organic psychology of consciousness. On the one hand, he admits to having Jesus always "for his darling." On the other, he cannot help perceiving A Kempis's Imitation of Christ as a mere confirmation of his own "abject self-referring quietism" of "the sparrow alone upon the housetops" that makes him replace "the plenitude that he (MacGreevy) calls 'God' with a pleroma to be sought only among my own feathers and entrails."[4] The notion of the "self-referring quietism" and of the immanent organic "pleroma" clearly communicate the preference for existential naturalism. If Beckett's characterization of MacGreevy as "Titchener of the modern lyric" is ambiguous, this ambiguity is self-defeating, for the premises of Beckett's existentialism seem to be not far from the naturalistic concepts of the English psychologist.

His critical allusions in "Recent Irish Poetry" point, in fact, to what differs MacGreevy from the core of Beckett's attitude, i.e., to MacGreevy's essential opposition to the description of human consciousness as the floating immanent dynamism. The focus on reality does not mean in his case the reduction of the perceived existence to biological and social determinism. His consciousness tends for contemplative fulfilment, and when it cannot be achieved irony or dismay points to the spiritual void.

References

Arp, Hans, Samuel Beckett et al.. 1971. Manifesto. Poetry is Vertical. In *Collected Poems of Thomas MacGreevy*. ed. Thomas Dillon Redshaw. 73-74. Dublin: New Writers' Press.

Beckett, Samuel. 1984, *Collected Poems*. 1930-1978. London: John Calder.

Beckett, S. 1929. Dante... Bruno. Vico ... Joyce. In *Our Exagmination Round His Factification for Incamination of "Work in Progress."* ed. Samuel Beckett. 3-20. Paris: Shakespeare & Co.

Beckett, S. 1974. First Love. In *First Love and Other Shorts*. 9-37. New York: Grove Press.

Beckett, S. 1971. Humanistic Quietism. In *Collected Poems of Thomas MacGreevy*. ed. Thomas Dillon Redshaw. 11-13. Dublin: New Writiers' Press.

Beckett, S. 1983. Recent Irish Poetry. In *Disjecta: Miscelaneous Writings and a Dramatic Fragment*. ed. Rudy Cohn. 70-76. London: John Calder.

[4] Quoted in Kennedy's *"Beckett Reviewing MacGreevy: A Reconsideration,"* p. 283.

Bouchard, Norma. 1997. Rereading Beckett's *Dream of Fair to Middling Women. Beckett Today 6*, no.11: 137-147.

Kennedy, Seán. 2005. Beckett Reviewing MacGreevy: A Reconsideration. The Irish University Review, September *2005:* 273-288.

Knowlson, James.1996. *Damned to Fame: The Life of Samuel Beckett.* Grove Press: New York.

Libera, Antoni. 2004. Przedmowa. In *Molloy i cztery nowele.* Samuel Beckett. Transl. Antoni Libera. 5-31. Kraków: Wydawnictwo Znak.

MacGreevy, Thomas. 1929a. Anna Livia Plurabel. *Irish Statesman,* 16 February: 475.

MacGreevy, T. 1945. *Jack B. Yeats An Appreciation and an Interpretation.* Dublin: Victor Waddington Publications Limited.

MacGreevy, T. 1960. *Art Criticism and Science. 161-166. The Capuchin Annual. 1960.*

MacGreevy, T. *1991. Collected Poems of Thomas MacGreevy: An Annotated Edition.* ed. *Susan Schreibman. New York: Anna Livia Press & The Catholic University of America Press.*

MacGreevy, T. 1943. Saint Francis de Sales. Father Mathew Record June 1943:2.

MacGreevy, T.. 1929b. The Catholic Element in Work in Progress. In Our Exagmination Round His Factification for Incamination of "Work in Progress." ed. Samuel Beckett. 121-127. Paris: Shakespeare and Company.

MacGreevy, T. 1931. *Thomas Stearns Eliot: A Study*. London: Chatto and Windus.

Polt, Richard. 2006. The Emergency of Being. On Heidegger's Contribution to Philosophy. Ithaca: Cornell University Press.

Schreibman, Susan. 1991. Introduction. In Collected Poems of Thomas MacGreevy: An Annotated Edition. edited and annotated by Susan Schreibman. ix-xxvi. New York: Anna *Livia Press & The Catholic University of America Press.*

Schreibman, S. 1994a. *Richard Aldington An Englishman Thomas MacGreevy An Irishman. In Richard Aldington: Essays in Honour of the Centenary of his Birth.* ed. A. Blayac & C. Zilboorg. 113-127. Montpellier: Université Paul Valéry.

Schreibman, S. 1994b. The Penman and His Bleaters. In *Principle, Beckett is Joyce.* ed. F. Rathjen. 1-19. Edinburgh: Split Pea Press.

False Memories, Forged Identities and Murders: *Macbeth* for the Twenty-First Century

Edyta Lorek-Jezińska

Abstract Deborah Levy's Macbeth—*False Memories* (2000) addresses the transitional millennial crisis of identity and integrity, fuelled by postmodern uncertainties and deconstructions. Its major concern is the concept of authenticity and forgery, the original and the copy, partly conveyed through the play's problematic relation to Shakespeare's *Macbeth.* In her version, Levy seems to be interested in exploring the cultural fragmented memories of Shakespeare's text, inaccurate, imprecise and detached from the original context. The author herself comments in the introduction to her play that her *Macbeth* "explores how a story told five hundred years ago travels into the concrete, pollution and speed of the twenty-first century". One of the essential transformations that occur in Levy's version is the redefinition of the motivation for murder. The original Macbeth's desire for power is replaced with Levy's characters' search for authenticity of experience and feeling which they hope to gain in the act of murder. The possibility of achieving authenticity through murder is destroyed by the play's intrinsic meta-and inter-textuality which in a self-referential metaphor equates murder with acting and compares a murderer to an actor, who is always a forger of somebody else. Levy's characters' imprisonment in fakeness and lack of authenticity is largely defined by their bondage to Shakespeare's text, which preconditions all their actions.

E. Lorek-Jezińska (✉)
Nicolaus Copernicus University, Toruń, Poland
e-mail: lorjez@yahoo.com

J. Fabiszak et al. (eds.), *Crossroads in Literature and Culture,* Second Language Learning and Teaching, DOI: 10.1007/978-3-642-21994-8_29,

1 Introduction

In many ways Deborah Levy's *Macbeth—False Memories* moves back into a well known territory of Shakespeare's tragedy. It contains a number of elements that can be identified as borrowed from the original play, including several repetitive quotations from the original text. At the same time, however, the play subverts the viewers/readers' expectation of referring to the original text in a recognizable manner, of respecting it and adapting or even revising it. As the director of the play writes, "With *The Tempest*, we took an X-ray of the play and created a New version from the elements. With *Macbeth* we wanted to go a stage further. In our workshops we raised the edifice of the Shakespeare play and all that means to us to the ground to build anew from foundations" (Philippou 2010)[1]. This radical deconstructive urge met with unfavourable reviews, such as Nick Curtis's criticism of the play as detached from the original, simplistic and clumsy: "Having 'deconstructed' the original to the point of extinction, Levy and director Nick Philippou have nothing to offer but technical tricks and archness masquerading as postmodern irony. This is an exercise in vacuousness" (Curtis 2010)[2].

In her re-writing of *Macbeth* Levy indeed seems more interested in moving into a different cultural dimension and exploring the cultural fragmented memories of Shakespeare's text, inaccurate, imprecise and detached from the original context. In a number of fragmented and non-chronological scenes combining live performance with film, the play presents two main characters—Bennet and his wife, who try to achieve authentic experience through murder and usurpation. The murderee—Lavelli, who is Bennet's Italian business partner, after death keeps haunting Bennet's wife, trying to train her to become an actress and a murderess at the same time. Deborah Levy comments that her play "explores how a story told five hundred years ago travels into the concrete, pollution and speed of the twenty-first century" (2000, p. 144). The concern with the idea of the millennial turn or passage situates the play in the context of millennial and postmodern theory conceptualizing the decadent, nihilistic and apocalyptic tendencies of the "*fin-de-millenium* mood of contemporary culture" (Kroker and Cook 1988, p. i). One of those postmodern tendencies is the redefinition of the concept of individual identity and authenticity. Of equal importance to the transposition of the original into the contemporary context is the use of film and mobile phone on the stage. The effect of speed is generated through film editing, fast changes of scenes, fragmentation and anachrony while the extensive use of a mobile phone problematizes the idea of spatial relativity and authenticity, replacing and preventing real actions that could be mimetically represented on the stage.

---

[1] From the director's Web page, http://www.nickphilippou.com/archive/2000/macbeth/macbeth_1.html.

[2] Curtis, "False alarm rings for Macbeth", http://195.234.240.57/theatre/review-583687-false-alarm-rings-for-macbeth.do.

The quintessential transformation in addressing the question of postmodern identity occurs in the redefinition of the motivation for murder, which is to feel something, for in the twenty-first century *Macbeth* the characters do not crave anything. What is missing in their lives is the feeling of being alive, of authenticity and individuality. Consequently, one of the central concepts that emerges in the play—both on the primary and the metatheatrical level—is the question of authenticity and forgery. The "murderee", whose death is supposed to give a sense of being alive and authenticity to the murderer, paradoxically believes that he himself is a forgery and as such lacks authenticity. The very possibility of achieving authenticity through murder is destroyed by equating murder with acting and comparing a murderer to an actor, who is always a forger of somebody else.

Deborah Levy in the introduction to her play presents her version as referring to the original *Macbeth* in three themes—children revenging their parents' death, a ghost of a murdered man appearing in a play and the motif of camouflage. These themes, present in the original, gain in importance in Levy's version, turning into the major structural and conceptual frameworks. Through the motif of camouflage Levy seems to explore the ways in which the original *Macbeth* itself exposes the idea of mutable identity and usurpation/forgery. The concept of camouflage, deriving directly from the original *Macbeth*, embraces human relationships with doubles, imitation, illusion, visual confusion and the surface, the ideas discussed, among others, by H. Schwartz in his *Culture of the Copy* (1998), a study in the millennial significance of replication, camouflage and forgery. These notions will be instrumental in the examination of the self-reflexive and intertextual exploration in multimedia theatre. The double, the twin, the Other, through which the act of usurpation and imitation is effected, reflects the conceptual doubling of identities within the millennial "doubled time of ends and beginnings" (Schwartz 1998, p. 47). The conceptual significance of the vanished twin, the absent part of the self, seems to represent the "paradoxical sense of exhaustion and incompletion at this fin de siecle" (Schwartz 1998, p. 47). The missing or fragmentary identity is represented as seeking completion in the Other through acts of usurpation and illusion as well as acting and imitation, each involving degrees of camouflage, forgery and repetition. Levy's characters' fragmentation is further conditioned by their relationship to the characters in Shakespeare's *Macbeth*. Levy suggests that they resemble the imaginary son and daughter of Shakespeare's *Macbeth*: "It is as if they have inherited the nervous tics and twitches of their parents" (Levy 2000, p. 144).

2 Usurpation and Illusion

The act of murder committed by Bennet is described in the stage directions as followed by three actions of usurpation and appropriation—taking over Lavelli's mobile phone, inheriting the eczema mark and eating his food:

> "Bennet stabs Lavelli in the stomach. [...] Blood pours out of Lavelli's nose and drips on to his salami roll. [...] Lavelli's mobile phone falls out of his pocket. It begins to ring. Bennet looks at it. Eventually, he answers it.[...] He puts his hand to his forehead. Bennet seems to have caught Lavelli's eczema in the shape of a number three. [...] He bites into Lavelli's bloody salami roll. Chews with gusto" (Levy 2000, p. 158, italics in the original).

In Levy's *Macbeth* desire for power that fuelled most of the characters' actions in Shakespeare's text has been transformed into postmodern consumerism that fails to bring satisfaction. The original Macbeth's greed for more power and authority is transposed into more literal corporeal voracity in Levy's play—committing murder by Bennet is accompanied by his repetitive acts of eating. Through murder and symbolic cannibalism Bennet hopes to be able to feel again, which, however, he fails to. The act of eating the murderee's sandwich sprinkled with his blood (bread/body and wine/blood) operates as a postmodern pastiche of the holy communion, which necessarily fails to bring any effects. The real experience that Bennet tries to grasp through consumption incessantly escapes him. Consumption gives Bennet a temporary illusion of experiencing something and thus being alive, which ceases instantly.

The eczema mark, the second usurped part of the murderee's identity, originally functions as a symbol of life energy and ability to feel. During a business meeting Bennet's deficiency of feeling is confronted with Lavelli's excess of emotions: "This is a good business partnership then? You need to feel more. I need to feel less" (Levy 2000, p. 153). Bennet's face is partly paralised and numb while Lavelli's eczema is according to his doctor a sign of him feeling too much. Through murder and symbolic cannibalism Bennet inherits the eczema mark but he still cannot feel anything—he is just a bad forgery of Lavelli's emotional energy.

The third element of usurpation, using Lavelli's mobile phone, is developed in the play particularly in the relationship between Bennet and Lavelli's daughter. By taking over Lavelli's mobile phone he becomes a foster/fake father to Lavelli's daughter. She keeps phoning him on Lavelli's phone to share most intimate experiences in her life as if only through the mediation of telephone communication could they become authentic. Paradoxically, the relationship that Bennet enters into, that is the relationship between Lavelli and his daughter, instead of authenticity evokes forgery that ultimately turns out to be a double deception. In one of the first scenes Lavelli, when talking to Bennet, explains his fascination with forgeries as deriving from the sense of having a fatal flaw—of being fake in relation to his daughter: "She has made of her father a forgery—a man who is more like the father she wants than the real thing. I'm fake" (Levy 2000, p. 148). Towards the end of the play it turns out that this fake father role over the mobile phone, the role undertaken by Bennet after Lavelli's death, was originally devised for Lavelli's second daughter, a five-year-old girl whose tripled video image in a Victorian costume functions as the three weird sisters from the original *Macbeth*. The misplaced role of the fake father makes Lavelli's alleged emotional authenticity and intensity only a game of appearances. The simplicity with which the replacement of the real father for the fake one is conducted is a comment on the redefinition of identity mediated through postmodern technology.

Mobile phone conversations expand the diegetic space of the play by creating the illusion of mobility and travel. At the same time, through constant availability and oppressive frequency, they limit conceptually the private space granted to Bennet. In this respect the use of a mobile phone in the play refers to the paradox of the world as a telephone box suggesting its simultaneous spatial openness and constriction (Levinson 2004, p. 86–87). Lavelli's daughter's accounts of her actions, journeys and experiences gradually change from probable and hypothetically authentic into very unlikely and ostentatiously fictional. The diegetic space created through conversations fails to overlap with the visual representation of the characters' movements; the closer the daughter is to her intended victim—Bennet—the more intimate her reports are (culminating in references to sexual experiences and defecation) and more exotic the spaces she apparently travels to. The last conversation takes place when she is standing outside his house, telling him about the existence of Lavelli's younger daughter. Despite their implied fictionality, Lavelli's daughter's reports of what she is doing are a source of experience for Bennet, the substitute of the life that he does not lead himself. The inherent authenticity of the medium naturalizes and verifies what remains speculative in the relation between Bennet and Lavelli's daughter. The communicative directness of telephone conversations, which according to Levinson does not involve Coleridge's suspension of disbelief (2004, p. 85), authenticates the daughter's accounts and reactions. The naïve trustfulness on the part of Bennet motivated by his craving for real experience makes him vulnerable to deception. The voyeuristic pleasure that he seems to derive from Lavelli's daughter's real time reports of what she feels and experiences turns out to be fake. It was used by the daughter to keep him busy while she was trying to locate him to avenge her father's death[3].

3 Acting and Imitation

The process through which Lavelli's ghost transforms Bennet's wife into an actress and murderess establishes a double metatheatrical perspective. On the primary level, the ghost's possession of Bennet's wife's body functions as an ironic reversal of the act of usurpation in the original play and as an equivalent to the famous sleep walking scene. Structurally, the ghost connects the metatheatrical levels in both plays—the life of the murderer as a badly acted performance and the

[3] To a certain extent Bennet is manipulated by Lavelli's daughter e.g. when she instructs him what flowers to choose for his wife—he repeats her words in conversations with his wife as if he was already haunted by the spirit of his future victim. In fact, Lavelli's daughter rematerializes for a moment just before the murder occurs—her ontological position as a mobile phone interlocutor does not differ much from her later status as the voice of the ghost.

murderer's identity as a series of usurpations and deceptions in the original and the complicated implications of the equation between acting and murdering in Levy's play. The metaphorical link between the sleep walking scene, the idea of camouflage and acting is established early in the play when Lavelli's ghost puts a white sheet over Bennet's wife dressed in a white night dress—against the white background she becomes completely camouflaged and invisible. From this moment Bennet's wife is subjected to the process of transformation into a murderess/actress, by which her identity is simultaneously usurped and appropriated by Lavelli's ghost. The relationship between Bennet's wife and Lavelli's ghost resembles the possession by a spirit of the dead which seems to define the essence of theatre and acting.

The idea of acting as possession by the spirit of the dead was developed by Tadeusz Kantor in his concept of the theatre of death. Acting was compared to so called lower usurpation, which can be illustrated by a Jewish legend of Dybbuk, a spirit that enters the body of another and speaks through this body. Kantor admits that the act of calling up the dead spirits is a suspicious and impure procedure—it is the act of hiring. An actor becomes an impersonator, a usurper of a personality. That is why he or she does not have to act. Every usurper or impersonator can never imitate a particular person; he or she always fails to do it; because he cannot transform himself into somebody else, he reveals his own personality (Pleśniarowicz 1997, p. 221). Life in art, as Kantor argues, can be conveyed "only through the absence of life, through an appeal to DEATH, through APPEARANCES, through EMPTINESS, and the lack of a MESSAGE" and the model through which this can be achieved is the notion of the mannequin (Kantor 1997, p. 222, capitalization in the original). When confronting the idea of acting as haunting and usurpation and an actor as a mannequin with the postmodern forgery and fakeness the relationships into which Bennet's wife and the ghost enter expose both the exhaustion of postmodern copy and surface imitation and the potential of this exhaustion to reveal the true identity not necessarily yielding to representation.

The interaction between Lavelli's ghost and Bennet's wife involves a double process of usurpation and transformation—the transformation into an actress/murderess and the ghost's attempts to enact and usurp her identity. Under the influence of Lavelli's concept of murder as acting and pretending, Bennet's wife in appearance and behaviour changes gradually into a film star, her final version being "a screen goddess" wearing a red dress and long gloves, combing her long golden hair crowned with a diamanté tiara. It is in those film star costumes that Bennet's wife is triggered by the ghost to utter and partly perform fragments of passages from the original *Macbeth*—the part about the importance of the promise and Lady Macbeth's readiness to harm a baby fed by her if she ever pledged to do so and the fragment of the sleep walking scene.

The former quotation from Lady Macbeth's part:

> "I have given suck, and know
> How tender'tis to love the babe that milks me;
> I would while it was smiling in my face
> Have plucked my nipple from his boneless gums

> And dashed the brains out, had I so sworn as you
> Have done to this" (Shakespeare 1971, p. 71)

fulfils two functions in Levy's version—it establishes the conceptual mother–daughter relationship between Lady Macbeth and Bennet's wife and hence indirectly between Shakespeare's and Levy's plays, but even more importantly, as the text charged with extreme affective intensity, it exposes the emotional exhaustion of millennial relationships. The concept of the mother's body is introduced in the play a couple of pages earlier in a dream Bennet's wife narrates to the ghost—the dream of the mother dressed up in a "gold fake-fur lionskin" clasping her and pressing against "her nylon fur breasts". Lavelli's ghost interprets the dream for her as a comment on the pretense and fakeness of imaginary paradise, of the unity between mother and daughter: "Your dream is telling you that paradise is synthetic. The breast your mother pulls you to is made from nylon. Paradise is a forgery" (Levy 2000, p. 155). When, later, he teaches Lady Macbeth's part to Bennet's wife, he refers to Lady Macbeth as Bennet's wife's mother: "Your mother said, I have given suck…", adding at the end of the quotation "and then she killed herself" (Levy 2000, p. 177). The terror of being pulled from the mother's breast and the horror of infanticide, although in a sense fake because never realized, are paradoxically more human and authentic than the fondness of the artificial breast. The two mothers, the one that threatens, rejects and then abandons her daughter and the other who offers artificial fulfillment again expose the problem of transposing the original text of excessive affectation into the millennial fascination with fake surfaces. The emotional exhaustion of Levy's version of Macbeth is problematised in Bennet's wife's inability to utter Lady Macbeth's words. When she tries to repeat after Lavelli's ghost what comes out of her mouth is a flatly, as the stage directions suggest, spoken stream of broken and fragmentary words: "I know how… I know how… ta ta tender it is ta ta to love… to love…" or "to love to love to love the bah bah to love love love the—" (Levy 2000, p. 168).

In the sleep-walking scene, Bennet's wife parrots the text prompted by the ghost, turning the original into parody and the grotesque:

> Lavelli: Stare out into the middle distance.
> *She does so.*
> Bring your hands together as if you are washing them.
> *She does so.*
> Say after me (*Thick Italian.*), all the perfumes of Arabia will not sweeten this leetle hand. Oh! Oh! Oh!
> Bennet's Wife: All the perfumes of Arabia will not sweeten this little hand. Oh! Oh! Oh!
> Lavelli: Yes, not a bad forgery. Oh! Oh! Oh! (Levy 2000, p. 171).

In contrast to rather failed attempts at enacting Lady Macbeth, Bennet's wife is fairly successful, with the help of the ghost's instructions, to act the part of Lavelli's daughter just before she is murdered. Repeating the intimate passages from Lavelli's Daughter's telephone conversations with Bennet, Bennet's wife finally manages to embrace the Mediterranean spirit of her victim and approach her "*like a screen goddess*" "*walking towards her prey*" (Levy 2000, p. 178). While quoting Lavelli's Daughter she takes hold of her and bangs her head violently against some object as if enacting the diegetic violence of Lady Macbeth's passage about feeding and harming the baby. In fact she seems to fulfill in this way Lavelli's plan to get rid of his elder daughter as Bennet's wife explicitly explains to Lavelli's Daughter: "I talk to your father because he is training me to be your assassin" (Levy 2000, p. 177). Lavelli performs both the role of the theatre director and a murderer, who by theatre training (instruction) and spiritual possession (haunting) merges the aesthetic and ethical notions into the idea of murder as a work of art.

The reversed process of usurpation and acting occurs in the ghost's attempts to imitate and embody the character of Bennet's wife. When training Bennet's wife to become a perfect actress and murderess, Lavelli is concerned with the surface transformation, the visual appeal, the façade. He is partly interested in the use of tricks that will create illusions of effects rather than the real or authentic transformations. For example, while burying both corpses—Lavelli's and his daughter's, Bennet's wife keeps drinking water from a bottle—which seems to suggest thirst and exhaustion but in fact is only a fulfillment of one of Ghost's instructions: "I give you a tip… always drink a lot of water when you bury the dead… it will help you sweat. The actress and murderess have something in common" (Levy 2000, p. 152). Thus an attempt to feel something through murder is doomed to failure as in the twenty first century murder is entangled with faked emotion, pastiche and spectacle and devoid of authenticity and terror. Sweating has to be produced to fake the authenticity of feeling.

At the same time, from the beginning of the play Lavelli's ghost tries to imitate and copy Bennet's wife's behaviour. Through the ghost's training the wife acquires the surface which is a mimicry of some other surface identity and as such yields to further imitation and appropriation, like a Barthesian "material which has already been worked on" (Barthes 1994, p. 15). In the screen goddess scene, he enters the stage dressed as an exact copy of Bennet's wife. Mimicking her flat manner of speaking the woman enacted by him tells a dream in which her mother dressed as a prawn wearing a Bette Davis hat (a parody of Bennet's wife's dream of the artificial mother) addresses her: "You are just a little version of me" (Levy 2000, p. 166). Upon these words Lavelli turns round and in his own voice addresses the same words to Bennet's wife. The horror of seeing her own copy, expressed in scream, is the primary charge of recognition and confrontation with the Other that, according to Kantor, is at the bottom of each theatrical experience—"the shock taking place at the moment when opposite a man (the viewer) there stood for the first time a man (the actor) deceptively similar to us yet at the same time infinitely foreign" (Kantor 1997, p. 224).

Apart from the reaction to this moment of recognition the scream as a sign of terror can be seen as one of the "nervous tics and twitches" inherited from Bennet's wife's conceptual and intertextual mother. Each of Lavelli's attempts to imitate the scream fails as he says he does not have “enough terror in [his] throat to make a good forgery of what is inside [her]” (Levy 2000, p. 167)[4]. The terror inside Bennet’s wife, which is, according to him, a prerequisite of being a murderess, seems to originally come from her relationship with her mother—Lady Macbeth: “A gaze of terror will do [...]. Your mother pulled you from her breast. She killed herself. Your blonde head is still pressed against the wall. Remember, you cannot murder without terror... so use what you have inside you...” (Levy 2000, p. 177). Thus what seems to be original, immune to imitation and usurpation originates from the appropriation of an earlier text. The only solid element of identity—the terror of a child pressed against the wall—which seems to motivate the authenticity of the scream, ultimately also turns out to be a forgery as after the final murder Bennet’s wife sarcastically mocks Lavelli’s instructions.

The final speechless pantomime of drinking tea, which according to stage directions is supposed to reflect the atmosphere of “inertia, heat, and nausea” brings together the ideas of postmodern excessive consumption incapable of satisfying one’s desire and the inability to feel anything. The instantaneous illusion of action and affection temporarily granted in the acts of murder fades the moment the action is terminated. Massaging the murderer’s stigma—the eczema mark—becomes an empty habit, a nervous tic of no consequence. The characters are finally caught in postmodern excess, putting sugar, lump after lump, into tea and vomiting, hardly aware of the murderees’ off-stage voices attempting to haunt them.

4 Technology, Forgery and Replication

The experimental multi-media quality of Levy’s performance becomes instrumental in exposing and complicating the notions of imitation, forgery and authenticity. Both the mobile phone and the film replace the textual diegetic function with mimetic representation or rather the illusion of such representation in the former case.

The use of film to present off-stage events is one of the accepted techniques of diegetic expansion of the stage replacing either the dramatic text or troublesome changes of the set in the performance. However, in the context of the play’s conceptual preoccupations, the film is employed to further question the possibility of authenticity and real experience. The interaction between the stage and the film is complicated by the lack of synchronicity between the two, despite their separate chronological developments. Lavelli’s ghost, for example, appears on the stage and is referred to as dead long time before his murder is shown in a film scene.

[4] Lavelli’s ghost: “It is difficult to become you. [He circles her]. Difficult because you have not yet become you. You have no centre for me to excavate in my mimicry. I can only imitate your surface. For this I am grateful. I do not have enough terror in my throat to make a good forgery of what is inside you” (Levy 2000, p. 167).

Likewise, Bennet presents a butterfly shaped comb to his wife before the scene of him murdering Lavelli's lover, an oriental waiter who received the comb from Lavelli, is shown on the screen. When confronted with the metatheatrical concerns of the unmediated stage events, the film situations strike as artificial and fictional. The effect is increased because of the explicitness of the camera work, which seems to respond too easily to the characters' words and ostensibly select visual stimuli, as if the situations have been directed by one of the characters. Apparently authentic and unrehearsed events are relegated to mediated representations by technological media, which seem to offer a temporary substitute for emotional excitement and fulfillment. The stage scenes left largely uneventful come to life only in self-referential or intertextual explorations of camouflage and forgery. None of these is capable of offering the characters the possibility of authentic emotional experience.

Relegating the role of the three weird sisters to several film close-ups of three identical faces of a five-year-old girl (Flavia—Lavelli's younger daughter), mouthing, whispering or singing the word "tomorrow" either in English or in Italian further problematizes the question of authenticity and illusion. Similarly to twins, triplets can be seen through the legend of the vanishing twin to be able to articulate in their oracular power the "profound uneasiness with postmodern confusions of identities and postindustrial contusions of the 'real thing'" (Schwartz 1998, p. 21)[5]. As technically multiplied ideal copies, the sisters are a perfect realization of mimicry and forgery, a fulfillment of a wish to become the Other through the process of becoming the Same and as such a mockery of the stage characters' concern with and horror of perfect imitation. The word 'tomorrow' is the condensed message of Macbeth's witches' prophecy parodied in its ambiguity and vulnerability to misinterpretation. The word cuts through the whole play in both temporal and conceptual reverberations referring to the ways in which the idea of 'tomorrow' functions in the original. However, the intertextual interpretation of the sisters scene together with its prophetic function turns out to be a misinterpretation by both Bennet and the reader/spectator. The multiplied Flavia's 'tomorrow' refers to the promise that her father made to her to visit her, which he never kept—the broken promise that originally stood behind the concept of identity as a forgery. The final nauseating pantomime is followed by Lavelli's ghost's conclusion: "Before I died, I was thinking that tomorrow is a promise that is always broken" (Levy 2000, p. 181). In the last film presentation of the three girls, one of them holds and fires a gun earlier found in Lavelli's older daughter's suitcase, upon which event Bennet massages his or rather Lavelli's eczema mark on his forehead, completing the play's intra- and inter-textual retroactivity.

[5] "The legend ultimately has less to do with science or medicine than with the oracular powers of twins who, together, appear to tell us most of what we want to know about being uniquely human and, apart, more than we want to know about feeling alone. The legend of the vanishing twin articulates our profound uneasiness with postmodern confusions of identities and postindustrial contusions of the 'real thing'" (Schwartz 1998, p. 21).

5 Conclusion

The metatheatrical and self-referential mode of the play is a source from which postmodern nihilism and exhaustion in Levy's play seem to derive—the emptiness that rests beneath the masks and roles played by the characters, inability to feel something to be someone, to discard the forged self. The conflation of acting and murdering places the events in the metatheatrical and intertextual perspective, with all actions undertaken turned into forgeries and fake.

Levy's play seems to embody both processes attributed to postmodernism's vacating of "the traditional self and stimulating of self-effacement" suggested by Ihab Hassan (1992, p. 196)—"a fake flatness, without inside/outside" and "its opposite, self multiplication, self-reflection", "diffusing itself in depthless styles, refusing, eluding, interpretation" (1992, p. 197). With the accompanying disturbance of temporality and instantaneous gratification, the play illustrates the Jamesonian depthlessness as "fixation with appearances, surfaces, and instant impacts that have no sustaining power over time" (Harvey 1992, p. 313). However, the disappearance of the notion of the self and individual, which according to Jameson, leads to repetitiveness and imitation of dead styles, also results in the narcissistic self-exploration, self-referential preoccupation with art itself, its exhaustion and failure (cf. Jameson 1993, p. 196). The disturbing and contradictory conceptualizations of the millennial crisis of identity, although linked with the negative notions of superficiality, exhaustion, forgery and fakeness, seem to reveal profound uncertainties and deconstructions of more constructive but illusive traditional concepts of the self.

References

Barthes, R. 1994. From *Mythologies*. In *A critical and cultural theory reader*, ed. A. Easthope and K. McGowan, 14-20. Buckingham: Open University Press.

Curtis, N. False alarm rings for Macbeth. http://195.234.240.57/theatre/review-583687-false-alarm-rings-for-macbeth.do. Accessed 15 April 2010.

Harvey, D. 1992. The condition of postmodernity. In *The post-modern reader*, ed. Ch. Jencks, 299-316. London: Academy Editions.

Hassan, I. 1992. Pluralism in postmodern perspective. In *The post-modern reader*, ed. Ch. Jencks, 196-207. London: Academy Editions.

Jameson, F. 1993. Postmodernism and consumer society. In *Studying culture: An introductory reader*, ed. A. Gray and J. McGuigan, 192-205. London: Edward Arnold.

Kantor, T. 1997. The theatre of death: a manifesto. Trans. V.T. and M. Stelmaszynski. In *The twentieth century performance reader*, ed. M. Huxley and N. Witts, 216-227. London: Routledge.

Kroker, A. and D. Cook. 1988. *The postmodern scene: Excremental culture and hyper-aesthetics*. London: Macmillan Education.

Levinson, P. 2004. *Telefon komórkowy: Jak zmienił świat najbardziej mobilny ze środków komunikacji*. Trans. H. Jankowska. Warszawa: Warszawskie Wydawnictwo Literackie MUZA SA.

Levy, D. 2000. *Plays* 1. London: Methuen Drama.
Philippou, N. http://www.nickphilippou.com/archive/2000/ macbeth/macbeth_1.html. Accessed 15 April 2010.
Pleśniarowicz, K. 1997. *Kantor: Artysta końca wieku.* Wrocław: Wydawnictwo Dolnośląskie.
Schwartz, H. 1998. *The culture of the copy: Striking likenesses, unreasonable facsimiles.* New York: Zone Books.
Shakespeare, W. 1971. *Macbeth.* Harmondsworth: Penguin Books.

Part VII
Extending the Boundaries of Genre and Medium

Beyond the Confines of Realism: Seeking New Metaphors for Memory in Kazuo Ishiguro's *When We Were Orphans*

Wojciech Drąg

Abstract This article examines the interplay between realism and non-realism in Kazuo Ishiguro's *When We Were Orphans* and argues that some of its departures from realism could be interpreted as illustrative of the workings of memory. The article opens with an outlining of Ishiguro's earlier novels' curious suspension in the middle ground between endorsing and challenging realism and with a brief discussion of their preoccupation with the work of remembering. What follows is an analysis of some of the most significant non-realist themes and motifs in *When We Were Orphans*, including the adoption of a naïve perspective of the first-person narrator, the relativity of time and space and the use of narrative coincidence. The delusion of the protagonist, his immersion in the past trauma as well as the ultimate disillusionment are shown as central to the construction of the novel's key metaphor—that of an orphan. In conclusion, the novel's problematisation of the workings of memory is discussed. Memory emerges from *When We Were Orphans* as an inherently unreliable mechanism prone to manipulation and fabulation and subject to the dictates of identity. Finally, the novel's meticulously woven innovative blend of realism and surrealism, redolent of the bizarre realm of memory, is considered as a unique artistic achievement staking out new territory for literature.

1 Introduction: Between Realism and Non-Realism

Part of Kazuo Ishiguro's acclaim as a novelist rests on his skilful use of various kinds of narrative experimentation. The innovative qualities of *When We Were Orphans*—his fifth novel, released in 2000—hinges on the incorporation of

W. Drąg (✉)
Wroclaw University, Wrocław, Poland
e-mail: moontauk@gmail.com

J. Fabiszak et al. (eds.), *Crossroads in Literature and Culture,* Second Language Learning and Teaching, DOI: 10.1007/978-3-642-21994-8_30,

evidently non-realist sections into an ostensibly realist narrative. Surface realism becomes interspersed with surrealist, expressionist, as well as patently fabulist elements. Ishiguro has received abundant praise for the highly original handling of the narrative, which has even been hailed as "extending the possibilities of fiction" (Carey 2000, p. 45). The novel's many departures from realism have attracted several interpretations foregrounding its vulnerability to Freudian analysis (Holmes 2005) or suggesting other metaphorical readings. This article shall argue that the non-realist passages in *When We Were Orphans* (*WWWO*) can be interpreted as illustrating certain mechanisms of remembering and its relationship with identity. A brief overview of the notion of realism and the significance of memory in Ishiguro's fiction will be followed by a closer analysis of these notions in reference to *WWWO*.

1.1 The Middle Ground—Realism and Non-Realism in Ishiguro's Early Novels

Ishiguro's first three novels were widely praised for their "historically grounded realism", each of them taking the form of a first-person narrator reconstructing their past (Holmes 2005, p. 11). His fourth novel, *The Unconsoled* (1995), is commonly regarded as a turning point in Ishiguro's oeuvre and his most challenging work to date. What puts that novel in opposition to his previous works is its radical rejection of realism and replacing it with a bizarre mixture of the dreamlike and the surreal. Realism, defined by Holmes (2005, p. 11) as "the attempt to use linguistic and narrative conventions to create a fictional illusion of social and psychological reality that seems plausible to ordinary readers", is broken in the very first scene of *The Unconsoled*, where the laws of time and space are revealed not to apply. Such is the extent of the novel's use of the improbable that Holmes labels the narrative as possessed of the elements of "fabulism", a term used to refer to the works that "violate standard novelistic expectations by drastic … experiments with subject matter, form, style, temporal sequence, and fusions of the everyday, the fantastic, the mythical, and the nightmarish" (Abrams in Holmes 2005, p. 12). The novel directly following *The Unconsoled*, *WWWO* also contains certain non-realist sections; they are, however, incorporated into an otherwise realist narrative modelled on a detective novel. It is the middle ground between a rejection and an endorsement of realism that marks a new quality in Ishiguro's output. His literary debut, *A Pale View of Hills*, could be considered a harbinger of that narrative strategy, insofar as it very subtly intersperses realism with certain passages redolent of a ghost tale (Lewis 2000, p. 61). Ishiguro's novels' complex attitude to realism is analysed by Holmes (2005, p. 11) in the context of postmodernist fiction, which frequently demonstrates the awareness that "what constitutes reality is a matter for speculation and debate".

2 Ishiguro and the Significance of Memory

Ishiguro has stated in an interview (Sim 2010, p. 120) that he is never interested in abandoning realism for the sake of sheer experimentation. The reason why he does choose to depart from realist narration is because adopting "non-realist methods allows for the depiction of themes of broader significance", unconfined by the historical context (Holmes 2005, p. 13). Elsewhere, in a discussion of generic experimentation, Ishiguro points out that he finds reworking genres for the sake of it a "tedious" task and expresses the belief in the purposefulness of literary experiment only as a vehicle for enhancing "the emotional dimension" of a particular theme (Sim 2010, p. 120). One of the most prominent themes that features throughout each of Ishiguro's novels is the notion of memory. Despite the great variety of narrative voices, settings and genres employed, each text can be summarised as a first-person account of an attempt to come to terms with one's past. As the narrators struggle to recreate their past from memory, the reader becomes increasingly aware of the unreliable nature of their narration.

2.1 Remembering and the Reconstruction of Identity

Each novel accentuates the inherently fallible character of memory and its susceptibility to manipulation and fabulation. Remembering emerges as a mechanism based on reconstruction rather than reproduction; in other words, the content of memory is exposed as constructed anew each time the subject remembers. It is therefore inevitably coloured by "afterwardsness"—the knowledge of what happened after the event. The project undertaken by the narrators can be interpreted as an illustration of Frederic Bartlett's (1995) principle of "effort after meaning". The term, regarded by the English experimental psychologist as the underlying feature of human memory, can be defined as an ongoing attempt to endow any material that enters memory with "the maximum possible meaning" in order to render it intelligible and coherent (84). Wong (2005, pp. 63, 95) observes that Ishiguro's remembering subjects struggle to "reassert meaning into past events" and "find an alternative and acceptable version of their desolate lives". In order to do so, they inadvertently manipulate and distort the facts to such an extent that remembering gradually transforms into an exercise in misremembering. However accurate its content, memory emerges as the foundation of the narrators' sense of identity. Their effort to remember can therefore be interpreted in terms of a search for a more stable identity.

2.2 *Memory in the Earlier Novels*

Attempts to underscore more effectively certain facets of the process of remembering by means of departing from realism can already be traced in Ishiguro's first novel, A Pale View of Hills (1982). In several scenes that dramatise haunting, certain elements of a ghost tale appear to override realism. The two ghosts, whose status in the novel as literal or merely figurative remains unclear, are the apparitions of the narrator's suicide daughter and a woman whose child was killed during the Nagasaki bombings. Their spectral presence in the text can be interpreted as the persistence of a traumatic event in memory or as a metaphor for the experience of loss and the work of mourning. The narrator's apparent invention of a fictional character from the past in order to alleviate her own sense of guilt is seen by Sim (2010, p. 32) as an exemplification of "how memory uses the past in order to address present needs and concerns". The Unconsoled, in turn, can as a whole be read as an extended allegory of memory. With its innumerable distortions, compressions and blends of the past and the present, the novel creates its own alternative reality governed by the logic of an uncanny dream. Lewis (2000, p. 128) sees this bizarre reality as "uncertain, quivering and subject to erasures and displacements"—a realm possessed of "the texture and timbre of memory".

3 *When We Were Orphans* and the Departures From Realism

WWWO is in many ways a continuation of Ishiguro's preoccupation with the theme of memory and his interest in challenging the confines of realism to achieve a particular end. It takes the form of a first-person account of a rise to fame of an English detective named Christopher Banks (depicted in parts one, two and three) and his subsequent return to Shanghai, the city where he believes his parents have been kept since their kidnapping, which put an end to his idyllic childhood some thirty years before. Parts four, five and six document the search for his lost parents, whereas the last section provides a generic dénouement and culminates in Banks's meeting with his elderly mother. The fictional reality conjured up in the novel can be described as occupying—in Ishiguro's own words—the terrain "between straight realism and out-and-out fabulism" (Lewis 2000, p. 128). What makes the text even more puzzling is its fickle and unsteady character. It begins as a realist detective novel and continues in that fashion until the end of part three. Parts four and five represent the gradual abandoning of realism, which reaches its peak in a highly expressionistic and considerably longer part six. The last section, in turn, reinstates realism.

3.1 The Naïve Perspective

Most of the sense of the unreal generated in part six, where Banks undertakes to make his way to a house where he believes his parents are being held hostage, is the result of adopting the narrative perspective of a person who remains immersed in childhood mentality. The reality is recorded through a distorting lens of childhood idealism and naivety. Banks as the narrating voice of this section appears to know the things he has learnt throughout his adult life and yet remains governed by a peculiar logic of striking immaturity. Banks's childish mindset seems increasingly at odds with his reputation as a highly intelligent detective. Brian Finney notes that this section appears to be dominated by a strong sense of wish fulfilment on the narrator's part (Holmes 2005, p. 16). Among the most notable instances of that mechanism is Banks's unfounded certainty that his parents are being kept in a particular house in Shanghai despite the lack of any evidence indicating that they are still alive. Even though several decades have elapsed since his parents' mysterious disappearance, Banks considers tracking them down to be a matter of weeks. He even makes enquiries as to which house he and his parents are going to occupy after they are reunited.

The events recounted by Banks strike the reader as curiously colluding with his naivety; other characters are likewise ready to act according to his every wish. Ishiguro notes in an interview (Holmes 2005, p. 15) that "a lot of the time the world actually adopts the craziness of [Banks's] logic". The owner of the house in which Banks wants to locate his parents is willing to give it up. Similarly, the Chinese lieutenant agrees to assist Banks in his journey to the house regardless of the dramatic situation of his regiment. The reality depicted in part six of *WWWO* can be labelled as expressionistic: it is radically distorted in a way that mirrors the state of mind of the viewer and represents the intensity of his emotion. It is, in other words, rendered through the naïve eye of the protagonist, whose "childhood vision of the world" has remained "frozen" in time and is now "applied to the adult world he encounters" (Wong 2005, p. 89). Banks's immersion in the distant past is labelled by Sim (2010, p. 69) as "a mummification of childhood", which effectively emphasises the resistance of his mindset to the passage of time.

3.2 Time, Space and the Use of Coincidence

The narrative produced by Banks is unconstrained by the flow of time and treats space and distance as relative. The house which is meant to be "just over there" (Ishiguro 2001, p. 224) takes forty-six pages of anguished wanderings to reach and is located at the heart of a maze-like war-torn area resembling a symbolic landscape rather than a district of the twentieth-century Shanghai. Despite their suspension of disbelief, the reader is increasingly puzzled at the amount of coincidence, which, as Lewis notes (2000, p. 149), "can only happen in dreams or in Dickens". The truly Dickensian narrative twists are no coincidence—the novel's great indebtedness to Great Expectations has frequently been pointed out by the critics as well as Ishiguro himself. One of the least likely instances of narrative coincidence in the novel is Banks's bumping into his old childhood friend Akira in the middle of his quest to find his lost parents. Although he has not seen Akira for decades, he is certain that the wounded Japanese soldier is indeed his old friend. The other man's astonishment at Banks's profession of friendship and his failure to share any of his childhood memories arouses the reader's scepticism as to his being the real Akira. Banks's hastiness of judgement can be viewed as yet another product of the logic of wish fulfilment.

3.3 The Deluded Protagonist

Set in the city where Banks spent his childhood and revolving around the fate of the characters who played a pivotal role in it (such as Banks's parents and his best friend), part six of the novel can be read as a symbolic portrayal of his mindscape rather than a depiction of external reality. The piling up of manifestly unrealistic elements generated by Banks's mind exposes the extent of delusion, which he has nursed ever since his parents disappeared. The naivety of his perspective becomes apparent when Banks reaches his destination only to discover debris and not a trace of his parents. The full degree of his delusion is revealed in the closing scene of part six, in which Banks learns from his uncle that what he believed about his parents all his life was a mere fantasy. The truth is exposed to be much more prosaic than Banks wanted to believe. He is therefore made to shake off the heroic scenario of his parents' kidnapping for political reasons and to come to terms with the fact that his father escaped Shanghai with a mistress and died soon after while his mother is a pensioner of a mental asylum. This represents a turning point for the protagonist as he finally needs to confront the rift between the objective reality and the distorted concept of reality, which he has clung to for decades. "I'm beginning to see now, many things aren't as I supposed," Banks reflects in a conversation with a Japanese colonel and adds, "[childhood] is where I've continued to live all my life. It's only now I've started to make my journey from it" (Ishiguro 2001, p. 277). This moment of reckoning puts an end to the non-realist content of the novel, which from then on leads to a classical dénouement in accordance with the generic expectations of a detective novel.

3.4 The Past Wound

Alongside revealing the extent of Banks's delusion, the non-realist section of the novel also indicates his inability to recover from the loss of his parents, which cast a shadow on his subsequent emotional development. Holmes (2005, p. 17) sees his "increasingly surrealistic narration" as "a neurotic result of his desperate need to heal the emotional wound inflicted by his parents' disappearance and to regain the Edenic innocence that he had known as a boy". The first part of the novel contains subtle intimations that the aura of confidence and satisfaction that Banks wants to project might be a mask concealing an emotional wound of some kind. Banks is troubled by an old friend remarking that he had always been "an odd bird" at school, which is incongruous with his own memory of having "blended perfectly into English school life" (Ishiguro 2001, pp. 5, 7). Elsewhere, another schoolmate's reminiscence of Banks as "a miserable loner" is dismissed as an instance of "self-delusion" on the friend's part, although, ironically, it is Banks's self-delusion that resurfaces here. The notion of a past wound that continues to ache and haunt in the present is one of the dominant motifs in Ishiguro's fiction. It is metaphorically present in A Pale View of Hills under the guise of a ghost and in The Unconsoled, exemplified by an uncanny wound on one character's arm, which refuses to heal. In his discussion of the latter novel, Sim (2010, p. 64) interprets Brodsky's fantastic wound as a metaphor for the Freudian notion of melancholia, a condition representing the subject's failure to accommodate the loss and leading to an unhealthy obsession with the lost object. Banks's persistent holding onto the illusion that his lost parents are there waiting to be found and the resulting obsession, which finds its culmination in part six, may invite the interpretation of Banks's condition as displaying certain aspects of melancholia.

3.5 The Inner Clash and the Elements of Expressionism

The primary function of the non-realism embodied in part six of the novel is to dramatise the clash of the two realities which Banks has been hovering between all his life. The first, objective reality is where Banks dwells professionally when solving the numerous crime mysteries and socialising with the London elite of the period. The other reality—internal and subjective—is the child's vision of a harmonious universe, in which good always triumphs over evil and what is lost can always be found. However, Banks's idealism, which springs from the latter reality, also appears to be at the heart Banks's professional life, as it inspired the very choice of career as a detective. The two realities co-exist until the realism of the former and the fantasy of the latter are painfully confronted in part six. The war enveloping Shanghai in the background may be seen as an expressionistic illustration of the violent clash which is about to take place in Banks's mind (Holmes 2005, p. 16). At one point during the quest for his lost parents' house, he can finally glimpse it in the distance. The house strikes him as miraculously "untouched" by the utter destruction that surrounds it. Banks remarks that the house resembles "an apparition from another more civilised world" (Ishiguro 2001, p. 267). This scene effectively encapsulates the dichotomy between the two realities—the "more civilised" world of his fantasy and the fragile and violent reality.

The chasm between the two realities resurfaces as the subject of a conversation that Banks has with "Akira" as the Japanese soldier emerges from a dream. The confession that he was dreaming of being a child elicits Banks's telling response, "it must have been a rude shock. To come from the world you were dreaming of into this one here" (261). In a brief passage delivered in broken English, "Akira" acknowledges his nostalgia for childhood, which he sees as an expression of a longing for "the better world" to come back (262). Rather than a true memory of a paradise lost, the childhood world as experienced by "Akira" in his dream fantasy seems to be more of a retrospective projection of the idealism and the values that the adult world has thwarted. On the literal level, it serves as an antithesis of the reality of war and destruction, which "Akira" remains engulfed in as a soldier; on the figurative level, it may be read as an antithesis of the disillusioned reality of adulthood.

The dichotomy between the two realities is yet again underscored in the scene that provides a kind of *anagnorisis* redolent of Pip's discovery of his true identity in Great Expectations (Sim 2010, p. 72). Banks's uncle reveals to him a series of painful truths, which destroy the illusory reality that he has clung to since childhood. Such is the extent of Banks's delusion that despite his intelligence and expertise in detective matters, he has been unable to see through the contrivance of the version of events that his mother and uncle had him believe in order to "protect" him (Ishiguro 2001, p. 286). "Your mother," the uncle explains, "she wanted you to live in your enchanted world forever. But it's impossible. In the end it has to shatter. It's a miracle it survived so long for you" (294). The moment of painful revelation comes only now: in the face of the overwhelming body of evidence proving the fallacy of his ideas and the futility of his hopes, Banks has to let go of his fossilised immaturity.

4 The Protagonist as an Orphan

The new identity that Banks is resigned to adopting is that of an orphan, as hinted at by the very title of the novel. He does not, however, become an orphan only in the literal sense. The broader theme embodied in WWWO, as well as in other Ishiguro's novels such as The Unconsoled and Never Let Me Go, is the process of becoming an orphan as a metaphor for gradually abandoning the belief that the world is benevolent and just. This notion is encapsulated in the most moving scene in part six of the novel, where the wounded Japanese soldier's dream of being a child again triggers concern about his little son, "When my little boy. He discover world is not good…. I wish I with him. To help him. When he discover" (263). What "Akira" hopes to have a chance to achieve is accompany his child in the inevitable journey of disillusionment and of becoming an orphan to the idealism and safety of the world of childhood. Assuming the identity of an orphan creates the acute sense of lack and emptiness, which is difficult to fill, as well as nostalgia for the time when the metaphorical parents were there. This longing may in some, such as Banks, arouse the sense of mission to revive the dead parents, which is as idealistic as it is unachievable. In the penultimate paragraph of the novel, Banks reflects in retrospect on the mission that the likes of him have the urge to undertake, "our fate is to face the world as orphans chasing through long years the shadows of vanished parents. There is nothing for it but to try and see our missions to the end, as best we can, for until we do so, we will be permitted no calm" (313). The unavoidable defeat of such quest does not, in Banks's view, either question its purposefulness or undermine its importance.

5 Conclusions: Flights from Realism, Metaphors for Memory

The last question that needs to be raised here is how WWWO as a whole addresses the issue of memory and its significance to the construction of identity. Throughout the narrative, Banks is at pains to emphasise the unreliable nature of memory and his uncertainty whether how he remembers certain events was indeed how they happened. He undermines the precision of his memories by recurrently describing them as "hazy" (Ishiguro 2001, pp. 68, 185, 189, 215, 239). "[T]his is how, admittedly with some hindsight, I have come to shape that memory" (87), he remarks elsewhere, highlighting the inherently reconstructive character of memory—as discussed in the earlier part of the article. Although aware of the minor distortions that are inevitable in the process of remembering, Banks remains ignorant of the larger manipulations embodied in his narration. The content of Banks's memories, distant and recent, emerges as subservient to his identity, which has been built upon an idealistic image of childhood. His memories are therefore prone to manipulation and fabulation. The degree of the fabulatory capacity of memory is epitomised in the expressionistic passages in part six, where distant memories resurface and become bizarrely intermeshed with the present. The memory content is shown as material for creative reconstruction, which may endlessly be reworked in order to serve present needs, such as the coherence and stability of identity. The novel as a whole also demonstrates the significance of the past in the present. "The past," notes Lewis (2000, p. 149) in his discussion of WWWO, "is peeping into the present like an under-painting beneath a weathered oil canvas". What shines most visibly through the canvas of Banks's present are past wounds as well as the lost paradise of childhood innocence and blissful ignorance.

The interweaving of realism and non-realism serves in the novel as a tool to effectively represent the mechanisms which oscillate between the domain of the rational and the unconscious, such as the complexities of memory, its intricate relationship to identity, and working through the experience of loss. The numerous departures from realism divorce the novel from its immediate context and setting of pre-war Shanghai, inviting allegorical readings. A highly innovative and compelling blend of realism and non-realism—bizarre as memory itself—the reality constructed in *WWWO* can be regarded as a literary achievement staking out a new territory for fiction.

References

Bartlett, F. 1995. *Remembering: A study in experimental and social psychology*. Cambridge: Cambridge UP.
Carey, J. Few novels extend the possibilities of fiction. This one does. *Sunday Times* 2 April 2000: 45.
Holmes, F. M. 2005. Realism, dreams and the unconscious in the novels of Kazuo Ishiguro. In *The contemporary British novel*, eds. J. Acheson and S. C. E. Ross, 11-22. Edinburgh: Edinburgh UP.
Ishiguro, K. 2001. *When we were orphans*. London: Faber and Faber.
Lewis, B. 2000. *Contemporary world writers: Kazuo Ishiguro*. Manchester: Manchester UP.
Sim, W. 2010. *Kazuo Ishiguro*. Oxon: Routledge.
Wong, C. F. 2005. *Writers and their work: Kazuo Ishiguro*. Horndon: Northcote.

Between a Fragment and a Whole. A Cognitive Analysis of the Gothic Fragment as a Literary Genre. A Case Study of Anna Letitia Aikin's 'Sir Bertrand: A Fragment.'

Anna Kędra-Kardela

Abstract Drawing on the principles of cognitive poetics as proposed by Peter Stockwell (*Cognitive poetics. An introduction.* Routledge, London and New York, 2002), combined with the theory of Current Discourse Space in the sense of Ronald W. Langacker (*Cognitive grammar. A basic introduction.* Oxford University Press, Oxford, 2008), the paper undertakes to analyse Anna Letitia Aikin's "Sir Bertrand: A Fragment", representing a literary form known as "the fragment", popular in eighteenth century English literature. Assuming that a literary work is "open" in the sense of Umberto Eco (*Lector in fabula.* Państwowy Instytut Wydawniczy, Warszawa, 1979/1994) and that reading is, effectively, a "gap filling" process performed by the reader (Kędra-Kardela. *Reading as interpretation, Towards a narrative theory of fictional world construction,* Lublin, 2010) or specifying the "places of indeterminacy" in the sense of Roman Ingarden (*O dziele literackim,* Państwowe Wydawnictwo Naukowe, Warszawa, 1930/1988), the paper accounts for the reading mechanism of the fragment which, apart from the gaps it contains as a literary work, also lacks a beginning and an end.

1 Introduction

The Gothic fragment is a literary form, beside the Gothic novel, popular in the last decades of eighteenth century.[1] What characterizes the aesthetics of the Gothic novel, whose representatives include Clara Reeve, Ann Radcliffe, Matthew

[1] Let us note here that in Romantic literature there appear a number of texts that are "fragmentary" such as Coleridge's "Kubla Khan","Christabel" and *Biographia Literaria*, Wordsworth's *Prelude* and *The Recluse*, Keats's *Hyperion* and *The Fall of Hyperion* or Shelley's *The Triumph of Life*—unfinished because of Shelley's death (Thomas 2005, p. 506).

A. Kędra-Kardela (✉)
Maria Curie-Skłodowska University, Lublin, Poland
e-mail: annakardela@wp.pl

J. Fabiszak et al. (eds.), *Crossroads in Literature and Culture,* Second Language Learning and Teaching, DOI: 10.1007/978-3-642-21994-8_31,

Gregory Lewis, and Sophia Lee, is the novel's length: owing to its multiple plots, the novel creates suspense and prolongs the sense of tension by introducing digressions and lengthy descriptions (Grove 1997, p. 1).

According to Allen W. Grove, the popularity of the Gothic fragment, virtually forgotten nowadays, matched that of the Gothic novel.[2] The characteristic feature of this form is—as explained by Grove—that it varied in length; it began "in the middle of an adventure" to "end abruptly" at an important moment of the action (Grove 1997, p. 2). As a literary form, the fragment is likely to get the reader involved in a dialogue with the text. In particular, confronted with this form, the reader is induced to provide a beginning and an end to the story, thus making an attempt to complete the "incomplete" text.[3] The aim of the paper is to provide a cognitive analysis of Anna Letitia Aikin's "Sir Bertrand: A Fragment"(1773)—one of the best known, if not the best known, Gothic fragments. The paper proposes a way in which this completion can be accounted for in the theoretical framework of cognitive stylistics as developed by Stockwell (2002), combined with the theory of Current Discourse Space in the sense of Langacker (2008).

2 The Gothic Convention

Written to illustrate Edmund Burke's aesthetic theory of the sublime, published together with the essay "On the Pleasure Derived from Objects of Terror", Anna Letitia Aikin's story was immensely popular at that time; it was published and imitated by numerous writers (Grove 1997, p. 2). While some critics, mentioned by Grove, treat fragments—"Sir Bertrand" included—as merely "a case study of aesthetics and form" or as "an exercise in atmosphere", or a "disconnected episode of terror", Grove sees in this genre a "quintessential unit that creates both the poetics and politics of Gothic fiction" in that it reveals "gender politics characteristic of the ancient romance," which is a typically male form (Grove 1997, p. 2–3). Achieved through the apparent fragmentation, "an omission from a tale of male adventure" (Grove 1997, p. 2), highlights, according to Grove (1997, p. 6), the silenced voices of repressed women.

There are several important questions that should be asked about the fragment: How is the fragment to be read? How does the reading of an incomplete text such as the fragment, which has the sense of incompleteness encoded in its structure, differ from the reading of a short story? Does the fragment contain enough

[2] To support this view, Grove (1997, p. 1) quotes Robert D. Mayo (1950), who managed to trace 20 magazines in which Gothic short stories were published. According to another scholar—Pritcher (1976)—also referred to by Grove (1997, p. 1), a large number of works of Gothic short fiction can be found in late eighteenth century pamphlets and pamphlet anthologies.

[3] For a discussion of generic implications of the fragment's incompleteness, see Thomas (2005, pp. 510–512).

information to enable the reader to complete the gaps or to provide the missing beginning and ending determining the fragment as a genre?[4]

The answers to these and other questions of this sort are important from the point of view of our analysis because they provide guidelines for the readerly interpretation of the text. Specifically, it is the knowledge of the Gothic convention that makes it possible for the reader to complete the gaps—the gaps which in this case appear before and after the story told by the fragment.

Because, as already has been mentioned, the fragment as a literary form is closely associated with the Gothic tale, it is important that the reader should know its characteristic features, which include typical settings, stock characters and the events unfolding.

In Gothic tales, elements of the romance may also play an important role. The pattern of lovers' meeting-separation-reunion is often followed in the development of the plot and is linked with such motifs as powerful love (often connected with feelings of uncertainty), which may be threatened by paternal control.

Being acquainted with the Gothic convention and with the convention of the romance, the reader is able to judge to what extent the story follows and to what extent it departs from the convention; she/he can also predict what may happen in the successive sections of the story. The Gothic fragment thus seems to be a perfect example of a literary text which, being highly underspecified, challenges the reader to complete its gaps.[5]

3 Current Discourse Space

In our analysis of the fragment we shall make use of Ronald W. Langacker's theory of the Current Discourse Space—CDS, i.e. all "interactive" aspects of a situation which are "shared by the speaker and hearer as the basis for discourse at a given moment" (Langacker 2008, p. 466) by means of which the speaker is held to exert some influence on the hearer (Langacker 2008, p. 460). The CDS subsumes "an

---

[4] The fragment is an "open" literary structure in that it contains, as any literary work does, gaps (Eco 1979/1994), or what Roman Ingarden (1930/1988) calls "places of indeterminacy". When interpreting the text, the reader makes attempts to close the text and thus "fill" those gaps. However, no matter how hard she/he tries, she/he will never succeed, for new places of indeterminacy will be opened. (Also see Kędra-Kardela 2010).

[5] Among the representatives of the so-called "reader response theory", which deals with the reader's interaction with the text and its "concretization", Wolfgang Iser undoubtedly deserves special mention. The process of reading, Iser asserts, is creative, consisting, in filling the "gaps" in the text by the reader, who establishes the "virtual dimension of the text", forms its *gestalt* and produces the ("unwritten") text. The word "virtual" is used by Iser to emphasize the fact that what the reader comes up with is not explicitly presented in the text; what thus has not been said—"the unformulated becomes concrete" (Iser 1972/1975, pp. 42). As it stands, Iser's reader response theory is similar in many respects to the reader-response approach advocated by cognitive stylistics.

immense body of background knowledge, but as discourse proceeds, it is continually updated as each successive utterance is processed. At any point, the CDS provides the basis for interpreting the next utterance encountered, which modifies both its content and what is focused within it (Langacker 2008, p. 281).

The CDS contains thus a series of so-called usage events, i.e. instances of language use which include the expression's full context-determined understanding. Generally, the CDS includes the Current Usage Event—CUE, the Previous Usage Event—PUE, and the Anticipated Usage Event—AUE. Although a usage event is never identical for the speaker and the addressee, a significant overlap between the speaker and the addressee usually secures successful communication (Langacker 2008, pp. 457–458; 465–466).

In Langacker's theory, the CUE is what the conceptualizer (i.e. the speaker/hearer) interacts with and what she/he may most readily refer to. The central element of CUE is the so-called "objective content", i.e. an entity: a thing or a situation which the conceptualizer perceives. The objective content is encoded by linguistic units, which are also said to serve as "instructions" for the conceptualizer to produce a particular conceptualization.

Let us suppose now that a literary text—analogously to linguistic structures—plays a similar role: it serves as a set of instructions issued for the reader of *how* to interpret the literary work. If so, then a literary text can also be viewed as a series of usage events, whose interpretation is going to be determined by the context, and by what are said explicitly and by what is inferred.

In the case of the fragment, the reader has to reconstruct not just what the story tells us that happened, but also what the story *does not* tell us that actually happened or that is going to happen—the Previous and the Anticipated Usage Event, respectively.

In order to perform the readerly "reconstruction" of the world represented, we first have to briefly address the problem of text boundaries—the text's beginning and ending. And this is exactly the point made by Bartmiński and Niebrzegowska-Bartmińska (2009, pp. 225–226), who say that the text is "framed" by these two opposite points—"delimitators", as they call it.

It should be clear now why the question of text boundaries acquires special importance in the case of the Gothic fragment. As is generally claimed, in narrative communication such signals have an impact on the way a text is going to be read, i.e. interpreted; the lack of a proper beginning/an abrupt beginning can also be regarded as an interpretative signal. The lack of beginning can be viewed as a particular kind of "opening formula." The same holds true for the ending: the lack of a proper closure/conclusion, the unfinished sentence is a particular kind of "closing formula."

4 Anna Letitia Aikin's "Sir Bertrand: A Fragment": Analysis

Aikin's "Sir Bertrand: A Fragment" begins with the following words:

> After this adventure, Sir Bertrand turned his steed towards the wolds, hoping to cross these dreary moors before the curfew. But ere he had proceeded half his journey, he was bewildered by the different tracks, and not being able, as far as the eye could reach, to espy any object but the brown heath surrounding him, he was at length quite uncertain which way he should direct his course. Night overtook him in this situation (Aikin 1773/2002, p. 3).

Opened in such an abrupt way, "Sir Bertrand" unfolds in a manner typical of the Gothic story in terms of the setting and events. The knight travels through mysterious dark surroundings, "desolate waste", "unknown pits and bogs" (Aikin 1773/2002, p. 3), guided by a dim twinkling light and sounds of a distant bell, which bring him to a moat surrounding a large antique mansion. The mansion has features of edifices described in Gothic fiction:

> The injuries of time were strongly marked on everything about it. The roof in various places was fallen in, the battlements were half demolished, and the windows broken and dismantled. A drawbridge, with a ruinous gateway at each end, led to the court before the building (Aikin 1773/2002, p. 3).

The interior of the building proves to be equally mysterious as its surroundings. The ominous and terrifying silence is interrupted by the echoing sound of the knocker used by the knight. As nobody responded to the loud knocking, the knight forced the door open: "the heavy door, creaking upon its hinges, reluctantly yielded to his hand" (Aikin 1773/2002, p. 4). The dim light goes off as soon as the knight enters the place, "all was still as death" (Aikin 1773/2002, p. 4), and the door would not be opened again. Moving wearily through the dark place, the knight experiences truly Gothic adventures. As he was groping his way through a staircase,

> A dead cold hand met his left hand and firmly grasped it, drawing him forcibly forwards—he endeavoured to disengage himself, but could not—he made a furious blow with his sword, and instantly a loud shriek pierced his ears, and the dead hand was left powerless in his (Aikin 1773/2002, p. 4).

Following his walk through the nameless mansion, Sir Bertrand reaches a lofty gallery, where "a figure appeared, completely armed, with a terrible frown and menacing gesture, and brandishing a sword in his hand" (Aikin 1773/2002, p. 5). He attacks the figure with a blow. The figure disappears, but before that happens a massive key falls on the floor, by means of which Sir Bertrand opens the door to a large apartment where he finds a coffin, whose lid is lifted to the sound of a tolling bell. Out of the coffin rises a woman veiled in black, who kisses Sir Bertrand's lips. The same moment the building shakes and collapses with a crash. Sir Bertrand is thrown into a trance, and when he wakes up, he finds himself on a velvet sofa in a magnificent room, where an impressive banquet is set. "A lady of incomparable beauty, attired with amazing splendour, entered", approached the knight and "falling on her knees thanked him as her deliverer" (Aikin 1773/2002, p. 5).

Dumbfounded, Sir Bertrand could not say a word. The story ends with the following words: "After the banquet was finished, all retired but the lady, who leading back the knight to the sofa, addressed him in these words:———" (Aikin 1773/2002, p. 6).

The woman's voice is silenced and her words remain a mystery. The ending to this story thus reflects the male dominance featured characteristically by the Gothic convention in fiction and points to the gender inequalities which are part of many stories of male adventure (Grove 1997, p. 5).[6]

Just like the beginning, the ending of the story is also missing. The abrupt ending of the story does not answer questions that have been raised, the reader's expectations remain therefore unfulfilled. According to Abbott (2002, pp. 53,188), closure is "something we look for in narrative:" it has to do not only with the resolution of the conflict, but it also is closely related to expectations the reader hopes to be fulfilled. Following Barthes's (1974) idea of codes put forward in his *S/Z,*[7] Abbott distinguished two levels at which suspense and closure obtain in narrative: the *level of expectations* and the *level of questions* (Abbott 2002, p. 54). In the case of the story under discussion, closure is missing at both the level of expectations (as readers expect the story to end somehow) and at the level of questions: we do not know what the interrupted episode described in the fragment resulted in. Nor do we get answers to the questions which in a natural way are asked in the course of reading "Sir Bertrand. A Fragment:" Did the knight and the lady, who considered him "her deliverer" get married and live happily ever after, which would be typical of the fairy tale or the romance? Or was some important truth revealed to the knight concerning, for example, the lady's identity, the relationship between the two ladies in the story, the reasons for her being put in a coffin? The convention of a trance/dream employed in Aikin's text provokes even more questions in the reader. The reader her/himself has to provide closure to the story, which can be attempted on the basis of her/his knowledge of the literary convention.

The part of the story directly available to the reader belongs to the CUE. But on what basis can a closure to the story—both in the case of the PUE and AUE—be provided by the reader? The answer is: through her/his knowledge of the convention of the Gothic.[8]

As already mentioned, the Current Discourse Space comprises three events: Previous Usage Event, Current Usage Event and Anticipated Usage Event and a

---

[6] For a more detailed discussion of the 'missing' ending in "Sir Bertrand. A Fragment", see Grove (1997, pp. 5–6).

[7] Two out of five Barthes's codes are particularly useful to Abbott's discussion of closure—the "proairetic code", which relates to "expectations and actions", and the "hermeneutic code", which concerns "questions and answers" (Abbott 2002, p. 53).

[8] Mishra (1994), quoted by Grove, observes that "Sir Bertrand: A Fragment" "is concerned only with the formal dimension of a fragment, with the 'materially' incomplete text." Mishra finds Aikin's text disappointing due to the fact that its Gothic quality is considerably limited by the fairy-tale quality (Grove 1997, p. 2).

literary text may be interpreted precisely in terms of this division. In the case of the text under scrutiny, the Previous Usage Event refers to what happened before the narrative telling of the story began. The opening sentence of the story: "After this adventure, Sir Bertrand turned his steed towards the wolds, hoping to cross these dreary moors before the curfew." (Aikin 1773/2002, p. 3) implies some events prior to the beginning of the plot. The demonstrative pronoun "this" in the expression "after this adventure" suggests that the events described in the story were preceded by an adventure which should be known both to the narrator and to her/his addressee. The time of the day when the adventure took place is implied by the reference to the curfew before which Sir Bertrand is hoping to cross the dreary moors. It also follows from the first sentence that the adventure may have delayed Sir Bertrand's trip and this accounts for his concern if he will be able to cross the moors before the curfew.

In the case of the fragment, the reader seems to get a set of highly incomplete instructions. The word "fragment" which appears in the title, foregrounds the text's "incompleteness" and suggests a surviving piece of a written work which is lost as a whole, or a piece of work which was left unfinished by its author (cf. *The New Shorter Oxford English Dictionary* 1973/1993). If the text is incomplete, the reader may reasonably judge, there must be a way of completing it; otherwise the author would not have written something that would look as if it was an unfinished literary work. Then the reader might reason that the author would want her/him to take up the game and complete the "puzzle."

Further, the reader might reason that the fragment was a popular literary form of the eighteenth century associated with the theory of the sublime. Thus, commenting on Edward Burke's theory of the sublime, Thomas (2005, p. 507) notes that the fragment is associated with the sublime in that it fails to represent a "concrete, finite or present object." What this means is that the Gothic fragment had been intended as an invitation on the part of the author to engage in the game whereby, as Burke puts it (quoted in Thomas 2005, p. 507), "the imagination is entertained with the promise of something more, and does not acquiesce in the present object of the sense."

When analyzing "Sir Bertrand" we notice many features typical of Gothic fiction (Harris 2008, also see Botting 1996/2007, pp. 21–60). In particular,

1. The story takes place in a typically Gothic setting, first in the moors and then inside an old mansion full of secret passages and corridors.
2. The atmosphere of mystery and suspense is created by the darkness of the night, desolation and the sense of the unknown.
3. Sir Bertrand falls into a trance from which he recovered in totally different surroundings.
4. Supernatural creatures mark their presence in the story when "in the midst of [a lofty gallery] a figure appeared, completely armed [...]. Sir Bertrand undauntedly sprang forwards [...]; instantly vanished, letting fall a massy iron key" (Aikin 1773/2002, p. 5).

5. A woman in distress approached Sir Bertrand and "thanked him as her deliverer" (Aikin 1773/2002, p. 5).
6. The hero's identity remains a mystery throughout the story. Albeit Sir Bertrand is presented as a deliverer, his identity is never revealed, perhaps because the story is interrupted in the middle.
7. The presence of a passion driven villain hero is suggested indirectly—if there is a woman for whom Sir Bertrand is a deliverer, there must be somebody she is delivered from. In the context of this Gothic story we are justified in thinking that a villain hero could be the cause of the mysterious lady's predicament.
8. The vocabulary used in the story expresses mystery, fear, and surprise: Sir Bertrand was bewildered; he dreaded, his "blood was chilled", his "heart made a fearful stop", "terror impelled him to make some hasty steps towards his steed" (Aikin 1773/2002, p. 4), "he could not speak for astonishment" (Aikin 1773/2002, p. 6).

Now, given these clues interspersed throughout the text, and using her/his more general knowledge of literature and culture, the reader can make an attempt to provide the missing elements of the story world of the fragment.

Thus, in regard to (1), the reader is exploiting PUE as well as AUE and, based on the Cognitive Narrative Frames, she/he "invents" a series of episodes in which the protagonists will be seen as wading through moors, covering the corridors and secret passages of other old mansions, etc.

With respect to (2), both the PUE and AUE will describe the fictional world in which the atmosphere of mystery and suspense will be upheld.

In regard to (3), the hero will, for example, be expected to encounter in another dream vision a legion of hostile individuals threatening him with the weapons of various kinds.

As far as (4) is concerned, more supernatural phenomena may be seen to be taking place, such as ghosts, apparitions or sighing portraits. The reference to "this adventure" (Aikin 1773/2002, p. 1) at the beginning of the story offers a number of options for the PUE, including encounters with ghosts, bandits or wild animals. AUE may also include the presence of ghosts, sighing statues and bleeding portraits, common in the Gothic convention.

In regard to (5), the PUE may introduce facts concerning the lady's persecution, from which she was delivered by Sir Bertrand. The AUE may include the happy ending and wedding bells.

The unknown origin of the protagonist referred to in (6) is likely to be explained in the AUE—introduced by the narrator as Sir Bertrand, the protagonist may reveal his family background and is very likely to turn out to be a nobleman, an heir to an estate, a distant relative to the lady.

As far as (7) goes, the identity of the villain hero, the persecutor who is referred to only indirectly in the story, is very likely to be explained in AUE—he may turn out to be a usurper, a violent man whose past and whose wicked deeds towards the lady delivered by Sir Bertrand (i.e. PUE) will be given account of.

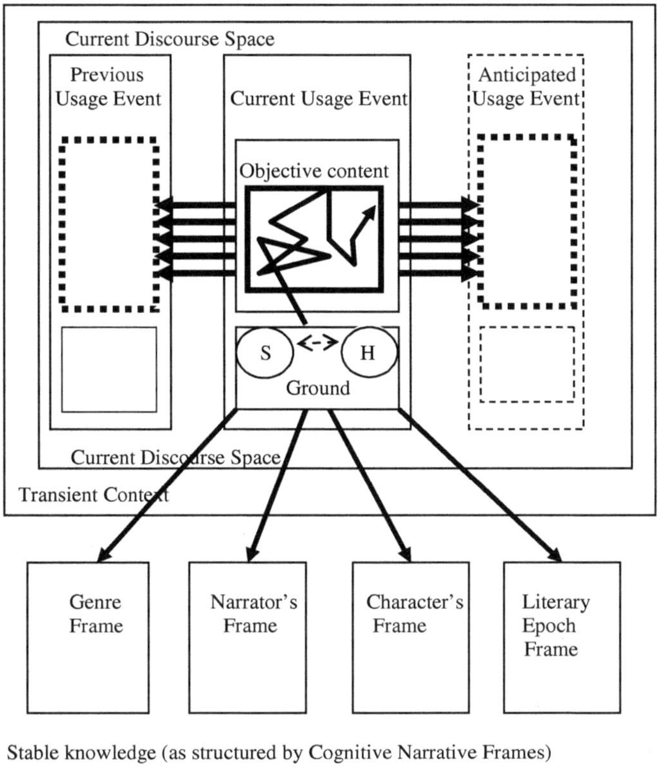

Fig. 1 Gap filling in the literary fragment

The mechanism of "filling in" the gaps in literary works like the fragment can be now diagrammatically represented as follows (Fig. 1, based on Langacker 2008, p. 466):

The reader (S/H), making use of the stable knowledge pertaining to the literary world as structured by the Cognitive Narrative Frames, scans the text of the fragment (squiggy arrow in the Objective Content box) and during the "gap filling" process, creates the text to the left and to the right of the CUE (dotted boxes in bold).

5 Conclusion

Reading is a creative process; it is a process of creating new, on-line senses. The reader makes use of her/his knowledge of the world, of the world of fiction, as well as the knowledge of literary conventions and, when reading the text, she/he fills in the gaps (or places of indeterminacy) the literary work contains. This she/he does in accordance with the basic cognitive principle of *gestalt-sehen*, namely the closure.

Particularly demanding in this respect is a literary form called the fragment, as it "calls upon" the reader to make guesses as to the "beginning" and the "ending" of a whole of which the fragment is a part. The cognitive analysis of "Sir Bertrand: A Fragment" proposed in this paper offers a principled account of how the missing initial and final chunks of the narrative are provided.

References

Abbott, H. Porter. 2002. *The Cambridge introduction to narrative*. Cambridge: Cambridge University Press.

Aikin, Anna Letitia, 1773/2002. Sir Bertrand. A Fragment. In *Gothic short stories*. ed. David Blair, 3-6. Ware: Wordsworth Classics.

Bartmiński, Jerzy, Stanisława Niebrzegowska-Bartmińska. 2009. *Tekstologia*. Warszawa: Państwowe Wydawnictwo Naukowe.

Barthes, Roland 1974. *S/Z*. New York: Hill and Wang.

Botting, Fred. 1996/2007. *Gothic*. London and New York: Routledge.

Eco, Umberto. 1979/1994. *Lector in fabula*. Trans. by Piotr Salwa. Warszawa: Państwowy Instytut Wydawniczy.

Grove, Allen W. 1997. To make a long story short: Gothic fragments and the gender politics of incompleteness. *Studies in Short Fiction* 34: 1-10.

Harris, Robert. 2008. Elements of the Gothic novel. Virtual Salt. http://www.virtualsalt.com/gothic.htm. Accessed 6 April 2010.

Ingarden, Roman. 1930/1988. *O dziele literackim*. Trans. by Maria Turowicz. Warszawa: Państwowe Wydawnictwo Naukowe.

Iser, Wolfgang. 1972/1975. *The implied reader. Patterns of communication in prose fiction from Bunyan to Beckett.* Baltimore and London: The John Hopkins University Press.

Kędra-Kardela, Anna. 2010. *Reading as interpretation. Towards a narrative theory of fictional world construction*. Lublin: Wydawnictwo UMCS.

Langacker, Ronald W. 2008. *Cognitive grammar. A basic introduction*. Oxford: Oxford University Press.

Mayo, Robert D. 1950. Gothic romance in the magazines. *PMLA* 65: 762-789.

Mishra, Vijay. 1994. *The Gothic sublime*. Albany: State University of New York Press.

Pritcher, E.W. 1976. Changes in short fiction in Britain 1785-1810: philosophic tales, Gothic tales, and fragments and visions. *Studies in Short Fiction* 13: 331-354.

Stockwell, Peter. 2002. *Cognitive poetics. An introduction*. London and New York: Routledge.

The New Shorter Oxford English Dictionary. 1973/1993. ed. Lesley Brown. Oxford: Clarendon Press.

Thomas, Sophie. 2005. The Fragment. In *Romanticism. An Oxford guide*, ed. Nicholas Roe, 502-520. Oxford: Oxford University Press.

Poe Goes Pop, or Adapting "The Fall of the House of Usher" in the 21st Century

Anna Krawczyk-Łaskarzewska

Abstract In one of the many articles commemorating the bicentennial of Edgar A. Poe's birth, it was argued that the writer "still does scare people" and his tales of horror "retain their grip on the imagination." Paul Lewis went as far as to suggest that Poe played a key role in the development of popular culture, which has always relied on "visceral experience"—a claim that is worth investigating particularly in the context of the influence the mass media continue to exert on its non-mainstream varieties. This paper will be devoted to the ways in which "The Fall of the House of Usher", one of Poe's most famous and enigmatic short stories, has been visualized, interpreted, transformed, and/or appropriated in several recent film and computer game adaptations. While the interpretive problems with the story will be touched upon, the focus will be firmly on how the mass culture market uses and abuses Poe's efforts, while at the same time reaffirming his enduring relevance and topicality. The examples of popular culture representations I intend to discuss in detail will include works by professional filmmakers and visual artists as well as fan-made videos.

1 Introduction

In one of the many 2009 articles commemorating the bicentennial of Edgar Allan Poe's birth, it was argued that the writer "still does scare people" and his tales of horror "retain their grip on the imagination" (Nuckols 2009). Numerous scholars suggested that Poe played a key role in the development of popular culture, and

A. Krawczyk-Łaskarzewska (✉)
The University of Warmia and Mazury, Olsztyn, Poland
e-mail: jlask@o2.pl

J. Fabiszak et al. (eds.), *Crossroads in Literature and Culture,* Second Language Learning and Teaching, DOI: 10.1007/978-3-642-21994-8_32,

websites such as "Edgar Allan Poe's Footprints"[1] trace and document his influence on movies, books, TV shows, comics, music pieces, or even advertising campaigns, a ubiquity which at times might seem puzzling, or at least indicative of the complexities and paradoxes of globalized modern culture.

This paper will be devoted to some of the ways in which "The Fall of the House of Usher",[2] one of Poe's most famous and enigmatic short stories, has been visualized, interpreted, transformed, and/or appropriated by visual artists and filmmakers for the sake of the contemporary, 20th and 21st century audiences. While the interpretive problems with the story will be touched upon, the focus will be firmly on how the mass culture market uses and abuses Poe's efforts, while at the same time reaffirming his enduring relevance and topicality. In order to illustrate this tendency, I will discuss several works by professional filmmakers and visual artists as well as fan-made videos.

The following remarks should perhaps be prefaced by an attempt to contextualize Poe's creative efforts and his "persistent struggle to influence the taste of the reading public" (Whalen 1999, p. 76). Far from being a naive participant in the literary marketplace of his time, and cynically aware of the rules governing it, Poe was keen on achieving not only the critical, but also commercial success, even if his attitude towards the mass audiences changed and evolved, ranging from pragmatic approval to idealistic snobbery.[3] Furthermore, as Mark Neimeyer (2002) succinctly points out, "Poe himself integrated the popular culture of his own day into his tales and poems" to the extent that his work "was of his time perhaps more than that of any other American author of the nineteenth century" (208).

According to Gerald Kennedy (2001), "no American writer of the antebellum period enjoys greater current popularity and recognisability" (3) than Poe, a contention confirmed by the fact that the author of "FHU" has several Facebook profiles, the most popular of which has, as of September 2011, attracted the attention and support of over two hundred and eight thousand fans (a truly impressive number, surpassing even the nearly one hundred and ninety five thousand Facebook users liking H. P. Lovecraft, who, as a writer, was clearly indebted to Poe). Alongside various other scholars, Kennedy tried to determine the reasons for Poe's undying popularity and mass cultural appeal and came to the conclusion that people read him "to experience the frisson of near-encounters with annihilation" and because "he scares us to death" (2001, p. 4). In addition, Kennedy located Poe's continuous importance "not only… [in] his projections of violence or insanity, but also… [in] his articulations of estrangement and doubt" (8): "the deepening spiritual uncertainty of the nineteenth century" (11). Of a similar nature is Mark Neimeyer's twenty-five page article in *Cambridge*

---

[1] http://www.poedecoder.com/Qrisse/footprints.php.

[2] From now on referred to using the abbreviated form "FHU".

[3] For a thorough discussion of Poe's status in the nineteenth century literary circles, see Terence Whalen, *Edgar Allan Poe and the Masses: The Political Economy of Literature in Antebellum America* (Princeton, NJ: Princeton University Press, 1999), especially the fourth chapter titled "Poe and the Masses" (pp. 76–108).

Companion to Edgar Allan Poe, in which the author enumerates various instances of adapting the prolific American writer to the needs of popular culture and offers several explanations for the persistence of the Poe phenomenon. His diagnosis seems clear enough: "Poe has largely been taken up by popular culture because of its ability to exploit his personal suffering and the sad, and sometimes strange, realities of his life as well as the even more fantastic myths that have grown up around him" (2002, p. 209).

The melodramatic quality of Poe's biography and legends accompanying it helps explain why his name is often mentioned in the context of the decadent poets and rock stars who lived hard and died young. Another probable reason for the writer's popularity is connected with the fact that most fans of Poe's short stories read them for the first time at a very young, impressionable age. In addition, one should certainly mention the slightly peculiar direction of the cultural transactions involving Poe: his transatlantic/cosmopolitan otherness may have contributed to his being accepted and idolized predominantly on the European continent.

Paradoxically, the themes of family curse, sexual transgression, premature entombment and revenge make the oscillation between the highbrow and the lowbrow perfectly understandable. Even though the most basic bibliography for "FHU" on the website of The Edgar Allan Poe Society in Baltimore contains seventy five items,[4] Anthony Magistrale recalls some critics' claim that "the story has no profound subtext whatsoever but is simply Poe's ingenious condensation of a gothic horror novel" (Magistrale 1997, p. 126). The fluctuating cultural status of the tale explains, to a certain extent, why those who adapt or appropriate it do not and cannot transpose its narrative and psychological complexities, and usually content themselves with sensationalizing its already gruesome content.

2 A Picture Can Tell a Short Story

Illustrations rarely attract scholars' attention. As Catherine Golden (2006) melancholically observes, critics usually "privilege a purely linguistic text over the material condition in which that text appears in a book or magazine" (62). The scarcity of studies devoted specifically to the visual reimagining of Poe's prose attests to the truthfulness of her claim.

Artists of various nationalities and using various techniques have shown a long-lasting fascination or even obsession with Poe's work. The ranks of the most distinguished Poe illustrators included Alfred Kubin, Odilon Redon, Edward McKnight Kauffer. Gustave Doré, Arthur Rackham, John Tenniel, etc. Thanks to Aubrey Beardsley's ascetic lines, dark, sombre, cloud-like shapes preferred by Alberto Martini, and Edmund Dulac's stunning colour palettes, clearly inspired by the pre-Raphaelite movement, the viewers have the privilege and the pleasure of

[4] http://www.eapoe.org/works/info/pt025.htm#bibliography.

dealing with intriguing, if at times scandalizing interpretations of Poe's controversial oeuvre. The mediation of the above mentioned artists between the nineteenth-century past and the more contemporary times provides a sense of continuous cultural dialogue and demonstrates their fearlessness in dealing with macabre, socially unacceptable themes.

2.1 *Harry Clarke and Alexander Alexeieff: The Joys of Subversive Interpretation*

Considered to be one of the most talented and influential Poe illustrators,[5] Irish-born Harry Clarke "stamped everything with his wry and exotic humor" (Pollin 1986, p. 4). His often reproduced 1923 colour illustration for "FHU" might be treated, among other things, as a tribute to bondage bordering on sadomasochistic fantasy.[6] At the same time it hints at the interpretive possibilities many literary critics and scholars refused to take into account while commenting upon "FHU" in Clarke's times. Judging by the closeness of crazed Roderick and the smug narrator in the upper part of the composition, a conclusion might be drawn that the two men have long conspired to control the lamentable fate of Madeline. The middle part of the illustration is occupied by hideous, grotesquely deformed lizard-like creatures, some of which are anthropomorphized. They are aggressively stalking and circling the tied and prostrate Madeline, whose position brings immediate associations with the motif of Christ's crucifixion, although on second look it becomes obvious that she broke the wooden lid of the coffin into which she had been thrown against her will. Madeline's armpit hair brings associations with some repulsive insects, and it is easy to discern blood trickling from the captive girl's naked body. Even more disquietingly, she has the same large, bulging eyes of a mad person as her twin brother.

Another, black and white illustration[7] by Clarke concentrates on Madeline's arrival from the crypt. This time she is imbued with a "badass" quality: no longer a passive victim, she seems to be screaming with anger and extends her unnaturally thin hands ended with razor sharp fingers, similarly to the claw-like feet (a motif often used by Clarke and augmenting the overall creepiness of his designs). Madeline's tattered dress resembles shreds of bandages on a mummy, but her state could not be confused with somnambulism: the contorted body of the avenger is ready to strike, to exact revenge on those who decided to bury her prematurely.

---

[5] In Pollin's view, Nicola Gordon Bowe's *Harry Clarke: His Graphic Art* (Mountrath, Ireland: The Dolnen Press, 1983) provides an indispensable introduction to his artistic achievements.

[6] http://www.flickr.com/photos/ajourneyroundmyskull/3564070123/sizes/o/.

[7] http://www.flickr.com/photos/ajourneyroundmyskull/3564869670/sizes/o/.

The set of twelve Usher-related aquatint illustrations made by Alexander Alexeieff in 1930[8] offers slightly more subtle, but no less demanding viewing pleasures. The Russian-born graphic designer and filmmaker is best known for his boldly innovative pinscreen technique in *Night on Bald Mountain* animation, where he demonstrated a creative use of the chiaroscuro, "sand dune" effect. Additionally, and with no small help from his female partners and collaborators, he produced animated commercials of great artistic value, a feat which makes him the right person in the right place as far as illustrating Poe's prose is concerned.

The introductory sentence of the tale in the above mentioned 1930 edition finds a fascinating visual equivalent which, apart from its aesthetic qualities, offers a very effective bit of foreshadowing. The heavy, depressing clouds referred to by Poe provide a background for a white tree which is partly reflected in the tarn of a water reservoir. The trunk has two boughs, one darker than the other and looming over it—this depiction could easily symbolize the intertwined lives of the Usher twins. The upper part of the trunk has a disturbing, torso-like quality, while its lower part resembles a fold of a dress; in fact, the whole tree assumes a shape of a human being. In addition, the boughs, especially the darker one, seem to invade the clouds, as if trying to destroy the natural order. All in all, the bizarre visual content effectively prevents the viewer from making unambiguous assumptions, but, as Christian Drost (2006, p. 272) aptly observes, "as a whole, the etching is suggestive of several keynotes struck in the narrator's account."

Another Usher-related illustration by Alexeieff shows sinister human-like shapes set against the background of numerous windows, deformed stars and thin, sharply delineated crosses of window frames. One of these shapes, perhaps a woman, has a large nose and a weird hat; the person is extending her predatory hands in a witch-like fashion, as if to threaten the hands of the remaining characters, whose identity is not at all obvious.

Finally, there is the brilliant aquatint featuring Roderick standing with legs wide apart and playing the lute.[9] The illustration incorporates the motto of the original story: "His heart is a lute strung tight; As soon as one touches it, it resounds" (a similar technique was used in the 1928 film adaptation of "FHU" directed by James Watson and Melville Webber). An ear is placed where Roderick's heart should be, and his limbs are partly cut off, making it impossible for him to control the instrument. It is as if someone else's hand (the fingers clearly belong to the right hand, but it is placed where we would normally expect Roderick's left hand) strummed the strings, and the lute covers the genital area of Roderick's body. With his eyes closed, Roderick embodies numerous

[8] The illustrations appeared in a Dutch edition of the story, published by Halcyon Press in 1930. The aquatints can be seen online, courtesy of *A Journey Round My Skull* blog, where the entry titled "Alexeieff's Fall" was published on the 19th of December 2009, at http://ajourneyroundmyskull.blogspot.com/2009/10/alexeieffs-fall.html. As of September, 2011, the blog entry has been redirected to the following address: http://50watts.com/#1614187/Alexeieffs-s-Fall.

[9] http://www.flickr.com/photos/ajourneyroundmyskull/4028229088/sizes/o/.

contradictions. He is an evil-doer and a victim of circumstances; his body seems full of energy, yet incapacitated at the same time; the shaded shapes on the lute vaguely resemble a crab or a flaccid penis, but the suggestion that they do not depict anything in particular would be equally valid. A 2009 commentary by blogger nicknamed London Archaeologist and Windowless Consultant describes the spirit of Alexeieff's Poe-related work very accurately:

> there's something about the use of monochrome layering to of shades on the one hand and clear outlines and sharp contrasts on the other that captures the obscure atmospheres of the tale while suggesting the searching light it shined on those depths; chilling, yes, but with analytical intent.[10]

2.2 Greg Hildebrandt and Francesco Spinelli, or the Diluted Power of Now

The exact nature of the relationship between Roderick Usher and his sister proved to be a major bone of contention in the scholarly circles, producing various, at times mutually exclusive analyses from the likes of Gerald Garmon (1972), Daniel Hoffman (1973), John Marsh (1972), Floyd Stovall (1969) or Renata Wasserman (1977), especially in the 1970s, a decade that saw the most radical instances of critical revaluation of "FHU". Curiously enough, the scholarly controversies did not find their counterpart in the realm of more contemporary visual arts. It seems that most of the truly imaginative depictions of Poe's poems and stories had been done much earlier, in the frenzy of new ideas and inspirations before World War II.

To give just a few examples of the more recent and usually safer strategies, Greg Hildebrandt's two "FHU" illustrations show his compositional and drawing excellence, but do not significantly contribute to how the story could be interpreted. The first one[11] relies on a palette of blue hues and shows the narrator of the story approaching the house of Usher, or, to be exact, a spacious castle. Gnarled tree stumps and branches do not produce an ominous effect; rather, they provide an aesthetically appealing frame for the impeccably dressed gentleman who will soon enter the Ushers' castle. The setting is reminiscent of a fairy tale, rather than a dark tale of transgression. In the second, black and white illustration[12] drawn by Hildebrandt, Madeline seems to have just left the crypt. However, she looks more like Mother Theresa of Calcutta begging for forgiveness than a revenge-seeking woman who was buried alive by her vicious brother. There are traces of blood on her semitransparent cloak, but Roderick, who is facing Madeline, seems too

---

[10] 22 October 2009, http://ajourneyroundmyskull.blogspot.com/2009/0/alexeieffs-fall.html.

[11] http://webling.at/atteap/gallery/gh_13.html.

[12] http://webling.at/atteap/gallery/gh_14.html.

orderly, too contained. Fortunately, this visual representation gains in expressiveness thanks to the hands of the characters (a trademark of Hildebrandt’s drawings) and cracks on the wall behind Madeline’s back which suggest the imminent fall of the Ushers’ dwelling.

Elegant, precise contours and rigorous composition are the most striking features of the Usher-related illustration by Francesco Spinelli, who posted his work under the nickname Galhad.[13] His Madeline is wearing a long, unadorned dress and a rather heavy make-up; her long, wavy hair and pronounced bust make her look like a model, rather than a victim of an unhealthy lifestyle. Though she is, quite literally, looming over the Usher house, whose reflection can be clearly seen in the lustre of the water, Galhad’s Madeline shares the space of the illustration with some of the most iconic symbols associated with Poe: a black cat on the left, and a raven on the right. Even Poe himself is shown there, at the bottom of the page, accompanied by an oil lamp and a bottle partly filled with wine, possibly amontillado. He faces the potential viewer emotionlessly, apparently focused on the book he is working on, while Madeline seems to be as bland and irrelevant as her dress: devoid of a significant presence, unable to function as a fully developed character.

Obviously enough, the above examples do not offer a full picture of more recent Poe-related visual efforts and should not be treated as a convenient point of departure for sweeping statements and/or broad generalizations. Nevertheless, at the risk of simplification, it could be argued that while the illustrators in the first half of the 20th century focused their creative energy on showing the unnameable, many of the late 20th and 21st century visual representations of “FHU” are technically accomplished but less edgy: they lack the spark of curiosity, the readiness to push the boundaries, to plunge into the abyss created by Poe and reveal the most ominous possibilities of the literary text.

3 It’s Getting Filmier

As of September, 2011, Edgar A. Poe’s entry on IMDb,[14] the most popular film data base available online, contains 246 credits as a writer (the list includes feature films, TV episodes, animations, video games and short films). Eighteen of those credits are connected with the process of adapting and/or appropriating “FHU”. Interestingly enough, the famous American’s message board is pretty active (ten to twelve threads per year—not bad at all for a dead writer), with possibly the most hilarious threads titled “Happy deathday!!!” and “Dating Nicole Kidman?”[15] The

---

[13] The image was posted by the author himself on the 7th of October, 2009, at http://galhad.deviantart.com/art/The-fall-of-the-house-of-Usher-139529998?offset=30.

[14] http://www.imdb.com/name/nm0000590/.

[15] See http://www.imdb.com/name/nm0000590/board/flat/37096347.

latter topic was initiated by a poster named "hulagurl is born" in February 2006 and has so far generated 80 responses, the last of which was written in January 2011. One feels almost tempted to conclude that if Poe were alive today, his eccentricity would probably enliven the rather dull and tabloid-unfriendly world of the literati; maybe he would even be stalked by some relentless Baltimore paparazzi or forced to publicly announce a rehab stint.

At least three of the above mentioned eighteen interpretations of "FHU" deserve to be mentioned in connection with innovative filmmaking. The year 1928 saw two highly idiosyncratic, black-and-white adaptations: a short film *The Fall of the House of Usher* by James Sibley Watson and Melville Webber,[16] and a feature film *La chute de la maison Usher*, by Jean Epstein.[17] The former serves as a representative example of Americans' fascination with German expressionism, while the latter is very often treated as an outstanding work of French impressionistic cinema, with a healthy dose of surrealist elements. Incidentally, Poe himself was considered by many to be a precursor of the surrealist movement and, generally, a source of inspiration for artists representing various trends and schools. Another groundbreaking take on Poe's story was possible thanks to the creative spirit of the Czech artist Jan Svankmajer, whose *Zánik domu Usheru* (1981),[18] a mixture of stopmotion animation and claymation, focuses, ingeniously and rather obsessively, on the dilapidated mansion and its gradual destruction.

However, it is Roger Corman's 1960 adaptation, titled *House of Usher,*[19] that got the greatest approval from mainstream audiences all over the world and continues to be best known and fondly remembered. It has, in fact, become a staple of popular culture imagination by virtue of its horror provenance, corny dialogues and a beautifully campy set design. Based on the script written by Richard Matheson, a celebrated author of science fiction stories, it, rather disappointingly, offered an ersatz of the original story in which the subtle subtext of sexual rivalry successfully replaced a much murkier content. The casting of Vincent Price may have been a good idea in terms of acting skills, but because of the age difference between him and his co-star Myrna Fahey the concept of a potentially incestuous relationship between the Usher twins got completely ruined. Cheap special effects, problems with maintaining the suspense in the story and hilarious casting decisions make it possible to suggest that from the point of view of the contemporary cinephile watching the film as a black comedy, rather than a "frightener", is perhaps the most reasonable strategy.

The House of Usher (2006),[20] the most recent feature film based on Poe's story, was made by Australian director Hayley Cloake and constitutes an example of a thoroughly modernized adaptation. Curiously enough, although it significantly

---

[16] http://www.imdb.com/title/tt0018873/.

[17] http://www.imdb.com/title/tt0018770/.

[18] http://www.imdb.com/title/tt0082357/.

[19] http://www.imdb.com/title/tt0053925/.

[20] http://www.imdb.com/title/tt0455537/.

departs from Poe's tale, it nevertheless bears a striking resemblance to the literary original both contentwise and structurally. The most radical change involves having a female narrator who is, in fact, the key character in the story, not only in terms of narrating it, but also having a genuine influence on the course of events.

Jill Michaelson is informed by Roderick Usher, her ex-boyfriend, that his twin sister died from an unspecified disease, and gets invited to the funeral ceremony. As can be inferred from Roderick's words later on, the three protagonists used to be connected in a bizarre erotic triangle. After the funeral, Jill decides to spend a few days with the clearly depressed Roderick. Mr Usher admits to her that he suffers from a weird illness whose symptoms involve chronic fatigue and hypersensitivity. Very soon they rekindle their romance, largely owing to Jill's sexual initiative. However, Ms Thatcher, Roderick's nurse, keeps on reminding her that a terrible curse has always crippled and threatened the Usher family, and Jill slowly begins to question Roderick's motivation.

The very title of the film, deliberately omitting the word "fall", promises that the resolution of the conflict proposed by Cloake will not be subjected to the logic of the original. There are no visible signs of decay as far as the Usher mansion is concerned, and the interior does not rely on typical Gothic accessories. Its alienating aspects are certainly alleviated and beautified by the presence of Jill, who spends a lot of time wandering around in a conveniently skimpy lingerie. At least the sinister looks cast by Mrs Thatcher invite the viewer to a familiar, properly gothicised territory. Initially she seems to be the principal villain of the story, but ultimately it turns out she is the only person who tries to put an end to the curse afflicting the peculiar family.

Cloake's adaptation did not score many positive reviews among the professional critics and if the discussions on its IMDb message board are anything to go by, even the staunchest supporters of B-horror movies were not satisfied with the script and its execution. Most importantly, they were disappointed with the final scene of the film and claimed that Jill's twin pregnancy does not put an end to the Usher curse and thus makes no sense from the point of view of the original story. Other debaters stressed the fact that at least Jill is not genetically related to the family of Ushers, so her twins will not be bound to repeat the sins of Roderick and his ancestors, although the possibility of their having an incestuous relationship in the future cannot be altogether discarded.

Ultimately, the most obvious problem with Cloake's adaptation is that it "fails on a basic level at doing exactly what Poe did best—creating palpable tension in the plotlines" (Anderson 2006). On the plus side, one should appreciate Collin Chang and Boyd Hancock, the authors of the script, for tackling the theme of sexual transgression in a straightforward manner, finding interesting contemporary equivalents for Poe's story and using a consistently biological/genetic point of reference. The tagline promoting the film, "Some Doors Should Never Be Opened," could, of course, be conceived as a brutally honest warning and an ironic reminder that adapting a work of a literary genius most often leads to a mediocre result. However, from the point of view of how contemporary culture always mediates the achievements of the past, filmmakers should not be afraid of reimagining demanding source material, even if the price for opening the doors might occasionally seem rather exorbitant.

4 The Usher Sims

Youtube, a video-sharing website that was created over five years ago and permanently changed culture consumption patterns, might be said to constitute the ultimate proof that Poe remains a vital part of popular culture. Apart from fragments of well known mainstream productions by professional filmmakers and actors, fans of his work upload and share numerous videos paying homage to the author of "The Raven". Among the clips available there, one can find comic strips animated by means of simple film techniques, short montages of Poe-related photographs and illustrations, musical performances and readings of his poems and short stories. The line between amateur production and professional work can sometimes be blurred, as is the case with the Youtube video I will be discussing in the following paragraphs.

The clip titled "Edgar Allan Poe: The Fall of the House of Usher"[21] was created by a user called simbiosi77, who in real life is Paola Amadesi, a published author of historical romances and a moderately experienced filmmaker.[22] Understandably, then, her animation cannot be treated as a work of a pure amateur.[23] Rather, it is a loving non-mainstream homage to Poe's most famous short story by a representative of the so-called knowing audience which is familiar with the original and draws considerable pleasure from its modern-day versions. What makes Amadesi's retelling of the story particularly compelling is the fact that she used the settings, characters and structural elements of the enormously popular computer/video game "The Sims", and that she was capable of exploiting its characteristic features to achieve a very impressive, by no means alienating effect.

"The Sims", a game developed by Maxis and first released nearly eleven years ago, belongs to the genre which has been classified as "life simulation" or, to be more specific, "a simulation of the daily activities of one or more virtual persons ('Sims') in a suburban household near SimCity."[24] Typically, the player creates a family of Sims, for whom s/he chooses names, biographies, gender, skin colour, age, hair and facial features, body type, clothing items, personality, etc. Each family has at its disposal a limited fund to design and build a house using various home building materials and interior design options. In addition, the Wikipedia

[21] The clip was uploaded on April 23, 2008, and can still be watched and commented upon at www.youtube.com/watch?v=_41ER3SL5f4&feature=related.

[22] The information is available on Amadesi's Italian-language webpage: http://www.paolaamadesi.it/, to which she provides a link on her Youtube account.

[23] "The Tell Tale Heart, THE Original Horror Story, from www.taletube.com", the clip uploaded by Taletube1 and available at http://www.youtube.com/ watch?v = j_yE0h75LvA, might serve as yet another example of the difficulty with distinguishing amateur and professional renditions of Poe's work on Youtube. The information concerning the file mentions "originally composed music and sound effects." They accompany seventy illustrations by Joel Rivers, and the five-part project was directed by Stephen J Conley and John Dur, with editing by Allan Wright and music by Felix Pando.

[24] See http://en.wikipedia.org/wiki/The_Sims.

entry for "The Sims" informs the readers that the game "technically has unlimited replay value, in that there is no way to win the game, and the player can play on indefinitely." The game continues to inspire numerous amateur video clips in which the Sim characters and settings created by the Internauts serve as a visual background for well known popular songs and/or as a means of telling new stories and adapting/appropriating the already existing ones.

It would be difficult not to see the ironies involved in adapting "FHU" to the Sim environment. Firstly, the inherent artificiality of Sim characters and the repetitiveness of their activities is not that strikingly different from what Poe reveals about his most famous fictitious siblings. One could almost argue that the Ushers' everyday existence is a simulation of life. Secondly, the technological limitations of the game design have exerted a significant influence not only on the plotline of the Italian Youtube user's appropriation, but also on the physicality of the characters: seen from the "simmed" narrator's point of view, Roderick and Madeline wander around stiffly, in a rather awkward fashion, even though it is possible to see in their stilted movements a timeless dignity. Thirdly, from the above quoted Wikipedia website one learns that Sims "may die, either by starvation, drowning, perishing in a fire, electrocution or by virus," and even the more vicious ways of killing Sims ("getting them into a pool and deleting the steps, or putting them into a room then deleting all of that room's doors") do not include the exciting option of a sisterly revenge. Therefore, the characters created by Amadesi cannot die, nor can the house fall.

Apart from subtitled fragments of the dialogues between the narrator and Roderick, Amadesi uses in her clip the beginning of Mozart's "Requiem". This decision greatly aids the emotional impact of her clip, but also complicates the already ambiguous relationship between a celebrated short story by a "mass-cultural writer" (Elmer 1995, p. 4) whose works have enjoyed academic prestige, and its "lowbrow" computer game adaptation. In the establishing shot of Amadesi's short film, the rather ominous Usher mansion is shown in torrential rain, but there are no fissures or other symptoms of decay here. The motif of the sentient house is not touched upon at all, although some unmistakably Gothic trappings of the setting are, indeed, present in the film: the "narrator" of the story enters the house and passes through several dark corridors in order to find and greet his friend in an equally dark, church-like chamber. Out of the limited stock of activities available to Sims, Amadesi chooses for Roderick playing the piano, painting, taking care of the fireplace, kneeling, praying and showing signs of utter dejection and despair. A truly emotional response to some of the sinister events in the story is not technically possible, e.g. upon hearing entombed Madeline's frightened scream the narrator and Roderick merely part their lips.

Characteristically, and inevitably, considering the structural components of *The Sims*, the scenes are almost never shown from a unified, consistent perspective. The scene in which Madeline "dies" or simply faints is shown from three different angles which prolong and fetishise the moment of her mysterious demise. In fact, Madeline remains a complete enigma throughout the story, an essentially passive character. The narrator meets her once, but they do not communicate with each other. In the penultimate scene of Amadesi's clip we see her "resurrected" body, or, to be exact, blood on her lower lip and chin. Madeline's loud, agonized scream is the only proof that she actually has emotions. Once again, the technical limitations of the game may have prevailed over the wish to accurately render Poe's vision. Judging by the specific scripting decisions made by Amadesi, she does not seem to be particularly interested in peeling off the layers of the original story and explaining its mysteries. Instead, she offers a visually beautiful, simple and highly entertaining tale in which she manages to maintain the suspense until the very end.

As has already been pointed out, the Usher mansion cannot collapse. We only get to see its blurry, coloured reflection in the last shot of the clip. Hence, maybe, the relatively happy endings for all the three protagonists, at least in the superficial sense. The narrator merely leaves the house and we do not get to see the deaths of Roderick and Madeline. However, it would be tempting to suggest that such an ending in fact helps to perpetrate the curse of the Usher family: the housing structure is left intact, the evil twins will be allowed to continue their incestuous relationship and maybe reproduce. The circularity of the electronic medium finds its logical extension in the adaptation proposed by Amadesi and, in addition, helps to retain the open-endedness of the story.

5 Conclusion

The first decade of the new millennium proves beyond doubt that Edgar Allan Poe's "The Fall of the House of Usher" continues to attract interest from graphic designers, painters, comic book creators and filmmakers. Their efforts have attested to the changing level of tolerance for taboo subjects. The most recent feature film adaptation of Poe' short story is currently being developed under the working title *The Ushers*, and according to the short description available on IMDb[25] the updated Roderick and Madeline "have lived in the same Martha's Vineyard house for a long time but begin to feel trapped there, and who then bring in a real-estate broker to help them resolve their crisis."[26] Apparently, to satisfy the current demands of the film market, this completely fresh look on a certain housing problem will be shot in 3D. And, as many film fans fear, it will probably turn out to be an unmitigated disaster. To use Neimeyer's pronouncement, it is very likely that the audiences will have to content themselves with one more item on the long list of Poe-related cultural artefacts which are "all strangely Poe and not Poe at the same time" (2002, p. 222).

[25] http://www.imdb.com/title/tt1440341/. Accessed 21 July, 2009. As of September, 2011, the project is still in development and the IMDB entry has been removed.

[26] The news about the film was published in May, 2009. See http://www.imdb.com/name/nm0443181/news?year=2009.

References

Anderson, T. W. 2006. The house of Usher. A review. 12 April 2006. http://www.filmthreat.com/reviews/8906/. Accessed 15 May 2010.

Drost, Christian. 2006. *Illuminating Poe. The reflection of Edgar Allan Poe's pictorialism in the illustrations for the* tales of the grotesque and arabesque. An unpublished doctoral dissertation. 15 February 2006. Hamburg. http://ediss.sub.uni-hamburg.de/ volltexte/2007/3279/pdf/Illuminating_Poe.pdf. Accessed 17 September 2011.

Elmer, J. 1995. *Reading at the social limit: affect, mass culture, and Edgar Allan Poe.* Stanford: Stanford University Press.

Garmon, Gerald. 1972. Roderick Usher: portrait of the madman as an artist. In *Poe Studies* 5.1 (June): 11-14.

Golden, C. J. 2006. Speaking a different story. The illustrated text. In *The yellow wall-paper by Charlotte Perkins Gilman: A dual-text critical* edition, ed. J. Shawn, 61-72. Athens: Ohio University Press.

Hoffman, Daniel. 1973. *Poe, Poe, Poe, Poe, Poe, Poe, Poe*. London: Robson Books.

Kennedy, G. J. 2001. Introduction. Poe in our time. In *Historical guide to Edgar Allan Poe*, ed. G. J. Kennedy, 3-17. Cary, NC: Oxford University Press.

Magistrale, A. S. 1997. *Poe encyclopedia.* Westport, CT: Greenwood Publishing Group.

Marsh, John. 1972. The psycho-sexual reading of "the fall of the house of Usher." In *Poe Studies* 5.1 (June): 8-9.

Neimeyer, M. 2002. Poe and popular culture. In *Cambridge Companion to Edgar Allan Poe*, ed. K. J. Hayes, 205-224. West Nyack, NY: Cambridge University Press.

Nuckols, Ben. 2009. Even at 200, Poe endures in pop culture (Bicentennial today). Yahoo – AP. http://www.freerepublic.com/focus/ chat/2167397/posts. Accessed 17 March 2010.

Pollin, Burton R. 1986. Review of Nicola Gordon Bowe's *Harry Clarke: His Graphic Art* (Mountrath, Ireland: The Dolnen Press, 1983). In *Poe Studies Association Newsletter* 14.1 and 14.2 (Spring and Fall).

Stovall, Floyd. 1969. *Edgar Poe, the poet: essays new and old on the man and his work.* Charlottesville: University Press of Virginia.

Wasserman, Renata. 1977. The self, the mirror, the other: "the fall of the house of Usher." In *Poe Studies* 10.2 (December): 33-35

Whalen, Terence. 1999. *Edgar Allan Poe & the masses: the political economy of literature in antebellum america*, Princeton, NJ: Princeton University Press.

Adaptation and the Idea of the Death of the Author: The Case of Samuel Beckett

Jadwiga Uchman

Abstract The article discusses the usefulness of Ronald Barthes' concept of the death of the author in reference to the output of Samuel Beckett. On the one hand, the playwright refused to provide any interpretation or explanation of his *oeuvre* and thus left the finding of its meaning to individual receivers. On the other, however, he rigorously insisted on a strict following of his stage directions, or, to put it differently, on respecting his original vision of the given drama. At present, the faithfulness to the spirit of his output is guarded by the Beckett Estate, represented by his nephew who has to cope with a difficult, if possible at all, task of protecting the authorial rights of his great uncle. The problems connected with the legal status of Beckett's writings are further complicated by the fact that, in most of the cases, there exist a number of his manuscripts, often introducing changes even in the same language and there are also differences between the French and English versions. As far as the productions of his dramas are concerned the situation is also similar—for instance, apart from the published version of Beckett's *Production Notebook* of *Happy Days*, registering his work on the Royal Court production in 1979, there are three annotated copies prepared by Beckett himself and records made by Martha Fehsenfeld and Alfred Hübner concerning still other productions on which Beckett worked.

J. Uchman (✉)
University of Łódź, Łódź, Poland
e-mail: jagodauchman@wp.pl

J. Fabiszak et al. (eds.), *Crossroads in Literature and Culture*, Second Language Learning and Teaching, DOI: 10.1007/978-3-642-21994-8_33,

1 Introduction

The aim of the present paper is to investigate the validity of Ronald's Barthes' theory of the death of the author in reference to the process of stage adaptations of drama. The discussion of the issue concentrates on Samuel Beckett as a playwright and a supervisor of the production of his plays. Special attention is paid to his authorial and directorial practice concerning the staging of *Happy Days*.

2 Adaptation

The term intertextuality, first introduced by Julia Kristeva, has been widely applied, defined and redefined.[1] While discussing the passage from one sign system to another, she argued:

> To be sure, this process comes about through a combination of displacement and condensation, but this does not account for its total operation. It also involves an altering of the thetic *position* and the formation of a new one. The new signifying system may be produced with the same signifying material; in language, for example, the passage may be made from narrative to text. Or it may be borrowed from different signifying materials: the transposition from a carnival scene to the written text, for instance.... The term *intertextuality* denotes this transposition of one (or several) sign-system(s) into another but since this term has often been understood in the banal sense of 'study of sources,' we prefer the term *transposition* because it specifies that the passage from one signifying system to another demands a new articulation of the thetic – of enunciative and denotative positionality. (Kristeva 1980, p. 111)

The notion of intertextuality is perceived by some critics (Hu 1989, p. 4) as a characteristic trait of post-modern craftsmanship, whose features are enumerated by Ihab Hassan. I would like to concentrate on his ninth 'definien' of the term, namely "Performance, Participation": "Indeterminacy elicits participation; gaps must be filled. The postmodern text, verbal or non-verbal, invites performance" (Hassan 1991, p. 198).

The communication with modern art, that of Beckett undoubtedly included, presupposes an aesthetic attitude where the audience are simultaneously receptive and co-operative. Their approach should be the one propagated by Ingarden (1937, pp. 38–41 and 1988, pp. 409–436): they should locate "the fields of indeterminacy" and then remove them in a process of concretization. The playwright's dramas contain many fields of indeterminacy for the audience to fill in. Each interpretation will differ from others as the final responsibility for the coherence and meaning belongs to individual viewers. Keir Elam (1980, p. 93) finds that the theatrical communication depends to a great extent on the intertextual basis of the theatrical

[1] See, for instance: Culler 1981, pp. 102–118, Culler 1975, pp. 102–152, Głowiński 1986, Jenny 1988, Lachmann 1991 and Pfister 1991.

frame: "Appropriate decodification of a given text derives above all from the spectator's familiarity with other texts (and thus with learned textual rules)." In her article "The Pleasure of Spectator," Ann Ubersfeld (1982, p. 129) describes "theatre as sign of a gap-being-filled. It would not be going too far to say that the act of filling the gap is the very source of theatre pleasure." Decodifiation and interpretation are elements of the adaptation process, the next theoretical issue to be tackled.

3 Staging Drama

Kristeva uses interchangeably intertextuality and transposition, the two terms being connected with the idea of adaptation, that is shifting elements from one sign system to another and introducing necessary changes. The same is true of most critics dealing with these problems. In *Dictionary of the Theatre* Patrice Pavis discusses intertextuality and adaptation, in both cases indicating the connections between two different sign systems. She also argues that, as a matter of fact, each production of a play fixes intertextual connections between literature and the stage (Pavis 1998, pp. 21 and 208). In his book, *Transposing Drama. Studies in Representation*, Egil Törnkvist uses the second term introduced by Kristeva while writing about stage, TV and film productions of drama and the problems connected with translating the playtext into another language. He also argues he will be using the term "transformation" "in a technical sense, as meaning 'transposing a play from a verbal semiotic system to an aural, visual or audiovisual one'" (1991, p. 7). Producing a play in a theatre is a case of adapting it to the requirements of the stage, turning a text printed on the page into a live performance. Dudley Andrew (2000, p. 29) presents a noteworthy insight into what this kind of adaptation really is:

> The broader notion of the process of adaptation has much in common with interpretation theory, for in a strong sense adaptation is the appropriation of a meaning from a prior text. The hermeneutic circle, central to interpretation theory, preaches that an explication of a text occurs only after a prior understanding of it, yet that prior understanding is justified by the careful explication it allows. In other words, before we can go about discussing and analyzing a text we must have a global conception of its meaning. Adaptation is similarly both a leap and a process. It can put into a play the intricate mechanism of its signifiers only in response to a general understanding of the signified it aspires to have constructed at the end of the process.

The points raised here by Andrew concerning the necessity of interpreting before adapting as well as the creative aspect of the process are stressed by other critics. Hornby quotes (1995, p. 99) Susanne Langer's "opinion with regard to musical performance, that the player is as creative as the composer... 'Real performance is as creative an act of composition, just as the composer's own working out of the idea, after he has conceived the greatest movement and therewith the whole commanding form, is still creative work. The performer still carries on'."[2]

[2] Susanne Langer. *Feeling and Form*. New York: Scribner's, 1953, p. 139.

He (1995, p. 99) continues: "In other words, *interpretation and creation* are not opposites." Similarly, Linda Hutcheon (2006, pp. 7–8) argues that "First, seen *as a formal entity or product*, adaptation is an announced and extensive transposition of a particular work or works" and "Second, *the process of creation*, the act of adaptation always involves both (re-)interpretation and then (re-)creation; this has been called appropriation or salvaging, depending on your perspective."

For centuries drama was discussed, similarly to poetry and the novel as one of the three literary kinds. For some time now, the literary theory of drama seems to have nearly disappeared. There are critics, however, who still insist on its validity like, for instance, Heuvel (1991, pp. 2–3) who writes: "By *drama* I mean generally that form of theatrical expression that is constituted primarily as a literary artefact, according to particular 'dramatic' conventions and empowered as text. Dramas, of course do not remain merely literary and textual; they are *often* (emphasis mine—JU) performed." Similarly, Wiśniewski (2006), even though he discusses the theatrical elements of the playwright's output, in the very title of his book stresses the literary shape of Samuel Beckett's drama. Now the dominant approach can be defined as theatrical theories of drama.[3] As a result, what is now taken into consideration most often is not the written literary text, which is often referred to as a playtext, playscript or simply text or script (Hornby 1995, pp. IX–X), but its concrete realisation, be it in the theatre itself, in TV theatre (which often makes use of video techniques) or in the cinema. The people involved in generating the final result (primarily the director, and actors, and secondly, all those responsible for costumes, setting, stage props, lightening, music etc.) become its co-authors. Needless to say, the final product often greatly departs from the text printed on the page. The former has been replaced by its realisation, "*the drama text*, experienced by the reader" has changed into "*the performance* text, experienced by a spectator (or listener)" (Törnkvist 1991, p. 2), thus a kind of adaptation, a shift to another semiotic field is noticeable.

4 The Playtext and its Stage Version

The question arising in connection with adapting drama concerns the authority of the playwright and the validity of his vision in the final product. Raymond Williams (1968, p. 170) argues: "When a dramatist writes a play, he is not writing a story which others can adapt for performance: he is writing a literary work in such a manner that it can be directly performed." In this context Roman Ingarden's (1976, p. 149) distinction between primary and secondary text is important, according to which the former refers to everything in the play that is verbalised

[3] See, for instance: Styan 1969, a book which discusses dramatic elements from the perspective of the theatrical theory of drama as well as Abramowska 1976, Hutnikiewicz 1976, Górski 1976, Kleiner 1976, Raszewski 1976, Skwarczyńska 1976a, 1976b, 1976c, and Sławińska 1976.

during the performance (i. e. the dialogues and monologues) and the latter to the stage and acting directions. The attitude of the critics to the function of the stage directions as well as to the role of the playwright and director varies. On the one hand, Pfister (2000, p. 45) stresses the importance of stage directions for the primary text and argues that the doubling of verbal and non-verbal information strengthens the overall meaning of the drama. A similar opinion is voiced by Clurman (1972, p. 48) who insists on the necessity of "the initial director-playwright consultations" and argues (5–6) that "The playwright as playwright is in part a director. If he is truly a man of the theatre (some playwrights only suffer the theatre but do not feel they belong to it) the playwright 'sees' the play on the stage as he writes. His dialogue as well as his notations of stage behavior suggest movement and part of the total physical life that the script is to acquire when it is produced."

Jeanne Luere, the editor of the book *Playwright Versus Director. Authorial Intentions and Performance Interpretations*, in the introduction devotes some time to the relationship between the text and the performance and the importance of the stage directions. She (1994, p. 11) discusses, among others, the article of Gerald Rabkin "Is There a Text on This Stage?" subtitled "Theatre/Authorship/Interpretation"[4] in which he refers to "Umberto Eco's classification of closed versus open texts in his work, *The Role of the Reader*. Closed texts 'aim at pulling the reader along the predetermined path, carefully displaying their effects so as to arouse pity or fear, excitement or depression in the due place at the right moment' (144). Rabkin suggests Arthur Miller and Samuel Beckett as, perchance, in this category, and examines media coverage of their hostile reactions to directorial license." Luere (155) adds: "But Rabkin stresses that if we deny the author's 'patriarchal authority over his text' we must also recognize 'the importance of protecting his personal rights to his work'."

Luere (1994, p. 10) also discusses Patricia A. Suchy's article "When Words Collide: The Stage Directions as Utterance,"[5] in which the author "cites Aristotle's emphasis on plot, character and thought over diction to support Samuel Beckett's position that the stage directions rather than the text's 'diction' often *are* the play (73). (Beckett's *Acts Without Words I and II* are nothing but the stage directions.) Suchy reasons that 'in modern times, with the concept of the 'sub-text' or the idea that much of the play's meaning can be discovered beyond the surface of its language, the uttered lines of a play seem skeletal, only *suggestive* of dramatic life' (73)."

On the other hand, however, an opposite opinion is recently more and more often expressed. Bradby (2001, p. 4) points out: "The idea that the stage directions are as central to the play's integrity as the dialogue between characters is far from

---

[4] Gerald Rabkin. "Is There a Text on This Stage? Theatre/Authorship/Interpretation." *Performing Arts Journal* 9, no. 2, 1985, pp. 142–59.

[5] Patricia A. Suchy. "When Words Collide: The Stage Directions as Utterance." *Journal of Dramatic Theory and Criticism* 6, no. 1, Fall 1991, pp. 69–82.

self-evident." This has been caused by the emergence of "the artistic director" whose function is greatly different from that of a director-coordinator from the past. When Wells ventures the opinion: "No matter how detailed the instructions a playwright may give, he remains a participant in an essentially collaborative act. The hints that he gives have to be transmitted to the audience by people other than himself,"[6] he has the same on mind as Patrice Pavis does. She writes (1992, p. 28): "*mise en scène* is not the performative realization of the text." She notices elsewhere (1982, p. 141) that the stage directions are comparable to a recipe for baking a cake: "As with the cake, some will prefer to scrupulously follow the recipe and others will add a 'pinch of salt' or will substitute their own culinary technique." In this context, it is worth returning to the already mentioned book edited by Jeanne Luere in which she presents a survey of criticism as well remarks of playwrights and directors on the subject. She (1994, p. 3) starts her discussion by mentioning the literary assumptions about text expressed by Ronald Barthes:

> Using a semiotic ground, Roland Barthes qualifies the old assumptions about literary text, professing that "the meaning of a work (or of a text) cannot be created by the work alone; the author never produces anything but presumptions of meaning, forms, and it is the world which fills them."[7] Texts, then, resemble "the links of a chain of meaning, but this chain is unattached," "so that someone else—in time, the reader—must "fasten it, give it a definite meaning" (XI).

Barthes' opinion brings us to the validity of his theory of the death of the author, especially relevant in reference to theatrical adaptations of drama. His theory has had a great influence on interpreting not only literature but also art in general.[8] His "Death of the Author" essay was published in English in an American journal *Aspen* no. 5–6 in 1967 and in French in the magazine *Mantea* no. 5 in 1968.[9] Whereas it seems justified to argue that a literary work exists on its own right and individual readers can perceive it differently, it is questionable whether people responsible for producing a play are justified to introduce drastic changes as far as the stage directions go.

5 Beckett's Standpoint as a Playwright

In this context the case of Samuel Beckett seems to be worthy of attention. On the one hand, as far as the interpretation and explanation of his writings are concerned, he seemed to support the idea of the death of the author. On the other, however, as

---

6 Stanley Wells. *Literature as Drama*. London: 1970, 7. Quoted after Törnikvist 1991, p. 13.

7 Ronald Barthes. „Literature and Signification." *Critical Essays*. Transl. Richard Howard. Kingsport, Tenn.: Kingsport Press, 1963. XI.

8 For its discussion, see, among others: Burke 2008, Staiger 1989.

9 http://en.wikipedia.org/wiki/Death_of_the_Author (16. II. 2010).

far as the production of his plays is taken into consideration, he expressed an opposite attitude.

Samuel Beckett was extremely unwilling to provide any commentary concerning the meaning of his works. While asked about the identity of Godot, he answered: "If I knew who Godot was, I would have said so in the play."[10] He also rejected the religious interpretation saying: "If Godot were God, I would have called him that"[11] and argued "Christianity is mythology with which I am perfectly familiar, and so I use it. But not in this case."[12] In the correspondence with Alan Schneider concerning *Happy Days* he tried to specify the identity of the "old Greek" but, not being able to find his notes on the pre-Socratics, he could not state it exactly:

> The leading Sophist, against whom Plato wrote his Dialogue, was Protagoras and he is probably the "old Greek" whose name Hamm can't remember. One purpose of the image throughout the play is to suggest the impossibility logically, i.e. eristically, of the "thing" ever coming to an end. "The end is in the beginning and yet we go on." In other words the impossibility of catastrophe. Ended at its inception, and at every subsequent instant, it continues, ergo can never end. Don't mention any of this to the actors! (Harmon 1998, p. 23)

Even while providing the director with an explanation and interpretation, Beckett still insisted on his not telling the actors.

The playwright sometimes modified minor elements of the original conception of his dramas while working on their productions in the theatre. At the same time, however, he was not eager to accept any alterations of the original playtext (stage directions included) by others. And so, for instance, originally *Breath* was given to Kenneth Tynan to be included in the New York production of *Oh! Calcutta!* but, as the latter changed Beckett's version, introducing naked bodies on the stage, it was not presented during the successive productions of the musical (Bair 1978, p. 603). In 1957 *Waiting for Godot* was produced in Teatr Współczesny in Warsaw. When Libera met Beckett in 1978, he mentioned having seen the play at that time. Beckett remembered that production as one of the first ones behind the iron curtain and also the photographs which had been sent to him. They impressed him, the only exception being the tree, which had too many leaves. Libera explained to the playwright that this resulted not from the scenographer's fantasy but from the faulty translation into Polish in which the word "*quelques*" was omitted in the stage directions (Libera 2009, p. 380–381). The fact that, despite the passage of so many years, Beckett remembered the departure from the original secondary text clearly indicates his hostility to the introduction of any changes of his original idea.

It seems noteworthy to mention the opinion of Pfister (1993, p. 45) concerning Beckett's stage directions:

---

[10] Samuel Beckett to Alan Schneider and many others. Bair 1978, p. 382.

[11] Samuel Beckett to Harold Hobson. Bair 1978, pp. 382–383.

[12] Samuel Beckett, November 17:1971. Bair 1978, p. 186.

> From the audience's point of view, the existence of "identity" means that, in the majority of dramatic texts, familiarity with the primary text is sufficient in itself to ensure a reasonable measure of comprehension.... Thus, identity always occurs when stage-directions are implicit in the *primary text*... of course, it is rare that identity, this redundant doubling-up of verbal and non-verbal information, occurs in such a pure form as in the following passage from Samuel Beckett's play *Happy Days*:
>
> WINNIE (*Pause. She takes up a mirror*) I take up this little glass, I shiver it on a stone—(*does so*)—I throw it away (*does so far behind her*)[13]

The virtual identity of primary and secondary text in this passage makes the doubling-up of verbal and non-verbal information clear even from the printed text. This creates the impression that in doing so Beckett was trying to produce a pure and extreme example of one particular dramatic technique in order to draw the audience's attention to that technique and to turn it into an implicit theme in itself. It should be also noticed that Beckett's stage directions, apart from being very precise are also characterized by high artistic value and are often also contrasted with the primary text, a point noticed by Wiśniewski (2006, pp. 21–33).

6 Beckett in the Theatre

While writing his dramas Beckett had a concrete vision of a theatrical production on mind. Beckett was truly a man of the theatre, he often co-directed his plays, introducing changes in the playtext and writing his theatre notebooks which not only specify the details of production but also often shed light on the meaning of the pieces. James Knowlson who edited *The Production Notebook of Samuel Beckett* of *Happy Days* (1985, pp. 19–20) in his notes to the text makes references to a number of different editions of the text and its annotated copies: the Faber and Faber English 1963 edition; the Grove Press 1961 edition; *Samuel Beckett inszeniert Glückliche Tage. Probenprotokoll von Alfred Hübner*, Frankfurt, Suhrkamp Verlag 1976 (1971 Schiller-Theater Werkstadt production); the Editions de Minuit French edition of 1981; an annotated copy of *Happy Days* prepared by Beckett for the 1974-5 National Theatre production; Beckett's production notebook of the Old Vic in London production of 1974-5 directed by Peter Hall, the rehearsals being attended by Beckett; an annotated copy prepared by Beckett for the 1979 Royal Court Theatre production; the Suhrkamp Verlag edition in German, Frankfurt 1968 which was used by Beckett when he directed the play at the Schiller-Theater Werkstatt in Berlin in 1971 and Beckett's production notebook for the production of *Glückliche Tage* at the Schiller-Theater Werkstatt in Berlin in 1971. The published notebook was written by Beckett during the Royal Court 1979 production. As may be noticed there are differences between the published versions of the play and also in the productions. This brings us to yet another problem to be discussed, namely "the legal status of Samuel Beckett's work and associated

[13] Samuel Beckett. *Happy Days*. London: 1963, p. 30.

issues of authority, authorship, and ownership, of quotation, appropriation, and transformation" the investigation of which "is overdue" (Ackerley 2006, p. 309). The entry in the *Companion* continues: "The legal status of the work is intimately bound up with its cultural status, and whatever Samuel Beckett's opinions and those of his Estate one cannot assume that its value is unassailable. Edward Beckett, Samuel Beckett's nephew and literary executor since the death of Jérôme Lindon has legal control over Samuel Beckett's work." His "stance accords with copyright's letter-based approach to textual essence. He has argued: 'There are more than fifteen recordings of Beethoven's late string quartets in the catalogue, every interpretation different, one from the next, but they are all based on the same notes, tonalities, dynamic and tempo markings. We feel justified in asking the same measure of respect for Samuel Beckett's plays'" (Ackerley 2006, p. 311).[14] The entry continues (311): "A text may be followed in letter or in spirit, and the debate is rife with references to Beckett's spirit. The estate claims that the best way to preserve that spirit is by strict observation of Samuel Beckett's text. The goal is to reproduce the 'spirit' unique to Samuel Beckett's work." The critic also stresses the fact that "Ample evidence demonstrates the complexity of Samuel Beckett's texts *even in print*—multiple versions of copyrighted, authorized texts exist" (311). The validity of the last statement has been earlier demonstrated in connection with *Happy Days*, the texts of two language versions written by Beckett himself and his numerous notebooks. It seems, however, that no matter which of these a director chooses he should follow instructions included by Beckett both in the primary and secondary text. The author in this case is really dead but when he was alive he did object to production changes which altered the vision presented by his dramas.

7 Conclusion

The answer to the question concerning the idea of death of the author in reference to Samuel Beckett is provided by Billie Whitelaw (1995, p. 147) in connection with *Footfalls*. For 26 years she worked with Samuel Beckett, acting in his plays which he most often directed or co-directed, she knew his intentions precisely well and was also his friend:

> Beckett didn't write this play for actors to "experiment" with. The plays were, in themselves, experimental. He wanted them to be done as he wrote them. To him, the speaking of the lines was only a small part of the whole work. If you throw out his detailed stage directions, you lose the play. An actor and director should have faith in what's on the page, and that comprises both the text *and* stage directions.

---

[14] *Guardian* 24 March 1994, p. 25.

References

Abramowska, J. 1976. Literatura – dramat – teatr. *Wprowadzenie do nauki o teatrze*. t. 1 – *Dramat – teatr*. Ed. J. Degler. Wrocław: Wydawnictwa Uniwersytetu Wrocławskiego:245-251.

Ackerley, C. J. & S. E. Gontarski, eds. 2006. *The Faber Companion to Samuel Beckett. A Reader's Guide to His Works, Life and Thought*. London: Faber & Faber.

Andrew, D. 2000. Adaptation. In *Film Adaptation*, ed. James Naremore, 28-37. New Brunswick, New Jersey: Rutgers University Press.

Bair, D. 1978. *Samuel Beckett. A Biography*. New York and London: Harcourt Brace Jovanovich.

Bradby, D. 2001. *Beckett. Waiting for Godot. Plays in Production*. Cambridge: Cambridge University Press.

Burke, S. 2008. *The Death and Return of the Author. Criticism and Subjectivity in Barthes, Foucault and Derrida*. Edinburgh: Edinburgh University Press.

Clurman, H. 1972. *On Directing*. New York: The Macmillan Company.

Culler, J. 1981. Presupposition and Intertextuality. *The Pursuit of Signs. Semiotics, Literature, Deconstruction*. London and Henley: Routledge and Kegan Paul:100–118.

Culler, J. 1975. *Structuralist Poetics. Structuralism and Study of Literature*. London: Routledge and Kegan Paul.

Elam, K. 1980. *The Semiotics of Theatre and Drama*. London and New York: Methuen.

Głowiński, M. 1986. O intertekstualności. *Pamiętnik Literacki* no. 4:75–100.

Górski, K. 1976. Literatura i teatr. *Wprowadzenie do nauki o teatrze*. t. 1 – *Dramat – teatr*. Ed. Janusz Wrocław: Wydawnictwa Uniwersytetu Wrocławskiego:233–244.

Harmon, M., ed. 1998. *No Author Better Served. The Correspondence of Samuel Beckett and Alan Schneider*. Cambridge, Massachusetts, London, England: Harvard University Press.

Hassan, I. 1991. Pluralism in Modern Perspective. *The Postmodern Reader*. Ed. Ch. Jencks. London: Academy Editions:196–199.

Heuvel, M. V. 1991.*Performing Drama/Dramatizing Performance. Alternative Theater and the Dramatic Text*. Ann Arbour: The University of Michigan Press.

Hornby, R. 1995. *Script into Performance: a Structuralist Approach*. New York, London: Applause.

Hu, S. 1989. *Tom Stoppard's Stagecraft*. New York: Peter Lang.

Hutcheon, L. 2006. *Theory of Adaptation*. New York, London: Routledge.

Hutnikiewicz, A. 1976. Czy dramat jest dziełem literackim? *Wprowadzenie do nauki o teatrze*. t. 1 – *Dramat – teatr*. Ed. J. Degler. Wrocław: Wydawnictwa Uniwersytetu Wrocławskiego:77–84.

Ingarden, R. 1976. O funkcjach mowy w widowisku teatralnym. *Wprowadzenie do nauki o teatrze, t. I, Dramat – teatr*. Ed. J. Degler. Wrocław: Wydawnictwa Uniwersytetu Wrocławskiego,

Ingarden, R. 1937. *O poznawaniu dzieła literackiego*. Lwów: Wydawnictwo Zakładu Narodowego Imienia Ossolińskich.

Ingarden, R. 1988. *O dziele literackim*. Warszawa: PWN.

Jenny, L. 1988. Strategia formy. Transl. K. and J. Faliccy. *Pamiętnik Literacki* no. 1:265–295.

Kleiner, J. 1976. Istota utworu dramatycznego. *Wprowadzenie do nauki o teatrze*. t. 1 – *Dramat – teatr*. Ed. J. Degler. Wrocław: Wydawnictwa Uniwersytetu Wrocławskiego:47–51.

Kristeva, J. 1980. *Desire in Language. A Semiotic Approach to Literature and Art*. Ed. L. S. Roudiez. Transl. T. Gora, A. Jardine and L. S. Roudiez. Oxford: Basil Blackwell.

Lachmann, R. 1991. Płaszczyzny pojęcia intertekstualności. Transl. M. Lukasiewicz. *Pamiętnik Literacki* vol. 82, no. 4:209–215.

Libera, A. 2009. *Godot i jego cień*. Kraków: Wydawnictwo Znak.

Luere, J., ed. 1994. *Playwright Versus Director. Authorial Intentions and Performance Interpretations. Contributions in Drama and Theatre Studies* no 54. Westport: Greenwood Press.

Pavis, P. 1982. *Languages of the Stage. Essays in the Semiology of the Theatre*. Transl. S. Melrose et. al. New York: Performing Arts Journal Publications.

Pavis, P. 1998. *Słownik terminów teatralnych*. Tłum. S. Świontek. Wrocław – Warszawa – Kraków: Zakład Narodowy imienia Ossolińskich.
Pavis, P. 1992. *The Theatre at the Crossroads of Culture*. London and New York: Routledge.
Pfister, M. 1991. Koncepcje intertekstualności. *Pamiętnik Literacki* vol. 82, no. 4:183–208.
Pfister, M. 1993. *The Theory and Analysis of Drama.* Transl. J. Halliday. Cambridge: Cambridge University Press.
Pfister, M. 2000. *The Theory and Analysis of Drama.* Transl. J. Halliday. Cambridge: Cambridge University Press.
Raszewski, Z. 1976. Partytura teatralna. *Wprowadzenie do nauki o teatrze*. t. 1 – *Dramat – teatr*. Ed. J. Degler. Wrocław: Wydawnictwa Uniwersytetu Wrocławskiego:105–138.
Skwarczyńska, S. 1976a. Dramat – literatura czy teatr? *Wprowadzenie do nauki o teatrze*. t. 1 – *Dramat – teatr*. Ed. J. Degler. Wrocław: Wydawnictwa Uniwersytetu Wrocławskiego:223–232.
Skwarczyńska, S. 1976b. Niektóre praktyczne konsekwencje teatralnej teorii dramatu. *Wprowadzenie do nauki o teatrze*. t. 1 – *Dramat – teatr*. Ed. J. Degler. Wrocław: Wydawnictwa Uniwersytetu Wrocławskiego:139–147.
Skwarczyńska, S. 1976c. Zagadnienie dramatu. *Wprowadzenie do nauki o teatrze*. t. 1 – *Dramat – teatr*. Ed. J. Degler. Wrocław: Wydawnictwa Uniwersytetu Wrocławskiego:53–75.
Sławińska, I. 1976. Główne problemy struktury dramatu. *Wprowadzenie do nauki o teatrze*. t. 1 – *Dramat – teatr*. Ed. J. Degler. Wrocław: Wydawnictwa Uniwersytetu Wrocławskiego:85–104.
Staiger, J. 1989. The Death of the Reader. *The Cinematic Text Methods and Approaches.* Ed. R. B. Palmer. New York: AMS Press:353–367.
Styan, J. L. 1969. *The Elements of Drama*. Cambridge: Cambridge University Press.
Törnkvist, E. 1991. *Transposing Drama. Studies in Representation*. London: MacMillan Education Ltd.
Ubersfeld, A. 1982. The Pleasure of Spectator. *Modern Drama* vol. 25, 1 March:127–139.
Whitelaw, B. 1995. *Who He*? An Authobiography. London: Sceptre.
Williams, R. 1968. *Drama in Performance*. Harmondsworth: Penguin Books.
Wiśniewski, T. 2006. *Kształt literacki dramatu Samuela Becketta*. Kraków: TAiWPN UNIVERSITAS.

From Commodification to Communal Art: "Above Sex… Above Politics"

Ewa Kębłowska-Ławniczak

Abstract The subject of art, often in a self-referential mood, is raised by a considerable collection of modern plays including Tom Stoppard's *Artist Descending a Staircase, Indian Ink, Arcadia, The Invention of Love*, Arnold Wesker's *Their Very Own and Golden City*, Terry Eagleton's *Saint Oscar* or David Storey's poetic *Life Class*. These and other art-oriented plays comprise a broad spectrum of approaches to the justification of artistic activity where the auto/biographical appears to be the most obvious. The aim of the present paper is to explore the manner in which two plays, David Storey's *Life Class* (1974) and—more importantly—Lee Hall's *The Pitman Painters* (2008), hardly experimental, focus "im Zeichen der Kunst" on the way art is capable of reaching beyond the confines of gender, politics and economic commodification governing the practice of art appreciation, examination, observation and even contemplation in search for meaning. The process of disrupting prevalent systems of evaluation, shown in both plays, leads via expressionistic and surrealist concepts of art as, among others, involuntary activity, towards a re-integration of art in Hall's insistence on communal participation and in Storey's "invisible event" and performance. Both, temporally remote, plays can be classified as committed or "engaged."

> *There is another story, a commoner one, according to which Remus, by way of jeering at his brother, jumped over the half-built walls of the new settlement whereupon Romulus killed him in a fit of rage, adding the threat, 'So perish whoever else shall overleap my battlements'. (Livy, I,6).*

E. Kębłowska-Ławniczak (✉)
University of Wrocław, Wrocław, Poland
e-mail: freeway.bohemia@gmail.com

J. Fabiszak et al. (eds.), *Crossroads in Literature and Culture,* Second Language Learning and Teaching, DOI: 10.1007/978-3-642-21994-8_34,

1 Introduction

In an interview for *Financial Times* Lee Hall tells Sarah Hemmings that *The Pitman Painters* (2007) was written "more out of despair than hope" adding that "[it] seemed a suicidally unfashionable play to write—about miners, about nostalgia for socialism, about abstract ideas" (Hemmings 2009 ES[1]).The play asks questions concerning the role culture has to play in the lives of individuals as well as in the contemporary society: in an age of dumb-down, overwhelming commodification and the "product" theatre-makers (like Mark Ravenhill in his play entitled *Product*) increasingly talk about suffering from entrapment in market economy. In his project Hall sets out to examine and celebrate the past by dramatising the real life story of The Ashington Group (1934–1984), of unprofessional painters, in order to imagine a better political future. However, what might have seemed a potentially "terrific piece of genuinely popular theatre" (Allen 2009 ES) developed into a precarious exercise the immediate context of which form such plays as David Storey's *Life Class* (staged 1974) and Arnold Wesker's (1966) *Their Very Own and Golden City* with the real life story of Centre 42. This, perhaps unintended, reference to the roots makes us realise that Lee Hall looks back nostalgically, indeed, trying to revive the once powerful tradition of political theatre—"committed" or "engaged"—so as to re-introduce a sense of *inspiration* through aspiration for High Art. A thus formulated task brings back the conceptual paradigms of political theatre with its strategies: the realistic *reflectionist* on the one hand and the modernist *interventionist* and Brechtian on the other (Patterson 2003, p. 24). Both, though in different ways, conceptualize and fence off dramatic territories. The political language the strategies generate precludes a "natural community" the *un-professional* Ashington Group recognized as their mode of existence. Hence it appears that in *The Pitmen Painters* Hall, somewhat unexpectedly, endeavours to employ paradigms his "material" seeks desperately either to transcend or to circumvent. In some measure, an analogous conflict pervades Storey's *Life Class*, a dramatised *ocular-centric* academic exercise the playwright strives to transform moving away from drama towards performance he terms an "invisible event." The aim of the following is to explore the manner in which Lee Hall, picking up the seemingly unpopular though recently, I believe, returning tradition of political discourse, seeks to address the current feeling of dissatisfaction with the state of culture.

2 Beyond Sex and Politics

In their dreams of New Jerusalem as escape from fragmentation blamed on capitalist methods of production and the resulting social order, both Wesker and Hall turn to William Blake's "Jerusalem" ("Introduction" to *Milton*) although the latter

[1] ES refers to electronic sources where pages are not marked.

playwright, ironically, brings the poet into close connection with the collector and wealthy patroness, Helen Sutherland. With a Blakean image of disintegration and fragmentation[2] in the background (Wesker 1966:34)—implying the dissolution of "universal man"—of "the dreadful and dreary industrial town"(*Their* 24)[3] of Satanic mills, Wesker's protagonist, Andy Cobham, reaches for the poetic Bible to build "Jerusalem in England's green and pleasant land" (*Their* 33). His intervention, following into the poet's footsteps in its curious reliance on capitalist production/profit, assumes a typically architectonic shape in Wesker's play—a complex of Garden Cities—while in the playwright's real life it fails to materialise as Centre 42. Andy Cobham, not a gentleman's architect (*Their* 17) but a Bleakean inspired craftsman (*Their* 15) and a wizard (*Their* 15) appears to be moderately successful in persuading the authorities in favour of his "real community project" (*Their* 56), the six Golden Cities. Launching the idea of free access to High Art, Wesker imagines his cultural centre as a place of contemplation but fails in calculating the consequences. Emphasis on access to art appreciation, only a point of departure for The Ashington Group, remains his goal. As a result, the working class, in Wesker's project, competes for a philistine position among art critics and collectors confirming thus their identity in the constricting mechanisms of production and consumption reducing culture, once again, to product.

The Pitmen Painters, being hardly experimental, fails to stake out new territories in the theatrical world of avant-garde and the postdramatic but, interestingly, crosses the boundaries of territories hitherto defined as political drama by indicating possible transgressions, which enables the playwright to signpost some cultural policy for the future. Storey's *Life Class*, with theatre as a metaphor for community, remains a significant background for Hall's concepts of the collective and the individual.

Access to High Art and education facilitated by The Workers' Educational Association commences the Ashington discussions on culture in *The Pitman Painters*. Indeed, the historical WEA, founded in 1903, organised extra-mural lectures and courses for adults interested in learning, for example, music, drama, elocution, geology or experimental evolution. By showing a group of pitmen and a dental mechanic from Ashington who, having completed courses in "[b]iology, geology and the rest of it" (Feaver 2009, p. 15), decide to start Art, Hall tells a real-life story based on Wiliam Feaver's historical account. As Feaver and Hall tell us, the course in Art Appreciation ordered by a group of almost 30 workers was conducted, starting with 1934, by Robert Lyon, Master of Painting at Armstrong College, Durham University. According to education norms, the course—also in the theatrical adaptation of the story—becomes a series of regular lectures in art history where a show of slides illustrates methods of examination and search for "meaning" in the paintings (Hall 2008:14). Initially disappointing to its participants, the course undergoes a rapid evolution as the students decide that meaning resides in the

[2] Arnold Wesker, *Fears of Fragmentation* (1970:110).

[3] Abbreviation used in reference to Arnold Wesker's *Their Very Own and Golden City*.

onlooker rather than in the object, that is the canvass. Further on, going beyond a professional critic's detached observation, beyond *mimesis* (*PP* 27)[4] and academic examination (*PP* 15), the workers decide that meaning "happens" in their heads, either in the conscious or in the subconscious (*PP* 25). The passive lecture form, with its rigid boundary between stage and audience, teacher and student, gives way to a communal exercise of dispute. Soon the group decides that, following Marxist injunctions, to understand "'why' an artist makes the choices he does"(*PP* 16), to acquire an "insider's" view, they have to start painting. Harry, one of the participants, explains the necessity by adapting rigid political discourse:

> **Harry** Praxis. Marx, isn't it? 'Hitherto, the world has been interpreted as an object of contemplation, not as sensuous human activity,' *Theses on Feuerbach*, isn't it? It's ne good just looking at the world, you have to change it. Theory in action. Practico-critical activity. (*PP* 17)

Though the WEA regulations considered unconstitutional any activity "which could be used for financial gain or practical employment" (*PP* 17), Robert Lyon persuades the regional committee that the works of the students would provide material for discussion. Indeed, the subjects given by Lyon were "straight from the teacher's manual" and had no intention to either encourage depiction of Ashington life or stimulate "proletarian art" (Feaver 2009, p. 23). Interestingly, the reversal of roles between Lyon and his students had an additional practical dimension: the experience of teaching pitmen helped Lyon in the completion of his MA on *The Appreciation of Art Through the Visual and Practical Approach* and ultimately earned him the future job. Except for Harry Wilson's, one of the painters-characters, distinctly Marxist voice, Hall captures the uniqueness of a group of working pitmen focused on self-improvement, the result being the workers' aesthetically minded, not "proletarian" art. Hall avoids the common preconceptions of life in a pit village Feaver defines as the "Lawrentian conceptions of the 'primitive'" or the "Orwellian images of hard graft"(Feaver 2009, p. 74), where the Ashington group might easily become a victim of what used to be termed "Britain in Pictures," "topographical artists," "archetypal English communities" or "Art by the Miner" genre. Going beyond these paradigms of cultural politics, conceptual territories within which the miners, as envoys from a nether culture, could be safely accommodated, the Ashington Group shifts swiftly from art appreciation to art *appropriation* transforming and familiarising the "manual" subjects as in, for example, *The Deluge* by Oliver Kilbourne:

> I started after the foreshift, I took the paints and that, but I had no Idea what I was supposed to be doing. I just looked at this blank bit of card. I didn't knaa what a deluge was supposed to look like. I mean, I have seen a storm – but a deluge. And I was completely struck, you know. So I just started painting these lines. I wasn't thinking – I was just painting. Throwing on lines just to de something – and then it all clicked, and it just came pouring out – all this feeling – the little houses being swept away, the teeming water – it all just came out. And when I stopped to look what I have done, suddenly I

[4] Abbreviation used in reference to *The Pitmen Painters* by Lee Hall.

> realised it was light – it was morning – time for work – I thought it'd been an hour or something – I'd been on the whole night. And I was shaking –literally shaking - 'cos for the first time in my life, I'd really achieved something – I had made something that was mine – not for someone else – not for money – not for anything really. And I felt like for those few hours there – I was me own boss. (*PP* 28)

The Biblical subject of the deluge remains alien to and confuses William Scott, called here "Jimmy," so he confesses in the play that although he "started deeing a deluge [...] it turned into a dog", a "Bedlington terrier" (*PP* 28–29),[5] the painting which became a signboard of the Group. Janet Adam Smith, Arts Editor on the *Listener*, interviewed Scott when she was writing an article on the Ashington phenomenon. Having listened to a simple explanation of why all the details were present on canvass, including the dog, she heard that "Miners are keen on Bedlingtons, so the terrier must be painted in the mining village," which confirmed the impression of originality and independence.[6]

The course, as the audience notices, promptly abandons the historical, mythological and iconographic frames to concentrate on the paintings and drawings judged "on their own merits." In what is probably the best part of *The Pitman Painters*, Hall produces a series of scenes dramatising the process of self-discovery within the group (*PP* 35). The discussions converge and spring from the participants' individual, but also significantly communal, experience of painting, their encounters with realism, avant-garde concepts, expressionism and abstract art—the famous "blob"—triggered off by the diversity of ways they understood or misunderstood the decisions made by other artists as well as a recurrent though abstract insistence on a political message (*PP* 32–33). In the process of purification, leaving behind the subsequent conceptual crutches, the painters arrive at understanding art as relative relevance, as meaning located democratically "with whoever's looking" (*PP* 33), which brings us back to Oscar Wilde's proposition that criticism is "a mode of autobiography" (Wilde 2008, p. 2).

Hall's, but also Feaver's, account of the pitmen's progress foregrounds the unique cultivation of *communitas* "integrating the study of art with [their] life" (*PP* 35-6). Still, the personal discoveries of Oliver Kilbourne soon reveal the constricting social and cultural context of institutions exerting pressure on the unprofessional artists: commissions, sponsorship, collectors and critics. The interest of Helen Sutherland, a patroness and art collector, is used by Hall to reveal the politically utopian and otherworldly existence of the group. Although the painters begin to exhibit and to sell their work, the profit is shared by the group to cover the costs of material and other activities, for example travel. Kilbourne, exceptionally talented, receives the offer of a stipend. Still, he rejects it even though he is strongly persuaded by Sutherland to accept the opportunity in order to give up his regular job of a miner (*PP* 62–63) and thus to become an individual,

---

[5] William Scott painted *The Bedlington Terrier* in 1937 and it was his only "Ashington Group" picture. Lee Hall calls him "Jimmy," perhaps to bring onto the stage a more prolific member of the group, James ("Jimmy") Floyd, the author of *Pigeon Creese* (1938) and *The Onsetter* (1942).

[6] *The Listener*, 28 April 1937. *The Listener* was the weekly Published by the BBC.

contracted artist like Ben Nicholson. The latter, it appears, is well aware of his own painful dependence on sponsorship and market. The "tabled artist" warns Kilbourne that he should "avoid all comestibles." Nicholson misses freedom from patrons, public opinion and press the pitmen still enjoy (*PP* 89). Soon it turns out that though Sutherland introduces herself as more of an enthusiast (hence perhaps her connection with Blake) than a collector (*PP* 42), her purchase of their work boils down to an attempt to place the *unprofessional* in the order of her and other collections. In the order of exhibition a painting does not mean anything "specific in the traditional sense" but –as Helen admits—it "is" (*PP* 55). In her case it is an illusory autonomy of an exhibited object whose value is commercially defined. "Jimmy," here the historical Jimmy Floyd, appears in the collection side by side with *her* Henry Moore's *A Woman Reclining* (1933) and Ben Nicholson's *White Relief* (1935), as an aspiring artist of the promising *other*. It is ultimately the meaning-less white canvass that costs £200 (*PP* 55) and thus reminds us of discussions on value in Yasmin Reza's *Art*. The gradually revealed mercantile interests of the collector encourage her to switch to "lovely" philistine ceramics (*PP* 102), to praise the Nicholsons that start to "fetch some handsome prices"(*PP* 105) and to criticise Kilbourne, the pitmen painter, who remains indifferent to the market demand for sex, women and desire the critic finds even in Henry Moore's "good grasp of hips"(*PP* 54). Unable to find the current "developments" in the work of the Group, Sutherland discards their work for a lack of sentiment, absence of sex, women, sensuality and, generally, desire (*PP* 104), which add to the market value.

Indeed, the group refuses, on what seem to be both moral and political grounds (for instance, exploitation), to study "the human form" (*PP* 40) in a life class. Allott, a teacher in David Storey's play explains pertinently the frequently noted futility and difficulty of the exercise as the students "are not [there] to seek sexual stimulation [but] "to peruse a beautiful and seemingly mysterious object" (Storey 1998:218–219). The academic "event", a "special treat" for Lyon (*LC* 39)[7], which employs a traditional theatrical regime transforming the "artists" into regular critics and the posing sitter into object is dismantled in both Hall's and Storey's projects. In *Life Class* the interventionist disruption of the academic rules rejects criticism, that is "contemplation" or "appreciation" masquerading as and confused with artistic "meditation" (*LC* 215). Additionally, in more explicit terms *Life Class* rejects sets of principles ensuing from academic practice: geometrical and gender entrapment (*LC* 218), focus on technique (*LC* 215), idealization (*LC* 212) and, ultimately, the whole of the visible. A radically inquisitive study of "human form" is ridiculed in Foley's, a character in *Life Class*, heap of broken statues and an avant-garde collection of his own urine in bottles (*LC* 247). From a traditional theatrical regime of the fourth wall and proscenium arch, Storey moves violently, staging a mock rape, towards a liberating performance he calls "invisible event" on which, however, he bitterly reflects: "How do you sell an event that no one will admit is taking place?" (*LC* 244). Violation of rules and values

[7] Abbreviation used in reference to *Life Class* by David Storey.

becomes, in *Life Class*, a "prerequisite" of art (*LC* 279) while art itself remains indefinable (*LC* 270), a form of celebration, initiation, and evocation (*LC* 234). Once again, in Allott's words, art "is wholly without meaning... it exists, merely, because it is" (*LC* 215). Hence it is not the question of market value. What Storey stages is an alchemical process of removing impurities—including politicization and commercialization—to revive the concept of art as an aesthetically conceived purposeless being (*LC* 216). Although the proposed concept of High Art and theatre violates the prevailing values, it distinguishes between the "real" artist who sings his song and the "*manufacturer* of events who, in his twentieth-century romantic role, sees art as something accessible to all and therefore the prerogative not of the artist—but of anybody who cares to pick up a brush, a bag of cement, an acetylene welder... anyone, in fact, who can persuade other people that what he is doing is creative" (*LC* 215). This formulation of artistic territory excludes, at least to some extent, the Ashington Group.

Hall marginalizes the concepts Storey's life class scene poses as metaphorically central. In a turn from Marxism to metaphysics, with Van Gogh in the background, in *The Pitman Painters* Art remains shared, communal experience excluding individual privilege and ownership: "Art doesn't really belong to anybody," says Harry; "Not the Artist," adds Jimmy (*PP* 72). Giving a lecture at the first exhibition of working-class painters, Lyon argues that talent is discovered once someone is given a paintbrush. Further on, painting is defined as "experience," "a vision of the everyday"(*PP* 71), "a gift" (*PP* 72) and "a force. The force that flowed through the flower—flowed through us" (*PP* 73). The pitmen artists openly refer to "spirituality," "religion or creativity or energy," "inspiration" in the etymological, medieval sense of "breathing life"(*PP* 73), suggesting and prompting (for example, Dante Alighieri). The climactic scene in act 2 becomes quasi-religious: a descent of creative spirit on the disciples who become an in-spired medium rather than a subject:

> The slide show is finished. The group are caught in the 'headlights' of the slide projector. They are looking out into the audience. In contrast to their high-flown rhetoric, they look vulnerable. They stare out uncertainly. (*PP* 75)

An analogously "inspired" group of unprofessional painters and miners was active in Upper Silesia in the years 1920–1960 and included the famous Teofil Ociepka and Erwin Sówka. They called themselves unprofessional painters and belonged to the occult circle of Janów. Unclassifiable as worker-painters, they created an oasis immune to both totalitarian and communist regimes. The film *Angelus*, directed by Lech Majewski, like Lee Hall's play, derives from a need to defend the poetic, metaphysical sensibility against various forms of materialism.

3 Conclusion

What follows in act 3 of *The Pitman Painters* is a disappointing dissolution of the previously formulated idea and a physical dispersal of the engagement the Ashington Group has promised. Exhibited and called "proper art" the artists-suddenly began to fear being patronised (*PP* 103), redefined (*PP* 92), eulogized (*PP* 109–10), exploited (*PP* 112), and turned into an object of academic analysis in the form of a dissertation (*PP* 108). The fear of discursive fragmentation they experience is additionally fuelled by a false desire for individualism which encourages them to contemplate their own art instead of practising it so that they become their own critics in search of the meaningless avant-garde blob they have previously rejected. To preserve the communal status they consider being useful and political (*PP* 119–121) and begin to write down rules conceptualising their territory—the *organisation* of the group. This marks an end of their flight into metaphysics encouraging them to fence off rather than jump walls.

References

Allen, Paul. 2009. Northern lights: How theatre undermined the coal industry. 16 September 2009. *Guardian Theatre blog*. http://www.guardian.co.uk/stage/theatreblog+series/northern-lights. Accessed 17 October 2009.

Feaver, William.2009. *Pitmen Painters. The Ashington Group 1934–1984.*Ashington: Ashington Group Trustees.

Hall, Lee.2008. *The Pitmen Painters.* London: Faber and Faber.

Hemmings, Sarah. 2009. *Financial Times* (January 10). http://www.ft.com/home/europe. Accessed 2 March 2010.

Patterson, Michael. 2003. *Strategies of political Theatre. Post-War British Playwrights.* Cambridge: Cambridge University Press.

Storey, David. 1998 [1975]. *Life Class. In: Plays. 3.* London: Methuen, 185-280.

Wesker, Arnold.1966. *Their very Own and Golden City.* London: Jonathan Cape.

Wesker, Arnold. 1970. *Fears of Fragmentation.* London: Jonathan Cape.

Wilde, Oscar. 2008. "Preface" to *The Picture of Dorian Grey.* http://www.gutenberg.org/catalog/world/readfile?fk_files=782464&pageno=2. Accessed 10 June 2010.

Part VIII
Countries and Cultures at Crossroads

The Subversive Power of Father Matthias. The Poetry of Maciej Kazimierz Sarbiewski as Vehicle for Political Propaganda in England of the 17th Century

Krzysztof Fordoński

Abstract Maciej Kazimierz Sarbiewski (1595–1640) was considered a pious Jesuit father, neo-Latin poet, equal if not better than Horace. However, when almost immediately after his death his poetry crossed the English channel, it gained an unexpected subversive power. It gained a new territory not so much for what it actually was but thanks to a new purpose for which it could be used. In Great Britain controlled by the forces of the Parliament acquaintance with the works of Sarbiewski became a telling sign for the Royalists. Consequently, several Royalist poets started to write and publish translations from Sarbiewski which departed from the originals in such ways which allowed the poets to express their true sentiments and bypass Parliamentary censorship. Others would quote excerpts from his poems in their original works for similar purposes. The present paper traces the way Sarbiewski's poems were used—translated, adapted, quoted, emulated etc.—by such Metaphysical poets as Richard Lovelace, Henry Vaughan, Abraham Cowley, and Sir John Denham, to mention but a few. It also presents an analysis of the most important testimony to Sarbiewski's popularity in the days of the War of Three Kingdoms and the Commonwealth period, a volume of translations by George Hils.

1 Introduction

In the two decades immediately following his death in 1640, the poetry of Maciej Kazimierz Sarbiewski, known on the Continent as Matthias Casimir Sarbievius and in Great Britain more likely simply as Casimir, gained an immense popularity

K. Fordoński (✉)
University of Warsaw, Warsaw, Poland
e-mail: k.fordonski@uw.edu.pl

J. Fabiszak et al. (eds.), *Crossroads in Literature and Culture,* Second Language Learning and Teaching, DOI: 10.1007/978-3-642-21994-8_35,

in the British Isles which was to last intermittently for almost two more centuries. Due to the complex political and social situation in the British Isles as well as due to several of its inherent qualities, the poetry was used in a number of ways to oppose and subvert dominating political discourse, reaching thus a territory which seemed completely alien to a Jesuit poet. A number of English poets chose either to express their political opinions in the disguise of translations or to seek consolation after the defeat in the poetry of the Polish Jesuit. The two aims combine in the translations produced after 1649 when the very need for consolation sought in neo-stoic poetry after the King's beheading quite obviously revealed the political position of the translators.

The present study is more likely to offer for consideration a number of cases rather than a coherent vision of the greatly varied phenomenon. Two elements of this variety should be mentioned here. The first of them are differences among the translators and their attitudes both to Sarbiewski and current situation. Different attitudes can be expected from soldiers fighting a still victorious war and from authors seeking a way of expressing their grief after the death of their monarch, still different ones from those who chose exile—either retiring to the country or going abroad. The second element and a source of problems is the circulation of translations in the period. The translations were either published in books or circulated in manuscripts. Some of the latter were prepared for print at a later date, some survived in hand-written commonplace books. However, many either were not published or appeared in numerous pamphlets which have only partially been researched yet. This makes an analysis quite difficult and currently existing attempts of approaching the Royalist publications of the War of the Three Kingdoms and the Commonwealth period miss the presence of Sarbiewski almost entirely (cf. Potter 1989). Sheer amount of material left by the so called manuscript culture flourishing in the 17th century, which has quite recently come under the scrutiny of scholars, makes a thorough search for texts which in a variety of ways (translations, adaptations, emulations, allusions) refer to Sarbiewski's oeuvre an almost impossible task. Consequently, the present article must concentrate on texts available in published form even if they may be only a fraction of all those created.

2 Maciej Kazimierz Sarbiewski and England

Even though the main hero of the present paper was once so famous that he was recognizable by his middle name alone (sometimes also as the divine Casimire) it seems quite necessary to start this presentation from a short biographical introduction.

2.1 The Life and Work of Sarbiewski

Maciej Kazimierz Sarbiewski (1595–1640) was considered one of the greatest Latin poets of his time. Pope Urban VIII awarded him with the poetic laurels and a gold medal in 1624. Sarbiewski was the author of 129 odes collected in four books of his *Lyricorum* (first published in three books in 1625 in Cologne, complete edition of four books in 1632 in Antwerp) just as the odes of Horace, 12 epodes, and numerous epigrams. He also wrote an epic poem *Lechias*, an imitation of *Jerusalem Delivered* by Torquato Tasso. His odes, characterized by erudition and a deep understanding of especially ancient poetry, were written under a clear influence of Horace. Many of them are paraphrases or parodies, as they were called then, of Horace's poems. Some take over Horatian forms, other continue Horatian motives often adapted to express Christian themes as Sarbiewski was a Jesuit. The Dutch humanist Hugo Grotius claimed that Sarbiewski not only equalled but surpassed Horace in his poetry, he was consequently known on the Continent as the Christian or the Sarmatian Horace.

2.2 Sarbiewski's Reception in England

England was the country where Sarbiewski enjoyed the earliest and the greatest popularity as measured by the number of known translations. In the period from 1644 to the mid-1800s there were at least fifty translators (it is impossible to give a more precise number as some translations are anonymous) who produced over 150 translations of 55 various odes and epodes.[1] Approximately one third of those translations were written in the period under discussion. Curiously enough, the fashion for Sarbiewski's poetry is largely unknown among British scholars as even such an important authority on history of translation as Robert Cummings claims that after the 1646 collection of George Hils Sarbiewski "is steadily represented, but with increasing embarrassment, evidenced by the focus on a limited number of poems" (2005, p. 500). Two out of five ratio may not seem very impressive but once we count out three groups of poems—Marian religious poetry, poems addressed to various Roman Catholic religious and political leaders as well as fellow Jesuits, and those commenting on current events of the 1620s, mostly connected with Polish wars against Turkey, which for various reasons could hardly appeal to an English reader, it seems that almost every poem of Sarbiewski which could find an audience in Great Britain was actually repeatedly translated.

[1] The translations have been collected in the volume *Casimir Britannicus: English Translations, Paraphrases, and Emulations of Maciej Kazimierz Sarbiewski*. Fordoński, K. and Urbański, P., eds. London: MHRA. The first edition was published in 2008, a new one, expanded and corrected, is expected in 2010

This popularity was brought about by three reasons. The first of them was his fascination with Horace's poetry quickly taken over by English poets of the mid-17th century, replacing to a degree an earlier interest in poetry fashioned after Pindar. The second reason was that his works were immediately admitted into the internationally established canon of Latin poetry, joining the works of Horace, Vergil, and Ovid, or, more locally, Milton and Buchanan (Kedslie McLeod 2004, p. 41). Ultimately, Sarbiewski's poetry was for the English a means of getting acquainted with the contemporary intellectual fashions of the Continent. It was filled by influences from a variety of different circles of cultural tradition such as Stoicism, Platonism, Hermeticism etc. (Urbański 2000, p. 187).

The most important of the latter was in the period under discussion Sarbiewski's interest in Stoicism which saturates a large part of his oeuvre. This philosophical current had been present in English literature from the early Elizabethan period, most notably through plays such as those of Seneca (cf. Monsarrat 1984, pp. 129–150), although it was never a dominating trend. In philosophy it was visible in the ongoing interest in the works of Cicero and Boethius. Neo-Stoicism propagated by contemporary continental thinkers such as Justus Lipsius and Guillaume du Vair reached England almost immediately. Lipsius' *De Constantia* published in 1584 was translated into English first in 1594 and three more times within the following century 1653, 1654, and 1670 (Urbański 2000, p. 194). Du Vair's works enjoyed a similar popularity to mention only a translation by Charles Cotton from 1664. Its political message was originally embraced by the supporters of the Parliament, the fortunes of the war, however, made the royalists more and more interested in the virtue of *Constantia* propagated by the ancient authors. Sarbiewski who was well-versed in both ancient and contemporary Stoicism was a natural choice in their literary pursuits. His poetry in the original offered solace, as a material for translation could be used for expression of hopes and intentions which could less and less be voiced as original poetry.

3 Translations During the War of Three Kingdoms

The two following examples of translations from Sarbiewski—a single poem translated by Richard Lovelace and a collection by George Hils—can be dated with some certainty as coming from the period of the First Civil War. Yet even if the dating may be doubted they share one common element which is absent from later translations—hope that the events can still bring about the success of the royalist party.

3.1 *Richard Lovelace and Lyr. IV 13*[2]

The first to use Sarbiewski in this way was probably Richard Lovelace (1618–1657) a Metaphysical poet, playwright, and soldier. In his lifetime he prepared for publication only one volume of poetry *Lucasta* in which he included, among others, his translation of Sarbiewski's Lyr. IV 13 entitled 'To his Dear Brother Colonel F. L. immoderately mourning my Brother's untimely Death at Carmarthen' (1649, pp. 110–111). Susan A. Clarke (2005, p. 125) argues that the translation was written immediately after the battle of Carmarthen in 1644 in which William Lovelace was killed. If we accept her theory, this is the earliest English translation of any poem by Sarbiewski.

The translation follows the original rather loosely, after the contemporary fashion. The biggest change comes at the very end. The original consists of eighteen lines which Lovelace splits into five quatrains and by adding two lines he gains "a fifth stanza [which personalizes] the 'Ode', counseling his brother to abjure sorrow in favour of victory on the battlefield" (Clarke 2005, P. 125).

But this way you may gain the field.
Oppose but sorrow and 'twill yield;
One gallant thorough-made Resolve
Doth Starry Influence dissolve.
(Lovelace 1649, P. 111)

The poem naturally first circulated in manuscripts yet the disguise of translation made it publishable in 1649 when Parliamentary censorship found nothing wrong with the alleged translation.

3.2 *George Hils and the Odes of Casimire (1646)*

The same Royalist sentiment saturates the collection of George Hils who seems especially willing to act out foreign identities and voice subversive opinions in his work. Very little is known about George Hils (1606–1655), apart from a volume of translations from Sarbiewski entitled *The Odes of Casimire* (1646) his literary output is limited to three liminary verses (two in English and one in Latin) included in editions of *Poems and* ca. by James Shirley (1646), and *Comedies and Tragedies* by Francis Beaumont and John Fletcher (1647) (Urbański 2000, pp. 204–209; Money 2006, pp. 160–174).

The feeling that the collection of twenty-nine odes and six epigrams, the largest ever published, is not exactly what it purports to be, appears first when we look at the frontispiece engraved by William Marshall which allegedly shows Sarbiewski

---

[2] Poems of Sarbiewski are usually referred to by their numbers and not their long Baroque titles—Lyr. IV 13 stands for the 13th ode of the 4th book of *Lyricorum libri IV*.

sitting on a rock and facing Horace. There is seemingly nothing wrong with the former dressed in fur (stressing his Eastern European provenance) but for the fact that Marshall's Casimire bears a striking resemblance to King Charles I. This resemblance may not be coincidental, as Marshall was well-known for his involvement in royalist publishing, most notably as the engraver of the frontispiece to *Eikon Basilike*, the King's Book of 1649 (Clarke 2005, p. 122).

If this is not enough to make a prospective reader wary (and it should be noted that very few people could actually know what Sarbiewski had looked like) in the English preface (there is also a Latin one) Hils writes:

> I confesse I have not been so precisely careful in every Ode, as to render line for line (a thing so strictly stood upon by some late translators) for indeed the exuberante torrent of Elegancie came so fast upon me, that I was forced to make my banks larger; choosing rather to make fault in the excesse then defect (Sarbiewski 1646, unnumbered page).

Changes introduced by Hils are clearly visible from the very beginning of the volume. Sarbiewski's Lyr. I 1, which opens the collection, translated as 'When the hatefull forces of the Thracians departed out of Pannonia'. The poem sets the royalist tone of a book published in the year the first Civil War ended. In the original the ode is a panegyric poem addressed to the pope Urban VIII which commemorates Hapsburg victory over Gabor Bethlen of Transylvania, Hils turns it into a message of hope addressed to royalists after their failure in the first civil war:

> The threats of cruell Warre now cease:
> In stead of them safety and peace,
> Banish'd th'unhallowed earth, doe please
> 'Returne in their white Waine.
> (Sarbiewski 1646, p. 3)

White was the Stuart colour and Charles I, who famously wore white coronation robes, was known popularly as the 'White King'. From his youth, he was associated with 'Charles his Wain', the constellation now known as the Plough or Ursa Major (Clarke 2005, p. 123). To add to poignancy of the passage the first line of the original is quoted from Seneca's tragedy *Tyestes* describing a civil war between kinsmen. It is a taken from a speech following a truce which should soon be broken when one cousin serves the other his own children as the main course during celebratory feast (Urbański 2000, pp. 97–114; Buszewicz 2006, p. 221).

The limits of space force me to move on without providing further examples of such alterations. They are the more striking as the volume is actually a bilingual one. Any educated reader (we may just say "any reader" bearing in mind the place the teaching of Latin held in Great Britain and especially among the target audience) could see clearly where Hils decided against following the original closely and where he got carried away by the "exuberante torrent of Elegancie" and his political views. Any reader acquainted with Sarbiewski's oeuvre could also tell that Hils pruned the original collection down to the neo-stoic poems only, removing any others.

4 Translations of Sarbiewski During the Commonwealth

The death of the king did not put an end to an interest in Sarbiewski's works which, however, for a while continued to be limited almost exclusively to the royalists. Hils was followed by other poets translating or alluding to Sarbiewski (sometimes through Hils' translation as is the case with Edward Benlowes' *Theophilia*), "indicating that Casimire had indeed achieved status as an accepted voice of royalism" (Clarke 2005, pp. 123–124). Even though some of them might have been written earlier (some of the dated poems included in Vaughan's *Olor Iscanus* come from the 1640s, however, none of the translations from Sarbiewski is dated) what they share is an altered mood. Former hope is replaced with the neo-stoic virtue *Constantia*.

4.1 Translations from the Volume Lachrimae Musarum

Striking examples of Sarbiewski subversive presence can be found in the volume *Lachrimae Musarum* from 1649. It is a volume of thirty-five elegies by various authors in English and Latin commemorating the death of the nineteen year-old Henry lord Hastings. The volume is remembered (if at all) because it includes the poetic debut of John Dryden and one of very few works of Andrew Marvell published in his lifetime. Its actual purpose was quite obviously rather different, the true hero of the elegies is the late king Charles I.

The volume includes two partial translations of the same poem (Lyr. IV 13 formerly translated by Richard Lovelace) by Sir John Denham and John Hall of Durham. Sir John Denham (1615–1669) was a poet, translator, courtier, and gambler, a fierce supporter of Charles I and the Royalist party. He is most often remembered for his descriptive poem *Cooper's Hill* (1642), long considered as one of the most famous poems in the English language, and one which uses a descriptive technique adapted from Sarbiewski's ep. 1 (Urbański 2000, p. 190). John Hall of Durham (1627–1656) was a minor poet and translator, the author of numerous short poems included in such volumes as *Poems* (1646) and translations such as *The Height of Eloquence* by Longinus.

Each of the two quoted elegies included in the volume opens with a passage paraphrased (without attribution) from Sarbiewski's Lyr. IV 13, in both cases the remaining parts of the poems do not resemble Sarbiewski's work. The passages thus include only one poetic image taken from Sarbiewski (an offer of paying with a pearl for every tear should it be proven that tears fight off grief) yet their purpose is not quite exclusively literary. As such, Denham's and Hall's contributions represent a particular kind of literary borrowing, one identified by Lois Potter as designed to generate and retain a coherent sense of royalist identity in the face of extreme pressure (Clarke 2005, p. 113).

Popularity of certain poems (Lyr. IV 13 translated by Lovelace, Hils, Denham, Hall, and Vaughan, Lyr. IV 28 by Sherburne and Vaughan, to mention only those quoted before) should not escape our attention. It is highly improbable that during the Civil War England imported large amounts of Sarbiewski's volumes of poetry (although the number of copies available today in library collections suggests that there were quite a few, Urbanski 2000, p. 203). Until 1684 when the first known English edition of Sarbiewski's poems published in the original appeared, access to his works was possible mainly within the manuscript culture. Consequently, rather few readers could access his complete works, most knew those which others had found interesting enough to copy and pass on. Quite obviously, the neo-stoic poems attracted more attention and especially authors living away from London (e.g. Vaughan and Sherburne discussed below) could know and translate only such works.

4.2 The Translations of Henry Vaughan

Henry Vaughan (1621–1695) was probably the most eminent of the translators of Sarbiewski in the period under discussion. Vaughan's studies ended when the Civil War broke out, he joined the royalist army and fought in the battle of Rowton Heath (1645). After the King's defeat Vaughan returned permanently to his native Wales where he began to practice quite successfully as a doctor although he had neither studied medicine nor ever claimed to have done so. His third volume of poetry, *Olor Iscanus* published in 1651, includes a selection of seven translations from Sarbiewski as well as the works of Juvenal, Boethius, Horace, and Ovid's *Tristia*. The common theme of all the translated pieces is that of banishment or retirement, suggesting that they should be read as a poignant political statement in the days immediately following the defeat of the royalists and execution of Charles I.

Vaughan makes this statement clear mostly through his selection of Sarbiewski's poems and not through the introduction of new elements, although he does domesticate and otherwise alter them to some degree. David Money (2006, p. 165) suggests that the language of Vaughan's translations is different from that of the originals, it is "the language of contemporary politics" an example of which is the use of the word 'prerogative' instead of 'power' in Lyr. II 8 (Vaughan 1651, p. 56), a word which appeared often in the debates concerning the mutual relations between the King and his Parliament in the early 1640s.

4.3 The Translations of Sir Edward Sherburne

Another royalist poet who took to translating Sarbiewski during an enforced retirement was Sir Edward Sherburne (1618–1702), a minor Metaphysical poet and translator. Just as Vaughan, Sherburne took an active part in the Civil War supporting the King, he fought in the battle of Edgehill. Following the execution of Charles I in 1649, Sherburne retired to his relatives' country estates and it was there that he studied and translated French, Italian, and Neo-Latin poets. The effect of this work was the volume of *Poems and Translations. Amorous, Lusory, Morall, Divine* (1651) which includes a selection of seven translations from Sarbiewski among which we find Lyr. IV 28 formerly translated by Vaughan.

Just as the formerly quoted translators and after the contemporary fashion Sherburne treats the originals only with moderate attention, adapting them to his taste as well as political and religious views:

To the Eternal Wisdom Upon the Distraction of the Times
O Thou Eternal Mind! Whose Wisdom sees,
And rules our Changes by unchanged Decrees,
As with Delight on thy grave Works We look,
Say; art thou too with our light Follies took?
...
O whilst for Wealthy Spoils these fight, let Me,
Though poor, enjoy a happy Peace with Thee.
(Sherburne 1651, p. 161)

The translation retains most of the qualities of the original, there are, however, two notable changes. The Eternal Wisdom is for Sherborne not the source of beauty but of decrees which rule human life. The original ends in a phrase "tecum ego rideam" ("I shall laugh with you") which changes into a plea for a "happy Peace". It is quite striking that Vaughan's translation includes almost identical changes:

But o good God!
While these for dust fight, and a Clod,
Grant that poor I may smile, and be
At rest, and perfect peace with thee.
(Vaughan 1651, p. 55)

4.4 *Émigré reads Sarbiewski: Abraham Cowley*

Abraham Cowley (1618–1667), one of the most famous English poets of his age, read Sarbiewski rather as an inspiration for his own odes which he originally fashioned after Pindar and only through Sarbiewski he came to know and adapt the Horatian model. In his only surviving translation of Sarbiewski ("The Extasie" after Lyr. II 5) he could not help altering the original ever so slightly by the addition of a whole new stanza:

Where shall I find the noble British Land?
Lo, I at last a Northern Spec espy,

Which in the Sea does lie,
And seems a Grain of th' Sand!

For this will any sin, or Bleed?
Of Civil Wars is this the Meed?
And is it this, alas, which we

(Oh Irony of Words!) do call Great Britanie?
(Cowley 1668, p. 41)

Cowley's attitude to translation as such is quite telling if we try to evaluate the discussed works. It is highly creative, his aim is to avoid slavish imitation but to write as the translated poet would write in the translator's times, country, and language (Urbański 2000, p. 188). This attitude was quite generally accepted and followed, and as a result, the discussed translators were all too ready to go one step further and make the following unspoken statement: had Sarbiewski lived in our times in England he would quite obviously make our sentiments more clear in his works.

5 Conclusion

Cowley's translation marks the end of Sarbiewski's "subversive political involvement" in the Wars of the Three Kingdoms. It should be noted here that also in the discussed period the role of Sarbiewski's poetry was not limited to his poems being a vehicle for propaganda. As it was suggested above, he was also a model followed in the emergence of the English ode as well as in the creation of the descriptive poem as a genre. After 1660 the general interest moved to less politically and more religiously charged of his works of which the manuscript collection of translations of Sir Philip Wodehouse (1608–1681), a politician and a poet, is an ample testimony. The collection preserved in the library of the University of Leeds includes translations of Sarbiewski's epigrams dating probably from the 1660s or the 1670s. Translations of Lucy Hutchinson, widow of an important Parliamentary officer colonel John Hutchinson (also surviving in her commonplace book, now in Nottinghamshire Record Office), suggest that in the late Commonwealth period the political value of Sarbiewski was gradually replaced by a more general popularity which was soon proven by translations of the Restoration period prepared by such authors as Lady Mary Chudleigh, John Norris, or Thomas Brown.

Sarbiewski's subversive power was present in three ways—the first was his very presence as a source of inspiration of either translations or allusions, it was enough to choose Sarbiewski as the material for translation to make a clear statement concerning one's political views. The second way was through the selection of texts used with a special stress on their neo-stoic undertones. Ultimately, the third way was using Sarbiewski's original poems as point of departure for alterations which brought the poems much closer to expressing the translators' political views rather than those of the distinguished Polish Jesuit. When the views the English translators presented in the disguise of translations once again became the dominant discourse and the need for such a disguise was gone, Sarbiewski temporarily disappeared from the English literary scene, only to come back in grand style when his poetry was rediscovered by the religious dissenters in the early 18th century.

References

Buszewicz, E. 2006. *Sarmacki Horacy i jego liryka*. Kraków: Księgarnia Akademicka.

Clarke, S. A. 2005. Royalists Write the Death of Lord Hastings: Post-Regicide Funerary Propaganda, *Parergon* 22(2): 113–130.

Cowley, A 1668 *Pindarique Odes, Written in Imitation of the Stile & Manner of the Odes of Pindar*. London: Henry Herringman.

Cummings, R. 2005. Post-Classical Latin Literature. In *The Oxford History of Literary Translation in English. Vol. 3 1660-1790*. eds. Stuart Gillespie and David Hopkins, 479-505. Oxford: Oxford University Press.

Kedslie Mc Leod, M. ed. 2004. *From Charlotte Square to Fingal's Cave: Reminiscences of a Journey through Scotland, 1820-1824, by Krystyn Lach-Szyrma*. Phantassie: Tuckwell Press.

Lovelace, R. 1649. *Lucasta: Epodes, Odes, Sonnets, Songs* etc. *to Which is Added a Aramantha, A Pastoral*, London: Tho[mas] Harper.

Money, D. K. 2006. Aspects of the Reception of Sarbiewski in England: From Hils, Vaughan, and Watts to Coleridge, Bowring, Walker, and Coxe. In: *Pietas Humanistica: Neo-Latin Religious Poetry in Poland in European Context,* ed. Piotr Urbański, 157-187. Frankfurt am Main: Peter Lang.

Monsarrat, G. D. 1984. *Light from the Porch: Stoicism and English Renaissance Literature*. Paris: Didier Erudition.

Potter, L. 1989. *Secret Rites and Secret Writing. Royalist Literature 1641-1660*. Cambridge: Cambridge University Press.

[Sarbiewski, M. K.] 1646. *The Odes of Casimire. Translated by G[eorge] H[ils].*, London: Humphrey Moseley.

Sherburne, E. 1651. *Poems and Translations. Amorous, Lusory, Morall, Divine: By Edward Sherburne Esq*. London: Thomas Dring.

Urbański, P. 2000. *Theologia fabulosa. Commentationes Sarbievianae*. Szczecin: Wydawnictwo Naukowe Uniwersytetu Szczecińskiego.

Vaughan, H. 1651. *Olor Iscanus. A Collection of Some Select Poems and Translations, Formerly written by Mr. Henry Vaughan Siluris*. London: Humphrey Moseley.

Early Modern Travel Writing and Thomas More's *Utopia*: An Attempt at Literary Interpretation

Olga Grądziel

Abstract The paper presents an analysis of the significance of association of Thomas More's *Utopia* with the Early Modern travel writing and the possible implications of this association for the interpretation of the highly controversial text. *Utopia* is here approached from the point of view of literary analysis, with reference to some of the theories proposed by modern scholarship. The main objective, however, is to investigate the likely impact of the adoption of the travel narrative convention for the perception of the text by the contemporary audience, which may make it possible to discover some of the author's assumptions regarding the responses that the text was supposed to evoke in the implied readers. Since the notion of foreignness inherent in travel accounts may be associated with distance and estrangement as well as fascination, the approach remains consistent with the recognition on highly complex and equivocal character of *Utopia*.

1 Introduction

Readers of Thomas More's *Utopia* are usually confronted with the question to what extent they should accept the vision presented in the book as the vision of the best commonwealth and to what extent they may feel justified in following their frequently ambiguous feelings about some of the solutions employed in Utopia as means to bring the society to the allegedly ideal condition. Conflicting theories have been developed regarding the interpretation of *Utopia* on the level of ideology, structure and literary conventions. In order to provide an answer to the

O. Grądziel (✉)
University of Warsaw, Warsaw, Poland
e-mail: olga.gradziel@gmail.com

J. Fabiszak et al. (eds.), *Crossroads in Literature and Culture*, Second Language Learning and Teaching, DOI: 10.1007/978-3-642-21994-8_36,

question which of these approaches should be favoured, it might be reasonable to consider how strongly are these theories based on responses conditioned by the critics' modern sensibility and how far they try to do justice to the response that *Utopia* might have evoked in contemporary audience.

2 Responses to *Utopia*: Selection From Literary Criticism

2.1 Utopia and Political Writing

A highly convincing interpretation was proposed by Paweł Rutkowski in the article *Polityczne korzenie Utopii Tomasza Morusa* (2001) in which the author argues that Utopia was closely connected with the contemporary tradition of political writing and was perceived as such by the sixteenth century thinkers and writers, who received it in an enthusiastic manner (Rutkowski 2001, p. 38). According to this interpretation, *Utopia* was a response to the contemporary need for a new form of political organization and lasting political structures (Rutkowski 2001, pp. 37–39), and rested on the same belief in the possibility of improvement as the numerous political treatises written and published at the time. The employment of the literary form was therefore not aimed to move the work into the sphere of mere fiction, but to give it the advantage of creating an ideal instead of merely optimal vision (Rutkowski 2001, p. 42) by means of transcending the limits of reality, according to Horatian and Platonic assumptions about the role and potential of literature (Rutkowski 2001, p. 42).

Convincing as it may appear, the theory does not address the problem of internal contradictions present in the text of *Utopia*, which strongly indicate that the book is not necessarily a straightforward presentation of an ideal state. Finally, it is difficult to recognize Utopia as a truly ideal model, even with the assumption that the literary form permits the author to ignore the limitations of reality, as one serious limitation has apparently not been overcome. The imperfection of human nature, which, although perhaps not incorrigible, as the Protestant writers opposed by More would like it to be, was still not free from fault. Thus, it seems, Utopia is not only entirely free from evil, but also from suffering, which must ensue whenever a wrong is committed.

2.2 Utopia as a Dialogue of a Statesman and a Traveler

The point on the imperfect character of the utopian model has been recognized by G.B. Wegemer, who interprets *Utopia* not as a political treatise but a literary text based on the structure of dramatized dialogue (Wegemer 1996, p. 91). The indication of a dialogical structure, reflecting the conflict between two competing philosophies of life and different conceptions of the ideal state that may be derived from them, is combined with an analysis of the likely sources of inspiration and references underlying More's text, as well as More's own ideals of statesmanship presented in other writings and suggested by his political practice.

Thus, Morus, the dramatic persona, is presented as a Renaissance civic humanist, connected with the entire body of tradition of writing and thinking of statesmanship recognized and reverenced by More the writer and statesman (Plato, Aristotle, Thucidydes, Cicero, the Church Fathers, English legal tradition). Meanwhile, his adversary, Raphael represents his opposite, a traveler, and a "free man" who refuses to serve the Prince and the state, and excels in the telling of stories about distant lands and unknown peoples (More 1516, p. 11). The confrontation of these two characters, representing conflicting models of life, in the dramatized dialogue, makes *Utopia* a "dialectical puzzle" (Wegemer 1996, p. 91), which must not be taken at face value, but should be interpreted according to the method advocated by More in the letter to Peter Giles included in *Utopia*, and intended as the "dialectical weighing of facts and perspectives" (Wegemer 1996, p. 83).

Therefore, while Morus expresses views characteristic for civic humanism and pre-Reformation Christian doctrine, grounding political order in property, free will, conscience, love of God and humility, and seeing the safeguard against evil in law based on collective reason and tradition (Wegemer 1996, p. 68) and the role of the statesman as the "guardian of the state, whose task is to appeal steadfastly to reason in an effort to bring about a consensus on the common good through slow development of law and tradition" (Wegemer 1996, p. 4), Raphael comes up with a utopian model that utterly disregards tradition and replaces the complex relation between individual virtue and collective law with a system of institutional arrangements that eradicates individuality along with private property. The utopian vision, according to Wegemer, is contradictory to two conceptions of a good republic, cherished by the tradition of civic humanism and Thomas More himself.

2.3 Utopia as Polyphonic Text

The dialogical form, so convincingly applied to *Utopia* by Wegemer, suggests that the decision about whether *Utopia* is or is not a vision of an ideal state may be a matter of individual preference and identification with the principles on which it is founded. Therefore, the possible responses may be as manifold and individualized as may be the responses of readers of literature to literary texts. This high degree of freedom of interpretation of *Utopia* is presupposed by the polyphonic theory proposed by Blaim(1984). Unlike Wegemer, this author allows more than two possible approaches to the text, structured along the lines of a complex exchange between figures whose views are not only conflicting but also complementing one another in the discussion of the ideal commonwealth.

What is more, the figures involved in that exchange not only voice different opinions, but also express them in different forms, the basic two being the dialogue and the travel narrative, complemented by the auxiliary form of the letter. According to the theory formulated by Artur Blaim, utopian writing should be separated from the traditions of political and social theory and approached as an independent literary genre (Blaim 1984, p. 12), which, obviously, has significant implications for the manner in which utopian texts are to be investigated. Viewed from this perspective, the meaning of *Utopia* becomes even more dependent on the question of identification with the solutions it proposes, both by the characters involved in the discussion and the implied readers.

3 *Utopia* and Early Modern Travel Writing: Consequences of the Employment of the Convention for the Interpretation of Thomas More's Text

3.1 *Book 2 of Utopia as a Travel Narrative: The Tradition*

Recognising the arguments provided by Artur Blaim in support of his interpretation, I would like to investigate the significance of the literary form selected by More for the presentation of his ideal commonwealth for the possible interpretations in the light of what has been said about the principle of identification as the primary criterion for the assessment of the utopian project by the implied reader. The classification of the second book of *Utopia*, presenting the utopian model as a travel narrative is generally recognized by critics. The book by More not only makes numerous references to the form and contents of Early Modern travel narratives, but also draws from the earlier traditions, which contributed to the development of travel writing as a genre. In this way, the book does justice to the entire body of associations that a sixteenth century reader would have in his mind while dealing with a travel narrative. Since the theme of travel was very strongly present in contemporary consciousness as a result of the Great Discoveries, the rebirth of utopian writing in Europe occasioned by Thomas More in 1516 in this form and not another may also suggest the strong link between the exercise of utopian imagination and social and ideological developments associated with the geographical discoveries. According to Lijlegren (1961) the emergence of Early Modern utopian fiction was a clear reflection of the process taking place in the contemporary consciousness in connection with the search for happier and better places, which took the form of ventures into the unknown.

In order to grasp the association of *Utopia* with the tradition of Early Modern travel writing, it is essential to remember about the earlier texts which still influenced the contemporary way of thinking about travel. Apart from the ancient classic texts which marked the very origin of travel writing in Europe, namely, Homer's *Odyssey* and Plato's account of the mythical Atlantis, the sixteenth century was still interested in Medieval travel narratives. In fact, the two most famous Medieval travel books; *The Travels of Sir John Mandeville* and Marco Polo's *Il Millione* were published in England only at the dawn of the 16th century (in 1499 and 1503). Mentioning *The Travels of Sir John Mandeville* it is impossible not to mention the fact that in the book, apart from the wondrous descriptions of imaginary creatures and lands, can be found an account of a state, which, though not named ideal, is designed as a political organism unquestionably more perfect than the countries situated in the known parts of the world. Since it was arranged along Christian principles, it could also be recognized as such by the contemporary readers. It is significant that the Land of Prester John was defined as a model place existing within the scope of the real world, accessible to living human beings, in contrast to the ideal place which was the Earthly Paradise; a place of essentially spiritual character, which could not be accessed by people. Significantly, *The Travels of Sir John Mandeville* appeared more credible to the Medieval readers than did the accounts of travels of Marco Polo, which suggests that the concept of ideal or nearly ideal places was definitely not foreign to the European sensibility before the year 1500.

3.2 *Utopia in the Context of the Early Modern Travel Writing*

The two books had their significance also for the Early Modern travels, as, according to tradition, Christopher Columbus was to have Marco Polo's book with him when going on his journey in search of India. As Lijlegren made clear in his book on the early utopian tradition, travelers and sailors taking part in the earliest of the Great Discoveries in a sense had to refer to the Medieval texts, as these constituted the only material which could guide them through utterly unknown realms (Lijlegren 1961, p. 124). Thus, the accounts written by Columbus in which he reported his alleged approach to the outer regions of the Earthly Paradise. This shows that the Medieval accounts, though largely mythical, served as a source of concepts and associations which helped describe and order the knowledge of newly discovered places, and incorporating it into the known world. Apparently, truth and fiction were sometimes difficult to separate from one another in the Early Modern travel texts. The presence of fictitious accounts was not always a matter of premeditated forgery. Facts could be adapted to reflect the popular myths in order to make them easier to understand or more appealing to the people, whose positive response was sometimes not only a matter of respect or fascination with the traveler's work, but also the prerequisite for financial support that the traveler could obtain for his venture.

It is important to remember that, in the sixteenth century, the main reasons for travel were utilitarian (Sherman 2006, p. 187). The travelers were usually merchants, pioneers of colonial ventures and conquerors. Travels for scientific purposes were undertaken in for the seventeenth and eighteenth centuries (Sherman 2002, p. 29). Therefore, scientific objectivity was not necessarily presupposed by contemporary travel accounts. On the other hand, however, it was in the Early Modern period that travel writing was developing into a convention based on the methodology, which would be continued and perfected in the subsequent centuries. The early descriptions were subordinate to the interests of trade, colonial expansion and conquest. Nevertheless, the set of instructions issued for the sixteenth century surveyor Thomas Bavin is recognized as one of the first overt statements of methodological guidelines for travel accounts, which included all essential elements of a description of a foreign land, to be present in travel accounts from the Early Modern as well as later periods (Sherman 2002, p. 20).

The genre-specific characteristics of style (like the matter of fact narration, geographical orientation of the account) and structure of the text (dedicating separate portions of the text to the description of the natural conditions, material culture and the local people and their customs (Sherman 2002, p. 26)) coexisted with a specific character of the content, which was focused on ethnographic description and which would give rise to ethnography as a separate discipline in the centuries to come. This deep interest in foreign cultures expressed in ethnographic investigations resulted from the need to understand foreign cultures and peoples and relate their modes of life and principles to their European counterparts. Confrontation with foreignness might have been difficult, as sometimes the ways of the natives could seem more humane and better even from the European's point of view. This impression could have been evoked by Marco Polo's description of the Brahmins' disapproval of the killing of animals, which, according to critic Lijlegren, might have been the inspiration for the Utopians' attitude described by More in largely similar terms (Lijlegren 1961, p. 88). Moreover, Columbus' discoveries in America and the confrontation with the Indians were to give rise to the myth of the "gentle savage", crediting the natives with capacity for sentiments and inclinations compatible with those represented by the Europeans', albeit developed to a lesser extent (Lijlegren 1961, p. 150). Sometimes, the native people were conceived of as the inhabitants of the Earthly Paradise, whose gentle nature was preconditioned by the fact that they did not need to struggle for survival, as the life in harmony with abundant Nature grants them all they need (Blaim 1984, p. 43). Finally, major concern among the Europeans was raised by the observation of hitherto unknown religious practices of the foreign peoples, which evoked different responses and inspired different conclusions regarding these particular religions and religion in general (Rutkowski 1998, p. 39).

It is difficult to underestimate the influence of these sentiments and conventions on Thomas More's *Utopia*. Parallels can be noticed both in reference to the older tradition of travel writing and the contemporary developments in the genre. Numerous studies indicate the connection between *Utopia* and the various accounts of ideal places that were to be found in medieval legendary texts; in some ways, it is believed to resemble the Earthly Paradise, which is located somewhere on Earth, but remains inaccessible to mankind, ("To prevent enterprising explorers, the site must be doubtful and the way difficult" (Lijlergren 1961, p. 86)). Associations with Eden can be suggested, though they are definitely not so obvious; after all, Utopia is a place and a product of political and social arrangements, not a spiritual realm created by God. There is a profound similarity between Utopia and the mythical Fortunate Islands (*Insulae Fortunae*), the place inhabited by heroes and exceptionally virtuous individuals, who may be transported there after death or when they are still alive (Lijlegren 1961, p. 16). Significantly, the tradition was present both in the Ancient world and the Middle Ages; the Fortunate Islands were to be the object of the quest of St. Brendan. Critics have also pointed to utopian elements present in *The Travels of Sir John Mandeville*, which contains references to the myth of the Earthly Paradise, but also presents the legendary Land of Prester John, which, being a country located within the accessible world and owing its strength and stability to the strict observance of Christian principles, seems to be closer to Utopia than the Earthly Paradise. Still, parallels have been noticed between the manner of presentation of the Earthly Paradise and Utopia. As the author of the *Studies on the Origin and Early Tradition of English Utopian Fiction* has observed, both places were distinguished by mild climate and favourable natural conditions, and located in the middle of dangerous seas and wild lands (Lijlegren 1961, p. 87). The critic reminds that it was a regular practice to locate the Medieval mythical and legendary places within the scope of real geography.

But for More, the world would already hold not only the well known lands and places derived from Medieval legends. The spaces opened by the Great Discoveries were in no way less intriguing than those to be found in works of literature. More was interested in the discoveries and was acquainted with the earliest accounts of the newly discovered continent of America, which can be ascertained even without going beyond the text of *Utopia*. But apart from the *Four Voyages of Amerigo Vespucci* referred to in *Utopia*, More might have known some of the pamphlet editions of the Letters of Columbus and Henry the Navigator, which conceived of the ventures into the unknown as a search for a happier and better life and reality (Lijlegren 1961, p. 11). If these were the assumptions of real explorers and discoverers of the time, there should be nothing strange in the fact that it found its reflection in literature. Therefore, there are interpretations of *Utopia* that present it as a "playful rendering of a travel narrative" which displays all features of the genre that have evolved until that moment, including a supplementary map and a rudimentary description of the Utopian language (Sherman 2006, p. 190), provided to complete the semi- ethnographic account of the life and customs of the Utopian people.

3.3 Raphael Hythloday: An Early Modern Traveler

An explicit reference to the travel narrative convention is the fashioning of Raphael Hythloday as an Early Modern traveler. From the description in *Utopia* we know that he looks like a sea captain (More 1516, p. 11) and that he traveled with Amerigo Vespucci (More 1516, p. 12) but departed from the famous sailor's expedition at some point to continue the exploits on his own. During his lonely travels he accidentally landed on the island of Utopia, where he stayed for five years, during which he got acquainted with the Utopian institutions and ways, but also enjoying a life offered to him by the ideal society. Thus, his account is delivered as first-hand knowledge; a travel narrative told (and then retold by Morus the persona, who, nonetheless asserts that his retelling is as exact as it might be (More 1516, p. 5)) by a person who has experienced all the things described. It may seem that what the reader can expect of Hythloday is even more than would be expected of a regular Early Modern traveler, as, after all, the aims of his travels are, as it seems, purely intellectual. And indeed, he delivers a systematic, well organized account of life in Utopia, very much in the fashion of ethnographic accounts.

But very quickly the reader may begin to wonder whether the very travels of Raphael are not of "purely intellectual" nature too. His very surname, Hythlodaeus, which means "speaker of nonsense" is probably telling enough. But critics went further into the analysis of hints of a potential fabrication; Wegemer observed that only two of the four voyages of Amerigo Vespucci were undoubtedly true, while the other two were probably a fabrication, unconfirmed by any reliable source (Wegemer 1996, p. 101). Moreover, he interprets the Latin original of the fragment referring to Raphael's claim that he took part in Vespucci's voyages as suggesting that he circumnavigated the world together with the great explorer; meanwhile, the first voyage around the world (obviously, undertaken by Magellan) took place in 1519, of which More must have been aware. Apparently, More the persona is also suspecting a possible fabrication, and warns Raphael that he does not wish to listen to any "Ulysses style stories" about "Scyllas, man-eating Lastrigonians and prodigious monsters" (More 1516, p. 11).

While making the attempt to answer the question whether the suspicions were justified, it might be of use to recall the image of the lone traveler present in the culture of the time, and the underlying implication that he may make up stories, like Marco Polo or Mandeville (depending on which of the two a contemporary reader would prefer to consider reliable), and still present their accounts as credible, as there would be no way to verify them against reality. Since there was hardly any way to verify many of the travel accounts and no method of reliable description was available, in Lijlegren's opinion, Early Modern travelers in a sense had the right to invent stories and this eventuality was normally taken into account by their listeners (Lilegren 1961, p. 111). If we look at how Hythloday uses the information about foreign nations inhabiting unknown parts of the world (apart from the Utopians, he mentions the Polylerites and the Macarians as well as named and unnamed neighbours of Utopia), which are invoked as an illustration of his theses, one is inclined to believe, after Wegemer, that perhaps Utopia is also an illustration of a more complex conviction, this time not uttered as a verbal proposition, but expressed by the totality of Raphael's mode of life. If so, then we return to the dialogical and dialectical confrontation of the two points of view, represented by More the persona and Raphael.

4 Utopia as a Foreign Land

4.1 Morus the Persona Assesses the Utopian Order

What follows from these considerations for the very vision of Utopia and the way it might be approached, is that its interpretation is indeed impossible to conceive of in any finite and universally binding way; one may either approve or disapprove of it, depending on the degree of identification he is ready to assume in relation to this concept of the best commonwealth. More's response to Raphael's account is very telling in this respect. Having learned about the state that the teller so ardently believes to be the best of all possible states on Earth, More remains unconvinced and says that the system existing in Utopia "not only undermines all nobility, magnificence, splendor and majesty, which are (in the popular view) the true adornments and ornaments of a commonwealth" (More 1516, p. 134), but also, that it is absurd in its very foundation, which is the community of property (More 1516, p. 134). As some critics have suggested, the statement about the "true adornments and ornaments of a commonwealth" may be dismissed as ironic. But the accusation of absurdity reveals quite a lot about the attitude that More the persona adopts towards Utopia; namely, a kind of mental distance and essential non-identification, characteristic, significantly, for the attitude usually assumed with respect to all forms and manifestations of foreignness.

4.2 Reversal of the English Order?

It seems that assessing it as absurd, More makes it clear not only that he is unconvinced by the utopian order, but also that it is foreign to his sensibility, which precludes any possibility of identification. Yes again, his unwillingness to identify with the vision may be motivated by various factors. On the one hand, it is difficult to expect that a loyal citizen, statesman and father of a family that More was would identify with a vision, which, as Artur Blaim has observed, is a complete opposite of Elizabethan England (Blaim 1984, p. 19), as distant from it as it possibly be in terms of location, political system, customs and institutions (Blaim 1984, p. 35). The polar opposition can be ascertained between the two states because of the fact that Utopia has the same underlying structure as England; it is derived from a description of England (as opposed to the unstructured and unordered New World) and English standards regarding values by means of complete reversal of order and evaluation (Blaim 1984, p. 75). In the opinion of Elliott, the reversal of order and value judgments is here so prominent, that it may be linked with the satirical convention (Elliott 1970, p. 24). But a satirical tone is yet another expression of distance.

More the persona therefore seems to distance himself from Utopia. But More the author, embracing the travel narrative convention, distances himself from the overwhelmingly Anglo-centric perspective, which also existed at the time and figured prominently in contemporary literature. By the same token, his willingness to address the shortcomings and vices of the English order in Book 1 of *Utopia*, suggests that he is distancing himself from the uncritical praise of England undertaken by contemporary panegyric writers, deeply prejudiced against all that was "foreign" (Lijlegren 1961, p. 63). Therefore, if *Utopia* as a whole is to expose and address the imperfections of the English state, at the same time suggesting an impossibility of identification with a theoretically "best" but foreign order, the impossibility must reach deeper than just to the institutional arrangements and views on statesmanship.

4.3 Foreignness and Utopian Beliefs

What may be recognized as the ultimate point of reference when it comes to the consideration of possibilities of identification with a foreign culture, is the issue on which the Early Modern English people, and Thomas More perhaps even more than anybody else, would never compromise; his religious beliefs. Not only the outer manifestations of Utopian religious life, presupposing pluralism in worship, far-reaching toleration and an "abstract and impersonal" vision of God (Rutkowski 1998, p. 339) worshipped in formalized ritual envisaged more as ways of preservation of the state than cultivation of spirituality, are obviously unacceptable for the Catholic More. Significantly, in this respect, the views of More the author and More the persona are identical beyond any doubt, as we know with all certainty what religion was professed by the former, while the latter reveals his belonging to the Roman Catholic Church in the conversation in Book 1 of *Utopia*. Obviously, the definite nature of the Catholic faith, as opposed to the indefiniteness of the Utopian religion, implies that the beliefs of these two figures agree in all respects, allowing the reader to merge the two persons into one via the dogmatic creed they both profess.

Hence, the distance is also felt by both, and it extends from the outward religious practices onto the entire Utopian system, if only we recognize that its foundations depend on the character of Utopian religion. That they do can be easily inferred from the treatment of Utopian heretics. But the only dogmas that are defended by the Utopian state are the belief in the existence of God and immortality of the soul. As Paweł Rutkowski has rightly observed, these assertions can be made on rational grounds, which in fact is the assumption made by the Utopians, who identify their heretics with "madmen" (Rutkowski 1998, p. 342). This corresponds with the characterization of Utopia proposed by Chambers (1935), who argued that the only foundation of the Utopian order is reason (Gallagher 1964, p. 110), which also serves as the only basis for the Utopian morality. The thesis finds confirmation in the text of *Utopia*, where the account of the utopian customs and rules is summed up by the statement: "This is their view of virtue and pleasure; and in the absence of religious inspiration from heaven revealing something holier, they think human reason can discover no truer doctrine" (More 1516, p. 91). According to Chambers' elaboration of the statement, the utopian system is deficient in the knowledge of the Christian revelation and therefore also in all that follows from it; namely, the liturgy, sacraments, and, first and foremost, the Christian ethical code. Hence, as the author rightly observed, the moral foundation of the utopian order should be be identified with the cardinal virtues derived from the ancient philosophical tradition (Wisdom, Faith, Temperance and Justice), but deficient in the three Christian virtues of Faith, Love and Charity (Gallagher 1964, p. 110).

5 Conclusion

In the context of all that has been said, the distance felt by the Roman Catholic Thomas More towards the vision he constructed in his book, can be justified on the grounds of disparity between the essential principles underlying the order of Utopia and the order he was a part of. The disparity may be observed in a close analysis of the two systems, which led Donner(1945) to his famous conclusion which lay at the heart of the Roman Catholic interpretation of *Utopia* (When a sixteenth century Catholic depicts a pagan state founded on reason and philosophy, he is not depicting his ultimate ideal… The underlying thought of *Utopia* always is "with nothing save reason to guide them, the Utopians do this; and yet, we Christian Englishmen, we Christian Europeans…!" (Elliott 1970, p. 24)), but the adoption of the travel narrative convention may be treated as a more direct indication of the author's and contemporary readers' implied perception of the Utopian system.

References

Blaim A. 1984. *Early English Utopian Fiction: A Study of a Literary Genre*. Lublin: UMCS.

Chambers, R. W. 1935. The Meaning of Utopia. In *Utopia A Norton Critical Edition*, ed. R. Adams, 137-142. New York: W.W. Norton & Company.

Donner H. W. 1945. *Introduction to Utopia*. London: Sidgwick & Jackson.

Elliott R. C. 1970. *The Shape of Utopia: Studies in a Literary Genre*. Chicago and London: The University of Chicago Press.

Gallagher L. 1964. *More's Utopia and Its Critics*. Chicago: Scott Foresman and Company.

Lijlegren S.B. 1961. *Studies on the Origin and Early Tradition of English Utopian Fiction*, Uppsala, Copenhagen, Lund.

More, T. 1516. *Utopia*, *On the Best Form of a Republic and on the New Island of Utopia*. In *Utopia modern translation, ed.* C. H. Miller. 2001. New Haven and London: Yale University Press.

Rubies J. P. 2002. Travel writing and ethnography. In *The Cambridge Companion to Travel Writing*, ed. Hulme P., T. Youngs, 242-267. Cambridge: Cambridge University Press.

Rutkowski P 2001. Polityczne korzenie Utopii Tomasza Morusa. *Odrodzenie i Reformacja w Polsce* XLV: 33–44.

Rutkowski P. 1998. There is a certayne Godlie Powre Unknowen.: Abstract and Impersonal God in Utopia by Thomas More. *PASE: Papers in Literature, Language and Culture*. KUL: 339–347.
Sherman W. H. 2002. Stirrings and Searchings (1500-1750). In *The Cambridge Companion to Travel Writing*, ed. Hulme P., T. Youngs, 17–37. Cambridge: Cambridge University Press.
Sherman W. H. 2006. Travel Writing. In *The Oxford Encyclopedia of British Literature,* ed. Kastan D. S., 186-189. Oxford: OUP.
Wegemer, G. B. 1996. *Thomas More on Statesmanship*. Washington D.C.: The Catholic University of America Press.

Beryl Bainbridge's *The Birthday Boys*: Travelling Towards the Pole of Death

Bożena Kucała

Abstract Beryl Bainbridge's novel imaginatively recreates Robert Scott's ill-fated expedition to the South Pole, probing into the established heroic version of the events. Told in five monologues from the point of view of each member of the group, the narrative traces their arduous progress towards their destination, at the same time charting the evolution of their attitudes. The extreme conditions of the Antarctic expose hitherto suppressed aspects of their personalities. What the reader of this historical novel knows from the start becomes gradually clear to the characters—their journey is in reality a journey towards death. The juxtaposition of private and public personalities as shown in the monologues reveals the complex motives behind the enthusiasm for the Antarctic adventure. What started as a heroic exploit or a boys' game (the narrative implies that these aspects are closely related) becomes a journey of psychological and moral discovery, as the characters are forced to acquire self-knowledge, and knowledge about the others. The article analyses the correlations between the physical progress of the expedition and each man's journey towards self-knowledge and, ultimately, in the words of Peter Pan, towards the "awfully big adventure" of death.

1 Introduction

With the conclusion of the story all too well known—at the end, everyone dies—Beryl Bainbridge's novel focuses on how it happened rather than what happened next. The novel is composed of five monologues spoken by Captain Robert Scott

B. Kucała (✉)
Jagiellonian University, Kraków, Poland
e-mail: bkucala@o2.pl

J. Fabiszak et al. (eds.), *Crossroads in Literature and Culture,* Second Language Learning and Teaching, DOI: 10.1007/978-3-642-21994-8_37,

and the four men who died with him on the return journey from the South Pole. The expedition aroused much interest at the stage of preparation, but it was mainly after its tragic end that it captured the popular imagination and passed into the sphere of modern English national myth. In the eyes of the contemporary British public, Scott fitted well the criteria of imperial heroism: "His achievements took place in a remote realm; he displayed formidable courage, stamina and willpower; and, above all, he died a martyr's death in the service of his nation" (Barczewski 2007, p. 148). The outbreak of the Great War reinforced the perception of Scott and his companion as heroes, as their tragic fate seemed to foreshadow the death that British soldiers were suffering, and, more specifically, came to be seen as "an exemplar of the kind of death they *should* be suffering" (Barczewski 2007, p. 140).

Assessment of the polar expedition has gone through several stages, starting from the initial glorification of the lost heroes to debunking the myth and severe criticism of Scott's incompetent leadership, through a range of intermediate evaluations, balancing the natural adversities with human weakness and folly as the cause of the failure while acknowledging the men's heroism. It is the latter, even-handed attitude that prevails in Bainbridge's book. Her approach here, in contrast to her usual bleakly comic mode, is, in the words of Brett Josef Grubisic (2008, p. 136), "notable for its sympathetic rendering and refusal of comic caricature". An "admirable demolition job on Scott" had earlier been done by Roland Huntford, the author of *Scott and Amundsen* (1993), which replaced the image of Scott as the embodiment of heroic resolution with the image of a feeble creature driven by vanity (Linklater 1992, p. 25). Linklater (1992, p. 26) comments that although Bainbridge is more generous here towards her characters than she normally is, the novel resembles her other books in that the epic proportions of the Antarctic mission have been reduced to a "small-scale drama", characteristic of the writer's urban comedies.

2 Boys and Heroes

Bainbridge's own perception of the expedition was shaped by her discovery of a seemingly improbable friendship between Robert Scott and J. M. Barrie, the author of *Peter Pan*. But on closer inspection parallels do emerge between the creator of a character who refused to grow up and wished to inhabit Never Never Land, and the man who abandoned the comforts of ordinary life in order to explore the Antarctic. Gary Kirst (1994, p. 15), reviewing the novel for *The New York Times*, sums up Bainbridge's perspective as "a compassionate respect for the obvious courage of these men, on the one hand, but also a critical, somewhat ironic detachment". As she admits in an interview, the writer was both baffled by the folly of the enterprise and impressed by the undeniable heroism of the explorers: "I thought the men were daft going out into the snow, but once you read the diaries and the letters and the contemporary accounts you realize that they walked 1,600 miles in very poor shoes and with very poor equipment and on a bad diet. These were supermen" (qtd. Kirst 1994, p. 15). A particularly illuminating reading, which became the writer's key to capturing the Antarctic experience, was the line spoken by Peter Pan: "To die will be an awfully big adventure" (see Kirst 1994, p. 15).

Accounting for the enormous popularity of Peter Pan as a literary protagonist in Edwardian England, Stephanie Barczewski (2007, p. 151) points to the predominance of adventure stories featuring "teenaged boys in the guise of men". The psychologist Ann Yeoman (see Barczewski 2007, p. 152) speculates that Barrie's story touched "an unacknowledged nerve in the collective British psyche of his time". She is inclined to link the popularity of the hero to the "romantic idealism" of the elite public schools in Britain in the late Victorian and Edwardian eras. According to her, Neverland "suggests the exclusive boys' club, apart from parents, in a world of its own, with its singular population and hierarchy, moral code and set of values".

Bainbridge's novel tries to reimagine and render the men's attitudes as a peculiar combination of a boyish yearning for adventure and their awareness of the possibility of death, which becomes more and more real during the course of the journey. In the writer's interpretation of the expedition, both modes coexist from start to finish: the spirit of adventure is tinged with the flavour of mortal danger, while even the tragic experience of death smacks of defeat in a boys' game.

3 The Awful Adventure of Dying

The five monologues trace five successive stages of the journey, and in the course of each the speaker celebrates his birthday. The birthdays, apart from being occasions for festivities in the unlikely Antarctic conditions,[1] serve also as reminders of the passage of time and of the inevitable confrontation with mortality, bringing each birthday boy closer to his own death. Unlike Peter Pan, who, having immunised himself from time before he is one year old, never has a birthday, the grown-up boys in Bainbridge's book eventually have to face the dire consequences of their desire for adventure. The last event in the novel—the death of Lawrence Oates, whose remark "I am just going outside and may be some time" passed into the store of famous last words—coincides with his birthday. Severely ill, he voluntarily leaves the tent and freezes to death the day before his thirty-second birthday. Death finally catches up with him and all the other birthday boys, and, despite its tragic circumstances, becomes the ultimate, awfully big adventure.

The novel relies on a series of such dualities: the adventurous aspect of the expedition from the very beginning is outweighed by its deadly difficulty, the men's immaturity is repeatedly contrasted with their brave determination and

[1] One of the surviving photographs of the expedition shows Scott's last birthday dinner (see Preston 1997, illustrations between pages 80 and 81).

sense of responsibility. Hence, although the heroes are cut down to size and, as D. J. Enright (1992, p. 16) observes, the title of the book itself is deflatory, the explorers retain some of the heroic stature originally awarded them by public opinion. The final test is each man's direct confrontation with death, which they all pass successfully, in spite of, or perhaps even thanks to their quirks and limitations.

From the start the sinister prospect lurks behind all the public enthusiasm and imperialist propaganda accompanying the preparations, as well as behind the men's private motives for going to the Pole. The writer probes behind the veneer of heroism without destroying it; the public's all-too-easy glorification of the men is questioned as much as the attitudes of the explorers themselves. The men are widely recognised and hailed as heroes long before departure; they are also equipped with several flags to hoist at the South Pole, including a huge flag of the city of Cardiff, the sheer size of which makes the offer preposterous. However, although aware of the public's misconceptions about the nature of their hardships, the men to some extent go along with the pomposity of the farewell they are given. Taff Evans, whose monologue charts the preparatory stage, expresses pride at being one of the explorers: "I don't count myself as educated, but when I stood up [to give a speech] I had a feeling in my head—and it was nothing to do with liquor—that I was part of something special, something with glory in it" (Bainbridge 1993, p. 41). Scott, too, despite his scepticism about all the honours prematurely lavished on the team, in his speech to his subordinates casts the undertaking in an undoubtedly heroic mode: "We are all contributing to a great enterprise which is only just beginning. Each and every one of you will play your part" (Bainbridge 1993, p. 48). Yet, all the national feelings apart, each man sets out to fulfil his own private dream. Although Evans is inclined to ridicule those of his companions who follow their youthful ambitions, he has to admit that they all, himself included, are in search of their dreams: "His [Teddy's] eyes had a boy's light in them, guileless, shiny with hope, and I was sorry afterwards I'd made sarcastic with him. When we get back he'll become a Captain for real, and then he'll be down on the list in line for an Admiral. To each his own dream, and I know mine" (Bainbridge 1993, p. 21). Evans himself, coming from poor, cramped surroundings, goes to sea in order to leave behind the tedium of domestic life and feel again the enticing freedom of open spaces. His yearning for an adventure has its more practical dimension, too—on return he wishes to buy a pub by the sea, with an orchard and a meadow nearby, as a recreation of the idyllic scenes remembered from childhood. His far more sober-minded wife from the start opposes her husband's decision to accompany Scott, treating the whole enterprise as a mere game, and is mainly concerned about her income at the time when, as she puts it, "you and your precious Owner [Scott] are thousands of miles away playing at snowmen" (Bainbridge 1993, p. 27).

There are indeed occasions when her suspicion that the men will use the expedition as a chance to enjoy their freedom from the responsibilities of adult life are confirmed. Wilson's thirty-eighth birthday, for example, is celebrated with a wild war dance followed by a mock battle (Bainbridge 1993, pp. 60, 61); the men carry with them gifts to be unwrapped at Christmas time, and some of the gifts happen to be toys. The exhilaration of the expedition occasionally obfuscates the sense of danger. Wilson, who accompanied Scott on his earlier expedition to the South, obliterates the bad experiences in anticipation of more excitement: "... any doubts I may have had about the wisdom of coming south again have evaporated like snow under sunlight. After five weeks at sea I'm as fit as a fiddle and have actually put on weight" (Bainbridge 1993, p. 51). To Wilson and Birdie Bowers, the two most immature members of the expedition, travelling and confronting danger still has the appeal of a boyish challenge. During a side trip to a penguin habitat, Bowers is full of enthusiasm: "I've been five times round the world, and Bill [Wilson] quite as far in his mind, yet we still thought this an awfully big adventure" (Bainbridge 1993, p. 135), which is soon tempered by the realisation of how near death they have come. Wilson claims that "'the joy of being' incorporates a delight in annihilation" (Bainbridge 1993, p. 165).

From Wilson's perspective the experience most resembles a public school boys' game transferred into the realities of the adult world. Rules must be observed, each participant is expected to behave fairly and honourably towards the others, politeness conventions must never be abandoned, and the value of team effort is of paramount importance. Wilson's loyalty to Scott stems both from his personal admiration for the man and from his deep-seated respect for any fair, authoritative leader. Wilson explains to Cherry, who shares his upper-class mindset: "Men from our background [...] are at an advantage. They've been schooled to accept things, not to argue the toss once the umpire has made a decision. Abiding by the rules is a great help, you know... it does away with introspection, leaves one free to get on with the game" (Bainbridge 1993, p. 71). Wilson, however, prefers not to sustain the parallel with boys' games when Cherry wistfully wonders whether all they are going through now is nothing more than a game (Bainbridge 1993, p. 72).

4 Playing the Game

Cherry's question hovers over the expedition until the end. Apart from some research done into the animal life of the Antarctic, the journey seems to have no practical purpose. It fulfils the great national desire for imperial expansion as well as the men's private ambitions. In Bainbridge's subversive version, Scott's motivation is of course far more ambiguous than that the public endows him with. His personality still seems to be shaped by the carefully suppressed memories of his unhappy home and the constant need to assert himself against his despotic father. Scott's obsessive self-control both commands the admiration of his men and distances him from them. Scott is determined not to reveal his emotions, not to admit mistakes, not to consult with the others, which has the contradictory effect of both inspiring the men's confidence in the leader on the one hand, and, on the other hand, causes him to make decisions with fatal consequences. The code of behaviour which serves him well in Britain may be incongruous or even harmful in the Antarctic conditions. And so, for instance, Scott counted on the hut built during the previous expedition to be still habitable but the fact that the window had been left open ruins these expectations. Scott immediately blames another man for transgressing—what he calls—"all the boundaries of common courtesy": "Surely it's a mark of civilised human behaviour to leave a place in the condition one would wish to find it" (Bainbridge 1993, p. 82).

However, while he can within reason expect himself and his team to keep up the conventions of polite behaviour even in the extreme circumstances of the Antarctic, his tendency to expect the same of animals is downright ludicrous. One of the reasons why Scott makes the very unwise decision to rely on ponies rather than dogs is that he appreciates horses as supposedly humanoid companions while despising dogs for their brutal instincts. In the tragicomic episode when several dogs, dangling on a rope in a crevasse, aggressively turn against each other, Scott is shocked by their "uncivilised behaviour" (Bainbridge 1993, p. 100). This incident echoes an earlier one, involving Scott and Evans falling into a crevasse together. On that occasion, even when faced with death, they both managed to act with admirable self-possession and civility: Scott, concerned about the other man's safety, asked, "Are you all right, Taff?", to which the other replied politely, "Don't trouble yourself about me, sir" (Bainbridge 1993, p. 11).

The one occasion when Scott's self-possession fails him is when the news reaches him of the Norwegian Amundsen also heading for the South Pole. Scott surprises himself and the others by his impulse to forcefully prevent the Norwegians from getting there first. The question about the nature of the expedition recurs, this time asked by Wilson who tries to convince the team that the journey is more than "a vulgar scramble to reach the Pole" (Bainbridge 1993, p. 104). But that it is this, too, seems obvious enough. Scott's tentative declaration to his companions, "We must go on, without fear or panic, and do our best for the honour of the country" (Bainbridge 1993, p. 104), sounds unconvincing even to himself. Notwithstanding their denial, Scott and his men do treat the expedition as a boys' race. Since Amundsen kept his plans hidden until the very end, Scott feels cheated, accusing the other explorer of "duplicity" and a "lack of sportsmanship" (Bainbridge 1993, pp. 101, 102).

One of Scott's private preoccupations is the question of leadership—he constantly re-evaluates and adjusts his behaviour so as to suit his own idea of a good leader. Scott's conduct can be simultaneously noble and highly unreasonable. He does not show weakness himself but is also very intolerant of other men's weakness; he forbids his men to shoot the horses even though their frailty slows down the progress of the expedition. It is Scott's determination that made the journey possible in the first place and that now sustains the other members of the team; but it is also his old-fashioned code of values that to a great extent impedes the grand project. Bowers, too, entertains romantic yet quite impractical ideas about their race to the South Pole: "I think there's an unsporting element in the use of either motors or dogs. Far better to stride out, nation against nation, man against man" (Bainbridge 1993, p. 142).

5 Dying Like English Gentlemen

There is an episode in *Peter Pan* when Wendy addresses her adopted sons who are about to walk the plank: "These are my last words. Dear boys, I feel that I have a message to you from your real mothers, and it is this, 'We hope our sons will die like English gentlemen'" (Barrie 1995, p. 140).[2] Scott's expedition took place only a few years before the codes and values of the Edwardian world were to come to an end in the trenches of the First World War. Wilson has an intuition that the world he knows is rapidly disappearing, with the ideology of "love of country, of Empire, of devotion to duty" crumbling away. In a revealing episode preceding the final calamity the men sing hymns and ballads to keep up their spirits. Bowers's loud recitation of the requirements for the Gold Medal for cadets, which include "fearless devotion to duty and unflinching truthfulness" (Bainbridge 1993, p. 155) is interrupted by Cherry's derisive laughter and the collapse of their shelter. The explorers, therefore, may be seen at once as the best embodiment of Edwardian Britain, and as its epigones. Wilson reflects, "It often strikes me that Con and myself, Bowers and Oates, even Peter Pan Evans with his penchant for swinging one round by the seat of the trousers, are the misfits, victims of a changing world. It's difficult for a man to know where he fits in any more" (Bainbridge 1993, p. 64). This observation opens up an alternative perspective on the polar adventure, transforming it from a brave exploration of the frontiers of the world to a Peter Pan-like evasion of reality.

Yet, Peter Pan's refusal to become entangled in the real world entailed also an evasion of death. In stark contrast to the literary precedent, the men from the beginning envision and accept the very real possibility of dying, without trying to delude themselves by naive optimism, although the boundary between death as a big adventure and death as the tragic prospect of annihilation is quite fluid in their minds. Death is foreshadowed in several episodes until it becomes an inevitability. Some of the men have already had their brush with death during Scott's first expedition to the Antarctic. The death of Vince, one of Scott's companions in 1902, serves as a warning of the awful hardships of the present enterprise. It was also then that Scott and Evans fell down a crevasse together, and their mutual help in a moment of crisis sealed their loyalty to each other. Evans's declaration of readiness to "die with the man, and for him" (Bainbridge 1993, p. 14) is not a mere rhetorical figure. His motives in going south include not only the private dream to own a pub by the sea but also the desire to follow the man he looks up to as his leader. Evans's preparations for the deadly adventure are quiet and subdued,

[2] Among the numerous books, songs and pamphlets that appeared in the aftermath of the disaster was the booklet by Hodder Williams *Like English Gentlemen* (1913), specially prepared for schools. Scott is referred to as "the hero". The text reads: "This country of ours is so great, its Empire so wide, because, throughout the long years, Englishmen, everywhere, have fought and struggled and died to possess that unknown land, and to plant their flag on its topmost height. So this hero of heroes said, 'I'm going to find the South Pole. It will be a big adventure'". (qtd. Barczewski 2007, pp. 124–125).

free from pathos. Before departure, he toys with and eventually rejects the idea of making a final confession of sins to his wife, and it is only when drunk that he openly mentions the sinister prospect of dying (Bainbridge 1993, p. 42). Scott is more given to rhetorical pronouncements: "I will reach the South Pole, or I will never come back again" (Bainbridge 1993, p. 47).

Wilson's acceptance of death stems from his strong sense of purpose; his own principles require him to fill all the days of his life with meaningful activity, which his wife insightfully sums up for him as "You have such a feeling of the absolute necessity to be doing something, at any hour of the day or night… before the end comes" (Bainbridge 1993, p. 57). Like Tennyson's Ulysses, Wilson believes that the end of activity may only come with the end of life: "Death closes all; but something ere the end, /Some work of noble note, may yet be done…" (Tennyson 2000, p. 1,214)[3] Yet, despite his seemingly stoical acceptance of the inevitable, Wilson has a suppressed terror of death, which manifests itself only in his half-conscious vision of a bird-like creature, interpreted by him as a harbinger of death. Momentarily losing self-control, he experiences a foreshadowing of his demise in the snow: "… my body shook, and through chattering teeth I heard myself stuttering over and over, 'So cold… so cold… so cold'" (Bainbridge 1993, p. 58). When he later tries to dismiss the incident as a hallucination, his normally puerile companion Bowers unexpectedly urges him to face up to the truth that "there are some things we have to accept for what they are" (Bainbridge 1993, p. 61).

The complexities of the five men's personalities, oscillating between the polarities of infantilism and a mature response to mortality, are perhaps best shown in Birdie Bowers. His participation in the expedition reveals simultaneously cowardice and bravery. His shyness with women and social awkwardness are matched only by his courage in confronting the elements, which aligns his career with a Peter Pan-like escapism from adult life. Bowers's image of death on the Polar expedition is initially naively pleasant: "… the cold, from what I've read, merely numbs the mind and positively lulls one into sleep" (Bainbridge 1993, p. 50). His hopes that at least "the cold won't rot us" (Bainbridge 1993, p. 65) are soon disproved as the first two men die of gangrene caused by frost bite. Yet, confronted with a natural environment whose hostility surpasses his imaginings, Bowers rises to the occasion and is determined to go forward and reach the Pole. For all his talk about the need for spiritual growth, worldly ambitions and a deep conviction about divine providence, Bowers surprises Wilson again with the bleakly realistic assertion "All I know is, nothing matters a damn except that we should help one another" (Bainbridge 1993, p. 64).

The sequence of speakers is deliberate; it shows the fluctuations of mood from Evans's carefree expectations of a big—and profitable—adventure, to Wilson's and Bowers's naive belief in a great mission, to Scott's growing self-doubt, to

[3] The cross marking the site of the men's death bears an epigraph taken from "Ulysses": "To strive, to seek, to find, and not to yield". This line was discovered inscribed in a volume of Robert Browning's poetry found by Scott's body (Preston 1997, p. 229).

Oates's criticism of the folly of the undertaking. Yet the birthday boys continue to play by the rules, despite the severe disappointment of being pipped at the post by Amundsen, despite the leader's more and more apparent miscalculations and misjudgements, and despite death looming large. Bowers reflects: "Not once throughout our dreadful journey had a cross word come between us, nor had we forgotten the civilities, those please-and-thank-yous which can mean so much when everything else has gone by the board" (Bainbridge 1993, p. 151). Scott's final concern is for the lot of his subordinates; Oates never openly rebels. Some partial confessions are made: Scott eventually admits to himself that it is his relationship with his wife that is the actual focus of his life rather than his professional ambitions; he openly shares with the others memories of his unhappy childhood; Bowers identifies his loneliness as his main personal problem; Oates for the first and last time talks about his wartime experiences.

6 Conclusion

Although the approach of death brings a degree of self-knowledge and better knowledge of others, the prospect does not entail a radical readjustment of their perspectives. As Francis Spufford (1991, p. 24) notes, "death comes to silence the characters' half-understanding of themselves, not to complete it". Nor are the dreams and fantasies with which they set out completely abandoned. The explorers confront them for the last time after the anti-climax of reaching the Pole just after the Norwegians. Evans is the first to die, in his delirious state recalling the vision of a pub by the sea; Bowers continues to delude himself that life has some happiness in store for him, Oates still makes plans to change his life on return. In Oates's mind, his suicidal walk into the snow merges with memories of his nursery, and the childish fantasy of a man bringing a horse for him on his birthday. In dying, Oates becomes again the birthday boy he once was. Finally, an imaginary journey to a land of dreams is also Scott's strategy to prevent himself from going mad: "I start to think of her [wife] a dozen times a day, and then stop myself, for that way madness lies—it sends me into daydreams in which I sail into port, the bands playing and the flags fluttering, happy ever after in never–never land" (Bainbridge 1993, p. 127).

References

Bainbridge, B. (1991) 1993. *The Birthday Boys*. London: Penguin Books.

Barczewski, S. 2007. *Antarctic Destinies: Scott, Shackleton and the Changing Face of Heroism*. London and New York: Continuum.

Barrie, J.M. 1995. *The Admirable Crichton. Peter Pan* [play, 1904]. *When Wendy Grew Up. What Every Woman Knows. Mary Rose*. Ed. with an Introduction by Peter Hollindale. Oxford: Clarendon Press.

Enright, D.J. 1992. Just going outside. *London Review of Books* 30 January:16.
Grubisic, B.J. 2008. *Understanding Beryl Bainbridge*. Columbia: University of South Carolina Press.
Kirst, G. 1994. Antarctic antics. *New York Times Book Review* 17 April:15.
Linklater, A. 1992. No longer hero or villain. *The Spectator* 4 January:25-26.
Preston, D. 1997. *A First Rate Tragedy: Captain Scott's Antarctic Expeditions*. London: Constable.
Spufford, F. 1991. In Scott's footsteps. *Times Literary Supplement* 20 December:24.
Tennyson, A. 2000. Ulysses. *The Norton Anthology of English Literature*, ed. M.H. Abrams, 7th ed., vol.2, 1213-1214. New York and London: Norton.

Crossing Racial Frontiers in the Quest for Cultural Acceptance as Seen Through Selected Works by Jackie Kay

Aniela Korzeniowska

Abstract Jackie Kay, born in Edinburgh in 1961 of Scottish-Nigerian descent, was brought up by a white couple in Glasgow. Her first poetic work, a collection entitled *The Adoption Papers* (1991), is not only autobiographical in nature but also explores cultural identity and genetic origin. The highly expressive and telling story she recounts is presented from three different points of view: that of the birth mother, the daughter and the adoptive mother, three different voices additionally signalled by three different type-faces. Being black and also Scottish, as well as growing up in a predominantly white Glasgow, Kay knows only too well what a lack of acceptance actually feels like, and in a country that, through birth and upbringing, she felt she naturally belonged to. I would like to confront these two situations referring to such works as the above mentioned *Adoption Papers*, four poems from her 1993 collection *Other Lovers*: 'Gastarbeiter', 'In my country', 'Compound Fracture' and 'Colouring In', 'Old Tongue' from *Life Mask* (2005), as well as a very relevant piece of prose 'Why don't you stop talking' (2002). Another work addressed to young readers, *Strawgirl* (2002), completes the image. Writing consistently from her own experience and with her own subjectivity always coming to the foreground, Kay confronts the issue of race and the lack of acceptance of otherness, the feeling of being within as well as outwith a certain society or community.

A. Korzeniowska (✉)
Warsaw University, Warsaw, Poland
e-mail: a.korzeniowska@uw.edu.pl

J. Fabiszak et al. (eds.), *Crossroads in Literature and Culture*, Second Language Learning and Teaching, DOI: 10.1007/978-3-642-21994-8_38,

1 Introduction

Jackie Kay, born in Edinburgh in 1961 of Scottish-Nigerian descent, was brought up by a white working-class, leftist-oriented couple in Glasgow. Her first poetic work, a collection entitled *The Adoption Papers* (1991), is not only autobiographical in nature but also explores what the Scottish literary critic Roderick Watson calls 'constructed identity and genetic origin' (2007, p. 312). The highly expressive and telling story she recounts here[1] is presented from three different points of view: that of the birth mother, the daughter and the adoptive mother, three different voices additionally signalled by three different type-faces. Being black and also Scottish, as well as growing up in a predominantly white Glasgow, Kay has been over the years confronted in a multitude of instances with the situation where she has very clearly felt a lack of acceptance in a country that, through birth and upbringing, she felt she naturally belonged to.

In the following discussion I would like to confront these two situations referring to such works as the above mentioned *Adoption Papers*, four poems from her 1993 collection *Other Lovers*: 'Gastarbeiter', 'In my country', 'Compound Fracture' and 'Colouring In', 'Old Tongue' from *Life Mask* (2005), as well as a very relevant piece of prose 'Why don't you stop talking' from her 2002 collection of short stories so entitled. Another work addressed to young readers, the magical novel *Strawgirl* (2002), completes the image. Writing consistently from her own experience and with her own subjectivity always coming to the foreground, Kay confronts the issue of race and the lack of acceptance of otherness, the feeling of being within as well as outwith a certain society or community.

2 The Adoption Papers

A certain image of Scotland in the 1960s is already presented to us in the foreword to *The Adoption Papers*, that consist of three parts, each one devoted to a certain period in the lives of the three leading characters. When the adoptive mother realised she would not be able to give birth herself, she and her husband decided to apply to the appropriate authorities for adoption. She reveals that:

> even in the early sixties there was
> something scandalous about adopting,
> telling the world your secret failure
> bringing up an alien child,
> who knew what it would turn out to be
> (Kay 2005, p. 10)

[1] It was originally broadcast on radio which, with the interplay of voices, gave an even more effective impression than one receives through the written word.

To admit any sort of failure, even if it was no fault of one's own, gave off a whiff of scandal. At the same time, though, the waiting list to adopt a baby seemed to be endless, or the given agency did not want them on their lists because either they were not church-goers or simply did not live close enough to a church. This, however, changed the moment—in sheer desperation—she mentioned:

> Just as we were going out the door
> I said oh you know we don't mind the colour.
> Just like that, the waiting was over.
> (Kay 2005, p. 14)

Reflecting on her baby's childhood, she says in Chap. 7 of Part Two (1967–1979):

> I brought her up as my own
> as I would any other child
> colour matters to the nutters;
> but she says my daughter says it matters to her
> (Kay 2005, p. 24)

For the adoptive mother what had always been of utmost importance from the moment she and her husband had decided to adopt a baby, was *the baby*. Appearance or colour of skin were issues that were not even considered. This, however, did not seem to be the natural way of thinking according to the adoption agencies of the 1960s. It was only when, quite by chance and as if in passing, she mentioned that colour did not matter, it suddenly appeared that in fact there was a baby available for adoption. It just so happened the baby was simply a different colour of skin. This appeared to be a big problem for the agency for whom, according to the narrator, "she wasn't even thought of as a baby" (Kay 2005, p. 24), she was as if non-existent.

However, as this baby grew up, started school and was confronted by a world outside her loving family, she was soon made very aware of the fact that she was somehow different and it was this difference that was excuse enough for the other children and, what's more, the teachers, to bring this difference, as they saw it, very much to the foreground. What can be called cold-bloodied racism is what she encountered when called such names as *Sambo, Dirty Darkie* (Kay 2005, p. 24, original italics), the teacher siding with the white children against her because she fought anyone calling her names. She would react with her fists which in turn would lead to the teacher threatening she would be a juvenile delinquent in a few years time:

> Do you know what that is? Look it up in the dictionary.
> She spells each letter with slow pleasure.
> Read it out to the class.
> Thug, Vandal. Hooligan. Speak up. Have you lost your tongue?
> (Kay 2005, p. 25)

It is only natural that children will follow their teachers' example and if such names and emotionally-loaded epithets as the above originally come from the teacher herself: "the other kids are all right till she starts" (Kay 2005, p. 25), it was only natural that this would lead to leaving a lasting effect. When practising for the school show, for example, and having difficulties in learning the steps for the Cha Cha and the tellingly named Black Bottom,[2] she heard her teacher shout from the bottom of the class:

> …..……..Come on, show
> us what you can do I thought
> you people had it in your blood.
> (Kay 2005, p. 25)

The trait in human nature concerning curiosity in anything that is different, but at the same time experiencing a certain anxiety and fear of this difference and what it might bring to one's well-known and familiar world, is revealed in such often encountered reactions as when an extremely fair-skinned girl from the Highlands of Scotland—the future natural mother—went out with her Nigerian boyfriend:

> Olubayo was the colour of peat
> when we walked out heads turned
> like horses, folk stood like trees
> their eyes fixed on us—it made me
> burn, that hot glare; my hand
> would sweat down to his bone.
> (Kay 2005, p. 26)

The fear of the alien although that 'alien' may not be a threat in itself. This reference to the reaction of the passers-by to the relationship is well-known to us from the past, although also often still encountered today, especially in small, close-knit, mono-racial communities. It is hardly surprising that at the young age of nineteen and being left to bring up her child on her own—he wrote he couldn't leave Nigeria—she felt she was acting for the best by signing her baby's adoption papers. There were many decisions taken along the way that were to result in the child, teenager, then adult trying to find out what her roots really were and to confront the hostility she had so often encountered in the community she grew up in, a community that should have been naturally hers, but that kept questioning her identity only because the colour of her skin was different to that of the majority. What this majority perceived as 'otherness', in contrast to her white adoptive parents, was something they could not—and still often cannot—insert and absorb into their national collective, a collective in which there appears to be an incompatibility of being black and Scottish. This may seem all the more surprising when yet again this majority quite naturally accepts the very hybridity of the Scottish nation, the notion that one cannot in any sense of the word talk about national purity.

---

[2] A popular American dance from the 1920s.

3 Scotland Contradicting Itself

In reference to the above—although maybe a slight digression from Jackie Kay's writing as such—it might be interesting to quote another very famous contemporary Scottish writer, William McIlvanney, who delivered a memorable speech in December 1992 in Edinburgh. The city was host to the summit meeting of the European Community, during which Scots—McIlvanney among them—made the most of the situation to promote the long-standing Scottish battle for devolution. At a rally that amassed up to 30,000 people he said the following memorable words: "We gather here like refugees in the capital of our own country. We are almost seven hundred years old, and we are still wondering what we want to be when we grow up…". He then concluded in a tone of tremendous pride: "Scottishness is not some pedigree lineage. This is a mongrel tradition!" (in Ascherson 2003, p. 75). Neil Ascherson, a well-known Scottish historian, journalist and one of the campaign organizers, went on to comment this speech in his fascinating work *Stone Voices*: At those words, for reasons which perhaps neither he nor they ever quite understood, the crowd broke into cheers and applause which lasted on and on (Ascherson 2003, p. 75).

Apart from the fact that McIlvanney's own heritage is a mixture of Scots and Irish, if we go back in time and look at the diverse peoples who, over the centuries, have settled in what we know as the country of Scotland today, we can fully understand why he used the word 'mongrel'. It is indeed a beautiful national mix, which additionally has in very recent times—the last couple of decades actually—also started to be a 'mongrel nation' that is not only white in colour. The change, however, is not necessarily accepted by the majority and it is this lack of acceptance that Jackie Kay and many like her have experienced. To quote Kay directly on this subject:

What interests me particularly is the way in which you can be in a society that is cordoned off and oppressed[3] but also be oppressed within that society, or divided within that or not belong to the common group in exactly the same way. I do think that sense of being outside with being inside Scotland—with being very proud of the country and being very proud of being Scottish, and also being outside in terms of receiving a lot of racism from other Scottish people—is what fuels my sense of how and what I write (in Gish 2001, pp. 179–180).

It is this 'sense of being outside with being inside Scotland', or any other country for that matter, and different understandings of identity that can be observed not only in Jackie Kay's *Adoption Papers*, but also in such poems as her 'Gastarbeiter', 'In my country' or 'Compound Fracture' (1993/2001). 'Old Tongue' (2005) from a much more recent collection is concerned more with the

---

[3] Reference here is being made to how the English treated their northern neighbours over many centuries in the past.

importance and significance of one's own language in contrast to the acceptance of the foreign that often happens to be part of one's subconscious more than a conscious rejection of one's own 'old tongue'. In this sense, you not only are physically outside Scotland but linguistically, culturally as well.

4 The Issue of 'Otherness'

'Gastarbeiter' does not refer to Scotland as such, as the title itself clearly shows, but the issue of 'otherness', of trying so hard to please with it not making any difference to the consequences:

> … she learned the new tongue
> and spoke it like a faltering step
> wanting to please, …
> …..
> In one bed they all slept, rolled
> Tight, a bandage on an open wound,
> *Gaster, bite her*; sleep is always light
> When stars are the shapes of swastikas
> and the limbs of hate move clockwise.
> (Kay 2001, p. 22)

The poem ends with the house in which they were living being blown up. Strangers are simply not accepted.

The short and powerful poem 'In my country', on the other hand, brings us back to Scotland and the extremely uncomfortable confrontation concerning the difference between Kay's sense of belonging and how others may happen to see it. When out walking

> a woman passed round me
> in a slow watchful circle,
> as if I were a superstition;
> or the worst dregs of her imagination,
> (Kay 2005, p. 24)

and when she finally managed to open her mouth and say something, the typical question addressed to strangers cut harshly through the air: *Where do you come from?* (Kay 2005,p. 24. original italics), not expecting to hear in answer: 'Here. These parts.' (Kay 2005, p. 24).

Being asked in your own country by a stranger, whose looks could kill if given half the chance, where you come from simply on the basis of the colour of your skin cannot help but lead to bitterness and to what can spiral so quickly into hatred and so many unnecessary social misunderstandings. Here it was just one seemingly very ordinary question that happened to say so much. Just one adult addressing another. Speech, the tone of voice and the words expressed.

In 'Compound Fracture', however, Jackie Kay goes back to the experiences of childhood where the narrator was not only suffering physically from a compound fracture when her 'bone came through the skin' (Kay 2005, p. 25) but was soon to be additionally fractured emotionally, again caused by an adult's tone of voice and the angry and insulting words that were hurled at her. When finding herself in hospital, desperate and screaming for her mother who was 'in the next room repeating our names' (Kay 2005, p. 25), the nurse

> burst into the white casualty
> her eyes bulging with cruelty; a terse –
> *Now Now Now*, her voice hailstones
> pelting—*you won't be seeing your mother*
> *unless you button*
> *that thick lip*, and worse, worse...
> .....
> (Kay 2005, p. 25)

Instead of a soothing tone of voice and words of sympathy, which one would expect from a nurse, the child is made all the more aware of certain physical features that turn out to be an easy target and a highly convenient pretext for insult. The compound fracture has a much larger dimension and the hurt goes so deep that it cannot be erased no matter the number of years that have passed in the meantime.

5 Cultural Heritage

The last poem to be discussed here is 'Old Tongue' which does not concern racism as such, but to what extent language and how one speaks, how one uses one's own personal voice can change unconsciously—or sometimes consciously—in the light of the search for social acceptance. Here this is not stated explicitly but taking into consideration the poet's concern with cultural belonging and how important language is in reference to this, it can be read between the lines:

> When I was eight, I was forced south.
> Not long after, when I opened
> my mouth, a strange thing happened.
> I lost my Scottish accent.
> Words fell off my tongue:
> eedyit, dreich, wabbit, crabbit[4]
> .....
> and I turned my back on Scotland...
> .....
> Out in the English soil, my old words
> buried themselves. It made my mother's blood boil.
> I cried one day with the wrong sound in my mouth.
> I wanted them back; I wanted my old accent back,
> my old tongue. My dour Scottish tongue.

---

[4] The English equivalent here would be: idiot, drab (miserable), weak, cross (bad-tempered).

> Sing-songy. I wanted to *gie it laldie*.
> (Kay 2005, p. 50)

The last phrase is very important here. In English translation, it is the narrator who wanted to 'give it a thrashing', to punish it for disappearing, the result of which his/her heritage has been lost. At the same time, one never really becomes part of the new country and its culture.

In his article "The World and the Home", Homi Bhabha speaks about the ethics of co-existence emerging from "a social space which has to be communally shared with others, and from which solidarity is not simply based on similarity but on the recognition of difference" (1998, p. 51). It is the very lack of this 'recognition of difference' that can also be observed in Jackie Kay's story 'Why don't you stop talking' from 2002, which can be found in her collection of the same title. Emphasising the continuing presence of racist violence in the UK, "caused mainly by what Kay perceives as a general, large-scale regression to culture being apprehended as racially defined and enclosed" (Brown 2007, p. 226), this story portrays what she sees as this cultural regress and a moving away from any 'ethics of co-existence' whatsoever. One of this black narrator's main characteristics is her penchant for talking and for saying loudly and clearly what she thinks. Without going into the details of all the events that take place, we are confronted with what Matthew Brown, in his article "In/Outside Scotland: Race and Citizenship in the Work of Jackie Kay", calls the "transmutation of linguistic into physical injury" (Brown 2007, p. 226), with someone threatening to 'belt her one on that fat black mouth' (Kay 2002, p. 45) or a woman slapping her and telling her to 'shut [her] fucking trap' (Kay 2005, p. 46).[5] In the final paragraph, she turns on her own tongue with a razor, giving it what she says it deserves for always getting her into trouble:

> [I]t will be punished. Then it will learn. I pick up the razor and I cut my tongue. It is actually painful, but the pain feels good. The pain feels deserved. The blood is generous and very red and it pours down my face. I wish they could all see me now (Kay 2005, p. 50).

We can say that Kay's concern here actually goes further than just the occasional incident. It covers a British society where there is a "persistent official preoccupation with the identification and registration of 'difference', the conspicuous absence of a civic community and, finally, the ongoing transmogrification of social space from a realm of hospitality and empathy into one of violence and despair" (Brown 2007, p. 226).

[5] Recalling 'Compound Fracture', we can see how such phrases referring to a person's mouth or lips, which have their natural links with voice, keep returning in Kay's works as these are some of the most common insults hurled at another person whom one wants to stop talking, or screaming if it happens to concern a child in pain. The fact that the given victim has certain physical features that may be a little different to those of the majority instinctively evokes the aggressive use of such adjectives as 'thick', 'fat' or 'black'.

This issue can also be illustrated through the work and writing of such Scots as Robina Qureshi, for example, who, like Jackie Kay, was born and brought up in Glasgow, but to Pakistani immigrants. Because she feels that Scotland continues to be a deeply racist society, she is actively involved in challenging what she has personally encountered as institutionalised racism and the government's policy towards refugees (cf. Qureshi 2002, pp. 217–219). To quote her directly: "Nothing has changed, every day racism continues on the streets, in our faces and insidiously behind our backs. The refrains I heard in childhood—'Go back home, Paki', Go back to where you belong'—still remain ..." (Qureshi 2002, p. 218).

It can be perceived, however, that although pessimism appears to seep through Qureshi's views on the attitude of Scots to the 'other', she continues her battle, showing through it in-born hope for a better future for all races in the country of her birth. And it is this optimism—also subtly educational in tone—that appears in *Strawgirl* (2002), Jackie Kay's novel for younger readers.

In this moving story set in the Highlands of Scotland and full of magic realism, Kay moves from the all familiar story of racism at school—and all the heartbreak and unhappiness this involves—to acceptance, respect and optimism in the future. Through touching upon the subject that pervades so much of her work for adults, in this book for children, Kay shows how Molly Siobhan MacPherson (Maybe to her family)[6] manages to overcome seemingly unsurmountable hurdles to achieve much deserved happiness in the end.

The issue of national identity first appears when her teacher is doing family trees in class. Molly (11 and a bit years old) explained that her mother was from Lochinvar (Scotland) but her mother's mother had come from Galway, hence her mom would say "she feels just as Irish as she does Scottish" (Kay 2002a, p. 20). This Maybe said proudly, which is a reflection of the present acceptance of Scots of Irish origin.[7] She was not so happy, however, when asked about her father—a member of the Nigerian Ibo tribe—as "she didn't like it when everybody had to be reminded that she was different" (Kay 2002a, p. 20). Although Jamie MacPherson had been adopted by a Scottish couple at the age of eleven,[8] he insists Maybe never forgets her Nigerian—and Ibo—origins. This, however, is what she has certain difficulties with. She feels that "[i]f she forgets it, she can get by. Be like the rest of them. Fit in" (Kay 2002a, pp. 5–6). The wish for sameness, *not* to be different, to be like everybody else was the overpowering feeling she could not resist. She wanted the impossible: "to blend into the background" (Kay 2002a, p. 22).

Some of the children obviously did not help her in her blending wish:

[6] The nickname Maybe came from the fact that she was always undecided and her answer to everything was always 'maybe'. Yes and no answers simply did not appear. It is also interesting to note that the main character has the same background as Jackie Kay herself.

[7] See the above reference to William McIlvanney, for example.

[8] As we can see, the issue of adoption reappears here yet again.

> 'These kids at school keep calling me 'Ibo" and things like that, Maybe said, her head down. Irene's mouth fell open.
> 'You know Ibo is not a bad word, don't you?' Jamie said. 'You should be proud of being an Ibo.'
> 'They say like it's an insult! They chant it.' Maybe said, her voice rising. (Kay 2002a, pp. 29–30)

This feeling of wanting to be the same as everybody else in school suddenly changed when her father was killed in a car accident. In her overwhelming grief she suddenly became interested in everything to do with him. "One day she wanted to save up and fly to Nigeria .... She wondered whether his family would like her, or whether she would just be too different, too Scottish" (Kay 2002, p. 83). Here we have Maybe suddenly realising that for her Nigerian relatives, she also may be different, in this case too Scottish to be easily accepted.

Reaching the end of the story and after many adventures involving magic and the participation of Strawgirl, visible only to Maybe herself, our main character is yet again in the classroom, in which a discussion on difference is taking place. This time, Maybe is speaking about the Ibo and the Scots, about their pride and the similarities between the two peoples. She is full of confidence, so different to what we perceived at the beginning of the story:

> It seemed a whole lifetime ago that Maybe had felt embarrassed or ashamed about who she was. She would never again be ashamed to stand up in her class and say, 'My dad was an Ibo, and that makes me Ibo too,' she added proudly. (Kay 2002a, p. 246)

6 Conclusion

In conclusion, I would like to refer to Fiona Wilson and her recent article "Scottish Women's Poetry since the 1970s" in which she draws our attention to how Jackie Kay uses dramatic technique in her poetry to present her subject matter and when "writing her self" has the "opportunity to stage questions about identity" (2009, p. 27). Her dramatic technique is used to the full as the above poetic representations have illustrated and it is these questions about identity in our contemporary world that still require attention and discussion. This leads in its natural course of events to the acknowledged fact that our world is not uniform, that there is 'difference' or 'otherness', if we like, and that this is a form of enrichment, never of impoverishment.

This has been substantiated by Kay herself with the optimistic note she leaves her reader with in *Strawgirl.* The necessity to speak about difficult and controversial issues, to fill in gaps in communication through poetic expression as performed by Jackie Kay or through the work of Scottish-Pakistani activist Robina Qureshi, for example, continues to be necessary in the quest for cultural acceptance when crossing racial frontiers.

References

Ascherson, N. 2002/2003. *Stone voices. The search for Scotland*, London: Granta Books.
Bhabha, H. 1998. The World and the Home. In *Dangerous liasons: Gender, nation, and postcolonial perspectives*, eds. A. McClintock, A. Mufti and E. Shoat, 445-455. Minnesota: University of Minnesota Press.
Brown, M. 2007. In/outside Scotland: Race and citizenship in the work of Jackie Kay. In *The Edinburgh companion to contemporary Scottish literature*, ed. B Schoene, 219–226. Edinburgh: Edinburgh University Press.
Gish, N. 2001. Adoption, identity, and voice: Jackie Kay's inventions of self. In *Imagining adoption: Essays on literature and culture*, ed. M. Novy, 171–191. Ann Arbor: University of Michigan Press.
Kay, J. 1991/2005. *The adoption papers*. Northumberland: Bloodaxe Books.
Kay, J. 1993/2001. *Other lovers*. Northumberland: Bolldaxe Books.
Kay, J. 2002a. *Strawgirl.* London: Macmillan Children's Books.
Kay, J. 2002b. *Why don't you stop talking*. London: Picador.
Kay, J. 2005. *Life mask*, Northumberland: Bloodaxe Books.
Qureshi, R. 2002. Being Scottish. In *Being Scottish. Personal reflections on Scottish identity today*, eds. T. Devine and P. Logue, 217-219. Edinburgh: Polygon.
Watson, R. 1984/2007. *The literature of Scotland*, Vol. II, Basington & New York: Palgrave Macmillan.
Wilson, F. 2009. Scottish Women's Poetry since the 1970s. In *The Edinburgh companion to contemporary Scottish poetry*, eds. M. McGuire and C. Nicholson, 21-35. Edinburgh: Edinburgh University Press.

The Flying Dutchman's Mimetic Desire. Crossing Geographical and Moral Frontiers in Frederick Marryat's *The Phantom Ship*

Joanna Mstowska

Abstract The chapter investigates the figure of Captain William Vanderdecken, one of the protagonists of Frederick Marryat's sea novel entitled *The Phantom Ship*, who is not only an incarnation of the Flying Dutchman but may also be regarded as a Faustian character. In his enormous pride and overwhelmed by discovery passion, William Vanderdecken wants to usurp the Absolute's demiurgic power and regards discovering lands, crossing frontiers, both geographical and moral, and staking out new territories as an imitation of God's creative acts. The Captain's refusal to accept his limitations and powerlessness, as well as his diabolic determination to go round the Cape of Good Hope at all costs and to master the sea may symbolize his craving for the participation in God's omnipotence. Blinded by the desire to be God's equal, the Captain does not hesitate to give his soul to devil and to commit blasphemy against God. Therefore, it is also the author's intention to interpret the legend of the Flying Dutchman as a modern version of the biblical archetype of the fall of man as a result of succumbing to Satan's temptation. As the Faustian archetype, the Adamic myth and the legend of the Flying Dutchman are all marked by the presence of Satan, who is the source of negative inspiration, the paper will also explore the relation between Captain William Vanderdecken, Satan and God applying René Girard's theory of mimetic desire presented in *Deceit, Desire and the Novel* (1965).

J. Mstowska (✉)
Nicolaus Copernicus University, Toruń, Poland
e-mail: joannamstowska@o2.pl

J. Fabiszak et al. (eds.), *Crossroads in Literature and Culture,* Second Language Learning and Teaching, DOI: 10.1007/978-3-642-21994-8_39,

1 Introduction

According to folklore, the Flying Dutchman is a ghost ship doomed to sail the seas forever and to fight with the ferocious sea. It is also the name of the Dutch Captain, a daring and courageous traveller who is condemned to be forever tossed about by the waves around the Cape of Good Hope, willing death to bring an end to his suffering. He cannot reach salvation and escape his fate as other seamen seeing the ghostly ship on the horizon are petrified and quickly change their course in order to escape the phantom ship's fatal influence. The legend of the Flying Dutchman was already a well-known tale with numerous versions in nautical folklore before Frederick Marryat incorporated it into his text. In his novel set at sea entitled *The Phantom Ship,* Marryat presents one of the most extensive versions of the legend and stresses the Dutch Captain's crossing not only geographical but especially moral frontiers. Thus, the aim of this paper is to investigate the figure of Captain William Vanderdecken, one of the protagonists of Marryat's novel, and to interpret him not only as an incarnation of the Flying Dutchman but also as a Faustian character. In his enormous pride and overwhelmed by the passion for discovery, William Vanderdecken wants to usurp the Absolute's demiurgic power and regards discovering lands, crossing frontiers, and staking out new territories as an imitation of God's creative acts. I also want to explore the legend of the Flying Dutchman as a modern version of the biblical archetype of the fall of man as a result of succumbing to Satan's temptation. As the Faustian archetype, the Adamic myth and the legend of the Flying Dutchman are all marked by the presence of Satan, in my paper I will elaborate as well on the relation between Captain William Vanderdecken, Satan and God, applying René Girard's theory of mimetic desire presented in *Deceit, Desire and the Novel* (1965).

2 The Dutch Captain as an Aquatic Incarnation of Faust

Captain William Vanderdecken of *The Phantom Ship* may be regarded as a Faustian figure representing both individual and universal features. Being extremely gifted and enormously proud, Faust is beyond doubt an exceptional individual rising above mediocrity. On the other hand, however, his cognitive insatiability and his desire to know, not to be lost in conjecture, is common to all humans. Overwhelmed by his passion, he wants to touch the intangible and to know the unknowable, and is not able to accept the limits of cognition, which result from the weaknesses of human nature. This active and dynamic genius, whose life is characterized by constant eagerness, does not agree to accept his helplessness and insignificance within the arrangement of the universe. Rejecting the limitations of human condition and rebelling against the inscrutability of knowledge, Faust lives on the verge of normality and insanity. In his enormous pride, Faust is convinced that he is able to scrutinize the inscrutable and to be

God's equal. The Renaissance legend of Faust becomes the myth of human sin usurping God's rights.

In his book devoted to the fall of man as presented in Genesis, Marian Grabowski states that in Eden the serpent promises Adam a participation in divinity, that is, in God's omnipotence and omniscience. The serpent's lie is tempting for Adam because it presents the absurd promise of achieving perfection, which is in fact inaccessible, as being within human reach (Grabowski 2006, p. 217). The serpent always takes advantage of the feeling of something lacking, as well as of the fact that for each person the desired object is a mixture of what is real and imagined. Being the source of negative inspiration, the serpent presents the illusory and the impossible as reachable and possible, making a human being overwhelmed by desire and hardly able to resist the temptation. Faust's pact with the devil may be regarded as the repetition of the fall of Adam in Eden. Faust rejects his helplessness and desires to manifest his unusual power. Much like Adam, he wants to try a fruit from the tree of the knowledge of good and evil, to gain knowledge in order to unravel all the mysteries of the world and to decide what is definitely good and what is not. Mefistofeles gives an unfulfilable promise to satisfy Faust's desires for the price of his soul.

3 The Legend of the Flying Dutchman as a Version of the Archetype of the Fall of Man

The legend of the Flying Dutchman also may be interpreted as a romantic version of the Biblical archetype of the fall of man as a result of succumbing to Satan's temptation. The desires and temptations of William Vanderdecken are similar to Faust's constant eagerness and his cognitive insatiability. Vanderdecken's blasphemy against God, which took place 17 years prior to the action of *The Phantom Ship*, is not presented by a third-person omniscient narrator but appears as a recollection on the part of his wife, Catherine. In a flashback, she describes a tempestuous and horrifying night when the ghost of her husband suddenly appeared in front of her eyes to confess his dreadful deed. Catherine's story carries with it a sense of credibility as she heard the reminiscence of the fatal cruise directly from her husband's mouth and also in the form of a letter, left by the ghost of Captain Vanderdecken.

3.1 Hatred of Inscrutability

In a short retrospection, William Vanderdecken is presented as a bold and first-rate sailor who resembles a typical romantic hero. During his cruise to India with a valuable cargo, he attempted to round the Cape of Good Hope. Prevented by adverse winds and currents, he was not able to fulfil his daring undertaking. Not discouraged by misfortune and continually overwhelmed by the passion for discovery, Vanderdecken struck a mutinous pilot who then fell overboard and drowned. It seems to me that his irresistible desire to round the Stormy Cape at any cost, which soon becomes a kind of obsession, may be compared to Ahab's quest for the white whale. In his essay devoted to *Moby Dick*, Michael T. Gilmore investigates Ahab's aversion to inscrutability and his desire to eliminate all that is beyond his understanding, at the same time stressing the fact that the quest for knowledge is both fascinating and associated with unimaginable destruction. For Captain Ahab, the search for knowledge has a touch of obsession much like the extermination of Moby Dick, which for him represents everything that is evil in the world, and becomes the raison d'être of the Captain's existence. Similarly, Vanderdecken, another fanatic for knowledge and a truth-seeker, hates inscrutability and cannot accept the fact that some lands and territories still have the status of an enigma for him. Also the Captain's insatiable passion for discovery turns into an obsession as he is ready to risk his life and to destroy himself in quest for scrutability. Regarding the sea as a malicious adversary which is preventing him from going round the Cape of Good Hope and testing the limits of his power, Vanderdecken wants to emerge victorious from the battle with the salty element at any cost.

3.2 Succumbing to Satan's Temptation

Not only the Faustian archetype and the Adamic myth but also the legend of the Flying Dutchman are marked by the presence of Satan, who is the source of negative inspiration. In *The Phantom Ship,* Satan takes advantage of Vanderdecken's feeling of unfulfillment, and his readiness to risk his life or to give up his soul to satisfy his desire for conquering at all costs and being powerful enough to master nature. It is Satan who kindles the Captain's passion for discovery to an enormous extent and tempts him with the promise of achieving God's perfection. As Vanderdecken refuses to accept his powerlessness and insignificance in confrontation with the Almighty and is convinced of his own greatness, the Captain gives credence to Satan's empty words. At the moment of committing blasphemy, the Captain rejects God and, like Adam, reaches for the fruit of the tree of the knowledge of good and evil. In other words, he resorts to Satan as the mediator who will help him to satisfy the desire for God's omniscience and omnipotence.

4 The Flying Dutchman as a Victim of Triangular Desire

René Girard's controversial idea of mimetic desire may be applied here to interpret the nature of Vanderdecken's pact with the devil. Girard's theory of the triangular nature of desire, in which he employed Plato's and Aristotle's concepts of mimesis and fused them with his critical outlook on Sigmund Freud's and Jacques Lacan's investigations of desire, was presented in *Deceit, Desire and the Novel.* Even though Girard's concept of mimetic desire does not bear directly upon Marryat's novel, his idea of the triangular nature of desire, however slightly modified, is applicable to Captain Vanderdecken's relationship with Satan (who is not mentioned *expressis verbis* in the novel) and God (also not explicitly named). The fact that the ultimate object of Vanderdecken's desire is metaphysical, as he does not want to have God's attributes but to be his equal, is an essential but lacking element in Girard's theory. What the Captain desires is not an external object but he wishes to discover the enormous divine potential within himself. He is convinced that the better he imitates God, the closer to perfection he comes. The Captain's desire to be God's equal is spontaneous only to a certain extent and appears as a result of his enormous pride and his conviction of being exceptional. To a greater extent, however, his desire is initiated by Satan, who tempts Vanderdecken with illusory promises of power and knowledge. As Satan also imitates God, or rather mocks Him, and desires His omniscience and omnipotence, he becomes the Captain's mediator, having a profound influence on his actions. Nevertheless, Vanderdecken is convinced that his relationship to the object of desire is independent of the mediator. Living in the illusion of autonomy and spontaneity, the Captain believes that his desire to be God's equal is thoroughly original and certainly not rooted in the Other. In his unawareness of any mediation and apotheosis of the Self, Vanderdecken does not realize that he places his faith in a false promise from the outside. The hero's illusion of the possibility of having God's omnipotence is only partially created by his imagination; to a large extent it is fostered by Satan who maintains the Captain's illusion of spontaneous desire. Vanderdecken borrows his desire from the Other (Satan) and completely confuses it with the will to be Oneself. He does not realize that behind his seemingly spontaneous desires, there is someone else's suggestion.

Vanderdecken would never reach for God's power if he were not imitating Satan and succumbing to his suggestions. Having no idea about the real nature of God, he copies the imaginary characteristics of the model. Thus, Vanderdecken's desire for transcendence is deviated. Vanderdecken is a man of extremes who often indulges in his own emotions and is a subject of flights of imagination. Blinded by his immense pride, his youthful ardour and his imaginative nature, the Captain perceives himself in a very unrealistic way. Vanderdecken's conviction of his greatness, uniqueness and exceptionality is the source of the Captain's strength and exhilaration in the face of danger and the unknown. Unfortunately, the vision of himself that the Captain projects lacks any basis in reality and Vanderdecken's distorted image of himself has a destructive influence on his actions. The Captain's pride and boldness make him particularly prone to Satan's temptations. His suggestion of the possibility of Vanderdecken's being God's equal, slips into the Captain's consciousness. His powerful imagination transforms Satan's false promise into a passionate desire for transcendence.

In the chapter entitled 'Men Become Gods in the Eyes of Each Other', Girard explores a metaphysical aspect of desire. He stresses that the desire for transcendence is usually initiated by an individual's faith in a false promise from the outside. The protagonists analyzed by Girard desire to absorb the being of the mediator, as they are convinced that only the mediator is able to help them achieve the divine inheritance, from which they feel excluded. However, the protagonists' desire for transcendence becomes deviated, as it is based on a lie. The mediator, toward whom the hero passionately turns, in fact does not enjoy the divine inheritance but only pretends to participate in it. Directing their existence toward the Other, the protagonists explored by Girard choose human models and treat them as substitute gods. The denial of God leads to the imitation of the model who seems to have God's attributes and to the deviation of transcendence. Girard also emphasizes that the characters who obey the suggestion from the outside and begin to regard themselves as gods, usually are aware of their nothingness and their inability to equal the God they desire to substitute. However, Girard points out that the heroes are too proud to admit their failure. Although the protagonists analyzed by Girard are bitterly disappointed with the deviation of the desire for transcendence, hiding this disillusionment becomes the essence of their existence. Girard calls their idea of transcendence deviated towards human beings.

Girard's statements may also be applicable to the interpretation of the nature of William Vanderdecken's desire for transcendence. The Captain's craving for transcendence is also deviated but, to paraphrase Girard's words, in the direction of Satan. Like the protagonists analyzed by Girard, Vanderdecken gives credence to a false promise from the outside. In Marryat's novel it is Satan who tempts the Captain with the promise that divinity is within his reach. Satan's idea of transcendence is also based on a lie, as he is not God's equal and does not enjoy God's omnipotence and omniscience. He only wishes and pretends that he does. Vanderdecken, like the heroes investigated by Girard, chooses a mediator and is convinced that Satan is able to help him achieve divinity. Unlike them, however, the Captain lacks awareness. In the apotheosis of the Self, he is not aware of the existence of the mediator whose suggestions he follows. What is more, he is so blinded by pride that he is not able to realize his insignificance within the arrangements of the universe and his nothingness in confrontation with the Almighty. He also does not accept the present situation and constantly looks to the future, associating it with his final triumph and his participation in divinity. Yet another difference between the Captain and the characters investigated by Girard lies in the fact that they consciously reject God and move towards the human mediator, whereas Vanderdecken wants to imitate God at any cost, even following Satan's deviated suggestions. In other words, he does not choose substitute gods as models but his model is the Almighty.

Another striking difference between the protagonists of Dostoyevsky's novels explored by Girard and the Captain is associated with the way they are punished for their desire to participate in God's transcendence. While Dostoyevsky underlines that God no longer exists and confronts his heroes with the ungoverned universe, the world presented in Marryat's novel is marked by the Almighty's presence. Captain Vanderdecken is punished severely for his enormous pride and his desire to be God's equal by the Almighty against whom he has trespassed. Both in his pride and in his final failure, Vanderdecken resembles the Old Testament king, Belshazzar. The fragment of the Bible describing the King's sin reads as follows:

> 1 King Belshazzar gave a great banquet for a thousand of his lords, with whom he drank. 2 Under the influence of the wine, he ordered the gold and silver vessels which Nebuchadnezzar, his father, had taken from the temple in Jerusalem, to be brought in so that the king, his lords, his wives and his entertainers might drink from them. 3 When the gold and silver vessels taken from the house of God in Jerusalem had been brought in, and while the king, his lords, his wives and his entertainers were drinking 4 wine from them, they praised their gods of gold and silver, bronze and iron, wood and stone *The New American Bible* (Dn 5, 1–4).

Much as his father, Nebuchadnezzar, Belshazzar feels powerful enough to equal God. Extremely proud and self-confident as he is, Belshazzar, the Chaldean king, regards himself as having God's attributes and the right to judge what is good and what is bad. The fact that during a great banquet the king allowed his lords, wives and entertainers to drink from the vessels stolen from the temple in Jerusalem is significant. It may be interpreted as an act of initiation. Up to this point the king has felt excluded from full participation in divinity, being convinced that there are certain mysteries that God reserved only for Himself. Having drunk the wine from the vessels that belonged exclusively to God, Belshazzar believes that he managed to wrest all secrets from the Almighty. The act of profanation of the vessels can be interpreted as the King's symbolic transition from partial exclusion to full inclusion in God's omnipotence and omniscience.

Similarly, Captain Vanderdecken also equals God in his own eyes. He swears by the fragment of the Cross that is hanging round his neck that he will circle the Cape of Good Hope against God's will 'in defiance of storm and seas, of lightning, of heaven, or of hell, even if [he] should beat about until the Day of Judgment' (Marryat 2004, p. 11). Vanderdecken is convinced that going round the Cape will prove his godlike mastery over nature. The Captain's blasphemy, like Belshazzar's profanation, can be regarded as a symbolic satisfaction of the desire for penetrating all God's mysteries and for participating in His divinity.

In both cases, God does not remain indifferent to the transgressions which Belshazzar and Vanderdecken commit against Him but immediately reacts and inflicts punishment. The fragment from the Old Testament continues:

> 5 Suddenly, opposite the lampstand, the fingers of a human hand appeared, writing on the plaster of the wall in the king's palace. When the king saw the wrist and hand that wrote, 6 his face blanched; his thoughts terrified him, his hip joints shook, and his knees knocked *The New American Bible* (Dn 5, 5–6).

The mysterious inscription that appears on the wall consists of three words: MENE, TEKEL, PERES. As nobody is able to decipher the meaning of the writing, Daniel who knows 'how to interpret dreams, explain enigmas and solve difficulties' (Dn 5, 12), is asked by the King to resolve the mystery. Daniel interprets God's handwriting as a harbinger of imminent punishment for the King's appropriation of God's attributes and as a warning against Belshazzar's impending fall. Indeed, God keeps His promise and that same night the King is killed.

Likewise in *The Phantom Ship* it is God's handwriting that announces the punishment. The words UNTIL THE DAY OF JUDGMENT are written in flame and appear in the centre of a huge cloud surrounded by immense darkness: 'The hurricane burst upon the ship, the canvas flew away in ribbons; mountains of seas swept over us, and in the centre of a deep o'erhanging cloud, which shrouded all in utter darkness, were written in letters of livid flame, these words—UNTIL THE DAY OF JUDGMENT' (Marryat 2004, p. 11). Unlike Belshazzar, the Captain is not punished with death but is sentenced by God to go round the Cape of Good Hope till the Final Judgment. For the adventurous and daring seaman, such eternal and monotonous sea-faring always in the same direction is the greatest suffering imaginable. God transforms Vanderdecken into the Flying Dutchman, a character whose ontological status is uncertain, as he is neither a human being nor a ghost. It is striking that the Flying Dutchman is the exact opposite of what Captain Vanderdecken used to be. Once enormously proud and bold, the Captain regarded the discovery of new lands and active fighting with the sea as the essence of his existence. Now Vanderdecken's pride is stifled and his passion for adventure is replaced by overwhelming passivity.

Unlike in Marryat's novel, in Dostoyevsky's fiction a non-existent God is not able to punish the protagonists for their pride and for regarding themselves as substitute gods. The punishment for them is a bitter confrontation of their imagined divinity with their obvious nothingness. As Girard puts it: 'And in [splendid palaces of imagination] the Self entertains itself, indescribably contented, until the day when the treacherous magician—reality—brushes against the fragile dream buildings and reduces them to dust' (Girard 1976, p. 28). Extremely disappointed as the heroes of Dostoyevsky's novels are, in their pride they are not able to admit the failure and they suffer throughout their lives pretending to be perfectly satisfied with the deviated transcendence. The distance between the desiring subject and the mediator is in Marryat's novel even greater than in the works of Dostoyevsky. In this case it is not physical space that measures the gap between the mediator and the subject, but the distance between the two is primarily spiritual and will always remain insuperable. Therefore, there is no rivalry with the mediator. In *The Phantom Ship,* the mediator, being a spiritual creature not only much more intelligent and perfect than a human being but also capable of transcending the boundaries of time and space, remains beyond the universe of the hero. Girard defines such a relationship as external mediation, stressing the fact that it appears in those works in which 'the distance is sufficient to eliminate any contact between the two spheres of possibilities of which the mediator and the subject occupy the respective centres' (Girard 1976, p. 9). Girard further claims that the hero of external mediation admits the truth about his desire. Although Vanderdecken does not worship his model but openly desires to imitate His power and knowledge, the divergence between God's transcendence and his limited nature makes the Captain's mimetic desire as absurd and grotesque as that of Dostoyevsky's protagonists.

As has been shown in this paper, Captain Vanderdecken may be regarded as an aquatic incarnation of Faust. The Captain's refusal to accept his limitations and powerlessness, as well as his diabolic determination to go round the Cape of Good Hope at all costs and to master the sea, may symbolize his craving for the participation in God's omnipotence. Blinded by his desire to be God's equal, the Captain does not hesitate to succumb to Satan's temptations and to give up his soul to the devil.

To conclude, not only in Dostoyevsky's novels but also in *The Phantom Ship*, the protagonists' desires for transcendence are initiated by the false promise from the outside. The fact that in both cases transcendence is based on a lie, leads to its deviation, either in the direction of the human, or in the direction of Satan. As a result, the protagonists of Dostoyevsky's fiction, as well as Captain Vanderdecken, become bitterly disillusioned with their deviated transcendence. In a world devoid of God, in Dostoyevsky's novels, the protagonists sentence themselves to lifelong suffering caused by their pretending to be perfectly satisfied with the deviated transcendence. By contrast, in Marryat's novel it is God who imposes punishment for man's desire to be His equal and to participate in His divinity.

References

The New American Bible. 2002. http://www.vatican.va/archive/ENG0839/_INDEX.HTM. Accessed 14 July 2010.

Gilmore, M. T. 2003. *Surface and Depth*. New York: Oxford University Press.

Girard, R. 1976. *Deceit, Desire and the Novel*. London: The Johns Hopkins University Press.

Grabowski, M. 2006. *Historia upadku* [History of the Fall]. Kraków: Wydawnictwo WAM.

Marryat, F. 2004 [1839]. *The Phantom Ship*, the Project Gutenberg eBook. http://www.gutenberg.org/etext/12954. Accessed 14 July 2010.

Forays into the Scientific Mindset: The Two Cultures in Ian McEwan's *Saturday* and *Solar*

Stankomir Nicieja

Abstract The paper discusses two recent novels, *Saturday* (2005) and *Solar* (2010) written by a distinguished contemporary novelist Ian McEwan in the context of C. P. Snow's famous lecture *The Two Cultures*. As the paper suggests, McEwan should be listed among the narrow group of British literary intellectuals living today who follow C. P. Snow's recommendations and try to bridge the gap between the worlds of science and the humanities. Both *Saturday* and *Solar*, the author of the paper argues, successfully demonstrate how science may become a rewarding and rich source of inspiration for a novelist.

1 Introduction: C. P. Snow's Two Cultures Lecture

In 2009, many academic societies and associations on the both sides of the Atlantic as well as in other parts of the globe celebrated the 50th anniversary of the lecture delivered at Cambridge University in May 1959 by a popular novelist, scientist and public servant, Charles Percy Snow (usually referred to as C. P. Snow). This memorable address entitled *The Two Cultures and the Scientific Revolution* was part of the annual tradition of the Rede Lectures dating back to the sixteenth century and named after Sir Robert Rede, the Chief Justice of the time.

It may seem rather peculiar that a single lecture is still commemorated more than 50 years after its delivery. Cambridge University hosts literally dozens of similar special events every year customarily delivered by distinguished academics, artists and political leaders of the world. These occasions, however, are rarely celebrated or even

S. Nicieja (✉)
University of Opole, Opole, Poland
e-mail: stann@uni.opole.pl

J. Fabiszak et al. (eds.), *Crossroads in Literature and Culture,* Second Language Learning and Teaching, DOI: 10.1007/978-3-642-21994-8_40,

remembered after many years. With due respect to his achievements, C. P. Snow could not be comfortably listed among the grandest figures of his time. Furthermore, the enduring fame of the lecture may appear even more puzzling form the contemporary perspective, when one learns that among C. P. Snow's key recommendations was the call to emulate the example of the Soviet Union in educational policy, especially in regards to teaching science. Naturally, this was not the main point of Snow's address, and certainly not the reason for the lecture's enduring fame.

Instead, thanks to its resonant theses, Snow's address became one of the most important points of reference in the on-going debate concerning the significance of science in modern culture and its relation to other disciplines taught at universities (especially the liberal arts). In the *Two Cultures and the Industrial Revolution* Snow described Western intellectual life as damagingly split into two different and antagonistic camps, hence "the two cultures" in the title.

Borrowing imagery and spirit form Churchill's famous "Iron Curtain" speech delivered 13 years earlier, Snow argued that the representatives of those "two cultures" live in open contempt and hostility toward each other. As he graphically put it: "Literary intellectuals at one pole—at the other scientists, and the most representative, the physical scientists. Between the two a gulf of mutual incomprehension—sometimes (particularly among the young) hostility and dislike, but most of all lack of understanding. They have a curious distorted image of each other" (Snow 1961, p. 4). Thanks to his experience as a chemist and author of popular novels, Snow, as William Todd Timmons (2007) observed, "thought himself uniquely qualified to comment upon the communication barriers he found between the culture of the literary elite and the culture of the scientist".

Throughout the lecture, Snow tried to assume a pose of an impartial arbiter; nevertheless, his speech became a potent accusation directed against the community of humanists or literary intellectuals (as Snow preferred to call them), for their apparent inability to embrace the inspirational power of science and the benefits of the Industrial Revolution. Particularly resonant was Snow's comparison of literary intellectuals to new Luddites, who remain hostile to the ground-breaking progress in human knowledge. In other words, Snow accused his colleagues from the liberal arts departments of snobbery, ignorance and staggering inflexibility in their assessment of science and tangible benefits of industrialisation.

The potency of C. P. Snow's words was unintentionally strengthened 3 years later when one of the greatest humanists living in Britain at the time, F. R. Leavis, also from Cambridge, responded to Snow's criticism in the 1963 Richmond Lecture. In this now notorious polemical address, F. R. Leavis made one crucial mistake. Instead of focusing on rebutting Snow's points, he assaulted his adversary in a shockingly virulent and personal manner. With this *ad personam* attack, Leavis unintentionally confirmed some of the worst stereotypes about the prejudices among the literary intellectuals that Snow described 4 years earlier.

When we revisit the so-called Snow-Leavis controversy today, it may seem quite surprising that despite numerous cultural changes since the early 1960s, so many points of the dispute summoned by the both sides are still relevant. Snow's resonant argumentation, although not devoid of major simplifications and flaws still constitutes an indispensable point of reference in various analyses assessing the impact of science on contemporary culture. Many of Snow's original observations remain pertinent after more than 50 years since their presentation. The veil of prejudice and ignorance highlighted by Snow continue to mar the exchange of ideas between the worlds of science and the humanities.

2 Scientific Inspirations in McEwan's Prose

Saturday is narrated from the point of view of an accomplished neurosurgeon, Henry Perowne. *Solar*, on the other hand, tells the story of Michael Beard, a gaffe-prone Nobel Prize winner in physics. It seems that both novels explicitly embrace C. P. Snow's postulates, expressed more than 50 years earlier, of closing the gap between the two cultures. On the other hand, however, they contain intriguing insights into the divisive practices within the two cultures, which make such an exchange of ideas difficult.

Saturday and *Solar* exemplify McEwan's ambition, rather rare in modern literature, to celebrate the intricate splendour of the world and human life without resorting to mysticism, magical thinking or other supernatural phenomena. He also belongs to a narrow group of contemporary novelists who manage to produce compelling and simultaneously commercially successful fiction that does not evoke fairies, ghosts, vampires or ersatz mythologies. Instead, he demonstrates through his writing that mature and intense moral or ethical judgment is possible within strictly rationalist and naturalist framework. In terms of outlook, McEwan derives much inspiration from his Oxford friend Richard Dawkins, but does not adopt the latter's confrontational and often alienating style.

3 *Saturday*: Scientist as a Heroic Figure

Saturday may serve as a very good example of this approach. It is an intricately woven novel that engages with a number of intriguing issues and themes, from the meditation on the complexities of individual experience to the wider political and social issues, including the war in Iraq and the legacy of the New Labour government. Adam Kirsch (2010) aptly characterised *Saturday* as: "a novel in which a man of science is made to symbolize everything that is noble and worth defending about our civilization [...] It is a fable of reason against unreason, written at a time when London was still reeling from an attack by religious fanatics, and McEwan allows his hero to triumph, with strength but also with compassion". One of the most striking features evident in *Saturday* is its unashamed celebration of the pleasures of modern urban life and what McEwan presents as the vitality of Western liberal democracy. Most of the more serious contemporary literature fulfils a well-justified postulate to speak for the excluded, persecuted and underprivileged, McEwan on the other hand—much to the annoyance of some critics and scholars—follows the lives of the more prosperous and complacent members of English society. He therefore rejects the lure of easy fatalism colouring so many novels about the modern experience today. In a practical way, *Saturday* rejects the doctrine of decline permeating contemporary social imagination in the West. For example, McEwan repeatedly draws the readers' attention to what he sees as infinitely fascinating intricacy of the modern city. While there is a long and eminent tradition of decrying the cruelty and shallowness of urban experience in English literature, McEwan often constructs the city, and London in particular, as a triumph of human will and ingenuity. Thus for example, early in the novel pondering the panorama of the English capital Perowne comes to the conclusion that:

> [T]he city is a success, a brilliant invention, a biological masterpiece – millions teeming around the accumulated and layered achievements of the centuries, as though around a coral reef, sleeping, working, entertaining themselves, harmonious for the most part, nearly everyone wanting it to work. And the Perowne's own corner, a triumph of congruent proportion; the perfect square laid out by Robert Adam enclosing a perfect circle of garden – an eighteenth-century dream bathed and embraced by modernity, by street light from above, and from below by fibre-optic cables, and cool fresh water coursing down pipes, and sewage borne away in an instant of forgetting. (McEwan 2005, p. 5)

Taking advantage of the unique social placement of the protagonist, McEwan develops his version of a new humanistic sensitivity firmly rooted in scientific rationalism rather than religious axiology. He shows that, despite many stereotypes, rationalism does not have to lead towards nihilism, bitterness and decadence. On Henry Perowne's example, McEwan tries to demonstrate that ethical and moral stance is not produced by the fear of some supernaturally administered punishment but rather by a more elementary human empathy and sensitivity. Therefore, Henry Perowne despite some flaws is a uniquely ethical creature. He is not only a devoted husband and patient father but also a giver of life and a restorer of health. As a high priest of science, he produces miracles (or at least what looks like miracles from a layman's perspective) on daily basis.

Saturday is constructed around numerous antagonisms, clashes and collisions, sometimes understood in a very literal sense. The background of the novel features a huge demonstration against the war in Iraq, while the protagonist enters into several personal conflicts. Perowne argues with his colleague from work, later with his daughter and finally he has a fight with a local thug (the incident itself is preceded by a minor car crash). There is one conflict in the novel, however, that has received relatively little critical attention. In *Saturday*, McEwan arranges a symbolic confrontation between the two cultures, the two modes of understanding the world. They are personified by the protagonist, apparently representing scientific rationalism and modern science in general, and Perowne's father-in-law, an eminent poet John Grammaticus, the embodiment of modern highbrow humanism.

It is both interesting and significant that the narratorial sympathy in *Saturday* clearly sides with science. Grammaticus is presented as a famous poet and a prominent literary intellectual but not as an admirable human being. His ostensibly superior artistic sensitivity that secured him many prizes and a place in school textbooks is not sufficient to transform Grammaticus into a more agreeable human being. He is simply a loathsome figure, constantly wallowing in the sense of being slighted or underestimated. He cannot even afford elementary benevolence towards his poetically-gifted granddaughter. Grammaticus is vain, egocentric and misanthropic but also endowed with other, simpler human weaknesses like alcoholism and a predilection for fits of anger. As an allegory of modern humanities, he is both intriguing and disturbing.

Patronising manner in which Grammaticus treats his son-in-law, openly demonstrated contempt for Perowne's lifestyle and achievements, the mutually reinforcing feelings of self-importance and deep insecurity almost perfectly embody C. P. Snow's idea of the sins committed by the literary intellectuals. The asymmetry between Grammaticus and his son-in-law is so staggering that it cannot be mitigated by Perowne's exaggerated philistinism. *Saturday* was hailed as a remarkable literary achievement and one of the most interesting novels of the decade. Nevertheless, there were several prominent voices of criticism. Most famously, Banville (2005) in the *New York Review of Books* felt offended by what he saw as unbearable smugness and complacency characterising Henry Perowne and unrealistically rendered relations Perowne had with his children.

4 *Solar*: Scientific Picaresque

It seems that in many respects McEwan's latest work of fiction *Solar*, constitutes an indirect response to those critical voices. While *Saturday*'s Henry Perowne was jarringly reliable and predictable, Michael Beard the protagonist from *Solar* is one of the most grotesque figures ever created in McEwan's fiction. Although he is a world-acclaimed quantum physicist and sort of scientific celebrity (with his status confirmed with the Nobel Prize won several years earlier), in disposition and temperament he has more in common with John Grammaticus than Henry Perowne. Intellectually, Beard is completely burnt-out and focused exclusively on reaping benefits from his earlier achievements. The numerous sinecures, functions, honours and privileges he acquired over the years depended solely on the halo effect generated by his Nobel Prize in physics. Moreover, McEwan endowed Beard with an impressive portfolio of more elementary human flaws. *Solar*'s protagonist is an irreparable womaniser with the history of five broken marriages and countless affairs. On the top of that he is an obese junk food addict and drunkard not dissimilar to a notorious character of John Self from Martin Amis's seminal novel *Money* (1984).

However, *Solar* is not only a sardonic story about an overrated and ridiculous scholar similar in style to the campus novels by David Lodge, Malcolm Bradbury or Don Delillo (particularly his *White Noise*). It is a much broader take on the absurdities corroding contemporary culture. To fulfil that ambitious task, McEwan chose to explore a fashionable but artistically risky topic of global warming were such significant aspects of modern life as science, politics and economy intermingle and clash with unique intensity.

At the beginning of the novel, Beard is appointed head of the National Centre for Renewable Energy—a research unit created by the British government to address the problems of over-dependence on fossil fuels. However, the seriously-sounding name hides a "*Catch* 22 institution" which spends most of its sizeable budget on administration and the rest on dubious projects. They look well on paper, but in fact are useless publicity stunts to convince the public that the government is responding to the problem of climate change. The example of the Centre brilliantly illustrates how the attractive façade of science and apparently rational planning fosters accidental and arbitrary decision-making, and how politics contaminates ostensibly non-ideological scientific projects. Thus for example, Tom Aldous, the only truly dedicated scientist working in the Centre is detested by the rest of the staff, isolated and eventually dies accidentally killed by Beard. Intriguingly, Beard's grotesqueness and a dubious nature of the scientific institutions portrayed in *Solar* do not indicate McEwan's wholesale rejection of scientific rationalism. They are rather meant to work as a reminder that, like all human designs, science can easily fall prey to the grinding powers of bureaucracy, politicisation and inertia.

With *Solar* McEwan revisits the motif of the clash between the two cultures. In this case, however, he does not resort to a simple juxtaposition of two allegorical figures symbolising science and the humanities like it was in *Saturday*. Instead, the issue is rendered in a more complex manner and is spread out into several emblematic episodes. Indeed, at moments, *Solar* feels like a picaresque novel, with Michael Beard playing the part of the scholarly rogue. His encounters with the world of the liberal arts resemble a series of tragicomic adventures.

Particularly two episodes described in *Solar* seem important. First of them, at the beginning of the novel, is one of many publicity stunts that Beard happily takes part in, namely an expedition beyond Spitsbergen with a group of ecologically engaged artists and intellectuals. As a scientific celebrity, Beard goes to the North Pole to witness the melting ice cap and thus gain more personal insight into the problems connected with the change of climate caused by human activity. What he sees during the trip are pathetic and futile attempts of various artists to deal with climate change awareness by staging performances on snow or building ice sculptures. McEwan exposed here the sheer helplessness of the arts community in the face of a complex issue. The second scene depicts the clash between Beard and a prominent, outspoken feminist scholar, Nancy Temple during one of official committee meetings. When Beard recklessly remarks that according to some research female brains are inherently less suited for work in abstract science, particularly mathematics and physics, he triggers a media storm and public backlash.

The whole row, modelled on the real case of the Harvard president Larry Summers, ends with public disgrace and Beard's retreat from official duties. In this scene McEwan comically draws attention to the existing gap of understanding between scientists and the influential group of postmodern humanists. When in the *Two Cultures* lecture Snow was describing the contempt and sense of moral superiority demonstrated by many literary intellectuals towards their scientific colleagues in the academia, he was referring to late modernists, people for whom

Yeats, Eliot and Pound were the most venerated authorities. The group that Beard has to struggle with are moulded by a different set of theories and critical practices. Nevertheless, they display similar contempt and suspicion of science, which they interpret as just another discourse aimed at oppressing and subjugating the underprivileged and uninformed. This is brilliantly illustrated in the case of Nancy Temple, professor of "science studies" who after observing the group of geneticists at work, came to the conclusion that the lion's gene Trim-5 they were examining was:

> in the strongest sense, socially constructed. [...] The gene was not an objective entity, merely waiting to be revealed by scientists. It was entirely manufactured by their hypotheses, their creativity, and by their instrumentation. [...] [T]hat [scientific] description, that text, only had meaning, and only derived its reality, from within the limited network of geneticists who might read about it. Outside those networks, Trim-5 did not exist (McEwan 2010, p. 131).

5 Conclusion

It can be argued that McEwan by foregrounding scientific rationalism and castigating the weakness of the dominant modes of thinking within the humanities became an apostate seduced by the efficient simplicity of science. Nevertheless, that would be a rather superficial and in fact biased reading. It is far better to see him as a novelist who is disillusioned by the circularity and sterility of postmodernist thinking and who rejects simplistic apocalyptic scenarios. Although he never explicitly refers to C. P. Snow, it seems that of all contemporary novelists writing today, he is the most devoted follower of Snow's recommendations. McEwan successfully exemplifies how literature may be inspired by the achievements of modern science.

References

Banville, John. 2005. A Day in the Life. http://www.nybooks.com/articles/17993. Accessed 5 March 2010.

Kirsch, A. 2010. The Silppery Scientitst: Ian McEwan on the Gulf between Goodness and Intelligence. http://www.slate.com/id/2247911. Accessed 01 July 2010.

McEwan, I. 2005. *Saturday*. London: Johnathan Cape.

McEwan, I. 2010. *Solar*. London: Jonathan Cape.

Snow, C. P. 1961. *The Two Cultures and the Scientific Revolution: The Rede Lecture 1959*. New York: Cambridge University Press.

Timmons, T. W. 2007. Older Than Snow: The Two Cultures And The Yale Report Of 1828. A Journal of the Oxford Round Table. http://www.forumonpublicpolicy.com/archive07/timmons.pdf. Accessed 21 March 2010.

Trying to Cross Frontiers of Fortress Europe: Rose Tremain's Novel *The Road Home* (2007)

Anna Maria Tomczak

Abstract *The Road Home*, the novel about immigration by Rose Tremain which won the Orange Prize for Fiction in 2007, tells the story of a 43-year-old widower Lev, an Eastern European who goes to Britain in search of work. The country of his origin, though purposefully unspecified, lies somewhere in the former Eastern bloc and thus Lev represents a 'generic Eastern European', combining a set of characteristics which may be seen as symbolizing 'Fortress Europe's Other'. Tremain's novel does not try to challenge stereotypes and in its presentation of Eastern Europeanness exemplifies a well known truth that crossing intellectual and mental frontiers may prove much more difficult than crossing physical borders. The paper seeks to contribute to the critical debate around the issue of collective identity and its constructions.

1 Introduction

On May 1st 2004 the European Union expanded by admitting ten new member states—eight Central and Eastern European countries (the Czech Republic, Estonia, Hungary, Latvia, Lithuania, Poland, Slovakia and Slovenia) and two Mediterranean islands (Malta and Cyprus). The accession involved a prior fulfillment of certain political and economic criteria, a functioning market economy, among others. On the official website of the EU (http://europa.eu/abc/history/index_en.htm) presenting a brief history of the enlargement process, in the sections entitled "Europe without frontiers" and "A decade of further expansion", we may read: "With the

A. M. Tomczak (✉)
University of Białystok, Białystok, Poland
e-mail: atomczak@uwb.edu.pl

J. Fabiszak et al. (eds.), *Crossroads in Literature and Culture,* Second Language Learning and Teaching, DOI: 10.1007/978-3-642-21994-8_41,

collapse of communism across central and eastern Europe, Europeans become closer neighbours. … The political divisions between east and west Europe are finally declared healed when no fewer than 10 new countries join the EU in 2004". But the European community, rather than striving to achieve new forms of common citizenship, became open to, as Balibar (2004, p. 170) states, "the reactive development of all sorts of identitarian obsessions". Thus, to quote him further,

> European unification has introduced a qualitative change both in the symbolic register and in institutional practice. The status of European citizen … has begun to take on an effective content … But the fact that, with respect to individuals, this 'citizenship' is defined as the simple addition of the national citizenships of the member countries of the union, transforms the status of the foreigner. In each particular country the foreigner is only the national of another sovereign state, enjoying an equivalent 'belonging', which is the object of reciprocal recognition. But at the level of the newly instituted union, he or she becomes the object of an internal exclusion (Balibar 2004, p. 171).

The above quote comes from a collection of essays *We, the People of Europe? Reflections on Transnational Citizenship* by Étienne Balibar, a leading French political thinker and philosopher. The book discusses (among other issues) the unification of Europe. The author maintains that the concepts of 'borders' and 'territories' are inherently linked to the ideas of 'rights' and 'citizenship'. As Professor Read of Princeton University notes, borders and territories are no longer "descriptive categories, which can be empirically verified" (Read 2004). Thus 'open borders', although allowing an unrestricted movement of individuals, may still mean the presence of hidden internal borders.

When after the expansion of the UE Britain 'opened its borders', a wave of new immigrants/foreigners, mostly from the former Soviet bloc, could enter the country legally to seek employment in various areas of work—catering and construction among the most often chosen forms of occupation. Their cultural and economic background varied, as did their 'immigrant lot'. What they shared was a dream of a better future, a hope of success, and a strong determination to persevere at all cost. *The Road Home*, a 2007 novel by Rose Tremain, may be viewed as an attempt to portray 'the destiny' of such foreigners/new immigrants. Simultaneously, the book may be read as a text exemplifying, albeit unintentionally, what Balibar calls "the creation of a 'subaltern imperialism' in Europe" (2004, p. 170). The present article seeks to substantiate this claim by exploring the novel's narrative as an "ideological project" (Storey 1996, p. 33).

2 The Author and the Story

Tremain (b. 1943) is a well-known and prolific writer. In the past, a university teacher of creative writing, twice a judge for the Booker Prize (in 1988 and 2000), she published eleven novels, various collections of short stories, as well as radio and TV plays. Awarded a CBE in 2007, Tremain is perhaps best remembered for *Restoration* (1989), her "most celebrated historical novel" (Bradford 2007, p. 89).

The Road Home the winner of the 2008 Orange Prize—received a positive response from literary critics. Kirsty Lang, who chaired the Orange Prize judging panel, described the book as "a powerfully imagined story and a wonderful feat of emotional empathy told with great warmth and humour" (quoted by Singh 2008), while a Daily Telegraph reviewer called it a "finely balanced novel of urgent humanity" (Moore 2007). It may seem however that, for all its psychological insight, the book fails on one front—that of an honest, truthful portrayal of Eastern Europe. It is precisely in this aspect that the text reveals its "relationship to the ideological and historical conditions of its existence" (Storey 1996, p. 33).

A novel is, by definition, a work of fiction, but its semantic interpretation arises through the reader's knowledge of the external world and recognition of direct references to apparent reality. A literary work not only appeals to external knowledge but also helps in substantiating and forming such knowledge. (Markiewicz 1996, p. 183; Głowiński 1997, pp. 24–31).

Hardly an attempt to challenge stereotypes and break the stronghold of prejudice, *The Road Home* proves that crossing intellectual and mental frontiers may pose a task much more difficult than crossing geographical borders or, to refer to Balibar's analysis of the condition of Europe, it provides evidence of the existence of internal borders erected by the old Europe's (Fortress Europe's) anxiety over cultural difference.

The story is simple and the characters well defined. Lev, a forty-three year old widowed engineer, arrives in London seeking employment. Prepared to encounter obstacles and to work very hard, he sets himself a goal—to accumulate enough resources to improve the economic situation of his family back home—the aging mother and a five-year-old daughter. From a distributor of leaflets and a kitchen help to asparagus picker, waiter and chef, the man quickly rises in his immigrant status. When after a two-year absence the protagonist returns home with the money earned through long hours of back-breaking labour and a handsome check from one of the inmates of an old people's home, he opens a restaurant. The 2 years in England bring joy and sorrow. He befriends his landlord (a lonely Irishman), meets a number of his compatriots, has a short-lived love affair, is assaulted and robbed in the street, spends a few hours in police custody and even reads *Hamlet*. The bad news from home about the construction of a dam and the prospect of his village being flooded precipitates the decision to return. The title's "road home" may be long and slippery, the crossroads and bends not always easy to negotiate and surmount, but Lev arrives safe and sound at his destination to embrace his family and his new life.

3 The Portrayal of Eastern Europe

The novel's setting, for most part, is England. Lev's country of origin emerges only through a series of flashbacks as well as in the straightforward descriptions included in the opening and two closing chapters (on 44 out of 365 pages).

Significantly, it is not any recognizable state in the former Soviet block but a place which combines vague characteristics of the region, a kind of 'generic Eastern European country' depicted in the reality of communist days and post-transformation economy. Great pains are taken to create an impression of ambiguity. Each time it is necessary to identify a place or a person, the phrase 'our country' or 'our language' is used. Names of people and places are a mixture of Polish/Russian/Slavic-sounding words. Male names include: Stefan, Rudi, Jacek, Dmitri, Vitali, Vitas, Oskar, Pavel, and Karl; female: Ina, Lydia, Marina, Maya, Lora, Larissa, and Eva—'people from our country'. Auror, Glic, Yarbl and Baryn feature as geographical locations. A reviewer in The Daily Telegraph observes: "Tremain never specifies where her hero Lev is from; it could be Poland, it could be Slovakia" (Connolly 2007).

Lev's "own country" left behind, a seedy, dismal place, comes out as a mythical land where the temperature is given in Fahrenheit and distance in miles, where winters are cold and, accordingly, people wear fur hats and sheepskin coats, but it is never so cold as in England and thus the climate allows planting poinsettias on the porch. The currency (after transformation) is euro and the most common term of address (now as well as then)—'comrade'. The president in office, Podrorsky, has streets in his own name and the most renowned citizen—Maestro Pyotor Greszler, the well-known conductor—temporarily rehearses with the London Philharmonic Orchestra.

The novel's idyllic passages of rural simplicity and picturesque views, with "the fields of sunflowers scorched by the dry wind and the wild garlic growing green at the edge of the road" (Tremain 2008, p. 1), with storks on chimney pots and the fir-covered hills, are counterbalanced by descriptions of poverty and neglect. Pre-industrial and economically backward, it is a country where village children play "with the goats and the chickens in the dust" (Tremain 2008, p. 10) amid "broken tools, sacks of rotted potatoes, plastic buckets, [and] chicken bones left by the dogs" (Tremain 2008, p. 344), where the cityscape features mounds of debris (Tremain 2008, p. 351) and the railway station's "rusty urinals [are] never cleaned" (Tremain 2008, p. 86). Electricity is precious, hence constant power cuts and candle stubs stuck in broken saucers. An antique samovar bolted to the wall in a public place, village houses with rag rugs, a petrol station consisting of one rusty pump and wood cabins lit by paraffin lamps complete the picture. When it rains, in this country of pot-holed roads, pig farms and quarries, pedestrians clutch "cheap, flimsy umbrellas or [hold] magazines over their hair" (Tremain 2008, p. 116). Goods are in short supply so barter is the preferred form of economy; in a bazaar a shirt may be obtained in exchange for a wooden plane and some nails, and people are always happy to work for packets of Marlboro cigarettes and panty-hose (Tremain 2008, p. 296).

Although religion was banned until not so long ago and a local church has been changed into an indoor-swimming pool, many women would secretly place a gold icon above the fireplace and put candles around it to pray. Now, when religious practice is permitted again, employers may, very generously, let the workers "have the day of the Christmas festival off" (Tremain 2008, p. 126).

The inhabitants of this country who "mostly look the same" (Tremain 2008, p. 54) fall into two categories: representatives of the system and the destitute rest. Policemen "with muscled forearms and healthy complexions" have "hand-guns slung about their anatomy in clever places" (Tremain 2008, p. 20) and never look you in the eye, while the army personnel in stiff, round hats "like boxes of chocolates" clutch "their old Kalashnikovs to their chests" (Tremain 2008, p. 236). Corruption and all-powerful bureaucracy are rife, contempt for average citizens prevalent. Officials may tell you to remove your muddy shoes in offices. The old generation, superstitious, downtrodden people, who believe in wood-sprites, hang 'spirit rags' in the forest as offerings to the dead (Tremain 2008, p. 29). They still remember the times before the war, when villagers "wore wooden shoes" (Tremain 2008, p. 122) and express their hatred of aristocrats of an earlier era,

> the people who had brought about the suicidal stampede towards Communism—those long-dead members of the ancient nobility, sweeping forwards, always forwards, in their jewels and furs and pheasant-tail feathers, going towards lighted rooms, towards concerts and soirées, towards ten-course banquets, past the unseen poor ... (Tremain 2008, p. 201).

Adults of the younger generation try to make a living, struggling with difficulties and soldiering on. Many people around Auror and Baryn die of leukaemia as the nearby lake has been so much contaminated with chemical waste that even the fish are bright blue and shine at night with a morbid fluorescent light.

People drink vodka to excess, at eleven o'clock—for celebration, late at night—from despair and to still the senses; during fishing expeditions to wash down a meal of dumplings and at Christmas—for surrender and to alleviate the gloom. Vodka functions also as local currency. When Lev's friend, Rudi, sets out on an expedition to another town to buy an American car Chevrolet Phoenix, he carries eleven bottles in his suitcase, to help him make a bargain. The drinking habit travels with the nation. In the bar located in the country's embassy in London vodka is served at 10 o'clock in the morning. Before Lev departs for England, he remembers to tuck a small flask into his boot.

The narrator's realistic descriptions and the protagonist's recollections create an atmosphere of authentic despair and despondency. Wretched is the country of such squalor and unhappiness, its pitiful inhabitants deserving commiseration and sympathy. The use of selected lexical items (given in the text in italics), suggestive of the original language, enhances the impression of authenticity. An aged woman is called, a "babushka" (Tremain 2008, p. 152) and vodka—"vodichka"—translating into "darling little vodka" (Tremain 2008, p. 104). "Greys" rule the black market economy, "waiting for buyers who [offer] American dollars or motor parts or drugs" (Tremain 2008, p. 58) and all the evil and the misfortunes of the common man are blamed on the "system" (Tremain 2008, p. 10).

The new times bring high-rise flats and "the bright, flickering heartbeats of American franchises, ... a world slipping and sliding on a precipice between the dark rockface of Communism and the seductive, light-filled void of the liberal market" (Tremain 2008, p. 337). However, the economic transformations resulting in “the abandoned farms and silent factories … the deserted depots and lumber yards” (Tremain 2008, p. 337) cannot affect “the ugliness of the town suburbs” (Tremain 2008, p. 337) or submerge the stench of waste. The new does not offer a viable alternative to the old, nor does it affect the daily existence of the country. The construction of the dam will make many families homeless, but as yet life goes on. On the day of his return, Lev contemplates the familiar scene of “a straggle of mules being led through the purple afternoon carrying bundles of reeds on their scrawny backs” (Tremain 2008, p. 337). His fellow villagers have not yet embraced the change and the city people remain set in their ways—bribery, pessimism and the old mentality are the orders of the day.

The above descriptions are meant to ring a bell. After all, the story of Lev, an immigrant from Eastern Europe, is not without foundation. As Tremain discloses in the ‘acknowledgements’ section provided at the end of the novel, she drew inspiration from real life conversations. And so she dutifully expresses her gratitude to a certain Jack Rosenthal, for having been introduced to “his Polish field-workers who told true and invaluable tales from Eastern Europe” (unnumbered page, the probable page number 367). No other nationality deserves a mention. One might wonder which elements of these “true and invaluable tales” found their way to the 2007 winner of the Orange Prize. Probably the mules.

Clearly and overtly so, in its portrayal of Eastern Europe *The Road Home* may be considered an example of Orientalist thinking, Orientalist in the sense of a one-sided view leading to a wrongful attribution in which the difference is translated into a sense of superiority and the depiction of Other (i.e. Eastern Europe) serves as a self-affirming construct of the West. Nowhere is this attitude more evident than in the passages employing food references.

4 Food

In narrative fiction food may operate as a literary theme or motif on different levels, authentication being one of the most important functions. The food that the characters obtain, prepare and consume reflects their social position, age, financial status, upbringing and class, ethnicity and taste. Thus, food acts as a signifier, classifier and identity builder, for individuals as well as groups (Scholliers 2001, p. 7). The link between food and ethnicity often leads to stereotypes about whole nations (Farb and Armelagos 1980, p. 3), such terms as “frogs”, “curry eaters”, “krauts” or “macaronis” testifying to their wide currency.

Tremain is no stranger to the idea that countries and ethnic groups, by and large, have their specific culinary traditions, and no novice to meticulous descriptions of diets and menus. The credibility of food scenes in *Restoration* has been universally praised as a “magnificent example of researched minutiae” (Bradford 2007, p. 89). In *The Road Home*, food with its smell, taste, appearance and style of consumption functions as a tool of cultural identification and serves a clear purpose—to separate and marginalize. The novel’s protagonist, while talking to his former English employer, the famous restauranteur, GK Ashe, declares: 'My town, Chef, has never known food' (Tremain 2008, p. 281). Nor does he protest when Mr Ashe adds: "From what I've heard about your country, all anybody's eaten in the last century is goat meat and pickles" (Tremain 2008, p. 281); the opinion confirmed by Lydia: "Our people don't care about good food. They never have" (Tremain 2008, p. 300), and Panno, the owner of a Greek restaurant: "Not many people from your country are interested in good cuisine" (Tremain 2008, p. 275), the claim unreservedly supported by Lev: "No … That's because we've eaten Communist food for sixty years" (Tremain 2008, p. 275).

The reader gets a vivid impression of the notion of 'communist food' when Lev returns from England and needs to stop for a meal before resuming the journey that will take him to his home village. In a hotel dining room, on a stained tablecloth, he is served a "meal of tinned soup and unidentifiable stew …. [and] a carafe of inky red wine" (Tremain 2008, p. 338). Later, in a newly renovated and redecorated café, Lev's senses (adequately sharpened by the experience of working in London's catering establishments) become assaulted by an unmistakable "whiff of Communist food": "the smell from the kitchen … immediately familiar … the smell of beetroot soup, nameless stews, seaweed ravioli" (Tremain 2008, p. 345). The menu "encased in oily laminate" (Tremain 2008, p. 346) boasts of two specials: "rabbit cooked with juniper berries" and "cold venison served with boiled egg". The next day's visit will bring a variation – "rabbit cooked with mustard seeds" (Tremain 2008, p. 350).

Earlier references to food, scattered in different chapters, present mostly Lev's memories. Free of sentiment, nostalgia or longing for childhood's smells and mother's cooking, Lev's recollections create a vision of the country of scanty provisions, where poaching is a way of obtaining meat and thus supplementing home economy, and where eating out is a constant exercise in humiliation. In restaurants, waiters in stained aprons behave "like labour-camp guards, slamming down dishes of sinewy meat, sloshing out wine from dirty carafes, snatching [the] plates away before [the] meal [is] finished" (Tremain 2008, p. 39). Primary school children's lunch consists of "goat's milk and bread and pickled cucumber, with, sometimes in summer, wild strawberries from the hills above the village" (Tremain 2008, p. 24), or cold sausage and poppyseed bread. At home, after school, children may be given "goat's milk with cinnamon, raisin cakes and rose-petal jam" (Tremain 2008, p. 10). Also mentioned are: smoked herring for a picnic, carp as a coveted delicacy, roasted sunflower seeds customarily sold at an open-air fair, and a special Christmas treat – goose cooked with rosemary and chestnuts (Tremain 2008, p. 127). The staple diet is described as "the crude starch-laden stuff" (Tremain 2008, p. 101).

By comparison, the food that Lev cooks or sees and admires in England both looks and tastes beautiful. Sophisticated and elaborately decorated, it is "lovely" (Tremain 2008, p. 76), "delicate" (Tremain 2008, p. 78), "refreshing", and "bloody nice" (Tremain 2008, p. 78); so "delicious", it must be "rolled round in the mouth" (Tremain 2008, p. 76) and "savoured", not hastily devoured. The "little sandcastle[s] of potato gratin" (Tremain 2008, p. 77), "beautiful chicken broth" (Tremain 2008, p. 76), create dishes with vibrant colours and smells that awaken a "beautiful hunger" (Tremain 2008, p. 100); in sum, "the most delicious" combinations that "the human mind [is] capable of inventing" (Tremain 2008, p. 101).

The stark contrast that the novel offers cannot be easily and readily reduced to the difference between gourmet food and simple grub as, for example, between poached chicken legs with celery and carrots versus the slices of sausage wrapped in paper. On a symbolic level, food encapsulates the deep chasm between Western and Eastern Europe, not so much in a simplistic understanding as the gap between the culture of plenty and the culture of want, but rather in the sense of an established culture and tradition versus a non-existent culture, the culture wiped out by communism or never truly formed.

However, luckily for Eastern Europe, things are going to change. Lev has plans. In the country that has never known the value of good food, in the country of no culinary tradition and thus no civilization, Lev will open a restaurant, "the first one … where the food will be truly good" (Tremain 2008, p. 320). After all, he has arrived not only with some cash but also with something equally important if not priceless – the cultural capital. For the last two years abroad Lev had been learning. He observed and made notes. England has given him myriad new recipes and a bright novel idea: food matters. England's civilizing mission will soon take shape. At Number 43 Podrorsky Street, in a newly opened establishment, where "the cornflowers in slim vases" deck every table, Lev will serve "a chocolate tart … with a shortcrust so perfected, so astonishingly short, it melted on the tongue like fudge" (Tremain 2008, p. 363). The civilizing mission has borne fruit.

5 Author's and Readers' Comments

Interviewed in 2008, Tremain explained the philosophy behind *The Road Home* saying: "I believe that when we think about people in a collective way—as, for instance, 'immigrants', 'foreigners', 'outsiders' etc.—we tend to lack empathy with them… But the moment we become engaged with an individual story, empathy arrives and our attitudes alter" (in Singh 2008). Regrettably, Lev's story never becomes individual and thus Tremain commits the very sin she so openly warns us to beware of, that of creating a stereotypical portrayal based on collective identity, the identity of an Eastern European, an inhabitant of that part of the continent where, if we are to trust his evidence, communism prevented a development of any culinary tradition and culture, and where poverty and ignorance have outlived the market forces and processes of democratization, as the novel amply demonstrates. On the level of the plot, Lev is a foreigner in Britain, an alien, a "second-class resident … deemed to be of a different kind" (Balibar 2004, p. 122). As Balibar (2004, p. 122) states in *We, the People of Europe? Reflections on transnational citizenship*, contemporary Europe has transformed "a project of inclusion into a program of exclusion", where cultural difference may lead to stigmatization, a form of "'new racism', postcolonial and postnational", and Lev exemplifies this. On the level of the book's "unconscious" or "internal distanciation", to use Macherey's term, (in Storey 1996, p. 33), the evoked reading illustrates the paradox of the fall of communism described by Balibar in the following way:

> There was an 'Eastern Europe' and not only a 'Europe in the West.' And we are beginning to learn today that however dictatorial and regressive its political institutions and social forms may have been, Eastern Europe did have a culture and a 'civil society'. The paradox of the situation after 1990 in this respect is that the effects of the end of totalitarianism were handled as if there had never been more than one single society in Europe, as if the system of 'real socialism' had not only been undemocratic, contradictory, and a failure, but fundamentally imaginary (Balibar 2004, p. 166).

The word 'imaginary' perfectly describes the Eastern Europe of the novel, although apparently contrary to the novelist's intentions. In the author's statement included on the British Council website 'contemporarywriters', Tremain declares:

> I suspect that many writers deceive themselves about why they write. My self-deception is that I create in order to understand and that the final end of it all might be wisdom. This means that I deliberately seek out the strange, the unfamiliar, even the unknowable, as subjects for my novels and trust my imagination to illuminate them to the point where both I and the reader can see them with a new clarity... (in http://www.contemporarywriters.com/authors).

"The strange and unfamiliar" Eastern Europe has proven so "unknowable" to the novelist that her lofty aims—to comprehend, illuminate, and achieve wisdom—must be read as an exercise in hypocrisy or relegated to the sphere of wishful thinking. The clarity of vision becomes the sole domain of those readers who can detect the disparity between 'representation' and 'figuration' (Storey 1996, p. 34). Judging by an overwhelmingly positive readers' response expressed on one of the Internet websites (http://www.librarything.com), the capacity restricted to the very few. Out of over three dozen opinions, only one voice questions the novel's presentation of Eastern Europe. The person, who identifies herself as "Tess 22" writes:

> Lev is from an unspecified country in Eastern Europe, and in her acknowledgments Tremain thanks the agricultural workers she interviewed in England. However, she doesn't seem to have gone to Eastern Europe herself, which is disappointing. I am far from believing that literature's only purpose is to change public opinion, but can't help thinking that the descriptions of Lev's mother, his home town and his friends' experiences perpetuate the image of Eastern Europe as a backward, impoverished place which is out of touch with the modern world - some of the very ideas that other parts of the book try to challenge. Why doesn't Tremain give Lev's country a name? Is it to make his story more universal or is it to prevent offence? Offence, in my opinion, would be justified. While there are still some very poor rural places in Eastern Europe, her description is not true of the vast majority. Yes—it's harder to get world class cuisine—no, people are not blown away by American cars, women in their 60s are no stupider than anywhere else and goats don't generally wonder around the kitchen. It might have become clear that I'm not entirely objective. This is because I've lived in Central and Eastern Europe for a while now, teaching English.... (http://www.librarything.com/work/3248827/reviews).

6 Conclusion

Written in English and published in the UK, *The Road Home* is specifically addressed to the British readership. Its omissions, its gross errors and inaccuracies, largely go undetected, as the readers' comments provided on the Internet pages suggest. However, "[t]he task of a … critical practice is not … to complete what the text leaves unsaid, but to produce a new knowledge of the text, one that explains the ideological necessity of its silences, its absences, its structuring incompleteness" (Storey 1996, p. 32). It seems that the novel is a product of this type of ideological necessity that demands an existence of an insurmountable border/frontier which separates the old Europe (the cultured and civilized Fortress Europe of the West) from the new Europe (new member states in the EU). To quote Balibar again:

> [N]ot only has the East–West division in Europe not been abolished, but it has in a sense been reinforced. When Europe was cut in two by the geopolitics of the Cold War, it was possible to imagine that it was a provisional phenomenon that would come to an end with the settlement of the conflict. Now we can see that it is a part of the 'settlement' itself. It manifests the profoundly inegalitarian and dangerously conflictual character of that settlement… [T]he principal European nations [are] spontaneously rediscovering traditional imperialist postures … A veritable structure has thus arisen, a sort of the cold war after the Cold War … but one which is both plural and mobile, and which is hidden under a thick mythology about 'clashes of civilizations' between Eastern and Western Europe, supposedly inherited from religious tradition or from the form of state building, when not simply from the ethnic character of the nations involved (Balibar 2004, pp. 166–167).

The cold war after the Cold War is also a consequence of the stance adopted by writers and intellectuals. Balibar is right claiming that "European intellectuals do not sufficiently exercise their capacity to cross political and cultural borders, translate discourses (other than specialized ones) within and beyond the official limits of the European Union … They are not sufficiently acting as citizens of Europe, dare I say thinking European citizens" (Balibar 2004, p. 205). Rose Tremain's novel *The Road Home* and the critical acclaim and praise which it received in intellectual circles may be seen as a tangible proof of such an attitude. In fiction, as in reality, "it is possible that, seen from the East, Western Europe appears not as the example of cosmopolitanism it claims to be, but as a conglomeration of national egoisms" (Balibar 2004, p. 94).

References

Balibar, E. 2004. *We, the People of Europe? Reflections on Transnational Citizenship*. Princeton and Oxford: Princeton University Press.

Bradford, R. 2007. *The Novel Now. Contemporary British Fiction*. Oxford: Blackwell Publishing.

Connolly, C. 2007. Fresh off the bus. A review of *The Road Home* by Rose Tremain. http://www.telegraph.co.uk/news Accessed 30 Jan. 2010.

Farb, P. and G. Armelagos. 1980. *Consuming Passions. The Anthropology of Eating*. Boston: Houghton Mifflin Company.

Głowiński, M. 1997. *Narracje literackie i nieliterackie*. Kraków: Universitas.

http://www.contemporarywriters.com/authors Accessed 20 Feb. 2010.

http://www.librarything.com/work/3248827/reviews Accessed 10 April 2010.

Markiewicz, H. 1996. *Wymiary dzieła literackiego*. Kraków: Universitas.

Moore, C. 2007. An immigrant's attempt at a new life: a review of *The Road Home* by Rose Tremain. http://www.telegraph.co.uk/news. Accessed 22 Feb. 2010.

Read, J. 2004. Writing in the Conjuncture: Étienne Balibar, *We, the People of Europe? Reflections on Transnational Citizenship. Borderlands,* vol. 3, number 1, online http://www.borderlands.net.au/vol3no1_2004/read_balibar.htm Accessed 1 June 2010.

Scholliers, P. 2001. Meals, Food Narratives, and Sentiments of Belonging in Past and Present. In *Food, Drink and Identity. Cooking, Eating and Drinking in Europe Since the Middle Ages*, ed. P. Scholliers, 3-22. Oxford and N York: Berg.

Singh, A. 2008. Author Rose Tremain wins the Orange Prize for Fiction. http://www.telegraph.co.uk/news Accessed 22 Feb. 2010.

Storey, J. 1996. *Cultural Studies and the Study of Popular Culture*. Edinburgh: Edinburgh University Press.

Tremain, R. 2008. *The Road Home*. London: Vintage Books.

Carnivalesque Pop. Representations of the Commonplace in British Pop Art

Justyna Stępień

Abstract An image in the sixties was a site of what Bakhtin alludes to as hybridization that brought together exotic and familiar. This montage of variable discourses, which constituted visual material, highlighted fragmentation of time into certain visual de-centered segments. The breakdown of signifiers resulted in a direct experience of conditions of heterogeneous Swinging culture. As there was no former differentiation between art and mundane life, in the course of time artistic representations melted into industrial design becoming notions of new social formation of British culture. The chapter focuses on selected works by British pop artists (Peter Phillips, Richard Hamilton and Derek Boshier) in which individual representations gain a sense of communal feeling largely generated via mass culture mechanisms. The artists' works depict imagery filtered through the mass media and advertising industry. Inevitably, the analyzed paintings are intextuation of array of popular signs and symbols of social anxieties and visual pleasures. As John Fiske (1995, p. 83) points out, carnival is concerned with materiality of life that underlines and precedes individuality and spirituality. Thanks to unification of artistic modes of expression, pop representations are common visual experiences easily recognizable by a wider audience. As a result, we receive a level on which cultural representations are equal and cross all the former social and cultural barriers. In the condition of transgression, visuality becomes a valuable material around which consumer needs and pleasures met.

J. Stępień (✉)
University of Łódź, Łódź, Poland
e-mail: stepienjustin@yahoo.co.uk

J. Fabiszak et al. (eds.), *Crossroads in Literature and Culture,* Second Language Learning and Teaching, DOI: 10.1007/978-3-642-21994-8_42,

> The carnival is a fascinating substantive cultural vehicle for the examination and analysis of transgression...With the increasing politicization of cultural knowledge; with the increasing attention being paid to popular cultural forms, primitive, low-life, vulgar and marginalized cultural practice; and with the postmodern disassembly of traditional forms of cultural analysis, carnival has come to provide a new metaphor and a new style for reading the social (Jenks 2003, p. 164).

1 Introduction. The Concept of Carnival

Carnival has become an expression of social and cultural life's complexity and multiplicity. The elevation of vernacular elements to the realm of fine arts was a powerful stimulus for a new amalgamation of two antagonistic discourses, each struggling for an equal recognition. The voices and language of carnival destabilize and deconstruct boundaries marked by the frames of high culture, leading to a cultural recombination. For Clair Willis (1989, p. 133), "the power of carnival to turn things upside down is facilitated by bringing it into dialogic relation to official forms". Carnival's travesty of formal register, its mockery and masquerade mix with mimetic expressions to produce a cultural transformation, unfurling before us the quicksilver meanings and imagery of popular culture.

For Mikhail Bakhtin (1968, p. 120), "carnival was the true feast of time, becoming, change and renewal. It was hostile to all that was immortalized and completed". This disordering of carnival against ideological timelessness is a source of "eternal joy of becoming", an affirmation of everyday life's pungency and grain. By placing itself in opposition to static cultural forms, carnival unveils repressed pleasures and the continual, life-gluing seepage of the street into the cathedral. Bakhtin (1968, p. 130), referring to the carnivals of the Middle Ages, describes the seepage as combination of "the sacred with the profane, the lofty with the low, the great with the insignificant, the wise with the stupid". Through its celebration of human processes, it grounded a collective activity of people whose "ridicule fused with rejoicing" (Bakhtin 1968, p. 125).[1] The effect was a temporary disruption and removal of social hierarchies, feeding popular culture anew. Bakhtin's focus on carnival in early modern Europe contains a critique of modernity. According to Mary Russo (1986, p. 218), for Bakhtin as an observer of

---

[1] As John Storey (1993, p. 131) notes, Bakhtin insisted that "carnival was not just a retreat from medievalism; it offered the utopian promise of a better life, one of equality, abundance and freedom". This can easily refer to the counter-cultural movement of the sixties and its utopian ideas of the liberated and classless world. Similarly to carnival the Swinging decade had a temporal character and disintegrated rapidly after the 1970s, leaving the cultural repertoire that had an impact upon the next decades.

popular culture in a sociohistoric context, the culture of modernity is "as austere and bitterly isolating as the official religious culture of the Middle Ages, which he contrasts with the joy and heterogeneity of carnival and carnivalesque style and spirit".

Here we can draw an analogy between the Bakhtinian concept of carnival and the social and cultural changes of the sixties. This comparison is visible from two perspectives that are closely linked and that infuse the entire picture of the 1960s turmoil and revolution. First of all, a certain kind of popular festivity, which acted as a negation of a symbolic order, appeared during this time. The decade aptly illustrates a discontinuity in a would-be homogenic order and history, eluding, perhaps, what Bakhtin saw as the liberating mixture of languages in Rebelais. Here, ephemerality, identification with previously excluded voices and a fusion of high and low styles marked a new order of society, and also opened the gates to the postmodernism. Marek Krajewski (2005, p. 15) notes that the distinctive character of the popular culture of this decade is its dynamism and constant change of designation, rules and models of actualization. Its borders elude us, as does any universal definition that its essence is ungraspable and liquid. Secondly, the counter cultural movement, born as a reactionary force, grew into a revival of British cultural and social life. London became a locus of transgression, dissociation and hybridization and the unification of high and low arts. For Mark Donnelly (2005, p. 91), "the capital was seen to be at heart of the wider social and cultural loosening of the sixties, soaking up influences from the provinces and abroad and morphing them into an exotic motif of hedonism, modernity and affluent liberation".

1.1 A Society of Contradictory Impulses

Great Britain of the 1960s was a society of discourses struggling with each other and with an arch-discourse of tradition. These forces made the country a site of contest where different meanings, values and attitudes gained equal value. A rise in living standards and a better economic situation blurred divisions within the class system. Great Britain veered, at least as an ideal, towards a classless society in which "the long front of culture desired to remove the old hierarchies and wanted to achieve a radical alternative cultural life" (Hewison 1987, p. 81). As Mark Donnelly (2005, p. 50) points out, "at this point working-class empowerment was essential for national renewal in the late fifties to accelerate the changes within the cultural sphere". The social hierarchy was breaking down and an affluent society found its fulfillment in consumption of products offered by constantly developing culture industries. New patterns of consumption, and the ways in which they interacted with broader social and cultural changes, were evident in the everyday life of the sixties. All this contributed to a travesty of social order. In fact, spending power and a growing obsession with consumption were not the only acts on shackles of traditional society. Now it was easier for minority groups kick against traditional repressions. In such conditions, the young generation, not bound to a single class, not taking tradition at face value, played the greatest role in social transformations. Prosperity enabled young people to adopt certain identities to buy into other lifestyles and to lead a life dominated by a carnivalesque spirit. The culture of the sixties confused formal categories, opened doors to new material and choices, and celebrated chance. It subverted transformed the meaning of its acquisitions, and in doing so tripped off a rather boundless exploration of the commonplace.

The anti-establishment British artists of the sixties found new means of cultural and artistic expressions to illustrate a multi-levelled society. It is worth remembering that the majority of British pop artists were of working class background, its social taproot apparently sunk not in the establishmentù but in British music hall. There was awareness that it was impossible to rely on only one meaning for utterances and representations producing a plentitude of meanings that stemmed from this social and artistic flux. Following Stuart Laing (1992, p. 72), "culture started to mean all the signifying practices—from language through the arts and philosophy to journalism, fashion and advertising of a given society". These multi-layered combinations made use of parodies, both oral and written, commentaries on authority and decorum. The term "grotesque realism"[2] is often applied here to catch the open, protruding, extended body of becoming, process and change (Russo 1986, p. 219). This explosion of liberalism and liberty found its ultimate realization in Swinging London. As Chris Jenks (2003, p. 169) emphasizes, it was something like a carnivalesque that enabled young people to construct a "second world and life outside officialdom, a world without rank and social hierarchy".

1.2 Carnivalesque Pop Art

Pop art imagery, which appeared together with a loosening of post-war austerity the fifties, grew into its second during the cultural revolution of the sixties. Its works from this period illustrate a young generation's attitudes towards a constantly changing reality shaped by a mass culture industry. A montage of variable discourses rendered as visual material highlighted a fragmentation of time into certain visual de-centered segments. A breakdown of signifiers resulted in a direct experience of heterogeneous Swinging culture. Art and mundane life now merged artistic representations melting into industrial design and becoming both market imagery and records of new social formations. Mark Donnelly (2005, p. 99) recalls that the first British Minister for the Arts pointed out that "diffusion of culture is now so much a part of life that there is no precise point at which it stops. Advertisements, buildings, books, cars, radio and televisions, magazines, all carry

[2] This term is used by Bakhtin as a central category under which his reading of Rabelais as a carnivalesque text is organized.

a cultural aspect and affect our lives". The chapter focuses on selected works by British Pop artists in which the commonplace becomes a major object of inspiration. Or put another way, in which the carnivalesque enabled artists to grasp a sense of a process and a limitless attitude towards life. The morphology of ordinary imagery used by artists show that pop art apprehends reality as a flux of temporally unstable images. These artists introduced a subversive language independent of a symbolic order. Pop art becomes mediation between high and low culture, challenging, like as the carnivals of medieval times, the barrier between artist and viewer. In British pop art representations, carnival adopts strategies that invert existing social forms and cultural configurations.

2 Peter Phillips and a Carnivalesque Object of Fame

In Peter Phillips' works of art, a state of perceptual tension and uneasiness initiates a kind of game with the viewers. But his paintings do not express illusory beauty with utopian ideas. Instead, they "reach almost hysterical pitch with sharply-defined images placed against bright backgrounds" (Livingstone 1983, p. 10). The combination of dissimilar objects on the same canvas produces a surrealistic aura and mocks traditional representations. A student of the Royal College of Art, Phillips was one of the most prominent names on a rejuvenated artistic scene. A youthful working-class enthusiasm and playfulness infuse his paintings. Inspired by funfairs and games, Phillips invented his own language to reduce the distance between viewer and artist. He introduced certain visual games into his works "to share his experience of the world with the viewers and set in motion emotions and intellect" (Livingstone 1983, p. 11). For Phillips, the mood of urban life is best evoked by the secondhand imagery that floods the corners of the city, an imagery that mingles with certain artistic forms to produce an abstract dimension.

2.1 For Men Only—Starring MM and BB

The magazine images constitute Phillips' work of art from 1961 *For Men Only-Starring MM and BB*. Fascinated by exuberance, excess and inversion of traditional means of expression, Phillips builds a carnivalesque atmosphere by inviting viewers to participate in his artistic creation. A collage technique and the largeness of scale amount to a game, a technique used in Phillips's work. The frames of the painting, which resemble a pinball machine, are enticing us to play. Commenting on this kind of work Phillips say that "each game is a big image subdivided into little pictures, and you would play this, so what the thing became a sort of visual game where more or less anything could be acceptable" (Livingstone 1983, p. 19). The traditional borders between viewer and the painting are blurred; the division of the whole representation into segments is pushing us into a perspectival mobility.

We move from one frame to another as if following the movement of a ball, bouncing through a terrain of film-star images cut from glossy magazines. This work, thanks to its connection with entrainment imagery and movement, was meant to liberate the selected material "from conventions and established truths, from clichés, from all that is humdrum and universally accepted" (Bakhtin 1968, p. 34).

So both the format of the painting and its content are borrowed from popular imagery. The decision to link popular material with traditional techniques intensified the artist's involvement in a recombination and degradation of high culture. Here, "the collaged heads of Marilyn Monroe and Bridgette Bardot, the two great pin-ups of the time, taken from magazine photographs" (Livingstone 1983, p. 22) constitute the focal point of the work of art. The stars placed under the photographs and saturated red and yellow colours emphasize our impression that celebrities are cultural creations. "Phillips added the phrase 'Cert X' above the head of Bardot, highlighting the adult nature of her allure" (Moorehouse 2007, p. 92). The lower part of the work contains four images of a stripper, painted again from a secondhand source. This woman, presented in different sexual position, adds to the painting's kinetic quality. Additionally, the painted hare on yellow background is a sly pun on the women in the work. Letters spell out the message, "She is a doll" followed by the name of the stripper. This work of art implies that even sexuality has become a kind of game, a new form of leisure celebrated by consumer society. The scraps of popular imagery are enriched with the glutted headline "Elvis for Britain". Phillips is catching at a new condition of life, at a glittering junk which has the power to upset tradition.

3 Derek Boshier's Image in Process

Derek Boshier's paintings employ different modes of communication to transgress structural constraint. He was also an artist associated with the second phase of British pop art. Like Phillips's painting, his works deconstruct a high-art discourse and criticize the subjects presented on his canvas. "Boshier brought to British pop art a strong satirical edge which distinguished his work from other fellow students at the Royal College of Art" (Moorehouse 2007, p. 93). The majority of his paintings referred to political events and a buying public, emphasizing the manipulative forces of the mass media and a gradual loss of social identity. As Marco Livingstone (1983, p. 143) points out, "from the very beginning, Boshier proclaimed to eliminate a separation between the aesthetic and the non-aesthetic, thereby joining and reconciling art and reality". His continual interchange and fusion of high and low styles, his parody and his politics constitute a "shifting polyglossia"[3] (Russo 1986, p. 212) in which a viewer is transported into the artistic creation.

---

[3] The term polyglossia characterizes Bakhtin's definition of novelness and instances of language once can find in novels.

3.1 *Is It a Space Probe?*

We see this clearly in his *Man Playing Snooker and Thinking of Other Things.* This may also be a kind of metaphorical self-portrait, in which the artist tries to think over an idea for a future painting. Thus certain well-known images come to his mind, filling the empty space of his head. "A large head is shown in profile, its interior exposed as if to reveal an anthology of private images: the Union Jack, a comic-book cover with Superman, some samples of newspapers and several phrases are represented in thought bubbles" (Moorhouse 2007, p. 88). Is it a *Space Probe?* This question indicated in one of the bubbles, suggests a possibility of the themes used by Boshier. Here, in seeking a stylistic alternative to abstract works of art, his work epitomizes a central tenet of the pop aesthetic, searching for an answer through a morass of mass media imagery. Those figures taken from mass culture exert powerful effect on the viewer as they are woven into a daily life. As a result the flag, this through the continuous re-usage in variable social and cultural contexts, is also ascribed to the category of common objects used by the artist. The mixture of high and low discourses dialogizes the public realm, destabilizing the assigned mimetic order. As Christopher Finch (2001, p. 5) notices, "Boshier's idealism was combined with a keen analytical sense of how the power of the media, especially in its subservience to big business and government, could be used to shape public taste and opinion".

A sense of movement, which is accentuated by the format and open framing, adds to the carnivalesque dimension of the painting. The double framing here comes from the green space that surrounds a supposed snooker table. Parts of a hand sneaks into the colourful frame, so we are not only voyeurs, because we can follow the movement within the frame and somehow take part in the final creation of the work. And we intuit a dialogical laughter under the artistic process of thinking. This, combined with an intertext, and multiple references to popular imagery, minimize the distance between the viewer and the artistic world.

4 Pauline Boty: The First Lady of British Pop Art

Pauline Boty was the only female pop painter to work alongside other male students of Royal College of Art.[4] She appeared with Peter Blake, Derek Boshier and Peter Phillips in a 1962 Ken Russell television documentary *Pop Goes the Easel.*[5] She produced only a small number of paintings before her tragic death in

[4] I am referring here to the title of the monograph *Now you see her—Pauline Boty—First Lady of British Pop* written by Adam Smith in 2002.

[5] The film presents four major figures of the second generation of Pop artists, namely, Pauline Boty, Derek Boshier, Peter Phillips and Peter Blake with a wider audience, exploring their artistic routines and areas of interest. The four artists, called also "young meteors," embodied the typical

1966, and for years her work was in the shadow of the pop images of a male-dominated art world. It was almost three decades after her death that her work would receive new attention.

According to Christopher Finch (2001, p. 30), "what makes Boty's paintings resonate today is their proto-feminist rhetoric, which is all the more powerful because it pokes fun at the conventional eroticism found in the work of other pop artists like Jones and Phillips". Her works offered an alternative interpretation of the concept of femininity constructed by mass culture, and brought a romantic element into the fold. Boty's paintings, like the works of her colleagues, are an urban outlook on reality, exploring raw materials borrowed from the mass media. Her works showed in a new way how British pop art was a rebellion of the young against mannerisms and modes of the old masters.

4.1 Towards Interactive Art: My Colouring Book

"This is my book. An unusual book. This is my colouring book".[6] These are the words Boty placed at the very left top of her 1963 work *My Colouring Book*. Even though the inscribed letters are smaller than the other images within the frame of the painting, the line is a starting point for the viewer, marking the order of the visual material organized by the artist. The painting is divided into six blocks containing different images, which at the first glance are unrelated to one another. There is a male star and an interior design taken directly from some magazine, a faceless woman hugging a white silhouette; and a green necklace, sunglasses and blue heart are placed in the centre of the painting. All of these objects appear to be photographs pasted from magazines into scrapbook pages, which are additionally enriched with an ornamental design. Once the inscriptions have been read, the viewer may conclude that the images form a narrative path to follow. We move from segment to segment, reading textual material which situates a particular image as in film storyboard. For Sue Watling (1998, p. 16), *My Colouring Book* "is compiled like the verses of a blues ballad". The images revolve around a motif of romance, resonating with the theme of so many pop songs of that decade. Here banality mixes with abstract shapes furthering a love story in the style of a comic book. The painting is an exemplification of a specific woman's point of view, detailing a sudden sense of loss and loneliness. The lines: "these are the arms that

(Footnote 5 continued)
features of youth of those days. This is particularly visible while analyzing the initial scene of the documentary, the figures strolling along the funfair park to the accompaniment of pop music. Here the scene indicates the impact of mass culture on artistic personae and their dependence on the productions dictated by the commonplace.

[6] This inscription appears in the reproduction of the work of art published in a catalogue to the exhibition *Swinging London: A collection of Grabowski* organized by Museum of Art in Lodz in 2007.

held him and touched him and lost him somehow colour them empty now" strengthen a tension between written and painted material. This is the world of a teenage girl whose bedroom music and souvenirs construct her identity.

My Colouring Book embodies an interactive nature of art, modifying the rigid relation between viewer and the work through limitation of contemplative distance. As Lawrence Alloway (1962, p. 38) points out, "an awareness of the world as something that contains both the work of art and the viewer was at the core of the development of visual arts in London in the sixties". This dream to recover unity between spectator and work and to end an established geometry of body and space is made flesh in Boty's pictorial syntax and vocabulary. This work attempts to break down traditional hierarchies of representation and to reveal the old way's betrayal of its promise to deliver the truth. The format of *My Colouring Book* lures viewers to explore the canvas and arrive at their own meanings. The combination of images associated directly with high discourse and popular imagery helps to identify with a presented material and transgress the\ normative techniques. For Boty, art has to issue from our own messy, contingent experience, and to abandon what was once considered a 'pure' representation of the world.

5 Conclusion

Just as Bakhtinian carnival brings together crises of the past and present, the analyzed works create a transgressive language that comprises heterogenous vocabulary. The paintings recontextualize high discourse in the name of the low. Ephemera filtered through the mass media and advertising industry result in an eclectic blend of styles. Excess flows from an intextuation of an array of popular signs and symbols of social anxiety and visual pleasure, here immersing viewers into Swinging London scene. These works engage with, rather than reject, commercial advertising and consumer goods. Andreas Huyssen (1986, p. 141) notes that "pop became the synonym for the new lifestyle of the younger generation, a lifestyle which rebelled against authority and sought liberation from the norms of existing society". The carnivalesque elements of these works destroy the distance between artist and viewer, sucking us in, rubbing our faces in a reality that traditional art would prefer we ignore.

References

Alloway, L. 1962. Illusion and environment in recent British art. In *Art International,* February. 38–40.

Bakhtin, M. 1968. *Rebelais and his word.* Cambridge, MA: MIT Press.

Donnelly, M. 2005. *Britain: culture, society and politics.* Harlow: Pearson Education Limited.

Finch, Ch. 2001. London pop recollected. In *Pop art: U.S./U.K. connections, 1956-1966.* ed. by D. E. Brauer…[et all], 20–42. Germany: Hatje Cantz Verlag.

Fiske, J. 1995. *Understanding Popular Culture.* London: Routledge.
Hewisson, R. 1987. *Too much: Art and society in the sixties 1960-1975.* New York: Oxford University Press.
Huyssen, A. 1986. *After the modernism, mass Culture, and postmodernism: great divide.* London: The Macmillan Press.
Jenks, Ch. 2003. *Transgression.* London: Routledge.
Krajewski, M. 2005. *Kultury kultury popularnej.* Poznań: UAM: Wydawnictwo Naukowe.
Laing, S. 1992. The Politics of culture: Institutional change. In *Cultural revolution?: The challenges of the arts in the 1960 s.* ed. by B. Moore-Gilbert and J. Seed, 72–96. London and New York: Routledge.
Livingstone, M. 1983. *RetroVision: Peter Phillips.* Liverpool: Walker Art Gallery.
Moorhouse, P. 2007. *Pop art portraits.* London: National Portrait Gallery.
Russo, M. 1986. Female grotesques: Carnival and theory. In *Feminist Studies/Critical Studies.* ed. by T. de Lauretis, 202–220. The UK: Macmillan Press.
Storey, J. 1993. *An Introduction to cultural theory and popular culture.* London: Prentice Hall, Harvester Wheatsheaf.
Watling, S. 1998. Pauline Boty. The Only Blonde in the World. London: Whitford Fine Art& The Mayor Gallery Ltd.
Willis, C. 1989. Upsetting the public: carnival, hysteria, and women's texts. In *Bakhtin and cultural theory,* ed.by K. Hirschkop and D. Shepherd, 129–135. Manchester and NY: Manchester University Press.

Representation of Identities in the British Arts Policy of the 1980s

Katarzyna Kociołek

Abstract The Arts Council of Great Britain profoundly influenced the arts discourse in the 1980s Britain not only through the funding that it allocated to artists and art practices, but also through wide production of reports and documents that aimed at meticulous and detailed description of the British arts scene. In the turbulent period of the 1980s, when the concept of British art seemed to have undergone a major revision, these reports and various policy papers imposed neat division of artists and their works into ethnic and racial categories. Adopting a postmodernist approach to text analysis one may argue that these documents on the one hand describe the reality of the British art scene, which at that time was becoming increasingly more diversified and multicultural. On the other hand they shape this reality by provision of certain labels and extensive use of certain terms to refer to particular artists and their practices. Therefore, the paper features analysis of reports on ethnicity, diversity, and multiculturalism in the visual arts debate. It focuses on discussing the use of labels such as "ethnic", "black" or "British", as well as examines how exclusive or inclusive each of these terms was, and to which social groups they referred. The analysis aims at demonstrating that some arts policy documents covertly sustained the status quo which set Britishness against its ethnic or black "Other" even though on the surface it may seem that they showed black art as part of all inclusive, pluralist notion of Britishness.

K. Kociołek (✉)
Warsaw University, Warsaw, Poland
e-mail: kkociolek@uw.edu.pl

J. Fabiszak et al. (eds.), *Crossroads in Literature and Culture,* Second Language Learning and Teaching, DOI: 10.1007/978-3-642-21994-8_43,

1 Introduction

In his article "The long march from 'ethnic arts' to 'new internationalism'" artist and curator Gavin Jantjes observes that in the 1980s black artists in Britain were attached denigrating labels such as semi-professional or amateur. Although many were born in Britain, they remained ignored by the authorities, which would only subsidize them as "ethnic artists". He argues that the very term "ethnic art" was coined by the dominant culture to control access of non-white art practitioners to the mainstream. Thus, the discourse of ethnic arts represented black artists as "immigrants, outsiders, foreigners and guests" (Jantjes 2000, p. 267) with the "Other" being the common denominator of all these notions and concepts. Commenting on the British art scene in the 1980s, Jantjes notes that black artists remained outside mainstream art practices, and that their work would be labeled "amateur" regardless of its artistic quality. Political activism on the part of these artists led to the establishment of an ethnic arts subsidy, which not only failed to improve the existing situation but also reinforced inequalities along racial divide, as well as strengthened the system of binary oppositions in which the dominant white culture decided on the place and role of other cultures art in the arts policy. Therefore, the paper, which features analysis of reports on ethnicity, diversity, and multiculturalism in the visual arts debate, focuses on discussing the use of labels such as "ethnic", "black" or "British", as well as examines how exclusive or inclusive each of these terms was, and to which social groups they referred.

2 CDA (Critical Discourse Analysis) as an Anlytical Resource for Examining Identities in the Arts Policy

An effective analytical tools for close examination of the representation of identities in the arts policy is provided by one of the founders of critical discourse analysis, Norman Fairclough. As Fairclough observes in his book *Analysing Discourse. Textual analysis for social research* the term discourse renders language "as an element of social life which is closely connected with other elements" (Fairclough 2003, p. 3). Similarly to Fairclough, I view discourse as an integral component of social life. Consequently, I propose that discourse analysis be treated as a source of invaluable information on social phenomena, in particular the approach to and construction of national and ethnic identities.

Fairclough's definition of discourse as a "form of social practice (…) in which people may act upon the world and especially upon each other, as well as a mode of representation" (Fairclough 2007, p. 63) implies first of all that social structure constitutes discourse, i.e. that relations in a society such as those based on class, race, gender, etc. influence discursive practices. Secondly, it suggests that discourse constructs social realities "relations, identities, and institutions" (Fairclough 2007, p. 64). As Fairclough puts it, "discourse is a practice not just of representing the world, but of signifying the world, constituting and constructing the world in meaning" (Fairclough 2007, p. 64). The assumption that discourse plays as constitutive role in a social structure is crucial for my analysis.

Fairclough establishes a firm connection between assumptions and ideology by noting that assumptions improve "efficiency and adaptability" (Fairclough 2003, p. 58). He also observes that power relations are best sustained by means of some shared concepts and so he links "ideological work of texts" to "hegemony and universalization" (2003, p. 58). The pursuit of hegemony is voiced in the desire to express the particular as universal, which allows to maintain domination. According to Fairclough textual representation of identities which are made appear universal reflects the struggle over hegemony. By means of filling the text with a set of background assumptions "In many texts (...) one finds the whole vision as part of an assumed and taken-for-granted background" (Fairclough 2003, p. 46). There are three main types of assumptions, he notes, and these are: "existential assumptions: assumptions about what exists; propositional assumptions: assumptions about what is or can be or will be the case [and finally] value assumptions: assumptions about what is good or desirable" (Fairclough 2003, p. 55). There are also "bridging assumptions" i.e. assumptions that are necessary to create a coherent link or "bridge" between parts of a text, so that a text "makes sense" (Fairclough 2003, p. 57).

Following Norman Fairclough's approach to discourse, I examine various arts policy texts, and investigate how and to what extent these texts shape and influence social life, particularly in the area of identity formation. As "meaning-making depends upon not only what is explicit in the text but also what is implicit—what is assumed" (Fairclough 2003, p. 17), I focus primarily on the implied meaning of categories such as "ethnic minority arts", "black arts", "Afro-Caribbean and Asian arts", and indigenous, Western, European culture. What is more, I concentrate on what the documents fail to mention (e.g. the connection they establish between ethnic minority arts, Afro-Caribbean and Asian arts, and black arts) because, as Fairclough observes "what is said in the text is always said against the background of what is 'unsaid'" (2003, p. 17).

3 Analysis of Identity Constructs in Art Policy Documents

The Council Paper 762 on Ethnic Arts from the 1980 (The Council Paper 762, 1980) defines the term ethnic arts as referring to arts "associated with British populations of Black Commonwealth, Pakistani and Chinese origin". Thus inexplicitly the report views ethnic art as art produced by black artists. Although the report describes Britain as a "plural society" of "equal opportunities", which means that being British denotes openness, tolerance and multiplicity of identities, it seems to quite openly suggest inferiority of ethnic arts in relation to "Western culture", of which mainstream British culture is part. The report defends Arts Council's activities aimed at ensuring that ethnic minorities are "introduced to the best of Western culture". Additionally, point 14 of the report emphasises relative lack of professionalism on the part of ethnic artists by providing a sweeping generalisation that "(...) cultural traditions of ethnic minorities are much less reliant on professional artist than is the European cultural tradition, (...) the ethnic artist can often establish himself only with his own community". From the above quoted fragments one can infer that the report clearly defines ethnic arts as parochial, marginal and generally of lower artistic quality than European or Western art, for it is implied that such art is rarely appreciated outside the community of the artist. It seems that the main role that Arts Council performs in relation to ethnic arts is alleviating ethnic art practices to the standards of European culture. In the choice of such terms as European or Western culture one may easily trace the characteristics of universalism. The replacement of the term "British culture", which would seem much more appropriate in this context, considering that the report was written and published in Britain and for the British, may suggest an underlying assumption that British culture is a synonym of Western or European tradition. Even though point 17 of the report becomes more specific in that it refers to Britain's "indigenous population", it fails to clarify to whom exactly the phrase refers. It can only be inferred from an earlier definition of ethnic arts that Britain's "indigenous population" would be white British citizens regardless of their white ethic origins (English, Welsh, Scottish, Irish). Therefore, it seems that "ethnic arts" is a camouflage term that establishes divisions along the colour line.

The report entitled "Ethnic Arts in the East Midlands" from February 1981 (Ethnic Arts in the East Midlands 1981) defines the term "ethnic minority arts" as "the manifestation in this country of the rest of the world's culture". While the report makes it clear that "ethnic minority arts" are diverse and heterogeneous, one can easily infer from the above quoted definition that they simply denote the art of the Other. Points one and two of the report are based on the bridging assumption that links ethnic minority arts and folk arts, for they state that "ethnic minority arts" involve "traditional and 'pure' arts activities which are indigenous to other countries than Britain" or "cultural activities which represent part of the 'folk' tradition". The "Ethnic Minority Arts and Racism" section of the report represents Britain as a country with a long-established history of immigration, which for centuries not only hosted ethnic minorities, but also whose culture "owes everything to the steady influx of immigrants from the Celts onwards". Thus, Britain is represented as a melting pot and British culture comes to be established as a mixture of cultures and influences from all over the world. However, in this concoction of cultures not all ingredients seem to have blended equally well, the report suggests, and names Afro-Caribbean and Asian communities as "the more conspicuous minorities" who are victims of intolerance on the part of "present day British society". The "Recommendations" section of the report seems to further clarify the division into mainstream and "blackstream" arts in Britain, by noting that with regard to funds allocation "the present picture is one of disproportionate spending on traditional middle-class, white-oriented cultural activity (...) the facts are (...) that ethnic minority arts form a small but very important part of the wide spectrum of creative and artistic activities". Contrary to a broader definition of the term "ethnic minorities" which includes white minorities and which was proposed at the beginning of the report, the quotation seems to testify to the fact that "ethnic minority arts" are viewed as artistic activities of non-white, non-middle-class artists. Therefore, it may be inferred form the report that ethnic minority arts equal Afro-Caribbean and Asian arts, which in turn equals black arts. What is more, as far as visual arts are concerned the report stresses close relationship between ethnic minorities' visual arts and the category of craft. This may be interpreted as an attempt to deny ethnic minority artists access to mainstream art galleries, where craft work is not normally on display. Having introduced the assumption that ethnic visual arts belong to the category of craft activities, and that they are "different from the indigenous tradition" which means indigenous British tradition, the report suggests that the Arts Council should organise "appropriate exhibitions for touring in ethnic venues". By juxtaposing the above fragment with another that reads: "the working party recommends that the Association should try to put together over the next three years a series of exhibitions appropriate to small, non-gallery venues", one may discover that "ethnic venues" are more or less synonymous with "small non-gallery venues". All in all, despite acknowledging contribution of ethnic minority arts to the mainstream culture and the indigenous British tradition, the report relegates ethnic minority visual arts practice to some special places on the margins of the British artistic scene.

The report that sheds some light on the shift in the use of the term "ethnic arts" into "black arts" is the Great London Council (G.L.C.) Race Equality Unit Report from 12 December 1985 (Race Equality Unit Report 1985) . The very first point of the report states that the creation of Race Equality Unit within the G.L.C. contributed significantly to the "taking [of] black artists into the mainstream". Most importantly the report emphasises "professionalisation of black arts sector" which caused the separation of black arts from ethnic arts. "The most significant effect of G.L.C. funding on black arts has been the professionalisation of black arts and the emergence of clear distinction between what is described as 'ethnic arts' and 'black arts'". This seems to suggest that G.L.C. activities elevated black arts above the category of ethnic arts. The separation of the two and the context in which the division occurred clearly implies that unlike "ethnic arts" which remain in the domain of amateur arts, "black arts" approached the standards of professional art. This claim is supported by the following sentence: "There is now a clearly defined black art sector which is concerned with the development of contemporary art forms". In other words the report seems to indicate that "black art" exceeded the boundaries of "ethnic arts" characterised by tight connections with particular community and the culture of origin of the artist. "Black art", in contrast, entered the mainstream. "A number of highly successful key events have been organised in major art venues. This has been important in helping to establish black artists in the mainstream".

"Ethnic Arts Action Plan—Art" by Mike Sixsmith from January (Sixsmith 1987) deserves closer attention due to a frequent occurrence of the term "black artist" in the paper. The existential assumptions that one may easily spot in the text are: (1) that there exists a separate group of black artists, (2) that equal opportunities policies entail including the works of black artists in the gallery programmes. There are also the following bridging assumptions: (1) that the phrase "black artists" means artists of Afro-Caribbean or Asian origin; (2) that British mainstream art does not include non-European artists apparent in the fragment "(...) to show the contribution of non-European artists to 'mainstream' art in Britain" (Sixsmith). What is more, British born artist Sonia Boyce is classified as a black artist, as the opening sentence of point three which reads "Several Art Department Staff have attended race awareness courses, and have made special efforts to inform themselves by attending conferences, by visiting exhibitions and through discussion with black artists" (Sixsmith) is concluded by a following statement "Sonia Boyce has been appointed one of the purchases for the Arts Council Collection for the next fifteen months" (Sixsmith). The bridging assumption that logically binds the two fragments is that Sonia Boyce is considered a representative of the black art, with which Art Department staff attempt to become familiar. Section 5.1 features an existential assumption that black artists who have gone through "Western art school" are allowed access to "traditional gallery settings" (Sixsmith). At the same time a bridging assumption suggests that these artists do not create art that is in any way related to Afro-Caribbean or Asian art forms. "Most of the work shown is largely within a European context, from black artists who have been through the western art school system and who are now exhibiting in traditional gallery settings. So far we have found it difficult to attract proposals relating to Afro-Caribbean and Asian art forms" (Sixsmith).

"The Arts and Cultural Diversity. Symposium Report" form 1989 (The Arts and Cultural Diversity 1989) declares an attempt to promote diversity and incorporate black arts into the British culture. It explicitly states that the goal of the symposium was "the recognition of black artists as artists first of all, as individuals free to explore aesthetics, rather than as agents for solving political problems" (13). However, in order to account for the situation of black arts in Britain, the report does not flinch from appropriating rhetoric based on binary oppositions, whereby black arts is set against or juxtaposed to the British "mainstream tradition" (5). From a series of existential assumptions one can learn that in Britain there existed a separate area in the arts referred to as black arts, "A principal reason for the historical neglect of black arts has been the failure to place its achievements within the framework of a broad heterogeneous national culture" (7). Additionally, the report also suggests that black arts had been marginalised, underfunded and separated from the mainstream. In contrast, "the British way of life" (6) is represented as valuing heterogeneity, tolerance and openness, while "Britain's artistic life" is claimed to have been "immeasurably enriched by its cultural diversity" (2). It seems that one cannot fail to notice the dichotomy between the overtly positive image of Britain, which aims at moving from "a monocultural to a multicultural perception of the arts" (7) and the existential assumption which implies that there still prevails a division into the mainstream and the margin reserved for "people from other cultures" (5), which is expressed in a sentence "The Government is keen for people from other cultures to play a full part in the mainstream of British life without losing their own cultural roots and identity" (5). Blackness which surfaces in the report is fairly homogeneous and remains outside the mainstream tradition and by implication also outside the concept of "Britishness".

The Arts Council report entitled "Towards cultural diversity: the monitoring report of the Arts Council's ethnic minority action plan" from 1989 (The Arts Council 1989) includes a bridging assumption that "Black Arts" equals ethnic minority arts. What is more, the "Principal Recommendations" section features points whose logical cohesion depends on the bridging assumption that "Black Arts" stands for "cultural diversity", i.e. "that Council develop a strategy plan which creates (…) a place for Black Arts organisations and groups", and "that Council install a code of practice for monitoring cultural diversity". The entire report is based on the existential assumption that British national culture is heterogeneous, "Central to the philosophy of this report is the concept of a broad heterogeneous national culture, its make-up reflecting the diversity of cultural achievement issuing from contemporary society". The sentence not only implies that the concept of Britishness had lost its cultural homogeneity, but also that British national culture (and by implication also Britishness) no longer consists of a fixed cultural core to which other cultures become assimilated. On the contrary, as the report states, diverse cultural achievements are "not assimilated, but rather placed alongside all other achievements to construct through their diverse autonomies a new superstructure of cultural practice". Another existential assumption about British national culture is the assumption that it is plural: "These debates identified particular needs: to raise the national consciousness to the plural nature of the contemporary culture", that it is multicultural, and that still the representation of Britishness as a culturally diverse identity is prevented by racism. Throughout the report there appear phrases such as "black representation", "black artists", "black achievement", "black organisations", "black audience", in which the meaning of "black" is not further clarified. Bridging assumptions included in the text imply that the signifier "black" here simply refers to people, achievements, and organisations that are "culturally other" [section D of the report, point 1 (i)] and do not represent Western/Eurocentric consciousness, "The national cultural consciousness was Western/Eurocentric, even insular. Its interaction with the 'culturally other' was simply to relegate it to the margin and contain it at the periphery" [section D of the report, point 1 (i)]. By pledging its commitment to the idea of diversity, the Arts Council represents itself as fighting against insularity of the British society and culture, which is presented as hostile to any "foreign intrusion" [D, point 1, (i)]. Consequently, in the report "Black Art" and black identity appear to be foreign to Britishness, yet with the new policy of diversity this foreignness can be happily awarded a special space parallel to the mainstream arts. Hence, the report calls for the need to create "more black organisations", "black venues" and to employ more black staff. There can be no doubt that blackness and Britishness are clearly separated, where black means non-British.

4 Conclusion

All in all, the British government arts policy seems to have contributed significantly to the formation of cultural identities within the visual arts sector. The artists, who had to identify themselves as "ethnic", "black", or "other" in relation to the dominant British art and culture in order to secure financial support, would be forced to internalise identities that the policy discourse constructed. The analysis demonstrates that some arts policy documents covertly sustained the status quo which set Britishness against its ethnic or black "Other" even though on the surface it may seem that they showed black art as part of all inclusive, pluralist notion of Britishness. The picture of black identity that emerges from the documents analysed in this chapter stands in opposition to Britishness and is tightly connected with ethnic minorities. Surprisingly, Britishness is represented as a fairly stable identity, which yet can embrace diversity and readily accept difference. After all, as some reports seem to imply, at the very core of Britishness there lie openness and tolerance.

References

Council Paper 762. 1980. *Ethnic Minority Arts: Lawrence Mackintosh Papers 1980-1987 (ACGB/46/4).* Victoria and Albert Museum Archives, London.

Ethnic Arts in the East Midlands. 1981. *Ethnic Minority Arts: Lawrence Mackintosh Papers 1980-1987" (ACGB/46/4).* Victoria and Albert Museum Archives, London.

Fairclough, N. 2003. *Analysing Discourse: Textual Analysis for Social Research.* London: Routledge.

Fairclough, N. 2007. *Discourse and Social Change.* Cambridge: Polity Press.

Great London Council (G.L.C.). 1985. Race Equality Unit Report, 12 December 1985. In *Ethnic Minority Arts: Lawrence Mackintosh Papers 1980-1987 (ACGB/46/4).* Victoria and Albert Museum Archives, London.

Jantjes, G. 2000. The long march from 'ethnic arts' to 'new internationalism'. In *Black British Culture and Society*, ed. K. Owusu, 265-270. London: Routledge.

Sixsmith, M. 1987. *Ethnic Arts Action Plan – Art, January 1987, Ethnic Minority Arts Action Plan 1986-1989 (ACGB/46/106)*, Victoria and Albert Museum Archives, London.

The Arts Council in association with the Home Office. 1989. *The Arts and Cultural Diversity, Symposium Report.* Victoria and Albert Museum Archives, London.

The Arts Council. 1989. *Towards cultural diversity: the monitoring report of the Arts Council's ethnic minority arts action plan.* Victoria and Albert Museum Archives, London.

The Coenesque Zones: Alternative American Settings Beyond the Constraints of Everyday Logic in the Cinematic Works of Joel and Ethan Coen

Jakub Ligor

Abstract On the surface, all films of Joel and Ethan Coen unfold in strictly defined local American settings. Closer scrutiny reveals that these environments, despite numerous features concurrent with their real-world counterparts, are actually nonexistent and non-illusionist, alternative realms which parallel what Brian McHale refers to as "zones" in postmodernist literary fiction. Like many postmodernist and metafictional writers, the Coens delineate the boundaries of their own, meticulously constructed, parallel territories and create heterotopian spaces of fictional worlds by combining conflicting chronological frameworks, blending noncontiguous aspects of unrelated realities as well as fusing history and fiction. By crossing the ontological borders and constraints of time and geography, the directors destabilize and deconstruct common-sense contextual constructions forcing the viewer to review their understanding of what constitutes objective reality. I am analyzing how the Coens generate their universes by making different historic frameworks overlap, using anachronisms and juxtaposing incongruous aesthetics. I am also showing how seemingly average Middle American settings paradoxically function as "zones" both in literature and the cinematic works of the Coen brothers. As a consequence of the process of defamiliarization they undergo, America appears to be a heterogeneous amalgam of discontinuous solipsisms. The Coens' Arizona or Texas are mental rather than geographical entities, ultra-stylized, closed, synthetic film-spaces "two steps to the left of reality", projections of cultural myths, stereotypes and beliefs that complement the films' stories.

J. Ligor (✉)
State School of Higher Professional Education, Płock, Poland
e-mail: ligor@onet.eu

J. Fabiszak et al. (eds.), *Crossroads in Literature and Culture,* Second Language Learning and Teaching, DOI: 10.1007/978-3-642-21994-8_44,

1 Introduction

As Patricia Waugh (1985) observes, metafictional prose foregrounds the role literary conventions play in the simultaneous act of describing and creating objects in fiction and, therefore, it predominantly projects universes which are potentially possible but markedly different from the world as we know it. The 'truth' status of both literary fiction and what is believed to be 'reality' is examined and reviewed. According to Waugh (1985, p. 90) metafiction tends to "explore the notion of 'alternative worlds' by accepting and flaunting the creation/description paradox, and thus expose how the construction of contexts is also the construction of different universes of discourse." Commonsense contextual constructions of the everyday world are destabilized, the awareness of possible alternatives to this reality is increased and the reader is effectively prevented from settling into any given context.

Brian McHale (1987) draws our attention to a typically postmodernist predilection for building within a literary work a heterotopian space of a fictional world. McHale (1987, p. 45) calls it a "zone" and defines it as a setting whose space is "less constructed than *deconstructed* by the text, or rather constructed and deconstructed at the same time." In order to build a zone a number of strategies are employed: noncontiguous, unrelated realms are juxtaposed; an alien space might be introduced within a familiar space; two familiar worlds may paradoxically overlap creating a new, highly disorienting third reality; places are also linked with unsanctioned attributes which undermine conventional wisdom and automatic associations.

The settings in the cinematic works of Joel and Ethan Coen, frequently labeled as postmodernist filmmakers, often bear a striking similarity to such literary heterotopias, including Thomas Pynchon's zone in *Gravity's Rainbow* which projects an ontologically unstable, impossible, pluralistic reality where, as McHale (1987, p. 45) points out, "our world and the 'other world' mingle with increasing intimacy, hallucinations and fantasies become real, metaphors become literal, the fictional worlds of the mass media—the movies, comic-books—thrust themselves into the midst of historical reality." According to R. Barton Palmer (2004, p. 50), detractors of the Coens who accuse their artistic output of being merely a series of cynically shallow stylistic exercises, fail to recognize the fact that their "ideas and themes are developed in the context of cinematic re-creation, of the fabrication of derivative 'sealed universes' quite distinct from one another." Searching for a meaningful key to the interpretation of the directors' productions, which, as Palmer (2004, p. 52) observes, "do not constitute an oeuvre in the neo-romantic sense of *the politique des auteurs*" and show no intention to create any continuity from one picture to the next, one should focus not only on the obvious classic Hollywood models for their films but also acknowledge the fact that the Coens' influences are primarily literary rather than cinematic. Even though numerous critics, including Eddie Robson (2003) and Robert Ziębiński (2008), notice the impact of hard-boiled and southern gothic writers while Palmer (2004, p. 52) also highlights some similarities with the neon noir fiction of Elmore Leonard and Raymond Carver, what seems to go unnoticed is the staggering number of parallels between postmodernist prose or metafiction and the Coens' highly intertextual American narratives. One of them is the manner in which the Coens delineate the boundaries of their own, meticulously designed, idiosyncratic territories and create heterotopian spaces of fictional worlds by subverting stereotypical perceptions, combining conflicting chronological frameworks, blending incongruous aesthetics and inconsistent aspects of unrelated realities as well as fusing history and fiction in order to create self-standing alternative visions of what might be at first glance believed to be typical American locations. Analyzing the techniques employed by the directors to construct their filmscapes and being able to see how closely they correspond to relevant literary devices seems crucial in understanding the directors' deconstruction of American culture and the ensuing critique of American values and dreams.

2 The Big Lebowski

2.1 A Carefully Constructed Time Perspective

One of the themes of *The Big Lebowski* is the exploration of the Los Angeles society. What makes this setting particularly interesting to the directors is the fact that it is a place where some cultural and social patterns thrive although they have virtually died out elsewhere. The film is actually set in what Körte (2001, p. 200) refers to as "a peculiar time warp"—on the surface the viewer perceives it as the 1990s thanks to the narrator's introduction, the references to the ongoing Iraqi invasion of Kuwait; at the same time, however, we cannot fail to notice that the story has been placed in a carefully constructed yet highly confusing time perspective where, according to Körte (2001, p. 200), within the framework of the 1990s "the 1970s are struggling under the weight of the 1960s, and (...) the pop culture icons of the recent past are shot through with allusions to the literary noir of the 1940s."

The main characters of the Dude and Walter are direct throwbacks to the past refusing to leave an era which ended years ago and maintaining an anachronistic approach to the realities that surround them. Their appearance, the choice of clothing, emotionality, behavior, manner of speaking and life priorities are out of tune with the 1990s setting in which they operate. The Dude epitomizes the haze and hedonism of the hippy times by refusing to have an occupation, smoking joints and taking a daily dose of White Russians, listening to Creedence Clearwater Revival. The defining moments of his life, such as being one of the authors of the original Port Huron Statement, the 1962 civil rights manifesto of an American student activist movement, are seriously dated and clearly meaningless for anyone unfamiliar with the time context he adopts. His friend Walter Sobchack, like the demobbed characters of *Taxi Driver* or *The Deer Hunter*, explicitly manifests his Vietnam war veteran experience and trauma and cannot refrain from making oddly inappropriate and compulsive references to the conflict in all possible situations. In addition, he has an obsession with all things militaristic, he finds it difficult to control his anger and frequently makes use of strategies suitable for a combat soldier but painfully unacceptable in the 1990 s realm of *The Big Lebowski*. When one of the bowling game players does not admit to stepping over the line and insists that he should be awarded eight points, Walter simply takes out a gun and insists on following the rules, making it clear that his frequent bouts of political correctness are just a desperate attempt at keeping in tune with the times and culture where he does not belong.

The film's collection of chronologically displaced figures and themes is significantly more extensive. Sam Elliott as the Stranger, who is at the same time the film's narrator, speaks in heavy old-fashioned laid-back Western accent and, using his own anachronistic frame of reference of an American pioneer evoked in the first scene by the Western-like image of rolling tumbleweed and male voices gently singing Bob Nolan's classic cowboy song *Tumbling Tumbleweeds*, claims that the Dude's adventures are an example of "the way the whole durned human comedy keeps perpetuatin' it-self, down through the generations, westward the wagons, across the sands a time." Lebowski's elder daughter Maude is an eccentric, feminist artist whose expressionist vaginal art definitely brings to mind the 1960s Fluxus movement. Her vinyl collection includes a record by a fictional band Autobahn whose cover bears a striking resemblance to Kraftwerk's album *Man Machine* and the name pays homage to the German band's 1974 breakthrough vintage electronica masterpiece. The character of the adult movies producer Jackie Treehorn echoes porn industry moguls from the 1970s like Hugh Hefner while the old Lebowski in his wheelchair recalls General Sternwood in Howard Hawks' cinematic version of Chandler's *Big Sleep*. The private detective shadowing the Dude, who is himself a pastiche hippy version of Philip Marlowe, seems to come straight out of Chandler's novel *The High Window*. Walter claims that the paralyzed man Arthur Digby Sellers whose son might have stolen the Dude's car along with one million dollars was a writer of 156 episodes of the old TV series called *Branded*, which was an actual 1965 western show. Additionally, the man shown bowling in the picture on the Dude's wall is President Richard Nixon, an unlikely image to be displayed by anyone in the 1990s. The film's leading motif of bowling as a culture rather than sport was, as Joel Coen admitted in the interview by Ciment and Niogret (2000b, p. 168), an indication that all the characters are strangely embedded in patterns of bygone times and thus it was "important in reflecting that period at the end of the Fifties and the beginning of the Sixties. That suited the retro side of the movie, slightly anachronistic, which sent us back to a not-so-far-away era, but one that was well and truly gone nevertheless."

2.2 Discontinuity

A prominent feature of *The Big Lebowski*'s plot is the sense of fragmentation induced in the viewer. Like other locations in the Coens' films its setting projects itself as a heterotopian concentration of conflicting solipsisms and the story relies heavily on juxtaposing incongruous, self-contained episodes evoking a feeling of disorientation and discontinuity. The viewer, similarly to the film's main protagonist, travels through a multitude of worlds in their own right, individual solipsisms inhabited by disunited figures and is thus exposed to a collection of confusingly diverse and apparently unrelated puzzle elements and given a task of putting them together to come up with a definitive answer to the kidnapping mystery the Dude was commissioned to resolve. The efforts are, however, frequently subverted by the directors so as to enhance the overwhelming sense of disconnection and impenetrability reminiscent of plots in novels like Thomas Pynchon's *The Crying of Lot 49* as well as classic hard-boiled detective prose often mentioned by Joel and Ethan Coen as their primary source of inspiration. Erin A. Smith (2000, p. 82) states that the plots of hard-boiled stories, such as Raymond Chandler's *The Big Sleep* on which *The Big Lebowski* is indirectly modeled, were excessively complex and convoluted and lacked a coherent narrative trajectory relying on the labyrinthine combination of "loosely linked scenes of action (…) without being particularly concerned with the larger story (…)."As the literature of choice for working-class readers, the fiction tended to reflect their alienation in the increasingly confusing everyday reality, evoked the repetitive, episodic character of their jobs, expressed the sense of speed involved in the modern reconceptualization of time as a commodity and thus, as Smith (2000, p. 83) points out, rejected "the unities of bourgeois cultural organization and logic." According to Smith (2000, p. 11) "these stories are collective fantasies offering representations of the plots, characters and idioms through which readers of this class made sense of their world."

For O'Brien (1997, p. 16) hard-boiled literature, permeated by ubiquitous uncertainty, dissatisfaction, ugliness, horror and violence, offers a grotesque vision of capitalist America and its corrupted institutions set in a "dark world below the placid surface (…). This other America, when it is not a bleak rural wasteland inhabited by murderous primitives, is a glittering hell ruled by money and violence, flaunting images of beauty that are either deceptive or unobtainable." As O'Brien (1997, p. 77) notices, "what the Marlowe novels are finally about is loneliness in a sprawling city devoid of spiritual comforts. For a variety of reasons—greed, egotism, fear, selfish lust, or sheer thick-headedness—none of the people in the city can offer anything in the way of a human relationship." Hard-boiled prose, despite its proletarian virility, wit and frequent outbursts of comic energy, often appears to be set in an alternative, unreal realm of gloom and void which O'Brien (1997, p. 80) defines as "poised so precisely on the edge of the real that it seems to cancel itself out." The Coens draw upon their fascination on the genre to project their own distorted American settings beyond the constraints of everyday logic and use them as an appropriate background for their critique of the country's social, cultural and business systems as well as the expression of the overwhelming sense of isolation, discontinuity and confusion triggered by the degeneration of the American Dream and its mythical institutions.

3 The Hudsucker Proxy

3.1 The Disturbed Timescale

The sequence of events within the plot of *The Hudsucker Proxy* exemplifies the disturbed timescale of the film—undermining the principles of chronological realism the directors squeeze into one month of the plot's span not only the numerous ups and downs of the main protagonist but also the complete history of his brainchild being developed, marketed, sold, labeled a failure and eventually becoming a resounding success. The time framework of the film is complicated and ambiguous and the directors explicitly manifest it in the very first scenes by drawing our attention to the gigantic clock and the slogan "The Future Is Now" proudly gracing the façade of the impressive Hudsucker Industries building in addition to making us aware that these are the last moments of New Year's Eve at the end of 1958, a symbolically confusing moment of what was thought to be the future becoming a contemporary experience. Norville Barnes, the company's CEO, has just stepped onto the window ledge and contemplates a jump into the void below. However, we know neither the past background of the dramatic situation we have to confront nor its subsequent resolution and, as a result, we are devoid of any points of reference or grounds for logical reasoning. The opening frames challenge the viewers with a setting of the story placed in an indefinite time zone where conventional principles of chronology have been suspended or subverted.

The uncertainty is carefully reinforced by the Coens with multiple anachronisms and time inconsistencies. As Robson (2003) observes, Richard Hornung, the film's costume designer, insists that the choice of the men's clothes tends to reflect the 1930s and 1940s whereas designer Dennis Gassner's inspirations gave the film's architecture a 1920s look. The story, the characters and the style of dialogues are reminiscent of 1930s comedies while the setting and the central motif of the hula-hoop and the Frisbee as Norville Barnes's key ideas are emblems of the 1950s where the events apparently take place although these are clearly the alternative 1950s in which the hula-hoop has not occurred and Barnes is going to be its inventor. For Robson (2003, p. 141) this is a fantasy America of different eras fused into an unrealistic, exaggerated ideal "rooted in nostalgia for a more optimistic, outgoing age." He invokes Ethan Coen's claim that 1958 was chosen almost arbitrarily as the year in which the film is set whereas the period is actually indeterminate. The same concerns the place which is nominally New York although it should be perceived in more general terms as an overall vision of Metropolis, a hub of capitalist values presented as a place so neat and sanitized that it becomes virtually sinister. According to Robson (2003, p. 142), in *The Hudsucker Proxy* the directors "literally constructed a fully conceptualized environment" for their ironic re-telling of the American Dream story based on the diverse world-views depicted in a variety of its source films rather than any research into what the 1958 reality was really like.

In my view, the Coens seem to have adopted an approach to the constraints of chronology analogous to many postmodernist writers who, as Linda Hutcheon (1989) suggests, are keen to "de-doxify" even the most obvious and natural limitations. Hence, in Salman Rushdie's *Midnight's Children* the narrator overtly accepts errors in chronology in his narrative and inverts the order of events. Moreover, postmodernist narratives tend to be deliberately both historical and contemporary and yet offer no resolution of the ensuing contradictions. The narrator of Rushdie's *Shame* applies this double coded approach and complicates his narrative by transgressing boundaries between the present and the past and allowing his present knowledge of events to contaminate his representation of the past. Anachronistic intertextual references in Banville's *Doctor Copernicus* highlight postmodernist literature's intense self-consciousness about the act of narrating in the present the absent events of the past. The assumption is that although the past did exist, it can be inferred in the present only from its traces so, as Hutcheon (1989, p. 70) states, it "becomes a question of representing, that is, of constructing and interpreting, not of objective recording."

3.2 The Cognitive Clashes

The Hudsucker Proxy challenges the viewer also by exposing them to an abundance of references, most importantly allusions to Frank Capra's heartwarming fantasies of ordinary men standing up against the greed and corruption of the rich and powerful and Preston Sturges' mixture of the screwball comedy and social satire typified by juxtaposing sophisticated quick-fire dialogue with farcical situations. According to Harkness (2000, p. 127), another complication in the process of undisturbed straightforward reception of the film is also the fact that the viewer is faced with an unlikely incongruous clash between Capra's "romanticism of pure individualism (…) against the grinding power of big money and big politics (…) blind to its own darkest implications that untrammeled individualism is as much a piece of the capitalist monsters as it is of the heroes" and Sturges' "romantic wise-guy" approach with its focus on bizarre forms of tortured infatuation and his heroes battling with their own limitations against the background of the benevolent world. These two different perspectives and discordant artistic modes stand in such stark opposition to each other that they are essentially mutually exclusive, generating discomfort, confusion and a sense of inconsistency and improbability among film-goers who realize, more or less consciously, that what they see is simply an unfeasible combination. Thus, Capraesque dreams ultimately become Coenesque nightmares since, as Harkness (2000, p. 127) puts it, "[a] Sturges hero would be eaten alive in Capra's world (...)."

Another cognitive clash results from the fact that the Coens decided to place a Thirties story and characters in a Fifties setting generating the overwhelming feeling of inconsistency. As Harkness (2000) stresses, comedies in the 1930s relied on a degree of intimacy behind the story and the set, a realistic romantic rationale, the rat-a-tat style of dialogues, the exterior confidence of the stars and the urban smartness provided by New York writers. In the 1950s the acting style changed dramatically and became more introspective and tortured, dialogues switched to a casual drone-like patter and the scope of sets became much broader. The Coens utilize and jam together elements of the two conflicting aesthetics looking for the novelty potential of this puzzling juxtaposition but, sadly, the effect alienated majority of viewers and critics who seemed to share Harkness's (2000, p. 128) unfavourable opinion that "big empty sets are not conducive to snappy dialogues" and perceived the film merely as an unrewarding stylistic exercise leaving its audiences indifferent about the projected un-reality. Not surprisingly, the Coens' most expensive film with a budget of approximately 25 million dollars performed disastrously and the proceeds from the American box office were just under $3 million.

I must agree with Robson (2003, p. 134) that another explanation for this spectacular flop is that audiences are accustomed to watching films that take place in a comprehensibly recognizable environment, either a realistic historical or contemporary setting or a fantastic realm of the Earth in the future or another world of the magical past where realism is not expected and just about anything is possible. By convention, combining the two modes needs to be well-grounded: sometimes distinct fantasy elements, like the presence of a superhero in case of Tim Burton's *Batman*, justify a skewed image of the contemporary world, otherwise the aesthetic of visual realism is applied to ridiculous and contrived plots in the vein of James Bond movies to validate some unlikely or ludicrous concepts in the script. The Coens break this unspoken agreement and subvert these presuppositions, once again choosing to combine two conflicting aesthetics and make a uniquely flamboyant attempt at disconcerting their viewers by telling a fairly conventional story of a young simpleton striving for success in the world of big business in a stylized manner and in a fantastic setting with little consideration given to realism or credibility. The film's unrealistic ending involving stoppage of time and an act of angelic intervention provides an ironic signature commentary of the directors to the genres, styles and artistic sensitivities which they have subverted.

4 The Coens' America

4.1 American Zones

Brian McHale (1987) points at a peculiar phenomenon of a few favored geographical areas recurring as "zones" throughout postmodernist writing, including the low profile state of Ohio, present in Donald Barthelme's *Up, Aloft in the Air* or Guy Davenport's *The Invention of Photography in Toledo*, which represents a typically Middle-American average location. McHale (1987, p. 49) claims that numerous 19th-century writers, such as Fenimore Cooper, were fascinated by the American frontier—a prototype of a postmodernist zone, a powerful "ambiguous and liminal space" separating the world of civilization and the wilderness. Writers like Nathaniel Hawthorne, Herman Melville or even Edgar Allan Poe used its characteristic features to create a psychological borderline splicing the real and the imaginary. With the disappearance of the frontier and the absorption of the wilderness by civilization American writers had to reimagine and reconceptualize the notion of American space. While texts like William Faulkner's *The Bear* or Norman Mailer's *Why Are We in Vietnam?* attempted to recover the frontier, Frank Baum's *The Wizard of Oz*, whose Hollywood version subsequently acquired a cult status, reopened the frontier in Middle America by projecting "The Land of Oz"—a fantastic self-contained world encompassing several dissimilar realms and located within the state of Kansas. According to McHale (1987, p. 50), the external frontier along with everything it symbolizes in American ideology and mythology was thus relocated into the middle of the American interior and it was exactly this concept that captured the imagination of postmodernist writers for whom Ohio is "the 'Zone of the Interior'. Its strangeness and liminality are foregrounded by its being located not on the edges of the continent, but at its center. It is the historical descendant of the frontier zone, transposed to the flat, middling (in every sense) American heartland." This version of American space is a recurrent theme throughout postmodernist literature alongside other frequently used zones such as Latin America or Africa but its geographical continuity is usually disrupted and remodeled by means of distorted typography, the determination of the projected world by the play of the signifier or an alternative make-up of postmodernist heterotopian America which might be composed as a collection of what McHale (1987, p. 51) refers to as discontinuous and incongruous "enormous solipsisms", as in the case of Angela Carter's *The Passion of New Eve* where the country's inherent multiplicity and internal differences become emphasized.

Similarly, the Coens have so far set their films solely in a number of distinctively American settings. Most of their spaces are strikingly average and local microcosms contributing to the American patchwork, yet they are consciously defamiliarized by highlighting their internal heterogeneity, deranging and reshaping their characteristic features in addition to stressing their mutual incompatibility, for instance by exaggerating the differences in regional dialects. The directors do not seem to be interested in depicting the objective realities of different parts and regions of the country. Instead, they follow in the footsteps of postmodernist writers who, as McHale (1987, p. 55) puts it, try to reflect "our culture's ontological landscape, which allots a certain space to an unreal zone", whether it is New York, Los Angeles, Minnesota, the Deep South, Texas or Arizona.

4.2 *Raising Arizona*

Nominally located in Tempe, Arizona, *Raising Arizona* does not aim at providing viewers with an accurate portrayal of life in this particular location. Carolyn R. Russell (2001, p. 44) quotes Ethan Coen claiming: "It's all made up. It's an Arizona of the mind." In an interview conducted by Kevin Sessions (2000) the directors insist that their vision of the Southwest as a milieu for the movie is the result of looking at it as if it was an exotic, foreign culture and it was largely derived from literary portrayals of the area, supplied predominantly by Southern writers. At the same time, the film has a definite cartoonish feel to it and the studiously bad taste of the set and costumes discomfits the viewer. Russell (2001, p. 32) believes that the Coens managed to conjure up a highly subjective film-space embedded in an imaginary geography of "a minimalistic, ultra-stylized, closed universe wherein the earthy tones of the Southwest merge with the cotton candy hues of the fairground (…) which combines the architectural style of a theme park with the contours of a Warner Brothers cartoon. It is pure synthetic postmodern fabulism, a projected world where each image is metaphoric and highly allusive." Eddie Robson (2003, p. 63) points out that the choice of the suggestive desert backdrop as the location is a powerful, apposite metaphor both for the film's subtext of the apocalyptic fear and the central theme of infertility since Arizona is as dry and barren as Ed's uterus. Hi makes the connection himself in the opening monologue saying that "the doctor explained that her insides were a rocky place where my seed could find no purchase." According to Robson (2003, p. 65), through the picture's visual style and the language developed for the characters the directors managed to construct a highly distinctive aesthetic for this stylized comic movie that refuses to take its own fundamentally violent and distressing situations seriously. Hence, *Raising Arizona* creates "an internally consistent world for its story to take place in. (…) Those who do engage with it will appreciate its ability to transport the viewer into a world that seems two steps to the left of reality."

4.3 Texas in *Blood Simple*

The very first film by Joel and Ethan Coen foreshadows the subsequent cinematic zones constructed by the directors. *Blood Simple*'s Texas is a setting conceived and designed to match the film's theme of deadly passion presented in a story which, as Joel Coen insists in an interview by Ciment and Niogret (2000a, p. 59), "was supposed to be a slice of life, but still a fiction contrived to fit in an exotic place." Therefore, the Coens' Texas is an unrealistic, mythical entity, a projection of the public imagination triggered by similar cases that have occurred in this particular area and preserved in a collection of histories and legends. The opening scenes of *Blood Simple* which present deserted, faded images of rural Texas are not only the establishing shots but also the projected mindscape of Loren Visser—a depraved private detective and the film's initial narrator. According to Carolyn R. Russell (2001, p. 8), "Visser's mournfully stark and depopulated landscapes mirror his worldview, enunciated in the guttural slurred speech appropriate to a man who, throughout the film, cannot muster the enthusiasm to keep flies from landing on his head (…) and (…) who (…) will be unable to transcend his limited frame of reference (…)." When Visser tries to justify his ruthless and brutal understanding of what capitalism means, he compares the only environment he is familiar with to his private vision of Russia where "everyone pulls for everyone else." It is also Russia and the assumption that people there "make only fifty cents a day" that are used by Visser to justify his violent transactions and acts. Marty, on the other hand, invokes and imagines ancient Greece along with its culture to judge himself benevolent by its standards, claiming that "in Greece, they would cut off the head of the messenger who brought bad news." The fantastic condition of both places they refer to mirrors the similar status of their geography—as Russell aptly remarks (2001, p. 10), "it is a Texan dreamworld, a minimalistic, ultrastylized, synthetic Texas-of-the-mind."

5 Conclusion

Closer scrutiny reveals that the settings in the films of the Coen brothers, despite numerous features concurrent with their real-world counterparts, are actually nonexistent and non-illusionist, alternative realms which mirror McHale's notion of "zones". The directors parallel certain literary techniques and sensibilities and, by crossing the ontological borders and constraints of time and geography as well as emphasizing the internal inconsistencies and mutual contradictions of their environments, destabilize and deconstruct standard contextual expectations forcing the viewer to question the established perceptions of American reality, review their understanding of what makes up America and, in broader terms, what constitutes objective reality. Even though the films of the Coen brothers seem to be invariably set in typical American locations, these are historically and physically insecure, fictionalized realms producing a profound sense of ontological instability concurrent with the poetics of contemporary postmodern literature.

References

Ciment, M., and H. Niogret. 2000a. The Logic of Soft Drugs. In *Joel & Ethan Coen: Blood Siblings,* ed. P. A. Woods, 167–173. London: Plexus Publishing.

Ciment, M., and H. Niogret. 2000b. Interview with Joel and Ethan Coen. In *Joel & Ethan Coen: Blood Siblings,* ed. P. A. Woods, 58–67. London: Plexus Publishing.

Harkness, J. 2000. The Sphinx Without a Riddle. In *Joel & Ethan Coen: Blood Siblings,* ed. P. A. Woods, 126–129. London: Plexus Publishing.

Hutcheon, L. 1989. *The Politics of Postmodernism.* London and New York: Routledge.

Körte, P. 2001. The Big Lebowski. In *Joel & Ethan Coen,* ed. P. Körte and G. Seeslen, 191–208. New York: Limelight Editions.

McHale, B. 1987. *Postmodernist Fiction.* London and New York: Routledge.

O'Brien, G. 1997. *Hardboiled America: Lurid Paperbacks and the Masters of Noir.* New York: Da Capo Press.

Palmer, R. B. 2004. *Joel and Ethan Coen*. Urbana and Chicago: University of Illinois Press.

Robson, E. 2003. *Coen Brothers*. London: Virgin Books.

Russell, C. R. 2001. *The Films of Joel and Ethan Coen.* Jefferson: McFarland.

Sessions, K. 2000. Ethan and Joel Coen. In *Joel & Ethan Coen: Blood Siblings*, ed. P. A. Woods, 49–50. London: Plexus Publishing.

Smith, E. A. 2000. *Hard-Boiled: Working-Class Readers and Pulp Magazines*. Philadelphia: Temple University Press.

Waugh, P. 1985. *Metafiction. The Theory and Practice of Self-Conscious Fiction*. London and New York: Routledge.

Zięibiński, R. 2008. Alfabet braci Coen. *Newsweek* 43/2008:110–113.

Crossing Frontiers, Staking Out New Territories: Hollywood Remaking British Crime Locations in *Get Carter*

Agnieszka Rasmus

Abstract Artists have always borrowed from older works. The Hollywood film industry is no different in this respect. Its remaking practices, however, have met with much criticism. Remaking has been variably accused of cultural piracy, political imperialism and even American colonisation of national cinemas and cultures. Hollywood remakes of British films may be seen as exception to that rule. When language is no barrier and with the continued mutual cultural exchange between Britain and the United States, new elements of critical academic discourse arise, encouraging alternative means of reappraisal and comparison with film-makers, producers and stars alike inevitably becoming part of a binary discourse of compare/contrast. Such remakes often proudly disclose rather than hide their roots, containing parody, pastiche, and/or homage. In 1999 and 2000 a cult British film, *Get Carter* (1971) directed by Mike Hodges, metamorphosed into the America of today. In the original, crossing borders literally and figuratively plays a vital part as the London gangster travels to Newcastle. The paper examines two alternative approaches to rewriting the original's spatial and temporal relations and its geographically specific cultural identity into a new location on the basis of two remakes: *The Limey* and *Get Carter*, with the resultant different, sometimes even contradictory and conflicting, messages.

A. Rasmus (✉)
University of Łódź, Łódź, Poland
e-mail: rasmusag@yahoo.co.uk

J. Fabiszak et al. (eds.), *Crossroads in Literature and Culture*, Second Language Learning and Teaching, DOI: 10.1007/978-3-642-21994-8_45,

1 Introduction: Film Remakes, Hollywood and Britain

Artists have always borrowed from older works. The Hollywood film industry is no different in this respect. The Hollywood film industry is no different in this respect. Its remaking practices, however, have met with much criticism. Thus, remaking has been variably accused of cultural piracy, political imperialism, and even American colonisation of national cinemas and cultures. Such Hollywood films are cynically seen as safe financial investments in pre-tested materials with no artistic merit. Often, American remakes are viewed as unworthy and fake copies of a country's national treasures. By relocating the action of the originals to their new American setting and making alterations to the story lines in order to universalise messages, remakes might be perceived as a powerful agent of globalisation and American hegemony which obfuscates national boundaries and destroys what is unique and indigenous to a given culture.

Hollywood remakes of British films may be seen as exception to that rule. In recent years, British films from the 1960s and 1970s are enjoying new interest and makeovers as their settings repeatedly metamorphose into the America of today. When language is no barrier and with the on-going mutual cultural exchange between Britain and the United States, new elements of critical academic discourse arise, encouraging alternative means of reappraisal and comparison with filmmakers, producers and stars alike inevitably becoming part of a binary discourse of compare/contrast. With these works still engraved in public memory, updating them is motivated by factors other than the exploitation of unknown originals. As a result, remakes often proudly disclose rather than hide their roots, containing parody, pastiche, and/or homage. This, in turn, opens up a possibility of discussing remakes in a new light: as testaments to the originals' sustained cultural relevance.

In 1999 and 2000 the cult British film *Get Carter* (1971) directed by Mike Hodges was updated into a contemporary American context. In the original, crossing borders literally and figuratively plays a vital role as a London gangster travels to Newcastle. This paper examines two alternative approaches to rewriting the original's spatial and temporal relations and its geographically specific cultural identity into a new location, which results in different, sometimes even contradictory and conflicting, messages.

2 Mike Hodges' *Get Carter* (1971) Then and Now

After the prosperity of the swinging 1960s, full of hope for a better future, the 1970s in Britain must indeed have felt like a cold shower. As the state of the country was in growing disarray culminating in the so called Winter of Discontent in 1978–1979, Britain had to wake up from its swinging dream and face the harsh reality of the times. Politically, economically and morally, there was a lot to be critical of. The heroes and symbols of the times—the infamous Kray brothers, were now facing live imprisonment. The growing disillusionment with the 1960s awoke society to the fact that behind its swinging veneer, there may have been corruption, violence and often criminality involved.

Such ideas had already been explored and hinted at in the cult gangster film *Performance* made at the turn of the decade, which combines and blends the world of artists with that of criminals with amazing ease. Not surprisingly other titles swiftly followed that offered what we might call a reality check. The two significant British gangster films of 1971 include *Villain*, starring Richard Burton as the incarnation of both Kray brothers in one, and *Get Carter* with the swinging 1960s icon Michael Caine in the lead, here seemingly cast against the type as a Soho thug.

Although panned at the time for its high level of brutality, misogyny and lack of characters with any redeeming features, *Get Carter* has experienced a veritable renaissance since it was rediscovered by Lad culture in the 1990s and adapted as the mouthpiece of masculinity in crisis. Generically, the film seems to be rooted as much in American gangster film tradition (the credit sequence on the train shows Carter reading Raymond Chandler's *Farewell, My Lovely*) as in the tradition of 1960s British New Wave which largely focused on depicting the life of the working class of Northern England. What makes it unique therefore is its close resemblance to its American counterparts of the time, such as *Dirty Harry* (1971) or *Godfather* (1972), while simultaneously remaining deeply imbedded in its British context.

Unfortunately, *Get Carter*'s level of brutality, still shocking today, and its hero's cool cynicism seem to have come too early and to have overshadowed the film's important implications. The movie had to wait until the 1990s for its reappraisal as a work of social commentary whose bleak vision reflects the morals and mores of the late 1960s as well as providing a visual record of the post-industrial Newcastle cityscape that has since changed beyond recognition.

3 From London to Newcastle: Crossing Frontiers, Staking Out New Territories

Get Carter is about thresholds and crossing borders in more ways than one. On a literal level, the main character, Jack Carter, travels from London to his hometown which he deserted many years back for a more rewarding, liberating and exciting life in the capital. His unexpected appearance in Newcastle opens old wounds, brings back memories of unresolved past feuds with local outlaws and distorts the order of smoothly run criminal activities between London gang lords and their Newcastle allies. Now a reluctant traveller, Carter refers to his hometown as a "crap house." Still, he undertakes his journey against his better judgment urged on by a sense of familial duty and, as we find out later, feelings of guilt.

However, as we realise early on, there is no coming back and Carter's ticket is one-way only. As he plunges into the depths of Newcastle's criminal world, his gentlemanly manners and elegant London looks give way to outbursts of unchecked and often unnecessary violence. His maniac laughter at the film's closing when he fulfils his revenge suggests his affinities with the mad avenger of the Renaissance and Jacobean revenge tragedy. Needless to say, he has to die in order to fulfil the requirements of the genre. His death is also welcome or justified for one more reason. The film is based on the novel by Ted Lewis, *Jack's Return Home* (1970), whose title's melancholy tone creates associations with bringing home a dead body for burial.

The film is not only about one man's odyssey into the depraved underworld of Newcastle in search of justice, but also becomes, as Pamela Church Gibson and Andrew Hill (2001, p. 263) argue, "pivotal in understanding the transition from the 1960s to the 1970s." Thus, the movement from the South to the North bears profound significance. It marks a threshold between two geographically distinct locations as well as the chasm between the old and the new. The appearance of Jack Carter in his old town shows a stark contrast between the glamour of London and the poverty of post-industrial Newcastle. In the opening scenes of the film shot in a luxury London penthouse, Carter, his boss and his cohorts are watching cheap porn films produced for them up North. The first words that register the gulf between the two locations are uttered by one of the gangsters who comments on what he sees on the screen: "Bollocks naked with his socks still on?" "Yeah, they do it like that up North," replies another. Carter's own first words when he arrives in his home town show his airs as a sophisticated Southerner in a rough and uncivilised Northern environment. In a scene in a public bar, he orders a pint of bitter discarding his favourite London drink—Scotch. He snaps his fingers at a bartender and demands, "In a thin glass." This also explains why his brother's death caused by an overdose of whiskey seems suspicious. Newcastle's local alcohol is the less sophisticated and obviously cheaper brown ale.

The film thus contrasts the superficial sophistication and confidence of London represented by Carter and other fashionable looking gangsters from Soho with the tired looking faces of local people set against the desolate and grim landscape of Newcastle, its dirty streets, polluted beaches covered in grime and murky sea waters where coal spoil gets dumped and where Carter eventually meets his nemesis. Even places of entertainment such as pubs and clubs are imbued with a melancholy spirit. They reek of working class hardship and are a far cry from their swinging 1960s London equivalents.

Mike Hodges is therefore making numerous comments related to crossing frontiers and staking out new territories. Firstly, the film implies that the swinging 1960s only sealed the growing divide between the North and the South but also brought no benefits for the working class of Northern England. The only few permitted the luxuries associated with the swinging London scene are related to the criminal underworld with connections to the capital. Still, though they seem to be in a real party mood, Hodges soon turns this into the morning after. In one of the final moments of the film, we see the guests in front of the mansion of the

Newcastle gang lord played by John Osborn. As the police raid the residence, the visitors are brought out into the cold light of day. Their blank faces covered with layers of thick make-up register little emotion. The scene poignantly exposes that the swinging 1960s offered only momentary escape.

Secondly, the film shows the deformation of 1960s ideals. Instead of benefiting from the sexual revolution, "salt of the earth" Northerners lose their innocence and integrity. As Robert Murphy (1999, p. 125) observes, films like *Get Carter* show "a post-swinging Britain where permissiveness has curdled into pornography and violence." Sexual liberation has been turned into advantage of a home-grown pornographic business. Women's lib and the pill have resulted in only illusory freedoms with most women in the film shown victimised by men. The community and family ties that Northern England has always stood for seem to have loosened with mothers deserting their children, fathers unable to establish their authority in a world largely dominated by money and appearances, and teenagers losing their innocence for thrills and small change. 1960s values have been turned into profit in the hands of local and London thugs. Drugs and sex are shown as now commercialised and criminalised. The affluent South appears to be staking out a new territory for profits, exploiting the underprivileged North for shady business deals and treating it as a possible satellite and puppet to London.

Thirdly, Carter discovers that his niece, who really is his daughter, has become involved in the local porn business and thus his own actions have indirectly encouraged the system that took her innocence away and led to the murder of his brother. It appears then that his own corruption crosses borders that he thought were impenetrable and that Carter's past mistakes haunt the present. With his daughter's downfall, we see the next generation feeling the repercussions of the older generation's decision to live outside of the rules. In this sense, *Get Carter* could be read as a film about the consequences of 1960s fake utopianism. Thus, the idea of one gangster's journey from London back home to Newcastle presented the director with a chance to reflect upon the then passing era that seemed to offer unlimited possibilities and ended up as a bitter disappointment.

4 *Get Carter* (2000): From Las Vegas to Seattle

How important is the journey through diverse geographical locations and how is the idea of crossing frontiers explored in the 2000 remake starring Sylvester Stallone? The film appears to be a relatively faithful remake in terms of plot development and key ideas. However, as a star vehicle for Sylvester Stallone, it diverts from the original in a few crucial ways. As Christian Divine (2000, p. 18) comments in *Creative Screenwriting*, "The new script by David McKenna (revised by Stephen Kay) shows how Hollywood 2000 deals with moral ambiguity. In the first five pages, Jack Carter, now a Vegas mob flunkie, defends a woman's honor; helps lift an old lady's bag, plays ball with a child; and gives a big tip to a cabbie. There is no doubt this confused man is a good guy." Even though these scenes were not included in the final cut, the viewer is left with no ambiguity concerning Carter's moral integrity. With Stallone as the star, the main character can no longer be an anti-hero with psychopathic traits but rather a nice guy who somehow got involved with the wrong people. In accordance with Hollywood's new vogue for representation of masculinity in the 1990s, Stallone's character combines the lethal quality of his 1980s films with a softer and more caring U.S. manhood, one that is, as Susan Jeffords (1993, p. 197) pinpoints, "capable of change." Further on, she observes that in these movies "families provide both the motivation for and the resolution of changing masculine heroism" (200). Back home in Seattle, Carter comforts his brother's widow and tries clumsily to counsel his niece. Unlike his 1971 counterpart, he is not responsible for the girl's downfall. Needless to say, he does not die at the end of the film because he never exhibited double moral standards: he only kills bad people and helps out the good ones. In this respect, the film does not offer any character development and it is impossible to speak of his trip from Las Vegas to Seattle as representing his moral descent figuratively.

On the level of the plot, however, the idea of a journey through geographically and culturally distinct locations is more obviously preserved. Upon hearing the news of his younger brother's death, Carter travels to Seattle by train suspecting foul play. However, due to Sylvester Stallone's impenetrable gaze, it is hard to tell if the trip brings back memories of past traumas. Whereas Michael Caine's Carter had a very good reason to leave his home town for London as it seemed to be a move forwards, a rebellion, and a gesture discarding the old ways for newly found liberation, Stallone's character's decision to desert his home town remains largely unclear unless we take seriously the comment of one of the Seattle gangsters about his Las Vegas tan. After all, the film visually exploits Seattle as the rainiest place on earth from the moment the hero arrives to a symbolic spell of dry weather at the film's closing when he is shown leaving, determined to change his life. But could the weather be the only factor?

Unlike 1970s Newcastle, present-day Seattle with one of the highest levels of literacy in the country, liberal views and respect for green issues is a place of prosperity, computer technology, cultural development and political progression. This is definitely not the "crap-house" from the original. The only thing in common is its geographical location as the northernmost coastal city. Las Vegas, which substitutes for London, has a history of racism and criminality and combines adult entertainment with the highest number of churches per capita. Thus, Carter's journey from Las Vegas to Seattle seems to be of little consequence and as such produces conflicting meanings. In contrast to the original film, his choice of relocating to Las Vegas is not a progressive but rather a backward move indicating, perhaps against the scriptwriter's intentions, that he does not feel at home with liberal politics and progressive society but yearns for a more conservative locale. This may explain why his relatives are astounded by his old fashioned Frank Sinatra suit, strangely out of place in the present-day modern city.

It appears then that the idea of crossing frontiers and staking out new territories so important to the original film has not been explored in 2000 *Get Carter*, where the two cities, Las Vegas and Seattle, do not really affect one another and appear to exist independently. Although the authors of the updated script made a point of choosing two distinct geographical locations with their own unique weather conditions, this seems to be a choice largely dictated by the requirements of photography than social or political commentary. Consequently, the hero's journey through space does not push the former discourse into any new directions or offer new readings of social or cultural nature. The original story clearly serves as a springboard for exploiting Sylvester Stallone's star persona. Despite the fact that 2000 *Get Carter* makes a reference to the source by casting Michael Caine as one of the two main villains of the film, the actor's appearance seems to be an empty gesture rather than a comment on the original-remake divide or their mutual impact. As such, this tribute is hardly consequential to the reading of this or the previous version of *Get Carter*.

5 *The Limey* (1999): From London to Los Angeles

In contrast, Steven Soderbergh's 1999 *Limey* shows that remaking can enter into a form of negotiation with the original, a link between the old and the new, the past and the present, and a continued cultural dialogue between Britain and America. According to Verevis (2004, p. 94), "The ever-expanding availability of texts and technologies, and the unprecedented awareness of film history among new Hollywood filmmakers and contemporary audiences, is closely related to the general concept of *intertextuality*." Seen in this light, Soderbergh's film is not a self-contained structure, but a text that transforms and repeats other structures. It not only revives the old work and proves its present relevance but uses it creatively to push the gangster genre in new directions as well as comment self-reflexively on the whole concept of remaking itself. By incorporating elements from pre-existing films, it pays respect to the old masters while ascertaining its own place in the development of the genre. As such, it is little wonder that *The Limey* uses much more effectively the idea of crossing frontiers and journeying across time and space.

To start with, "Limey" is an old American slang term, often pejorative, for the British, originally referring to sailors. *The Limey*, as the title immediately indicates, is thus about cultural diversity and journeying across continents from Britain to America. It stresses foreigners, alien elements and divisions. The main character, Wilson, played by 1960s icon Terence Stamp, travels from London to Los Angeles to investigate his daughter's mysterious death. His Cockney slang, for example, "tea leaves" for "thieves", "china plate" for "mate", "butcher's hook" for "look", is repeatedly misunderstood, forcing him to translate for his American interlocutors. Freshly out of prison, he tries to find his way about. His dark suits and scrutinising gaze make him look out of place in sunny California where a smile is the order of the day.

In the original *Get Carter*, Londoners look down on Newcastle and make jokes about its lack of culture. In *The Limey*, Californians mock England as a country half the size of Wyoming where the cops do not even carry guns; thus, Wilson's London gangland origins hardly impress. Moreover, with LA now more in accordance with the London of the original, we can immediately read it as a location of perceived development and progress. It is similarly associated with "swinging" and its flipside of criminality, violence and false impressions. Its main representative, Terry Valentine, played by 1960s American counterculture icon Peter Fonda, is an aging music producer who still lives in awe of the past. Intent on "swinging" forever, he gets involved with drug dealers to boost his waning finances and continue his fading lifestyle. Therefore, it appears that both locations in the original and remake serve equivalent purposes. Equipped with the knowledge of the original film, the viewer easily sees through the superficiality of Californians' groovy lifestyle.

Looking at two antagonists played by 1960s key actors, another similarity between *The Limey* and *Get Carter* becomes visible. In the character of Wilson, we see the remnants of "The figure of the hedonistic playboy … one of the central media myths of the 1960s culture [with] working-class male stars like Connery, Michael Caine and Terence Stamp … essential to its creation. Its main features were a dedication to a conspicuous consumption and a freerer attitude towards sexual morality, combined with a new form of democratic accessibility" (Shail 2008, p. 154). Jenny ran away from London to California fed up with her father's immaturity and unremitting criminal tendencies. Ironically, in Los Angeles, she ends up in a liaison with a man her father's age, Terry Valentine, who, as it turns out, is descending into criminality himself. Played by Fonda, known for his rebellious attitude towards the establishment, Valentine's recurrent reminiscences of the 1960s echo Wilson's own persistent attachment to his old ways and his inability to leave the past behind (he still refers to his fellow cronies as "lads"). When Jenny discovers that her lover is involved in drug trafficking and money laundering, she pretends to be calling the police, repeating a trick she would use on her father when she was little. Valentine panics, pushes her against the wall and accidentally breaks her neck. It appears then that Wilson' lifestyle choices indirectly caused his daughter's death. In this respect, *The Limey* aptly mirrors the melancholy and sadness of the original story about mistakes that cannot be repaired. The film shows yet again how one generation's pursuits of freedom and living outside borders of society has repercussions passed onto the next generation.

Moreover, despite initial impressions of cultural divides and miscommunication, the fact remains that Wilson finds friendly helpers in his quest in California, such as his daughter's acting coach, Elaine, and her fellow actor-trainee, Eduardo. Consequently, Soderbergh might also be exploring cultural exchange, parallels and influences across borders rather than focusing on divisions. In the opening sequence of *The Limey*, we hear The Who's song "The Seeker," which not only mirrors Wilson's situation but provides a British comment on British-American cultural exchange:

> They call me The Seeker
> I've been searching low and high

I won't get to get what I'm after
Till the day I die
I asked Bobby Dylan
I asked The Beatles
I asked Timothy Leary
But he couldn't help me either (…)

The lyrics do not distinguish between American or British cultural icons but group them all together under one common heading: counterculture. Taking this into consideration, the continued cultural exchange between the two countries explains why Wilson's daughter left London for Los Angeles in pursuit of fame just like so many other "Limies" of 1960s Britain relocated to America—John Lennon, The Rolling Stones, Eric Clapton, and The Animals front man Eric Burdon—all of whom ironically reintroduced America to the Afro-American music it had itself created. When we further consider the song chosen as musical introduction to Wilson's archenemy, Terry Valentine, The Hollies' "King Midas in Reverse," we can see that the two crucial moments in the film rely heavily on 1960s British groups associated with the so called British Invasion: The Who and The Hollies, mirroring the eponymous character's own unexpected "invasion" of Los Angeles.

Finally, the idea of a journey across frontiers is not only limited to the film's opening scenes, as in the case of 1971 original, but thanks to editing permeates the entire picture and creates the impression of fluidity of time and space. The narrative follows a basic cause and effect structure but the shots are often repeated and scenes are arranged in a non-linear order. We repeatedly see a close-up of Wilson's face on a plane, but it is difficult to establish if they are flashbacks or flashforwards, if he is going to Los Angeles or coming back to London. This unusual editing technique reflects Wilson's jet lag, his experience of the new continent, his efforts to piece together scraps of information to solve the mystery as well as his many reminiscences of the past that seem to haunt him. It might also be read as a comment on the art of remaking as a work in progress where the clear divide between the original and remake is not only blurred but too simplistic. Moreover, Verevis's discussion on remake in terms of intertextuality captures the essence of the non-linear editing strategy adopted by Soderbergh which not only "disrupts the linear unfolding of the text impelling the reader toward a non-linear (tabular) intertextual reading," but also "enrich[es] meaning and salvage[es] the very same narrative linearity that was initially compromised (2004, pp. 94–95)."

To further establish the fluidity of time and space, Soderbergh even incorporates footage from Ken Loach's film, 1967 *Poor Cow* starring Stamp. This device firmly roots the film's main character in 1960s London, but more importantly, promotes the idea that British and American film can creatively feed off each other without obfuscating national borders, which again becomes an interesting comment on remaking itself: no longer seen as parasitical but symbiotic in nature. As such, *The Limey* shows the shift in attitude towards appropriation which according to Bourriaud moved from an "ideology of ownership" towards "a culture of constant activity of signs based on a collective ideal: sharing" (2002, p. 9).

Seen in this light, *The Limey* with its many cultural references, its choice of actors, soundtrack, and multiple homages to cult British films and bands turns out to be part of this ongoing cultural exchange between American and British artists without making either one culture subservient or dominant. In this way, it follows in the footsteps of Hodges' *Get Carter*. Looking at the 1971 film, we can notice how Hollywood and British cinemas have creatively borrowed from one another. In Hodges' film, Carter reads *Farewell, My Lovely* in a tribute not only to Chandler's book but primarily to its first screen adaptation, Dmytryk's *Murder, My Sweet*, a noir classic from 1944 shot in Los Angeles. 1975 Hollywood film *Farewell, My Lovely* starring Robert Mitchum is a remake of *Murder, My Sweet*, but its opening scene pays homage to the pre-credit London sequence of Hodges' *Get Carter*, as Graham Fuller (2000, p. 36) aptly points out. Finally, *The Limey* revives 1971 *Get Carter* but reintroduces it back to *film noir* original setting: Los Angeles. The above examples demonstrate that "the intertext, the precursor text, is never singular and never a moment of pure origin" (Frow, review of *Play It Again, Sam*, quoted in Verevis 2004, p. 98)

6 Conclusion

Rewriting geographically specific cultural identity into a new location frequently results in producing diverse, sometimes even conflicting, messages. 2000 *Get Carter* follows the script rather faithfully but seems to offer little if any social commentary related to the topic of crossing borders and staking out new territories so crucial to the original script. Although it updates the story to the contemporary US setting, it uses its locations mostly for photography's sake. Its main focus is the star of Stallone, which results in important changes to the plot: the hero cannot die at the end and cannot be morally dubious.

Viewed in terms of mutualism rather than parasiticism or competition, the source and update appear to complement each other and benefit equally in the case of *The Limey*. Such remakes not only revive interest in old films and prove their continued cultural relevance but more importantly provide an invaluable comment on the past as viewed from a contemporary perspective.

To conclude, *The Limey* points out that while Hollywood has always been said to haunt British gangster genre, the British are haunting this American gangster film, its soundtrack and its American heroes, showing the mutual exchange that lies at the heart of thoughtful film remaking and proving that the border between "the original" and "remake" is often quite fluid. Soderbergh's decision to move the original film's pivotal locations across the continent has widened the scope of discussion. No longer is it a film about a journey through distinct cultural sites with their own unique implications but also a deeply self-reflexive work that comments on filmmaking, remaking, cultural exchange as well as the rejuvenated presence of British artists, actors, directors and musicians in the present-day Hollywood.

References

Bourriaud, N. 2002. *Postproduction*. New York: Lukas and Sternberg.
Church Gibson, P. and A. Hill. 2001. 'Tutte e marchio!': excess, masquerade and performativity in 70s Cinema. In *The British Cinema Book*, ed. R. Murphy, 263-269. London: BFI.
Divine, C. 2000. Script comments on *Get Carter*. *Creative Screenwriting* Sept/Oct Vol. 7 No. 5: 18-19.
Fuller, G. 2000. *Get Carter* 2000. *Film Comment* Sept/Oct: 35-37.
Hodges, M. 1971. *Get Carter*. UK. Metro-Goldwyn-Mayer British Studios.
Jeffords, S. 1993. The big switch: Hollywood masculinity in the nineties. In *Film Theory Goes to the Movies*, eds. J. Collins, H. Radner and A. Preacher Collins, 196-208. New York, London: Routledge.
Kay, S.T. 2000. *Get Carter*. USA. Morgan Creek Productions.
Murphy, R. 1999. A revenger's tragedy – *Get Carter*. In *British Crime Cinema*, eds. R Murphy and S. Chibnall, 123-133. London and New York: Routledge.
Shail, R. 2008. 'More, much More… Roger Moore': a new Bond for a new decade. In *Seventies British Cinema*, ed. R. Shail, 150-158. London: Palgrave Macmillan, 2008.
Soderbergh, S. 1999. *The Limey*. USA. Artisan Entertainment.
Verevis, C. 2004. Remaking film. *Film Studies*. Issue 4: 87-103. Manchester: Manchester University Press.

From the New Right to the *Da Vinci Code*: Transformations of the Conspiracy Theorist After the Second World War

Michał Różycki

Abstract It is over 50 years since Richard Hofstadter penned his famous essay "The Paranoid Style in American Politics". The work, as Mark Fenster, a later scholar of conspiracy theories, claimed, simultaneously "created and cleared the field", elevating fringe, radical political movements and accompanying conspiracy theories to a topic worthy of an academic debate. A remarkable element of Hofstadter's work is a definition of the figure of conspiracy theorist, or "paranoid spokesman", as it is called in the text. Despite the author's concentration on right-wing examples and his use of the word 'paranoid', the image presented provides a basis for the analyses of conspiracy-related cultural phenomena even today. Echoes of Hofstadter's vision of a person who sees the world as a stage for "apocalyptic" struggles between the insidious, decadent conspiracy and the forces of Truth, of which he is a paragon, reminds us of such various and diverse figures as populist politicians, filmmakers like Michael Moore or even the writings of Dan Brown, who actively salvages and rewrites conspiracy theories as plot elements in his works. My paper will be an attempt to analyse how Hofstadter's definition of a conspiracy theorist withstood the test of time and the rampart proliferation of conspiracy narratives in the second half of the 20th century and the first decade of the 21st century. It will concentrate not only on the academic interpretations of the topic, though one cannot fail to mention the impact of semiotics on the field. Rather, it will be concerned with both fictitious conspiracy theorists, like Fox Mulder of the *X-Files* (1993–2002) and Jerry Fletcher, protagonist of the movie *Conspiracy Theory* (1997), as well as historical conspiracy theorists, such Bob Lazar or David Icke, and test them against Hofstadter's theory.

M. Różycki (✉)
University of Warsaw, Warsaw, Poland
e-mail: m.rozycki@uw.edu.pl

J. Fabiszak et al. (eds.), *Crossroads in Literature and Culture*, Second Language Learning and Teaching, DOI: 10.1007/978-3-642-21994-8_46,

1 Introduction

We all know at least one conspiracy theorist. But, in a sense, there is a conspiracy theorist in every one of us. In some moments, rare or abundant, we begin to fall for the lure of a belief in a conspiracy—a belief that seems to answer all our doubts about how the world works. Yet almost all of us stop at some point, brought to the realization of the amount of logical inconsistencies we have just spawned.

There are some, however, who are not as self-conscious. Those Richard Hofstadter dubbed the "paranoid spokesmen". The term, originating from his influential essay "The Paranoid Style in American Politics", printed in 1964, referred to a specific group of people and politicians contemporary to Hofstadter—the New Right movement and the followers of Senator Barry Goldwater or Joseph McCarthy. Delivered as a lecture a year earlier, the text "had grown out of Hofstadter's long-standing apprehensions about the rise of American right-wing extremism after World War II" most visibly seen in the form of "McCarthyite hysteria of the early 1950s" and organizations such as the John Birch Society (2008, p. xi). Taking these phenomena as the starting point, the essay strived to identify what Hofstadter called the eponymous paranoid style, a "chronic, rancid syndrome" that spawned radical movements adhering to a recognizable paradigm of beliefs (2008, pp. xi–xii).

This paper will strive to prove that, despite the passage of time and the bias of "Paranoid Style…", the definition of a paranoid spokesman remained not only valid, but also can refer to a much wider group than originally intended by its author. From the onset of the Cold War the concept of a conspiracy theorist was in an almost constant process of reinterpretation—an enemy of democracy became the last defender of American values, a potentially subversive individual a collector of curiosities. Today, as some critics claim, a conspiracy narrative can "be used to promote oppression as it can be made to advance democratic and emancipatory politics" (Fenster 2008, p. 287). For space reasons, this text will concentrate only on a selected body of conspiracy narratives. This list, subjective almost by necessity, strived to use the most notorious and recognizable texts, both from the point of view of the academia as well as the popular culture. Likewise, this text will not try to debunk any of the theories presented within, nor its aim is to provide a different interpretation to those, rather controversial, issues. Rather, each conspiracy narrative will be taken as is and allowed, to quote Yeats (1994, p. 19), to "go their way unoffended or defended by any argument of mine".

2 Definition of a "Paranoid Spokesman"

For Richard Hofstadter, the defining trait of a conspiracy theorist is his belief in the absolute power of a conspiracy. This fact differentiates him from those who seek to uncover true conspiracies, which after all do happen and knowledge of them cannot be seen as pathological;[1] Hofstadter (2008, p. 36) himself admits that "paranoid writing begins with certain defensible judgments". To a paranoid spokesman, however, a conspiracy is the "motive force in historical events" (2008, p. 29), a power group that can influence everything and everyone. Furthermore, the workings of a conspiracy are interpreted in apocalyptic terms, with the present moment being always a turning point (2008, p. 30) in the theorist's struggle against the conspiracy. This struggle is always Manichaean in nature—there is no room for discussion or political compromise, for "what is at stake is always a conflict between absolute good and absolute evil" (2008, p. 31). This unwillingness to compromise stems from the portrayal of the "enemy"—the "totally evil and totally unappeasable", "amoral superhuman: sinister, ubiquitous, powerful, cruel, sensual, luxury-loving" (2008, pp. 31, 32). This "demonic agent" is deemed to be beyond and above the normal limitations imposed by social order, often to the point of controlling some of its aspects—the press, the educational system or has access to brainwashing techniques (2008, p. 32).

What is to be adduced to the silhouette of the conspiracy theorist sketched thus far is what made him so. His place in the vanguard of the resistance may be due to his having been influenced by the malevolent power, or was even himself a part of the conspiracy. Here Hofstadter's examples range from an ex-Mason, a runaway nun or an apostate priest, to an ex-Communist who, to use the colloquial term, "saw the light" and came to appreciate the merits of democracy. Such a figure "brings forth with him or her the final verification of the suspicion" and embodies hope that the defeat the conspiracy is indeed possible (2008, p. 35).

3 Conspiratorial Motifs During the Cold War

Despite Hofstadter's warnings, the conspiratorial mode of thinking thrived during the Cold War, and consequently legitimized the paranoid spokesmen. This came as no surprise, as the world indeed had become a battleground of two opposing forces, either of which presented its opponent as a model of malevolence and tyranny. From the US perspective, therefore, one saw the conflict as between the peace-loving and democratic America, and, to quote president Ronald Reagan's famous 1983 speech, the "evil empire", the battle with which being "the struggle between right and wrong and good and evil".[2] This battle, neither overt nor conventional, had to be, by necessity, clandestine—and it was presented as such in popular culture. One needs only mention the works of such popular authors like Ian Fleming or Tom Clancy, who presented small, elite groups of individuals as

[1] The interpretation of conspiracy theories as a danger to democracy seems to be one of the most controversial elements of Hofstadter's analysis. For him, the paranoid spokesman is disruptive to the democratic process, as he provides fantastic information to the public (2008, p. 40). This, as can be seen below, began to be challenged in the 1990s.

[2] Online at http://www.presidentreagan.info/speeches/empire.cfm.

the true movers and shakers of world events. The notorious Agent 007 faced, both in the original books as well as their movie adaptations, conspiracies that could easily influence the whole world through their actions. Likewise, the very plot of *The Hunt for the Red October* (1984) revolves around a conspiracy of soviet officers commanding the eponymous submarine.

A more specific example could be brigadier general Jack D. Ripper from Stanley Kubrick's *Dr. Strangelove* (1964). His paranoid belief that a communist conspiracy has poisoned the US water supply leads him to ignore the chain of command and start a nuclear war against the Soviet Union, an event that, to summarize the plot, sends the no less paranoid leaders of the two superpowers on the path of mutual self-destruction. It is a curious coincidence that the first test screening of the movie was scheduled on the 22nd of November, 1963—the day of President John F. Kennedy's assassination, and that its original, uncut version included references to Dallas, while one of the characters exclaimed "Gentlemen! Our gallant young president has been struck down in his prime!"[3] Such coincidences can be seen as an indication of a much wider trend, which made the society not only perceive this "mirror world of secret power" with both a sense of fear and fascination, but also made the fantasy and reality mingle, to create a paradox of covert operations conducted "for show" (Knight 2000, p. 29).

Real events, such as the assassination of President Kennedy, furthered the need for conspiratorial thinking. As Timothy Melley remarked, the cornerstone of the interest in the John F. Kennedy's death was "a competition between the lone-gunman and grand conspiracy theories" (2000, p. 135). The latter, it would seem, is even today more satisfying than the findings of the Warren Commission. On the 40th anniversary of Kennedy's death some polls had shown that over two thirds of the American population believes that a governmental "cover up" was involved.[4] 12 years prior, the success, and notoriety, of Oliver Stone's movie *JFK* (1991) also contributed to the perpetual public rejection of the findings of the Warren Commission, a level of distrust that would later be matched only with the reaction to the 9/11 Commission Report.

Finally, one can briefly mention an even more influential fact. Not long after the first publication of Hofstadter's essay those believing in secret governmental schemes were indeed proven right. The cases dubbed "Watergate" and the Iran–Contra affair (also known as "Irangate"), of 1972 and 1986 respectively, have shown without a shadow of doubt, that true government conspiracies indeed take place, and do so on a grand, international scale. While this paper will not analyse those events, their very existence and media exposure is symptomatic. If it were not for investigations, first journalistic, then congressional, one could treat those historical facts as nothing more than conspiracy theories (McConnachie and Tudge 2008, pp. 190–195, 201).

---

[3] As can be seen on the documentary "No Fighting in the War Room Or: Dr. Strangelove and the Nuclear Threat", which was a part of the 40th Anniversary Special Edition DVD.

[4] Online at <http://www.cbsnews.com/stories/2003/11/20/national/main584668.shtml>.

4 Academic Reinterpretations of the Paranoid Spokesman

The post Cold War acceptance of the paranoid style, and with it the theorist, stems not only from the clandestine manner of practicing politics, but also from the fact, that by the end of the 20th century the whole world indeed became interconnected, and on many levels. All we need is a quick browse of the Internet to make the connection, to acquire data. As a consequence we are, to some extent, required to make connection between seemingly unrelated events, a hallmark of the conspiratorial way of thinking. For some scholars a perfect illustration of this fact is the internet—as Kathleen Stewart claimed, "the Internet was made for conspiracy theory: it is a conspiracy theory: one thing leads to another, always another link leading you deeper into no thing and no place" (1999, p. 18). The two entities became so intertwined that the phrase "reading wild conspiracy theories on the Internet" became almost common parlance. The Web's greatest impact lies not only in providing an accessible forum for any discussion, however, radical, but in blurring the boundary between the mainstream and the fringe reports (Barkun 2006, p. 20). Sites providing conspiratorially tinted new reports, one example being prisonplanet.com, are constructed to mimic those of leading TV stations, like the BBC or CNN.

Yet another aspect is the fact that, as Peter Knight observes, the "secure paranoia" of the Cold War's "continuous but often unspecific threat" not only began to slowly erode since the late 1960s, but finally gave way to an "insecure paranoia in which there is no longer a single recognizable enemy" (Knight 2000, p. 175). As a result, the American society began to search for what Slavoj Žižek, in his essay "'I hear you with my eyes': or, the Invisible Master", called "invisible masters": the true, but unreal controllers of the social and political reality (1996, p. 92)—the "doppelganger government" of unaccountable secret agencies spawned by the National Security Act of 1947 (Knight 2000, p. 28). In the postmodern social reality their alleged existence became not a danger or source of anxiety, but a "comfort" of what could be called self-imposed anxiety. In essence, it is a cure for the "breakdown of the symbolic universe", which is the true "illness" (quoted in Willman 2002, p. 36). This sentiment was voiced by Frederic Jameson who saw conspiracy theories as "a populist form of cognitive mapping that attempts to represent the un-representable totality of (...) seemingly disparate yet interconnected social, political, and economic transformations" (quoted in Lewis and Khan 2010 p. 13).

Indeed, as Keeley (1999) pointed out in his essay "On Conspiracy Theories", we are aware that some conspiracies could be potentially true and are able to make that distinction, if only based on our gut feeling. As Keeley himself admits by the end of his essay, his attempt to create a stable, objective category of "unwarranted" conspiracy theories, a class of explanations that should never be used by definition, effectively failed. He admitted that "there is nothing straightforwardly analytic that allows us to distinguish between good and bad conspiracy theories" (1999, p. 126). Jack Z. Bratich went yet further by claiming that what makes a conspiracy theory are not some distinguishable traits inherent to the narrative, but the opinion of the public. He postulated a category of "conspiracy panics"—reactionary discourses that brand narratives that could subvert the accepted socio-political paradigm as unacceptable (2008, p. 167).

If anything, many of the above critics saw a more danger in the full dismissal of conspiracies, which would make the public turn to probably equally false "contingency theories", which claim that events happen, for the most part, due to random chance. Yet while this stance is totally opposite to that taken by Hofstadter, it does not invalidate his definition of the phenomenon of conspiracy theories, only its interpretation within the framework of a given society.

5 Post-cold War Popular Depictions

The forming reality of the 1990s saw the arrival of conspiracy theorists both real and fictitious. Among them, the aforementioned movie *JFK* or the first revelations of David Icke, a British ex-footballer and sports commentator, both of which attracted ridicule and outrage. Oliver Stone's work, however, was nominated for eight Academy Awards, won two,[5] and attracted many viewers to the cinemas despite its unconventional, for a mainstream Hollywood movie, length of over 3 h. Likewise, as the years passed did David Icke manage to attract an enormous group of supporters, even though, or perhaps because, his ideology evolved from sombre predictions of impending natural disasters and the end of the world—with Icke being a saviour and son of God, to a complex network of separate conspiracy theories linked by the common denominator of an extraterrestrial species effectively controlling humanity.

5.1 The X-Files

If one were to point to a single event that accelerated the changes in the reception of the conspiracy theorist in the 1990s, it probably would have to be the television series The *X-Files* (1993–2003), created by Carter (1993) and aired by the Fox network. More specifically, the catalyst would have to be the character of FBI agent Fox Mulder, portrayed by David Duchovny. Mulder was not a paranoid lunatic, who saw extraterrestrial infiltrators, secret government laboratories and mind-control machines at every corner. While fully fictitious, he is believable as a man who holds a certain belief, that his sister was abducted by extraterrestrial forces, and acts upon it. His quest to uncover the truth is personal and marks his as

---

[5] Based on a search in The Academy of Motion Picture Arts and Sciences, online at <http://awardsdatabase.oscars.org/ampas_awards/BasicSearchInput.jsp>.

an outcast, but is not obsessive. Like the mysterious Mr. X in *JFK*, Jim Garrison's source of information, he is a calculating man with an agenda, so unlike the raving, paranoid David Ferrie, who is portrayed in Stone's film as a man driven to the edges of sanity by the very plot in which he supposedly participated. Perhaps to cater the need for such characters, later in the show the "Lone Gunmen" are added, a group of conspiracy theorists whose beliefs, attitudes, and actions are more in line with the stereotype of a paranoid spokesman. Overall, what he does is not unlike what most of us might have done when confronted with a traumatic event in one's youth—through this, the character of Fox Mulder is both believable and likable despite fulfilling most, if not all, criteria of a paranoid spokesman.

What may have also decided about the popularity of The *X-Files* was the way the series was embedded into its media format. For the most part, each episode consists of a single story of the protagonists' investigation which very often gives them hints of the grander conspiracy behind the investigated event. As Mark Fenster remarks, Mulder and Scully are always denied the final answer, not only due to the power of the conspiracy, but "the formalistic and institutional framework of the hour-long television series drama". As Chris Carter, the series' creator, quoted in Fenster, claimed: "(...) it's hard to put handcuffs on aliens every week and throw them into the slammer" (Fenster 2008, p. 145).

The perpetually frustrated necessity to reveal the conspiracy is relegated effectively to the background, as it is not the focus of the series itself. This constitutes a trend that reappears throughout many conspiracy narratives. One reason, as Mark Fenster claims, is that the conspiracy theorist can never achieve a full, totalizing image of the conspiracy. There will always be data to accumulate, to reinterpret and to fit into the conspiratorial paradigm. Yet in her critique of Fenster's work Jodi Dean pointed out, that the public engages in such a quest but infrequently. They rely, rather, on simply knowing that there is an "invisible master", that some events, seemingly random, are indeed connected. "The pleasures of the wired world are pleasures of knowing, of no longer being duped by the system"; all one needs is an apparent access to "fringe" information (Dean 2002, p. 103). In essence, we do not really need to know the full truth—the very act of linking, combining, clicking, and viewing seems to suffice, with the need of it intensified by the presence of the conspiracy theorist.

5.2 *Jerry Fletcher and Conspiracy Theory*

The FBI agent found his mirror opposite in the protagonist of the 1997 movie entitled simply *Conspiracy Theory*, directed by Richard Donner. In its very first scene we meet the taxi driver Jerry Fletcher, portrayed by Mel Gibson, whose main purpose in life seems to be bombarding his customers with wild tales of conspiracies and clandestine government machinations. This stereotypical image, complete with obnoxious, antisocial behaviour, gradually fades as we not only learn that Fletcher was indeed experimented upon, but also witness an unfolding romance with a Justice Department employee Alice Sutton (Julia Roberts). By the end of the movie Fletcher ceases to be a subversive, potentially dangerous, element within the society but a victim of the cultured Dr. Jonas (Patrick Stewart), whose good manners and a British accent seem to give the character a nefarious air.

Gibson portrayed, with almost cartoonish exaggeration, a fully domesticated paranoid spokesman; one who is not potentially dangerous, but hurt by some vaguely defined force and in need of healing. While alluding a well-known conspiratorial motif: MK-ULTRA, the very real US military program, the conspiracy itself is downplayed. Dr. Jonas seems to be working alone, not as a part of some larger cabal, while the secret agents sent after both Fletcher and Jonas are more like stern but sympathetic civil servants than the dehumanized men in black. Indeed, as Peter Knight puts it, Donner's movie "operates less as a political thriller and more as a romantic comedy", with Mel Gibson character's ordeal at the hands of a conspiracy "the plot device that brings him closer to Julia Roberts, and (…) merely a conventional part of the thriller's scenery" (Knight 2000, pp. 52–53).

5.3 The Da Vinci Code

Ten years after first episode of The *X-Files* this trend was enhanced even more by the publication of Dan Brown's The *Da Vinci Code* (2003). Even more so than the television series, Brown's books are about conspiracies, but do not reveal them. Likewise, the recurring fictitious character of professor Langdon is even less of a stereotypical conspiracy theorist than Mulder. In *Angels and Demons* (2000), the prequel to Code there is in fact no conspiracy at all. The whole murderous plot was orchestrated by a single man who, through his knowledge and influence in the Vatican, is able to make both the characters in the book and its readers believe that the actual perpetrators were the Illuminati, who in Brown's narrative are a secret society supposedly antagonistic to the Catholic Church. We believe in conspiracy theories, the books seem to tell us, because the popular culture made us consider them possible, not because there is some tangible proof.

This theme reappears in the *Da Vinci Code* in the character of Teabing, a theorist so obsessed with the Grail conspiracy that he creates a conspiracy to reveal and punish the secret society, the Priory of Sion, which supposedly knows the true nature and location of the Grail. As the plot progresses, Professor Langdon discovers what the Grail really is, a metaphor for the line of descendants of Jesus Christ, as well as the most guarded secret of the Priory, the tomb of Mary Magdalene, wife to Jesus. This revelation, however, incites no world-shattering events. Langdon, and with him the readers, simply walks away, content with solving the puzzle, making the right connections—and, to refer to Jodi Dean, "knowing".

6 Conspiracy as Explanation: The 9/11 Truth Movement

The New Testament adage "the truth will make you free" (John 8:32) became closely associated with conspiratorial thinking during the last two decades. It effectively became a creed for such filmmakers as Michael Moore, who in his documentaries, specifically *Fahrenheit* 9/11 (2004), claims he seeks to inform the public. This movie, though dubbed "conspiracy theory" by some (Bratich 2008, p. 144) is only just avoiding Hofstadter's definition of what a paranoid spokesman believes—it provides facts without the totalizing links, seemingly allowing the viewer to make his or her own judgment, despite the movie's overt thesis—that of president George W. Bush's incompetence. As such, the notion of whether a conspiracy did or did not exist is not an issue for Moore. If such an idea is introduced, it is used either to emphasize the supposed foul play on the part of the Bush administration, or to present the former president as a mere figurehead with no real say in any decisions.

Moore, however, paved way for other works, more radical and overt in their interpretation of conspiratorial themes surrounding 9/11. Works like *Loose Change* (2005) by Dylan Avery, *Zeitgeist* (2007) by Peter Joseph and a number of movies by Alex Jones—among them works with such ominous names like *Endgame: Blueprint for Global Enslavement* (2007), *Obama Deception* (2009) *and Fall of the Republic* (2009)—quite overtly present the existence of a global conspiracy. Avery's claim that the attacks on World Trade Center were "an inside job" pale in Jones' paradigm of a world dominated by a banker conspiracy for over a century.

Yet none of the above works would get as much publicity, internet distribution or not, if it were not for the unprecedented public reaction to the findings of the 9/11 Commission, more officially known as the National Commission on Terrorist Attacks Upon the United States. The so-called "9/11 Truth Movement", as this unofficial group came to be called, consists not only of conspiracy theorists or regular citizens, but also of respected academics, all of whom were concerned with the supposed inaccuracies of the 9/11 Commission Report. The body of texts describing various WTC conspiracy theories steadily grows since the 2002 publication of 9/11: *The Big Lie* by the French Thierry Mayssan and, as Mark Fenster recounts," millions of people, supposedly, have viewed conspiracy-related Web sites and films; thousands have purchased the bestselling books of the movement; and hundreds attend the largest conferences and gatherings" (243). The line between a paranoid spokesman and a regular citizen blurs, as some of the "truth movement writers (...) embrace the conspiracy theory moniker, notwithstanding its apparently debased status" and "claim to be part of the longstanding, honourable tradition of fearing conspiracy in American history" (Fenster 2008, p. 242).

7 The Conspiratorial Constant

One might say, that the conspiracy theorist became more aware of potential allies within the society, as in many cases he will not face ridicule. The rhetoric contemporary to Hofstadter pitted two all-powerful forces against each other, with the society as foot-soldiers of either. As Joseph McCarthy himself remarked: "Any man who has been given the honour of being promoted to general and who says, 'I will protect another general who protects Communists,' is not fit to wear that uniform, general".[6] Today, such division is to a large extent nonexistent in the accepted conspiracy narratives—instead of the cold war us versus them we have the informed versus the conspiracy battling over the uninformed. Alex Jones claims repeatedly throughout his documentaries that "the minute the public awakens to the fact that there is an agenda to manipulate them (…) the establishment begins to lose [the] edge they have over people, (…) so all I ask of the viewers is to think for themselves". This message is emphasized by guests who appear in his movies, one of the being the filmmaker David Lynch, who in *Invisible Empire* (2010), a movie produced by Jones, claims that "you don't have to believe everything in the documentary to still have questions". A similar sentiment is implied in *Loose Change*—its creators do not hide the fact that not only do they pose more questions than answers, but also do not any idea what the answer to some might be (McGuire 2010). Creating a totalizing vision of a reality subjugated by the conspiracy, the impossibility of which, according to Fenster, undermines any shred of credibility a theory might have, is no longer the aim of those theorists who want to attract an audience, or at least is practiced to a much lesser extent (2008, p. 100).

Coming back to the original question, we may arrive at the conclusion that it is not the conspiracy theorist that changed, by our approach to and tolerance of him. The paranoid spokesmen still see the world as a stage for the final battle between them and the conspiracy; as Alex Jones claimed in *The Obama Deception*: "Free people on the one side and the New World Order on the other (…), either they fall or we fall, either a worldwide darkness and a new age of tyranny and oppression with a scientific overlay, or a new age of liberty and freedom and a new renaissance". Like in the 1950s, the trap set to ensnare the American society grows ever tighter and complex, but never springs. The communist agents did not reveal themselves to pave way for a Soviet invasion, nor does the New World Order abolish states and partition the Earth according to their design. Nonetheless, to use the words of Robert W. Welch, Jr, "time is running out" (quoted in Hofstadter 2008, p. 30).

At the same time, the conspiracy has allegedly tremendous influence and power, effectively controlling every government they please, as well as keeping

[6] This passage was one of the many used by Edward R. Murrow in his *See It Now* show from March 9, 1954, which starkly criticized McCarthy. Transcript online at http://filepedia.org/files/Edward%20R.%20Murrow.pdf. Retrieved 01.07.2010.

the aberrant elements in check through technology or economical tools. The image of the conspiracy itself within some narratives is sometimes worse than during the McCarthy Era. Alex Jones' *Fall of the Republic* quotes one Dr. Webster Tarpley, who claimed that "if you think the bankers are the same as you, you're wrong. (…) to be one [of the bankers] means you are in a completely different world, with values which are the opposite of human values." An extreme can be seen in the works of David Icke, for whom the ruling elite are literally a different, extraterrestrial and hostile species, and they engage in satanic rites, which include drinking human blood. Finally, the theme of an outcast who turned against his conspiratorial overlords remained strong, as exemplified by Fox Mulder, an agent of the FBI, and Mr. X from *JFK*, who discloses insider knowledge to Jim Garrison.

8 Conclusions

Don DeLillo wrote in his novel *Running Dog* (1986) that his age was a time of conspiracies and secret, yet tangible, connections. One may find it ironic that the these words perfectly described the three decades that came after them—a rekindling of hostilities during the Reagan era, the "age of paranoia" of the 1990s, and the perpetual crisis of the post-9/11 world. In essence, this remains true today as well—even more so due to the advent of the great connector, the Internet, as well as our constant saturation with conspiratorial motifs. Not only the post Cold War political discourse, but also movies, novels, as well as the more popular genres of computer games and comic books made, make, and will make heavy use of various conspiracy theories that surfaced or were codified throughout the 20th century. Of course, one cannot forget the darker aspects of a conspiracy theorist—ones that lead to, for the lack of a better word, clinical paranoia and violence. The most visible example of such an undoubtedly pathological narrative would be Andrew Macdonald's, the *nom de plume* of William Pierce, *Turner Diaries* (1978), which not only presented a conspiracy with heavy racist undertones, but also vividly depicted a struggle of a protagonist's white supremacist group against it.

Since the 1990s the public, especially in the US, were subject to the aforementioned conspiracy panics, the term coined by Bratich to describe the "anxiety over the phenomenon of conspiracy theories". Events like the Oklahoma City bombings have painfully shown that a minority acting upon a conspiratorial belief can cause deaths and destruction. Yet if anything, people like Timothy McVeigh or the Unabomber Ted Kaczynski became the new radicals, the unacceptable element in the democratic society. They became a minority, last descendants of the 19th century conspirators, who sought, or were said to seek, change through violence and a total revolution. At the same time, those who fit Hofstadter's definition of a "paranoid spokesman" are seen by many as contributors to the democratic order, often controversial, yet necessary. While acknowledging the dangers, Mark Fenster claimed that the conspiracy theories "may well address real structural inequities, albeit ideologically, and they may well constitute a response, albeit in a simplistic and decidedly unpragmatic form, to an unjust political order, a barren or dysfunctional civil society, and/or an exploitative economic system" (2008, p. 90).

References

Barkun, M. 2006. A *Culture of Conspiracy: Apocalyptic Visions in Contemporary America*. Berkeley and Los Angeles: University of California Press.

Bratich, J. Z. 2008. *Conspiracy Panics: Political Rationality and Popular Culture*. Albany, NY: State University of New York Press.

Dean, J. 2002. If anything is possible. In *Conspiracy Nation. The Politics of Paranoia in Postwar America*, ed. P. Knight, 85-106. New York, London: New York University Press.

DeLillo, D. 1986. *Running Dog*. New York : Vintage Books.

Fenster, M. 2008. *Conspiracy theories. Secrecy and Power in American Culture*. Minneapolis: University of Minnesota Press.

Hofstadter, R. 2008. *The Paranoid Style in American Politics*. New York: Vintage Books.

Keeley, B. L. 1999. Of Conspiracy Theories. *The Journal of Philosophy* 96: 109-126. Jstor. http://www.jstor.org/stable/2564659. Accessed 22 July 2009.

Knight, P. 2000. *Conspiracy Culture: From Kennedy to The X Files*. Abingdon, New York: Routledge.

Lewis, T., R. Khan. *The Reptoid Hypothesis: Utopian and Dystopian Representational Motifs in David Icke's Alien Conspiracy Theory*. richardkahn.org/writings/culturalstudies/reptoidhypothesis.pdf. Accessed 29 April 2010.

McConnachie, J., R. Tudge. 2008. *A Rough Guide to Conspiracy Theories*. London, New York: Rough Guides Ltd.

McGuire, M. Web movie takes flight. Timesunion.com http://www.timesunion.com/AspStories/storyprint.asp?StoryID=506008. Accessed 01.07.2010.

Melley, T. 2000. *Empire of conspiracy* : *The Culture of Paranoia in Postwar America*. Ithaca, N.Y. : Cornell University Press.

Stewart, K. 1999. Conspiracy Theory's Worlds. *Paranoia Within Reason: A Casebook of Conspiracy as Explanation*, ed. G. E. Marcus, 13-19. Chicago and London: The University of Chicago Press.

Willman, S. 2002. Spinning Paranoia: The Ideologies of *Conspiracy and Contingency in Postmodern Culture. In Conspiracy Nation. The Politics of Paranoia in Postwar America*, ed. P. Knight, 21-39. New York, London: New York University Press.

Yeats, W. B. 1994. *The Celtic Twilight*. Gerrards Cross: Colin Smythe Limited.

Žižek, S. 1996. "I Hear You with My Eyes"; or, The Invisible Master. In *Gaze and Voice as Love Objects*, ed. S. Žižek and R. Saleci, p. 90-118. Duke University Press.

Filmography

Angels and Demons. Dir. Ron Howard. Perf. Tom Hanks, Ewan McGregor, and Ayelet Zurer. Imagine Entertainment, 2009.

Conspiracy Theory. Dir. Richard Donner. Perf. Mel Gibson, Julia Roberts, Patrick Stewart. Warner Bros., 1997.

The Da Vinci Code. Dir. Ron Howard. Perf. Tom Hanks, Audrey Tautou, and Ian McKellen. Columbia Pictures, 2006.

Dr. Strangelove or : How I Learned to Stop Worrying and Love the Bomb. Dir. Stanley Kubrick. Perf. Peter Sellers, George C. Scott, Sterling Hayden. Columbia Pictures, 1964.
Fahrenheit 9/11. Dir. Michael Moore. Lions Gate Entertainment, 2004.
Fall of the Republic: The Presidency of Barack H. Obama. Dir. Alex Jones. 2009.
The Invisible Empire: A New World Order Defined. Dir. Jason Bermas. 2010.
JFK. Dir. Oliver Stone. Perf. Kevin Costner, Tommy Lee Jones, Gary Oldman. Warner Bros, 1991.
Loose Change. Dir. Dylan Avery. Microcinema International, 2005.
The Obama Deception. Dir. Alex Jones. 2009.
"Pilot." The X-Files. Writ. Chris Carter. Dir. Robert Mandel. Fox Network, 1993.

The "Cultural Turn" and the Changing Face of the Humanities in Poland

Urszula Terentowicz-Fotyga

Abstract The paper focuses on the relation between the basic premises of cultural studies and the "cultural turn" that is beginning to affect the academic discourse in Poland. I shall argue that although the Birmingham model has not been embraced in Poland, it offers an interesting context for the discussion of the present condition of the humanities, especially as far as methodology and critical practice are concerned. The most valuable lesson that can be learnt from the present crisis of cultural studies concerns the pitfalls of binary thinking, whereby the high and the popular, poetics and politics, form and identity are analysed and discussed separately. The crisis also points to the need of exercising caution in the process of redrafting the disciplinary boundaries under the influence of new critical and theoretical paradigms. My primary focus, when considering the future of the humanities, will be on the position of English departments, or modern philologies more generally. I shall argue that in some ways we are in a liminal position, not only in between cultures but also between different schools of cultural study and that such position may be particularly apt for considering the strengths and weaknesses of different forms of cultural studies and possibly also for formulating a new paradigm that might be more in tune with the discipline's trans-national and trans-disciplinary ambitions.

U. Terentowicz-Fotyga (✉)
Maria Curie-Skłodowska University, Lublin, Poland
e-mail: uterentowicz@hektor.umcs.lublin.pl

J. Fabiszak et al. (eds.), *Crossroads in Literature and Culture,* Second Language Learning and Teaching, DOI: 10.1007/978-3-642-21994-8_47,

1 Introduction

The central question that I would like to raise and tentatively answer in the paper concerns the relation between the basic premises of cultural studies, the discipline that has redefined the humanities not only in Britain and America but across the world and has been instrumental in transforming such diverse fields as literary studies and geography, and the "cultural turn" that is gradually seeping into the academic discourse in Poland. I shall argue that although the Birmingham model has not been embraced in this country, it offers an interesting context for the discussion of the present condition of the humanities and the institutional changes that are beginning to redefine the face of our university departments.

My primary focus will be on the position of English departments, or modern philologies more generally. I shall argue that in some ways we are in a liminal position, not only between cultures but also between different schools of cultural study and that such position may be particularly apt for considering the strengths and weaknesses of different forms of cultural study and possibly also for formulating a new paradigm that might be more in tune with the discipline's trans-national and trans-disciplinary ambitions.[1]

In Poland, after the period of apocalyptic doom and gloom, of brooding over the death of the humanities, a more hopeful vision is beginning to prevail. There is a growing sense that in the crisis lie opportunities as well as difficulties and that what is really at stake is the new face of the humanities, a new post-postmodern paradigm that will bring the end of old securities and certainties but also open up the field to new areas and alliances. One of the important markers of the change is the growing role of cultural study. Voices postulating the study of culture as an umbrella discipline, capable of integrating various areas of the humanities into inter- or trans-disciplinary projects become more and more vocal.[2] While such fields as cultural theory, anthropology of literature, comparative anthropology or cultural semiotics offer different perspectives on the study of culture, they do so without aiming at over-embracing theory. The new projects tend to argue against unifying, metanarratives and opt for a pragmatic or functional approach to methodology.

In the discussion of the method, the development of cultural studies of the Birmingham tradition, and particularly the impact it had on other disciplines offers some interesting insight and room for thought. While its theoretical principles, as I shall argue, may not necessarily transplant well into the Eastern-European context

[1] The term "cultural study" is used to refer to the study of culture in a general way. The term "cultural studies" is used when talking about a particular school of cultural study that developed from the Birmingham tradition.

[2] See for example the papers presented at the first convention of Polish "Kulturoznawstwo" in October 2009, especially Anna Zeidler-Janiszewska's "Kulturoznawstwo jako project trans-dyscyplinarny" and Andrzej Szahaj's "Nauki o kulturze jako nauki podstawowe". For the "cultural turn" in literary studies see Markowski, Nycz (2006).

(not least because they are being questioned and undermined from within), the question of methodology remains a valid and useful point of reference. In fact, the most valuable lessons can be learnt by looking at the relation between the discipline's self-definitions and its actual critical practice, between the premises formulated by the founders of cultural studies at the moment of its inception and its subsequent, institutional development. Paradoxically, the lessons are most useful in the failures of the discipline rather than it its successes, in the proposed solutions to the present crisis rather than in the consistency of original formulations.

2 British Cultural Studies

2.1 From High to Popular Culture

Cultural studies, as a separate discipline, grew out of the polemic with the tradition represented by Matthew Arnold and F. R. Leavis, renowned literary critics, whose understanding of culture was polarized around the high/low distinction. Within this tradition, culture was synonymous with high culture or, as Matthew Arnold put it in his seminal book *Culture* and *Anarchy*, "the best that has been thought and known in the world" (1960, p. 70). High culture was understood as the one that meets the required standards of formal complexity, moral worth and difficulty; it was associated with the body of values superior to those prevalent in society.

Questioning these basic assumptions, the understanding of culture as elitist, national and associated with aesthetics (in particular with literary masterpieces) came to constitute the centre of cultural studies' revolution. The main objection was against separating one aspect of culture from other life-practices. In reality, it meant that the very concept of culture needed redefining. Raymond Williams in his seminal work *Culture and Society* identified four meanings of the term. Culture as "a general state or habit of the mind", "the general state of intellectual development, in a society as a whole", "the general body of the arts" and "a whole way of life, material, intellectual and spiritual" (1958, p. 16). The defining characteristic of cultural studies was the shifting of attention from the third to the fourth meaning of the word—from interest in the general body of the arts (the canon, the literary masterpieces) to the analysis of the everyday, the whole way of life of a particular group (Milner 1996, p. 11). The concept of culture was thus democratized and socialized to include "'the lived experience' of 'ordinary' men and women, made in their daily interaction with the texts and practices of everyday life" (Storey 1993, p. 56).

2.2 From Poetics to Politics

Having rejected the high/low dichotomy and proposed popular culture as a valid object of study, British cultural studies redefined culture as a space of struggle between those with and without power. The discipline has been highly influenced by the Frankfurt School, which drew attention to culture as the site of ideological struggle but also of "political critique and intervention" (Grossberg et al. 1992, p. 5). In effect, culture became "interesting not in itself, but for its role in social change" (Hartley 2003, p. 11). As Stuart Hall, the father figure of cultural studies put it:

> [Popular culture] is one of the sites where this struggle for and against a culture of the powerful is engaged: it is also the stake to be won or lost in the struggle. It is the arena of consent and resistance. It is partly where hegemony arises, and where it is secured. […] It is not a sphere where socialism, a socialist culture—already fully formed—might be simply 'expressed'. But it is one of the places where socialism might be constituted. That is why 'popular culture' matters. Otherwise, to tell you the truth, I don't give a damn about it. (qtd. in Hartley 2003, p. 24).

With such conceptualization of culture, in terms of the lived dominance of particular classes, cultural studies became an openly and self-confessedly a political project. Critical attention focused on the interplay of representations and ideologies of class, gender, race and nationality in cultural texts; on the questions of power and agency. As Easthope (1991, p. 71) put it, cultural studies reversed Leavisian paradigm, which placed work in the foreground and historical background in the background. It foregrounded "social and historical determination and bas[ed] analysis on sociological or Marxist model".

2.3 Critical Practice

While Raymond Williams's project aimed at expanding the subject matter of the study of culture, the praxis developed somewhat differently, largely as a result of a particular relation between cultural studies and literary studies. As Andrew Milner explains, cultural studies developed as an "antithetical, excluded other" of literary studies. Since literary criticism deliberately excluded popular fiction, the popular, almost by default, became the subject matter of "the new proto-discipline of cultural studies" (Milner 1996, p. 16). And while the new discipline "has defined itself as the study of 'all' texts, minority or popular […] it has become the study of popular culture in particular" and then, as a result, "it has increasingly tended to argue for the inherent value of the popular" (Milner 1996, p. 17). It has, Milner concludes, "tended to espouse a form of cultural populism" (1996, p. 17).

While equating cultural studies with cultural populism might be an oversimplification, the centrality of popular culture in the discourses of cultural studies remains unquestionable. There is no doubt that its focus on marginal and minority discourses has been accompanied by the marginalization of the study of canonical texts. In this way the study of the canon was left in the domain of literary studies while cultural studies focused on popular culture and other life practices.

2.4 Literary into Cultural Studies

The relation between the two disciplines and the exclusionary logic that underlines it is evident in Anthony Easthope's *Literary into Cultural Studies* published in 1991. At some level of generalization, Easthope's (1991, p. 164) celebratory book can be read as an attempt to define the identity of the two academic disciplines. Going from, what he calls "abstract and ideal types" to what seems a practical consideration of forms of identification and positioning, Easthope constructs the two fields as essentially antagonistic. Literary and cultural studies are defined through series of oppositions: transcendent vs. immanent, individual vs. collective, elitist vs. democratic, idealist vs. pragmatic and materialist, masculine vs. gender-neutral, and, most importantly, the former associated with conformism and consolidation of the established order, the latter with transgression and problematization. Easthope's vision of the humanities, though claiming to go beyond binary thinking, in fact underlines and consolidates the inner homogeneity and mutual antagonism of the two disciplines.

From his discussion of the "form of identification and positioning each [discipline] would inscribe for its pedagogic subject" (Easthope 1991, p. 166), one can reconstruct an ideal object of cultural studies, as contrasted with Leavis's. Against the difficult, elitist, modernist text, Easthope (1991, p. 165–176) sets a text that:

- takes the everyday as a point of departure;
- questions the borderline between high and popular literature;
- avoids fixing, unification and centralization; instead positions itself as relative, dispersed and decentred;
- welcomes interdisciplinary and contextual readings;
- problematizes questions of identity (gender, national, religious and class);
- speaks from the ethnic, national, gender or sexual periphery in a gesture of defiance aimed at the superego of the center;
- hints not at individual but collective production;
- dramatizes (or at least welcomes) the transgression of the everyday into the academic; and finally
- "challenges the category of the aesthetic and the canonical, stresses that experience always takes place through a means of representation, recognizes how readings of texts are discursively constructed" Easthope (1991, p. 174).

Such an abstracted and idealized object of cultural study bears a hardly veiled resemblance to a postmodern text. In fact, each of these points, in a slightly rephrased form, might be applied to define a postmodern novel. Camp recycling, radical eclecticism and questioning the borderline between high and popular literature can be read as turning away from modernist self-enclosure into a more democratic and collective terrain. Blurring the boundaries between literary and non-literary, the theoretical and the poetic, clearly welcomes interdisciplinary and contextual readings and dramatizes the transgression of the everyday into the academic. Polymorphous, playful and parodic intertextuality, undoubtedly acknowledges the fact that experience takes place through a means of representation and that readings of texts are discursively constructed. The positioning of the contemporary novel on the cultural periphery hardly needs explication. While the ethnic, national, gender and sexual periphery is given voice in the postcolonial, feminist and queer narratives, the centre is weakened by skeptical and desensitized anti-narratives.

2.5 The Persistence of Binary Logic

Despite the evolution of the literary form, which undermined the idea of textual unity and destabilized the split into high and popular culture and thus preempted the reasoning behind the antagonism of the two disciplines, the proposed "combined study of literary texts and those from popular culture" (Easthope 1991, p. 5) did not really materialize. The legacy of cultural studies of the Birmingham tradition remains deeply rooted in binary thinking.

Against the original ambitions to expand and combine, the critical praxis mostly shifted the focus from high to popular, from poetics to politics, from form to identity. The new paradigm gradually undermined and replaced the old one, it supplanted rather than completed the New Critical idiom. Yet, the binary logic has remained part and parcel of the picture.

The impact of the "cultural turn" has been both positive and negative. On the one hand, cultural studies has been instrumental in opening up the humanities onto new perspectives. On the other hand, there is an increased concern not only with the limitations of its theory but also with increasingly sloppy scholarship and undefined research method.[3]

3 The Polish Perspective

3.1 The Old and the New

In Poland, culturalist perspectives are gradually seeping into the academic discourse and changing the face of university departments. New courses in cultural

[3] The discussion of the crisis of cultural studies lies beyond the scope of the present paper. For interesting perspectives on the issue see Bowman (2003).

semiotics, popular culture, women's and gender studies are being opened and culturalist contexts are introduced or reintroduced into such diverse disciplines as literary studies, sociology and geography. However, the impact of the Birmingham model remains limited and the reasons behind its relative unpopularity seem to stem from the same line of thinking that has fuelled criticism of the discipline in the West.

One of the most contentious issues is the conflation of culture and ideology, the assumption that politics and ideology determine and overshadow other dimensions of the text. Such approach proves particularly problematic in post-communist countries, where the politicization of discourse and the obligatory application of the Marxist model are the nightmares that a number of academics have been trying to wake up from in the last couple of decades. Resistance to overtly political, ideological readings is an important part of the legacy of the politicized discourse of the communist years. In fact, as Michał Głowiński explains, from the early 1990s the natural tendency has been to avoid ideology and focus on the subjects hitherto neglected, such as metaphysics and religion. What is more, Głowiński argues, out of the three key concepts crucial for cultural studies, that is race, class and gender, only the last one seems naturally applicable in Poland (which is well reflected in the steady development of gender studies). Race is not as yet an important category in our social and cultural landscape, while the discussion of class has been exhausted and so discredited by the vulgar Marxism of the post-war years that it is hardly an inspiring path even for younger scholars (Głowiński 2009, pp. 55–58).

Equally important seems the legacy of semiotics. Communicative and semiotic models have always been at the centre of Polish "kulturoznawstwo" and of the humanities as such (See Dmitriuk 1990). At the heart of our thinking about culture is the question of the plurality of sign systems, heterogeneity of cultural processes. Therefore, many critics speak of a sense of discomfort with what Henryk Markiewicz calls the prosecutor's bias of the approaches aiming at denouncing the oppressive character of sign systems (2009, p. 47). In such readings, the conclusions are too tied up in the idiom to remain sufficiently sensitive to the complexity of individual texts and signifying practices. In effect, the discussion of social, cultural and historical situatedness of the text reduces rather than adds to the plurality of its meanings.

Yet, while the major premises of cultural studies may not translate well into the Polish context, there is a sense that the binary logic that lies at the heart of its paradigm and that has ultimately stalled the Birmingham model might be much more contagious. There is a manifest trend to see popular culture, both within the educational process and academic research, as a distinct phenomenon, separated from the canon and institutionalised culture. Warded off into separate journals and courses, married with the new media and technologies yet divorced from the discussion of the canon, popular culture seems to exist on its own terms and within its own context.

3.2 The Question of the Method

One of the reasons behind the separation is the fact that through all the different revisions and reconfigurations that are changing the face of the humanities we have not as yet worked out a method that would happily marry the old and the new, "high" and popular, poetics and politics, form and identity. But if the lessons of cultural studies can come handy they should do so primarily at the level of critical practice so that once popular culture storms our departments it does not do so on its own, enclosed and warded off in separate courses. If the question of the method is indeed to become the forefront of our thinking about the new humanities, the principal aim should be to supplant rather than reinforce the binary logic, making sure that our focus on identity complements rather than replaces the analysis of form, the discussions of poetics and politics, multidimensionality of the text and its political situatedness, though pulling us in different directions, share the common ground and a teaching room.

It is thus crucial that the proponents of cultural studies are read alongside its detractors, the successes of the discipline measured against its failures, that the conceptual frameworks that have defined the cultural turn in Britain, America and other part of the world are not transplanted automatically but read critically and against the local social, historical and cultural context.

Working by combining and including has got an additional advantage when the institutional aspect of the process is taken into account and again the changes that have taken place in Britain and America can be a valuable point of reference. The institutional success of cultural studies often came at the expense of other disciplines. While the "cultural turn" let the humanities get its second wind it also brought gradual morphing of English departments into either cultural studies or comparative literatures. As Mieke Bal puts it, interdisciplinarity often "has given university administrators a tool with which to enforce mergings and cancellations of departments that might turn out to be fatal for the broad grounding" (Bal 2003, p. 31) not only of cultural studies but of the humanities as such.

3.3 The "Cultural Turn" and the Future of English Departments

The "cultural turn" can clearly pose some dangers to the integrity of university departments but it can also be used to our advantage. As teachers at English departments we are in a liminal position, where on the one hand the historical and cultural experiences make us somewhat resistant to the charms of cultural studies and critical of its politicized, binary thinking. On the other hand, we teach courses, where such readings have been in the forefront of critical attention in the last couple of decades. This liminal position, however, may be seen as an advantage, particularly in view of a growing tendency to reach out across the disciplines.

From the point of view of modern philologies, one of the more interesting developments that have been evolving under the umbrella of the "cultural turn" has been comparative cultural studies, which Steven Tötösy de Zepetnek defines in the following way:

> field of study where selected tenets of the discipline of comparative literature are merged with selected tenets of the field of cultural studies meaning that the study of culture and culture products—including but not restricted to literature, communication, media, art, etc.—is performed in a contextual and relational construction and with a plurality of methods and approaches, inter-disciplinarity, and, if and when required, including team work. In comparative cultural studies it is the processes of communicative action(s) in culture and the how of these processes that constitute the main objectives of research and study. However, comparative cultural studies does not exclude textual analysis proper or other established fields of study

(1999, pp. 16–17).

Zepetnek points out at general inspirations and possible directions rather than offers a clear-cut definition of the discipline. His proposition represents a basis for discussion rather than a fully-fledged conception. However, its general framework seems to resonate well with the recent Polish debates over the future of the humanities, the "cultural turn", the changing face of comparative studies and the legacy of semiotics.[4]

Zepetnek's (1999, p. 15) proposition grows out of the need for methodological and theoretical "dialogue between cultures, languages, literatures, and disciplines" and the conviction that a comparative perspective, with all its pitfalls and shortcomings, offers a beneficial approach for the new humanities in the global world. Such perspective seems particularly well-suited for modern philologies, which are naturally comparative not only due to the liminal position of the academic staff, who, living within one culture and teaching about another, unavoidably work through comparisons but also because such departments often focus on supranational fields. In English we teach courses in British (itself comprising of English, Irish, Scottish and Welsh), American and Australian cultures.[5]

Under the umbrella of interdisciplinarity, Zepetnek (1999, p. 15) proposes to study: culture in its parts (literature, arts, film, popular culture, theatre, the publishing industry, the history of the book as a cultural product, etc.) and as a whole in relation to other forms of human expression and activity and in relation to other disciplines in the humanities and social sciences (history, sociology, psychology, etc.).

Such approach gradually becomes a necessity. In the global reality, ventures into other cultures and disciplines often prove unavoidable. While teaching a seminar on urban theory and representation at UMCS, I combine the study of different urban narratives, both canonical and popular, in literature and film with the reading of Russian semioticians, urban theoreticians, sociologists,

---

4 For the most interesting opinions, see Hejmej (2007), Kola (2008), Grochowski (2005), Nycz (2005), Markowski, Nycz (2006), Balbus (2004), Nowicka-Jeżowa (1998), Sławek (2008).

5 For a similar discussion of Slavic studies, see Kola (2008).

anthropologists and cultural geographers. And since it is difficult to talk about the changing forms of urban representation without considering Baudelaire and Benjamin or seeing such films as *Berlin* or *Metropolis* we often reach for texts from other cultures. In effect, while the primary texts selected for interpretation are from the English-speaking world, the historical, critical and theoretical context is set much more broadly.

While Zapetnek acknowledges and discusses a number of formal and institutional constraints and obstacles posed by such broadly-outlined area of study, he sees part of the solution in team work, in active cooperation of participants from several disciplines organized in inter-, trans- or multi-disciplinary projects. Such approach can offer an interesting option for different philologies, whose cooperation should not give up on disciplinary boundaries but be based on individual courses and projects built around themes, concepts and problems. Interdisciplinarity, as Ryszard Nycz argues, need not lead to trans-disciplinary integration but should act as a stimulus for critical self-reflection of individual disciplines and the humanities as such (2005, p. 176).

While not renouncing the importance of cultural studies and its political engagement, Zapetnek (1999, p. 16) reaches for the legacy of comparative literature and argues for the centrality of communicative processes.[6] He calls for renewed attention to the relation between cultural studies and comparative literary studies, particularly in the area of methodology. While he considers the systemic and empirical approach as the most advantageous methodology, he acknowledges that comparative cultural studies and its methodologies must be "implicitly and explicitly pluralistic" yet at the same time "based on scholarly rigor".[7]

A plethora of often conflicting answers to the problem of methodology suggests that the new trends in the humanities will evolve locally and gradually rather than as applications of a pre-defined theoretical framework. One may agree with Kola (2008, p. 56), who argues that in comparative studies tools and methods are not pre-given but each and every time must be either selected from the existing repertoire offered by different disciplines or created on an individual basis. It means that the path leading to an establishment of comparative cultural studies as a fully-fledged discipline is likely to be long and difficult. But the fact that the new areas evolve through critical praxis rather than out of a coherent theoretical or ideological position need not be a disadvantage.[8] In the contemporary academic world,

---

[6] In Poland, due to the importance of semiotic and structural method, this trend is even more evident. For Stanisław Balbus (2004, p. 12), for example, the terms "interdisciplinarity", "comparative cultural studies" and "inter-semiotics" are roughly synonymous.

[7] Critical engagement with Zapetnek's choice of methodology lies beyond the scope of the paper. In a parallel article I discuss the relevance of Lotman's semiotics of culture for contemporary study of culture (in progress). For a comparative analysis of cultural studies and semiotics of culture, see my article (Terentowicz-Fotyga 2007). For an insightful comparison of cultural semiotics and constructivism, see Blaim (1992a), (b).

[8] For an example of a successful venture into the new grounds see the series of comparative literary and cultural studies edited by Zbigniew Kadłubek and Tadeusz Sławek (2008).

distrustful of grand metanarratives, local and grass-root initiatives may prove most useful in providing answers to the crisis of individual disciplines and the humanities as such. As Grzegorz Grochowski (2005, p. 10) argues, at this point the most reasonable strategy lies in holding your breath and keeping an open mind:

> Rather than drawing up inventories of our assets and liabilities, we had better observe—with a degree of caution, but above all with a great deal of curiosity—the ideas of those who take up the risk of expanding the scope of our knowledge, who stick their necks out and venture, more or less successfully, into the new grounds (translation mine).

4 Conclusion

In the paper I have been trying to argue that an extensive and critical engagement with the theory and practice of cultural studies can offer some insight into the chances and risks posed by the "cultural turn" that is beginning to affect the academic world in Poland. The most valuable lesson that the present crisis of cultural studies offers concerns the pitfalls of binary thinking, whereby the high and the popular, poetics and politics, form and identity are analysed and discussed separately. It also points to the need of exercising caution in the process of redrafting disciplinary boundaries under the influence of new critical and theoretical paradigms. In the second part of the paper I discuss some premises of comparative cultural studies, a discipline which like other areas developing under the umbrella of the "cultural turn" is in the process of defining its identity and which can prove inspirational in the process of probing disciplinary boundaries of modern philologies.

References

Arnold, M. 1960. *Culture and anarchy*. Cambridge: Cambridge University Press.

Bal, M. 2003. From cultural studies to cultural analysis: a controlled reflection on the formation of method. In *Interrogating cultural studies: theory, politics and practice*, ed. P. Bowman. London and Sterling, Virginia: Pluto Press.

Balbus, Stanisław. 2004. Interdyscyplinarność—intersemiotyczność—komparatystyka. In *Intersemiotyczność: literatura wobec innych sztuk (i odwrotnie)*, ed. S. Balbus, A. Hejmej and J. Niedwiedź. Kraków: Universitas.

Blaim, A. 1992a. Constructivist aspects of cultural semiotics. In *Poetics of the text: essays to celebrate twenty years of the Neo Formalist Circle*, ed. Joe Andrew. Amsterdam: Rodopi.

Blaim, A. 1992b. The twain shall meet? Some observations on Soviet cultural semiotics and empirical study of literature and the media. *Poetics* 21. 1-2: 33-44.

Bowman, P. (ed.). 2003. *Interrogating cultural studies: theory, politics and practice*. London and Sterling, Virginia: Pluto Press.

Dmitriuk, K. 1990. *Współczesne polskie koncepcje kultury. Warszawa*: PWN.

Easthope, A. 1991. *Literary into cultural studies*. London and New York: Routledge.

Głowiński, M. 2009. Głosy do rozważań Johna M. Ellisa. *Znak* 10.2009: 55-61.

Grochowski, G. 2005. Blaski i cienie badań kulturowych. *Teksty Drugie* 1.2: 4-10.
Grossberg, L., Nelson, C. and P. Treichler (eds.) 1992. *Cultural studies*. New York and London: Routledge.
Hartley, J. 2003. *A short history of cultural studies*. London: Sage Publications.
Hejmej, A. 2007. Interdyscyplinarność i badania komparatystyczne. *Wielogłos: Pismo Wydziału Polonistyki UJ*. 1: 35-53.
Kadłubek Z. and T. Sławek (eds). 2008. *Wizerunki wspólnoty: studia i szkice z literatury i antropologii porównawczej*. Seria: Komparatystyka literacka i kulturowa. Katowice: Wydawnictwo Uniwersytetu Śląskiego.
Kola, A.F. 2008. Nie-klasyczna komparatystyka. W stronę nowego paradygmatu. *Teksty Drugie* 1.2:56-74.
Markiewicz, H. Przed teorią nie ma ucieczki. Z Henrykiem Markiewiczem rozmawia Łukasz Tischner. *Znak* 10.2009: 44-54.
Markowski, M.P., and Ryszard Nycz (eds). 2006. *Kulturowa teoria literatursy*. Kraków: Universitas.
Milner, A. 1996. *Literature, culture and society*. London and New York: Routledge.
Nowicka-Jeżowa, A. 1998. *Badania porównawcze. Dyskusja o metodzie. Radziejowice 6-8 lutego 1997r*. Izabelin: Świat literacki.
Nycz, R. 2005. O przedmiocie studiów literackich—dziś. *Teksty Drugie* 1.2: 175-187.
Sławek, T. 2008. Wprowadzenie do lektury. *Peribolos*. In *Wizerunki wspólnoty: studia i szkice z literatury i antropologii porównawczej* (eds.) Kadłubek Z. and T. Sławek. Katowice: Wydawnictwo Uniwersytetu Śląskiego.
Storey, J. 1993. An *introductory guide to cultural theory and popular culture*. New York and London: Harvester Wheatsheaf.
Terentowicz-Fotyga, U. 2007. The great divide over culture and politics: cultural studies and semiotics of culture. http://www.uel.ac.uk/ccsr/documents/TerentowiczGreatDivide.pdf. Accessed 14 July 2010.
Tötösy de Zepetnek, S. 1999. From Comparative Literature Today toward Comparative Cultural Studies. CLCWeb. http://docs.lib.purdue.edu/cgi/viewcontent.cgi?article=1041&context=clcweb. Accessed 14 July 2010.
Williams, R. 1958. *Culture and society 1780-1950*. London: Chatto & Windus.